ATOMIC FORCE MICROSCOPY

ATOMIC FORCE MICROSCOPY

Understanding Basic Modes and Advanced Applications

GREG HAUGSTAD

WILEY

A JOHN WILEY & SONS, INC., PUBLICATION

Cover image: Courtesy of the author.

Published by John Wiley & Sons, Inc., Hoboken, New Jersey.
Published simultaneously in Canada.

Library of Congress Cataloging-in-Publication Data:

Haugstad, Greg, 1963–
Atomic force microscopy : understanding basic modes and advanced applications / Greg Haugstad.
p. cm.
Includes bibliographical references.
ISBN 978-0-470-63882-8
1. Atomic force microscopy. I. Title.
QH212. A78H38 2012
620′. 5–dc23

2012003429

10 9 8 7 6 5 4 3 2 1

MIX
Paper from responsible sources
FSC® C013604

To Laurie

CONTENTS

Preface xiii

Acknowledgments xxi

1. Overview of AFM 1

1.1. The Essence of the Technique, 1
1.2. Property Sensitive Imaging: Vertical Touching and Sliding Friction, 6
1.3. Modifying a Surface with a Tip, 13
1.4. Dynamic (or "AC" or "Tapping") Modes: Delicate Imaging with Property Sensitivity, 16
1.5. Force Curves Plus Mapping in Liquid, 21
1.6. Rate, Temperature, and Humidity-Dependent Characterization, 24
1.7. Long-Range Force Imaging Modes, 28
1.8. Pedagogy of Chapters, 30
References, 31

2. Distance-Dependent Interactions 33

2.1. General Analogies and Types of Forces, 33
2.2. Van der Waals and Electrostatic Forces in a Tip–Sample System, 38
2.2.1. Dipole–Dipole Forces, 38
2.2.2. Electrostatic Forces, 41
2.3. Contact Forces and Mechanical Compliance, 44
2.4. Dynamic Probing of Distance-Dependent Forces, 51
2.4.1. Importance of Force Gradient, 51

2.4.2. Damped, Driven Oscillator: Concepts and Mathematics, 56
2.4.3. Effect of Tip–Sample Interaction on Oscillator, 60
2.4.4. Energy Dissipation in Tip–Sample Interaction, 64
2.5. Other Distance-Dependent Attraction and Repulsion: Electrostatic and Molecular Forces in Air and Liquids, 67
2.5.1. Electrostatic Forces in Liquids: Superimposed on Van der Waals Forces, 67
2.5.2. Molecular-Structure Forces in Liquids, 69
2.5.3. Macromolecular Steric Forces in Liquids, 72
2.5.4. Derjaguin Approximation: Colloid Probe AFM, 76
2.5.5. Macromolecular Extension Forces (Air and Liquid Media), 78
2.6. Rate/Time Effects, 83
2.6.1. Viscoelasticity, 84
2.6.2. Stress-Modified Thermal Activation, 85
2.6.3. Relevance to Other Topics of Chapter 2, 86
References, 88

3. Z-Dependent Force Measurements with AFM 91

3.1. Revisit Ideal Concept, 91
3.2. Force-Z Measurement Components: Tip/Cantilever/Laser/Photodetector/Z Scanner, 93
3.2.1. Basic Concepts and Interrelationships, 93
3.2.2. Tip–Sample Distance, 96
3.2.3. Finer Quantitative Issues in Force–Distance Measurements, 99
3.3. Physical Hysteresis, 106
3.4. Optical Artifacts, 109
3.5. Z Scanner/Sensor Hardware: Nonidealities, 113
3.6. Additional Force-Curve Analysis Examples, 118
3.6.1. Glassy Polymer, Rigid Cantilever, 118
3.6.2. Gels, Soft Cantilever, 123
3.6.3. Molecular-Chain Bridging Adhesion, 126
3.6.4. Bias-Dependent Electrostatic Forces in Air, 129
3.6.5. Screened Electrostatic Forces in Aqueous Medium, 131
3.7. Cantilever Spring Constant Calibration, 133
References, 135

4. Topographic Imaging 137

4.1. Idealized Concepts, 138
4.2. The Real World, 143
4.2.1. The Basics: Block Descriptions of AFM Hardware, 143
4.2.2. The Nature of the Collected Data, 149
4.2.3. Choosing Setpoint: Effects on Tip–Sample Interaction and Thereby on Images, 156
4.2.4. Finite Response of Feedback Control System, 162

4.2.5. Realities of Piezoscanners: Use of Closed-Loop Scanning, 167
4.2.6. Shape of Tip and Surface, 180
4.2.7. Other Realities and Operational Difficulties—Variable Background, Drift, Experimental Geometry, 182
References, 186

5. Probing Material Properties I: Phase Imaging 187

5.1. Phase Measurement as a Diagnostic of Interaction Regime and Bistability, 189
5.1.1. Phase (and Height, Amplitude) Imaging as Diagnostics, 189
5.1.2. Comments on Imaging in the Attractive Regime, 200
5.2. Complications and Caveats Regarding the Phase Measurement, 202
5.2.1. The Phase Offset, 202
5.2.2. Drift in Resonance Frequency, Phase Offset, Quality Factor, and Response Amplitude, 207
5.2.3. Change of Phase and Amplitude During Coarse Approach, 211
5.2.4. Coupling of Topography and Phase, 214
5.2.5. The Phase Electronics and Its Calibration, 221
5.2.6. Nonideality in the Resonance Spectrum, 230
5.3. Energy Dissipation Interpretation of Phase: Quantitative Analysis, 234
5.3.1. Variable A/A_0 Imaging, 235
5.3.2. Fixed A/A_0 Imaging, 240
5.3.3. Variable A/A_0 via Z-Dependent Point Measurements, 243
5.4. Virial Interpretation of Phase, 247
5.5. Caveats and Data Analysis Strategies when Quantitatively Interpreting Phase Data, 248
References, 255

6. Probing Material Properties II: Adhesive Nanomechanics and Mapping Distance-Dependent Interactions 258

6.1. General Concepts and Interrelationships, 259
6.2. Adhesive Contact Mechanics Models, 261
6.2.1. Overview and Disclaimers, 261
6.2.2. JKR and DMT Models, 263
6.2.3. Ranging Between JKR and DMT: The Transition Parameter λ, 266
6.2.4. The Maugis–Dugdale Model, 270
6.2.5. Other Formal Relationships Relevant to Adhesive Contact Mechanics, 273
6.2.6. Summary Comments and Caveats on Adhesive Contact Mechanics Models, 274
6.3. Capillarity, Details of Meniscus Force, 277
6.3.1. Framing the Issues, 278
6.3.2. Basic Elements of Modeling the Meniscus, 280

6.3.3. Mathematics of Meniscus Geometry and Force, 283
6.3.4. Experimental Examples of Capillarity, 287
6.3.5. Capillary Transfer Phenomena: Difficulties and Opportunities, 293
6.4. Approach–Retract Curve Mapping, 296
6.4.1. Motivation and Background, 296
6.4.2. Traditional Force-Curve Mapping, 298
6.4.3. Approach–Retract Curve Mapping in Dynamic AFM, 306
6.4.4. Approach–Retract Curve Mapping of Liquidy Domains in Complex Thin Films, 313
6.5. High-Speed/Full Site Density Force-Curve Mapping and Imaging, 315
6.5.1. Liquidy Domains in Complex Thin Films, 317
6.5.2. PBMA/PLMA Blend at Variable Ultimate Load, 319
6.5.3. PBMA/Dexamethasone Mixture at Variable Temperature, 320
6.5.4. Arborescent Styrene–Isobutylene–Styrene Block Copolymer Plus Drug Rapamycin, 322
6.5.5. Comments on "Force Modulation" Mode, 323
References, 324

7. Probing Material Properties III: Lateral Force Methods 330

7.1. Components of Lateral Force Signal, 330
7.2. Application of Lateral Force Difference, 336
7.3. Calibration of Lateral Force, 343
7.4. Load-Dependent Friction, 346
7.4.1. Motivations, 346
7.4.2. Load Stepping and Ramping Methods, 347
7.5. Variable Rate and Environmental Parameters in AFM Friction and Wear, 352
7.5.1. Motivations, 352
7.5.2. Interplay of Rate, Temperature, Humidity, and Tip Chemistry in Friction, 354
7.5.3. Wear Under Variable Rate and Temperature, 359
7.5.4. Musings on the Spectroscopic Nature of Friction and Other Measurements, 362
7.6. Transverse Shear Microscopy (TSM) and Anisotropy of Shear Modulus, 364
7.7. Shear Modulation Methods, 366
7.7.1. Motivations and Terminology, 366
7.7.2. Shear Modulation During 1D Lateral Scanning, 368
7.7.3. Diagnostics of Sliding Under Shear Modulation, 371
7.7.4. Complementarity of Shear Modulation Methods to TSM, 372
7.7.5. Shear Modulation Within Force Curves: Material Creep, 373
References, 375

8. Data Post-Processing and Statistical Analysis **379**

8.1. Preliminary Data Processing, 379
8.2. 1D Roughness Metrics, 383
8.3. 2D-Domain Analysis, 385
8.3.1. Slope and Surface Area Analysis, 385
8.3.2. 2D-Domain Fourier Methods for Spatial Analysis, 386
8.3.3. Fourier Methods for Time-Domain Analysis, 391
8.3.4. Grain or Particle Size Analysis, 394
8.4. "Lineshape" Fitting, 396
References, 398

9. Advanced Dynamic Force Methods **400**

9.1. Principles of Electronic Methods Utilizing Dynamic AFM, 401
9.1.1. Shifted Dynamic Response due to Force Gradient, 402
9.1.2. Interleave Methods for Long-Range Force Probing, 405
9.1.3. Interleave-Based EFM/KFM on Different Metals and Silicon, 408
9.1.4. KFM of Organic Semiconductor, Including Cross-Technique Comparisons, 412
9.2. Methods Using Higher Vibrational Modes, 414
9.2.1. Mathematics of Beam Mechanics: The Music of AFM, 414
9.2.2. Probing Tip–Sample Interactions via Multifrequency Dynamic AFM, 419
9.2.3. Contact Resonance Methods, 425
9.2.4. Single-Pass Electric Methods, 429
References, 433

Appendices **437**

Appendix 1: Spectral Methods for Measuring the Normal Cantilever Spring Constant *K*, 437
A1.1 Plan-View/Resonance Frequency Method, 438
A1.2 Sader Method, 441
A1.3 Thermal Method, 442
Appendix 2: Derivation of Van der Waals Force–Distance Expressions, 443
Appendix 3: Derivation of Energy Dissipation Expression, Relationship to Phase, 447
Appendix 4: Relationships in Meniscus Geometry, Circular Approximation, 449
References, 450

Index **453**

PREFACE

Since its invention by Binnig and coworkers in 1986 and the appearance of commercial instruments by the end of that decade, atomic force microscopy (AFM) has become an essential tool for materials and biological research. AFM is present in core facilities at all major research universities and in many single-investigator labs, and is a standard capability in central research labs at large corporations. In more recent times AFMs have been acquired by 2-year and 4-year colleges and occasionally by smaller companies. But the latter typically seek AFM services, or training for independent use in nearby facilities. In addition, there are now more opportunities for AFM demonstrations and even summer research experiences for pre-college students.

What seems almost universally lacking in the above research settings is rigorous AFM education and training. Unlike transmission electron microscopy (TEM), for example, for-credit curricular classes focusing on AFM are almost nonexistent. AFM training sessions in many core facilities are "bare bones," as confirmed in conversations with former graduate student users (i.e., later postdocs seeking self-described "remedial" AFM training). The trainer has some AFM experience but often little, or at most narrow, expertise in research applications as well as the plethora of artifacts, caveats and other interpretive issues that one encounters with the technique. There are also training sessions provided by AFM manufacturers, mostly relevant to the initial users of a newly installed instrument; but one often finds that these training sessions are not developed around fundamental AFM understanding (e.g., of interfacial forces) and broad research experiences.

Beyond AFM training issues, too often little knowledge exchange or "vetting" of data interpretations takes place between users of core facilities at universities or central analytical labs at companies. Together with the lack of formal education, a *culture of understanding* of AFM may not take root at a given institution, with the exception of a few single-investigator labs (with their own AFMs) whose research critically depends on the technique. This situation again differs from the learning environment around larger (i.e., more expensive) instruments such as TEM. Historically, TEMs have been heavily used by groups that take the time to fully understand the technique, even within a core-facility setting, because principal investigators cannot purchase "their own" instruments; and core-facility user fees are often high enough to preclude "casual" users. Similarly, groups at large research facilities such as synchrotron light sources, where usage proposals undergo peer review, develop a culture of understanding that is reinforced by "elbow-to-elbow" user interactions and "live-in" conditions. Finally, the peer review of journal publications or internal review of reports using AFM can be problematic. Typically the reviewers are experts in the material, biological, or technological issues of the samples studied, rather than knowledgeable of the AFM method. Often multiple techniques are employed such that AFM results are only tersely presented as part of a package of information, and thus correspondingly are often casually reviewed.

Then there is industry. As elucidating as advanced implementations of AFM can be, they have made only a meager foray into the R&D of technological and biomedical companies (with a few exceptions). It is this author's contention that academic research—especially that which is steeped in methods—must translate into advanced technological and biomedical R&D beyond academia. (Today's harsher political climate for public funding of universities can only underscore this contention.) Yet personnel from industry relate that they have received almost no training on AFMs for which they are charged to become local "experts." We *must* do better in translating volumes of AFM methods development, as published in scientific journals over a period of more than two decades, into analytical practice in industry. My observation is that a broad range of structures, properties, and phenomena in the materials technologies of industry and biomedicine beg insightful analysis with AFM. Yet the usage of AFM exhibits a troubling footprint: a diversity of in-depth and creative methodologies implemented in academia by *AFM-savvy research groups* (not casual users of core facilities), and relatively minimal usage in industry. It behooves academics who are hands-on active in AFM to reach out much more to industry. Collaboration and consulting are two avenues (familiar to the author), but I believe an AFM monograph also is in need: one which explains underpinning concepts and operational issues, and how these lead to advanced applications.

This monograph is intended to address the above issues associated with the use of the AFM. Its chapters target a broad audience of AFM users ranging from precollege students and technicians to PhD-level scientists. The materials derive in part from a diversity of curricular courses, training sessions, short courses and outreach programs at the University of Minnesota:

(1) on site and internet-based demonstrations for precollege students;
(2) a capstone fourth-semester characterization course within a Nanoscience Technology technical college (two-year) program;

(3) a lab course on structural characterization within a Materials Science 4-year university degree program (5th semester);
(4) a graduate course for students in their first or second year, taken primarily by Ph.D. candidates in chemistry, materials science, physics or chemical/mechanical/electrical engineering (for whom materials characterization is essential to later thesis research);
(5) training courses for researchers to use instrumentation in a core facility;
(6) short courses and demonstrations on advanced methods, primarily for industrial scientists and engineers.

I have taught more than a thousand curricular students and research trainees with a broad range of educational backgrounds and at various career stages, and have taught industrial users as well as academics from numerous institutions: not only research universities but also small colleges including visiting professors and undergraduates (e.g., Research Experiences for Undergraduates programs). In teaching the AFM technique, it has struck me that the knowledge required to understand its most basic operation and the simplest data (surface topography) is well within the grasp of this entire audience, given the intuitive nature of stylus profilometry and contact forces. Conversely, many of the less intuitive and more mathematical concepts (e.g., attractive and repulsive dynamic imaging regimes, raw versus processed data, piezoscanner nonidealities) often are murky for students and trainees at many levels of education, whether pre-college, 2-yr college, 4-yr college or graduate; even to professional scientists and engineers with PhDs. Often graduate students seek AFM training only to use it in a "zeroth-order" fashion, for seemingly simple measurements of surface roughness or step height. However, one cannot reliably determine even this simple information without addressing issues associated with attractive/repulsive imaging regimes, scanner cross coupling, nonlinearity and piezocreep, along with nanoscale contact mechanics; and issues of capillarity and tip-sample adhesion. In short: even if only the "simplest" information is sought from AFM, it does not follow that the simplest understanding of the technique will enable success.

Goals in the core facility that I manage, the Characterization Facility at the University of Minnesota, are to develop instructional materials

(i) with minimal mathematics leading to an *enabling understanding of AFM*, insofar as it impacts the measurement of basic surface characteristics (like roughness or step height);

(ii) that drill down into *user-selectable rigor* (calibration, physical origin of artifacts, signal/noise limitations, mathematical analysis of physical responses) to open avenues for in-depth experimentation and deeper understanding.

This monograph is thus written in a hierarchical fashion, rooted in basic understandings of distance-dependent intersurface forces, but drilling down into subtopics which are located in later subsections of early chapters or in late chapters and appendices. It seems that no existing books on AFM are based on such pedagogy. Often these (mainly edited) books consist of individual chapters derived from PhD or postdoctoral research projects, and thus assume a readership that is well into research applications and savvy of the difficulties associated with the AFM method. The writers typically do not introduce the technique in a way appropriate for a newcomer, and moreover do not get into the instrument realities with which one must wrestle to achieve success. These books usually emphasize advanced applications (e.g., viscoelastic properties of polymers, nanolithography, electrochemistry, single-molecule force spectroscopy) written by different authors with very different voices. Even if early chapters cover the "basics," there is often an assumption of a readership with in-depth knowledge of physics, interfacial science, etc. ... not the typical distribution of people who show up for AFM training! And in any case there usually is little in the way of "pointers" between chapters, because each is written by a different person.

Similarly, one can download application notes from AFM vendors, each application drilling down into a subtopic but without the context of an overall treatment rooted in basic principles. Some very good review articles exist, but newer ones are either about narrower special topics or are lengthy and exhaustively focused on a single methodology (e.g., force-distance measurements). Older review articles are shorter, shallower and broader in scope, but often contain flawed or at best outdated understandings. Finally there are basic tutorials at vendor and principal-investigator web sites, but the treatment of subtopics (e.g., frictional imaging) is usually minimal and sometimes physically incorrect when examined in detail.

A question that arises in any primer on AFM is at what point to move from quasistatic "contact mode" (or the newer force-distance mapping modes) to dynamic or "AC" (a.k.a. tapping mode) AFM. The former can be intuitively understood and tip-sample forces can be simply quantified, whereas the latter cannot be understood at even a base level without the physical concepts (and mathematics) of a driven oscillator with damping and nonmonotonic force-distance relationship. For some years now I have found that a majority of users need dynamic AFM because their materials are too soft, or because nano-objects are too weakly adhered to substrate, or because residual, mobile species are present at the surface (following film growth or surface modification or extraction from a vacuum chamber), preventing "clean" images. Although most of these users seek only topographic information, height is often sought at nanometer or subnanometer precision. Therein lies a problem, because one cannot accurately measure step heights or characterize roughness on this scale without carefully stabilizing the tip-sample interaction in the net repulsive or net attractive regime under delicate interaction conditions. Understanding regime bistability is essential even for casual users, and phase measurement is necessary for identifying and controlling the regime. But in the process of examining phase images, the novice stumbles across interesting contrast and naturally asks, "what does it mean"? Thus I do not relegate this topic to the last chapter on advanced methods, nor cover it early under superficial concepts. An introductory treatment must not, at the first stages of learning, mislead the user with oversimplified concepts such as "phase is equivalent to stiffness," which is incorrect in many cases. Nevertheless, the most mathematically rigorous understandings of dynamic AFM and phase imaging have been assigned to Chapters 5 and 9 and Appendix 3.

As already hinted, an important question for any introductory book is the degree of mathematical formalism and exactness to invoke. An AFM primer cannot shy from mathematics, because it is the language of physics; and a method with "atomic" and "force" in its very name cannot possibly be understood while avoiding physics! Yet a book should be organized so as to invoke mathematics as painlessly as possible. This book begins with an overview chapter that contains essentially no mathematics, yet introduces several advanced things that AFM can do. In the ensuing early chapters, some physical concepts are first developed with appeal to analogy (e.g., radar, tuning fork), then a bit more formally such that the reader first dips only lightly into mathematics. Then in later chapters, the book delves into greater physical/mathematical detail and rigor, presumably as the user gains experience with the technique in parallel with reading and possibly coursework. But words such as "introductory" and "lightly" should not mean incorrect. I endeavor to clarify approximations and possibly incorrect assumptions (e.g., treating tip shape as a hemisphere) during the introduction of simplified topics, rather than relegate to later sections or chapters (or cited literature) that the reader may not visit for months or years while becoming familiar with the basics and reporting results from this work ... even submitting publications! Later, the reader who pushes through all the chapters, methods, and physics hopefully will be enabled to conduct more rigorous and ambitious research, additionally aided by the references (and perhaps consulting the authors).

Throughout the book I also point out some unfortunate "realities." One is the variegated terminology for expressing a given concept (e.g., "tapping mode" and its synonyms), as easily found in journal papers, books, digital presentations, manuals, and so on; and in some cases I comment on appropriateness. So rather than invoke a single terminology and a pedagogical strictness, I attempt to make the reader aware of the diversity of terms in the "real world." Another reality is the prevalence of instrument shortcomings. These include (i) the "leakage" of optical interference, AC electrical signal and building vibrations into AFM data, generating variable background in images and plots; (ii) nonideal behavior in the split photodiode (central to instrument performance), resulting in misleading data trends; (iii) calibration issues arising from operation under liquid immersion; (iv) feedback oscillations due to tip contamination in dynamic AFM; and related topics. I endeavor to explain the fundamental origins of these shortcomings, and present manifestations of artifacts in real data and images, such that suggested mitigations can be better understood.

Another question involves the degree of emphasis on custom methods. I have rooted this book in what I consider to be core methods available on most research-grade commercial systems, needing no special attachments. This includes "secondary" probes like lateral force, phase in dynamic mode, *X-Y* mapping of force-*Z* curves, and interleave-based scanning (electrostatic methods). To this I have added examples of simple but useful custom methods, often requiring BNC signal breakout (commonly available on commercial AFMs for at most a small additional expense), using (1) a function generator and signal adder circuit (easy to build with an op amp and a few resistors), (2) additional lock-in amplifier (e.g., to analyze an external periodic signal such as used in shear modulation), (3) LabView computer with high-speed DAQ card running commercial virtual instruments (e.g., logger, oscilloscope, digital signal analyzer), and (4) Witec's Pulsed Force Mode because I consider it to be a hugely enabling add-on (especially the digital version), and one that dovetails with the pedagogy of force-vs-distance and intermittent contact AFM. Inexpensive function generators, lock-in amplifiers and LabView are often available even in 4-year colleges. Configuring these special measurements is a good way to reinforce a student's or technician's understanding of what is being measured and how, and what are the limitations (signal magnitude, time/frequency domain), while at the same time expanding capabilities.

A further educational tool that is missing from nearly all books on AFM is the use of digital files for the manipulation of real data and illustrating concepts. During my career as an experimental and analytical scientist, nothing has elucidated better than manipulation of numbers and seeing results rendered in graphical form. This is particularly true for AFM. I created Microscoft Excel workbooks available from the Wiley web site*, given the ubiquity of this software. (Many other programs that are more advanced in graphics and curve fitting, such as Origin, MATLAB, SigmaPlot, Kaleidagraph, etc., can open Excel files.) In addition to real AFM data, I have generated several theoretical/mathematical constructs to illustrate concepts, for example Fourier decomposition of surface topography and force-distance relationships within adhesive contact mechanics models. I also incorporate usage of the freeware and open-source AFM program Gwyddion for AFM data processing/rendering.

A final, related question is the number of Scanning Probe Microscopy techniques to cover in a single book. It is worth noting that this capitalized term and its acronym, SPM, is almost unknown to the diversity of first or second-year graduate students that show up for AFM training in a core facility (at least in the US), and to the range of personnel in companies who seek AFM services from their central analytical labs or from university facilities. Yet the majority of edited books that might be useful to these users (usually at a much later date), or conference symposia and workshops of potential interest, use the acronym "SPM." This reflects the inclusion of additional measurement modes and methods (some of which are not force based) such as scanning tunneling microscopy, conducting AFM (measuring current through tip under bias but in contact under force control), electrochemical AFM (in liquid under bias), scanning capacitance microscopy, scanning ion conductance microscopy, scanning near-field optical microscopy, piezoresponse force microscopy, and more. As already stated, this book focuses on core methods available on most research-grade commercial systems, methods requiring no special attachments, and which are force based (including force gradient), and are applicable to a broad range of material/biological systems and research questions. The reader is referred to other books on the previously listed SPM techniques.

* Excel workbooks can be found by entering ISBN 9780470638828 on booksupport.wiley.com.

For an industrial lab, the payback on investments into AFM systems and personnel come from the wealth of information obtained by broad-based applications of AFM to imaging, materials property characterization, in-liquid interfacial analysis, tribology, and more, by a staff spanning from technicians to PhDs. I hope that this book will aid in understanding the use of AFM in these applications, and will help different corporate laboratories to understand the various components of a complex technology (e.g., medical devices). Similarly, I hope that this book will fit the missions of research university core facilities, especially those serving a broad clientele. And of course I hope that graduate students or postdocs, whose research project benefits from a more penetrating application of AFM, will be aided by this book. Another important potential use of this book is as a textbook or supplementary reference in advanced undergraduate or graduate courses.

GREG HAUGSTAD

July 2012

ACKNOWLEDGMENTS

My introduction to atomic force microscopy occurred in late 1991 as a postdoc in an industry-collaborative project, with Wayne Gladfelter in the Department of Chemistry at the University of Minnesota and Mike Keyes of DuPont Medical Products in Brevard, North Carolina, in basic research relevant to medical X-ray film. Considering my then-minimal knowledge of chemistry and zero experience with AFM, silver halides and polymers (rather, a background in semiconductor physics and ultrahigh-vacuum synchrotron work), I must heartily thank these two scientists for opening the door to AFM. In a short time I fell in love with the technique and, relatedly, nanotribology and polymer/organic thin-film systems. (Some of this love derived from a love of physics, first nurtured at Gustavus Adolphus College through the teaching of Dennis Henry, Richard Fuller and John Bolkcum.) I also quickly experienced the open-endedness of AFM research within an industrial collaboration, repeated in several projects up to the present day. I am thus indebted to industrial collaborators who encouraged exploration and methods development: Libby and Rolf Weberg, Richard Jones, Andrew Avery, Ed Parsonage and Klaus Wormuth. (I further thank Klaus for valuable feedback on early partial drafts of this book.) My growth as an applied scientist/consultant, with one foot in academia and the other in industry, would not have been possible without the countless hours in discussion with these individuals. Nor would my broad interests in AFM have flourished.

I am also eager to thank several former and current PhD students for whom I have been privileged to serve as a thesis co-advisor, including Jon Hammerschmidt, Ron Schmidt, Craig Dykstra, Donna Staarup, Srini Somayajula, Kanan Puntambekar, Vivek Kalihari, Dave Ellison, Dabien Chen, Yanfei Wu, Pranav Agarwal, Govind Saraswat, Kirby Liao and Peng Li; also, postdocs Susheng Tan, Jinping Dong and Francois Ahimou, with whom it was a joy to work. Some of these research projects have contributed data and/or understandings of AFM methods contained in this book; in particular, those involving friction and wear, shear modulation, force curve mapping, transverse shear microscopy and Kelvin-probe force microscopy. In this co-advisor role I have benefitted from the collegiality and intellectual exchange with faculty collaborators Wayne Gladfelter, Mike Semmens, Paige Novak, Dan Frisbie, Raj Suryanarayanan, Ron Siegel, Tianhong Cui, Murti Salapaka, Andre Mkhoyan and Chris Macosko.

I further thank Wayne Gladfelter for championing the importance of PhD-level scientific staffing to core facilities, including roles in curricula, methods development, research (both independent and collaborative), and interaction with industry. I believe this to be a model for core facilities in a twenty-first century research university. To grow hands-on expertise in analytical techniques over a period of decades is in many respects a privilege, and a lot of fun. But it's also the right way to staff the research enterprise.

Finally I must thank my wife, for patiently awaiting returns from weekend and late-night work sessions, for allowing the author to quietly write from the passenger's seat, and for managing the home front including (among other things) sons' activities; and all the while, providing cheerful encouragement.

GREG HAUGSTAD

Minneapolis, Minnesota

1

OVERVIEW OF AFM

1.1 THE ESSENCE OF THE TECHNIQUE

Atomic force microscopy or AFM is a method to see the shape of a surface in three-dimensional (3D) detail down to the nanometer scale [1,2]. AFM can image all materials—hard or soft, synthetic or natural (including biological structures such as cells and biomolecules)—irrespective of opaqueness or conductivity. The sample is usually imaged in air, but can be in liquid environments and in some cases under vacuum. The surface morphology is not perceived in the usual way, that is, by line-of-sight, reflections, or shadows.[1] Rather, at each point or pixel within a 2D array over the surface, a measurement of *surface height* is made using a sharp solid force probe. One could thus say that AFM is "blind microscopy"; it essentially uses touch to image a surface, unlike light or electron microscopes. The force probe may move over a stationary sample or remain stationary as the sample is moved under the probe, as discussed in Chapter 4. Typically, one chooses to display the height measurements as colors or tints, some variant of dark-is-low/bright-is-high, with a gradient of color or grayscale in between. Thus, an *image of surface topography* is obtained for viewing purposes, as exemplified in Figure 1.1, for several surfaces relevant to hard and soft materials science, nanotechnology, and biology. The typical range of these measurements is several micrometers vertically with

[1] Or, as with scanning electron microscopy, by secondary electron emission enhanced or suppressed to give the perception of reflections and shadows.

Atomic Force Microscopy: Understanding Basic Modes and Advanced Applications, First Edition.
Greg Haugstad.

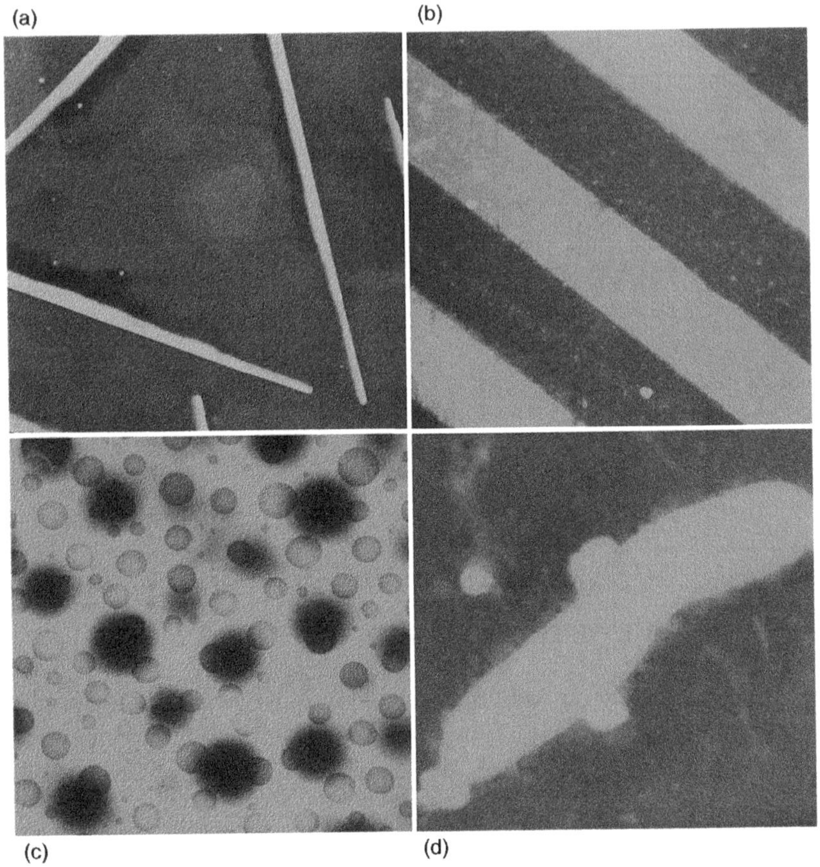

FIGURE 1.1 In-air surface topography images of (a) silver rods (15-nm tall) grown from a AgBr(111) surface by photoreduction, 5 × 5 μm [3]; (b) gold and aluminum lines (∼50-nm tall) lithographically created on silicon, 25 × 25 μm; (c) surface of a ∼1-μm thick polymer film (deepest valleys ∼100 nm) of a 75:25 blend of butyl and lauryl methacrylates (spin coated onto a silicon wafer), 8 × 8 μm; (d) wastewater bacterium (170-nm tall) on filtration membrane, 3 × 3 μm [4].

subnanometer height resolution and several tens of micrometers laterally, ranging up to ∼100 μm, with a highest lateral resolution of ∼1 nm (when not limited by the pixel density of the image, i.e., physical resolution as opposed to digital resolution).

Given that the image is constructed from height numbers, one also can measure *peak-to-valley* distances, compute *standard deviations* of height, compile the *distribution* of heights or slopes of hills . . . , and even *Fourier-analyze* a surface to identify *periodic* components (ripples or lattices) or dominant length scales (akin to a scattering technique). These *metrics* of topography can be relevant to technological performance or biological function, whether in microelectronics (e.g., roughness of layers or grain size, in deposition processes), tribology (e.g., friction and wear on hard disk read heads), polymer–drug coatings (e.g., surface contour area impacting

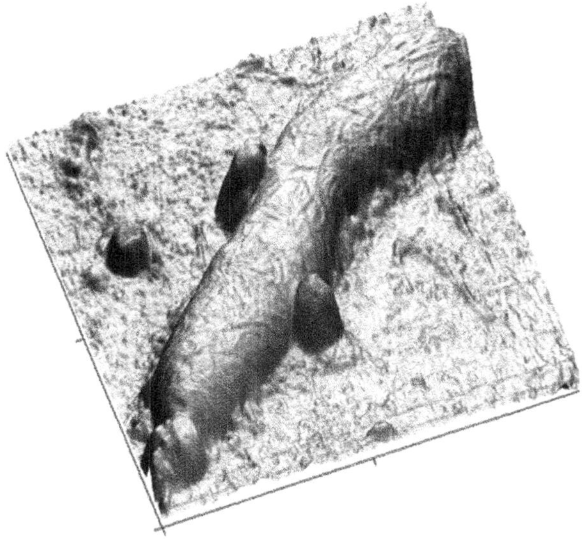

FIGURE 1.2 Wastewater bacterium (170-nm tall) on filtration membrane, 3 × 3 μm.

drug release rate), intrabody medical devices (e.g., shape of surface in contact with cells, tissues), cellular membranes and surface components (e.g., phospholipid bilayer, protein receptors), and much more.

As a bonus, with real height numbers in hand, one can render images in 3D perspective. The example in Figure 1.2 is an image of the dividing bacterium rendered in 2D in Figure 1.1d. *Computer-simulated light reflections and shadows* are incorporated to give the sense of a macroscale object and to enhance the perception of texture, even though the features may be nanoscale (i.e., below the resolution of real light microscopes). The angle of simulated illumination as well as the angle of "view" can be adjusted. The vertical scale has been exaggerated; the height of the bacterium is 180 nm, but is made to appear almost twice that high in comparison to the lateral scale. This is typical; often 3D-rendered AFM images exaggerate height by an even greater factor to bring out features for viewing.[2]

A bacterium, or for that matter anything hundreds of nanometers tall, is in fact a large object for AFM. With AFM's high precision, one can measure molecular or atomic crystal structures and indeed image striking, meandering steps. Figure 1.3 contains an image of five terraces on a surface of single crystal $SrTiO_3$, in ambient air. The steps between terraces comprise a "staircase" of increasing brightness from top right to bottom left. Also shown is a *histogram* representation or *population* of heights in the image: the number of pixels counted within narrow increments or "bins" of height (further discussed in Chapter 4), with the height scale increasing from left to right. One sees five well-resolved histogram peaks, spaced by 4 Å

[2] There is nothing wrong with this type of presentation, provided the scaling is made known to the viewer.

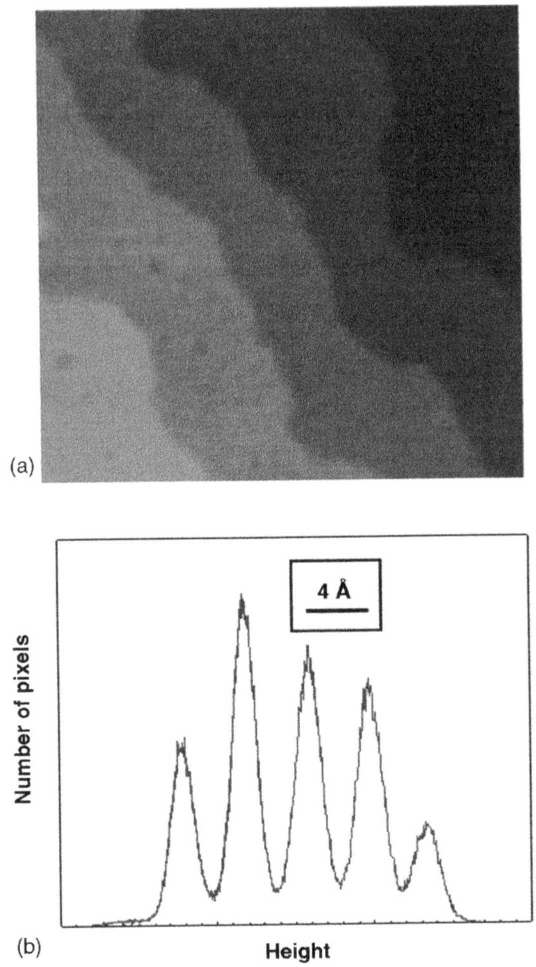

FIGURE 1.3 (a) 800 × 800-nm height image of $SrTiO_3(100)$. (b) Histogram of preceding image.

between adjacent peaks, the signature step size between adjacent (100) planes of $SrTiO_3$. The area under each peak—the total count of pixels—quantifies the relative surface area of each terrace within the imaged region. The shapes of step contours and extent of terraces are interesting for many reasons; for example, these may provide information on the *kinetics* and *thermodynamics* by which steps and terraces form during material growth processes [5].

How exactly does AFM determine the local height of a surface? By touching it with a sharp object, while measuring the *vertical or "Z" displacement* needed to do so. This "touching," however, can be very subtle; that is, the metaphor can be taken too literally. Moreover, heights are *indirectly* measured, as detailed in Chapter 4. In

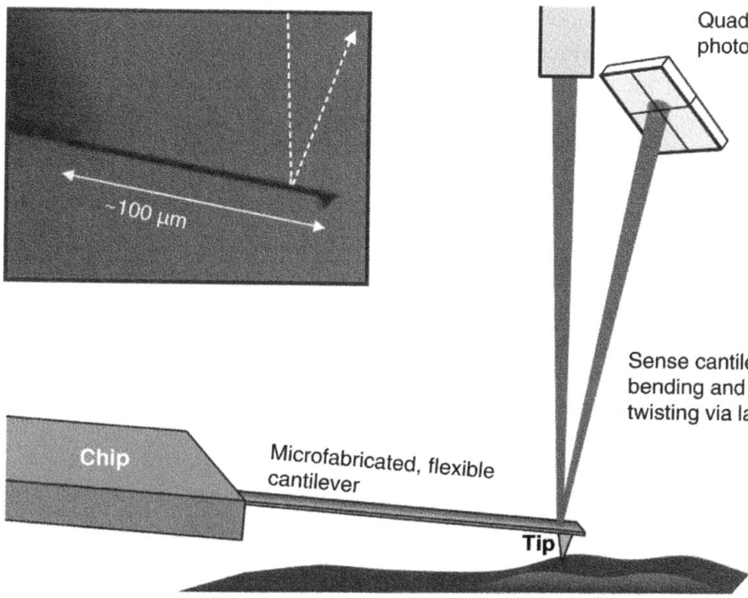

FIGURE 1.4 Schematic illustration of the core components of AFM: tip/cantilever/chip, focused laser beam, quad photodiode. Inset is light micrograph of a real AFM cantilever/tip viewed from the side; cantilever is 100 μm long, tip 10 μm tall.

most AFM designs,[3] and as depicted in Figure 1.4, the sharp *tip* (also known as stylus, probe, or needle) is attached to a flexible *microcantilever*—essentially a microscopic diving board—which bends under the influence of force. The behavior is that of a tip attached to a spring; a cantilever bent upward or downward is that of a compressed or extended spring. The bending is usually measured by reflecting a laser beam off of the cantilever and onto a split photodiode (a horizontal "knife edge"), the output of which gauges the position of the laser spot. The vertical tip movement, in turn, is quantified from this cantilever bending. Lateral forces that torque the tip, causing the cantilever to twist, can be measured via the horizontal movement of the laser spot (at a vertical "knife edge"). (We discuss lateral force methods in detail in Chapter 7.) The measurement typically will handle a vertical tip range of hundreds of nanometers, and with subnanometer resolution as detailed in Chapter 3 (including caveats). The vertical spring constants of cantilevers in common use range from 10^{-2} N/m to 10^{2} N/m (or nN/nm), resulting in a measurable force range from pico-Newtons to micro-Newtons.

In the simplest picture, one would bring the tip into contact with a surface, start moving or *scanning* laterally, and measure the vertical tip movement as the

[3] Several force measurement schemes are treated in Ref. [2], including the original method of Binnig et al. in Ref. [1].

cantilever bends up and down to gauge surface height while the tip slides over the surface. (Imagine the surface moving back and forth in Figure 1.4.) By doing so, over a 2D grid of locations across the surface, one could build up a surface *topograph*: height versus *X* and *Y*. But this scheme generally does not work very well because the up and down bending of the cantilever corresponds to higher and lower spring forces pressing the tip against the surface such that the tip or sample might be damaged due to high contact force atop the hills, and, conversely, the tip and sample might separate or *disengage* in the deepest valleys. Moreover, there is always some arbitrary *tilt* between a sample surface and the *X*–*Y* plane of the scanning device such that forces would continually grow while scanning in one direction (cantilever bending further up) and the surface would "recede from view" if scanning far in the opposite direction as contact is lost. The range of the split photodiode measurement may not be sufficient to gauge large excursions of the tip up or down anyway (i.e., large laser spot excursions). So AFMs normally employ scanning devices that displace not only *X* and *Y* but also *Z*, via *feedback*, to offset variations in height and keep the pressing force approximately constant.[4] This *reactive Z displacement* is, then, the sought measurement of surface height.[5] We will discuss in greater detail each of these components—tip/cantilever, laser, photodetector, scanner, and feedback circuit—as well as nonidealities and caveats associated with these components, plus the physics of the tip–sample interaction that affect topographic imaging—in Chapters 2–5.

1.2 PROPERTY SENSITIVE IMAGING: VERTICAL TOUCHING AND SLIDING FRICTION

AFM is, however, much more powerful as an analytical tool! One is touching the surface of an object that one wishes to understand. Using touch to measure height, but nothing else, seems unambitious. We all know that a piece of upholstery feels different from a piece of concrete. Food has a different texture if moist instead of dry. We wish to detect, even quantify, such differences with AFM. After all, a major goal of microscopy is to *differentiate* objects or regions. This may include materials such as metals, semiconductors, ceramics, minerals, polymers, or other organics—or biological entities such as cells, tissues, and biomolecules (e.g., proteins, polysaccharides, nucleic acids, lipids)—or, for that matter, may differentiate synthetic from biological. Also, one wishes to detect *changes* in a given material—say from *amorphous*, meaning atomically disordered, to *crystalline*—or from biologically *functional* to *denatured*. If we can touch at the nanoscale, and in a highly controlled way . . . , cannot we distinguish materials or biological entities based on unique

[4] This force is measured as the fixed vertical displacement of the tip, relative to its position when the cantilever is unbent as seen in Section 1.5, times the spring constant of the cantilever. The latter is approximately specified by the manufacturer or measured by the user as described in Section 3.7 and Appendix 1.

[5] Even this displacement is not directly measured, as detailed in Chapter 4.

properties, that is, how they "feel?" Understanding surface topography measurements by AFM is a first goal, but much of this book's subject matter relates to this second question: how to differentiate sample constituents and measure the properties of a given constituent. This encompasses changes in properties under variable environments including gaseous, liquid, and variable temperature, upon chemical treatment or with aging, and as a function of measurement parameters such as rate or applied force [6–8].

A common property metric is the rigidity or *stiffness* of a material, sensed as the resistance to the tip pushing in—the increase of *repulsive* force per unit distance of deformation.[6,7] *Rubbery* polymers, for example, derive their soft character from molecular composition, with further dependence on temperature and absorbed small molecules, such as water, residual solvent, or other such *plasticizers*, that tend to soften the material. Small changes in chemical structure or environmental parameters, such as temperature or humidity, can lead to dramatic changes in material properties. These properties are not only manifest in the 3D deformation of the sample as the tip pushes in but also at the *interface* between tip and sample. In what sense? AFM is exquisitely sensitive to the "grab" exerted by one material on another when we try to pull them apart or slide one past the other. The resistance to these motions depends in part on the strength of *attractive forces* between the materials constituting tip and sample. Most materials, when touching or very close together (~1 nm), experience *dipole–dipole* forces that produce *attraction*; in special cases in liquids, they produce *repulsion*. (This is discussed in Chapter 2.) Resistance to separation or sliding also can depend on *molecular motions* at the interface or internal to the sample. How? The motion of the tip itself can *activate* molecular motion or produce a stress that decreases the barrier to *thermal activation* of molecular motions at ambient conditions [9]. Once the tip and the excited molecules are far apart, there is no way for this motional energy to be given back to the tip. It is lost or *dissipated* as "heat" into the sample, in the most general sense of the term, meaning a large number of atomic and molecular *degrees of freedom* (e.g., bond vibrations); this heat, in turn, dissipates into the environment. Of course, these atoms and molecules already had motional energy prior to tip interaction; but in their "collisions" with the tip, this energy has on average increased. This is analogous to the kinetic energy of a car imparted to air (primarily N_2) molecules while driving down the road. Some molecules may actually collide with the back of the car to aid its motion, but on average the ensemble of collisions takes away kinetic energy (is dissipative for the car).

Thus, due to the "grab" exerted on the tip as manifest in adhesion and friction, as well as the finite mechanical stiffness of the sample, we have three differentiating measurements at our disposal. Figure 1.5 schematically depicts the raw

[6] This can be calibrated given that force is measurable as stated above and distance of deformation can be determined in comparison with force–distances measurements on a rigid reference sample; see Chapter 3.
[7] This is not to be confused with hardness, which formally refers instead to a resistance to mechanical yield, meaning plastic deformation, such as the creation of a permanent indent or hole. Some use stiffness and hardness interchangeably, but formally this is incorrect, just as using stiffness and density interchangeably would be incorrect. See Johnson, K.L., *Contact Mechanics*. 1985, New York: Cambridge University Press.

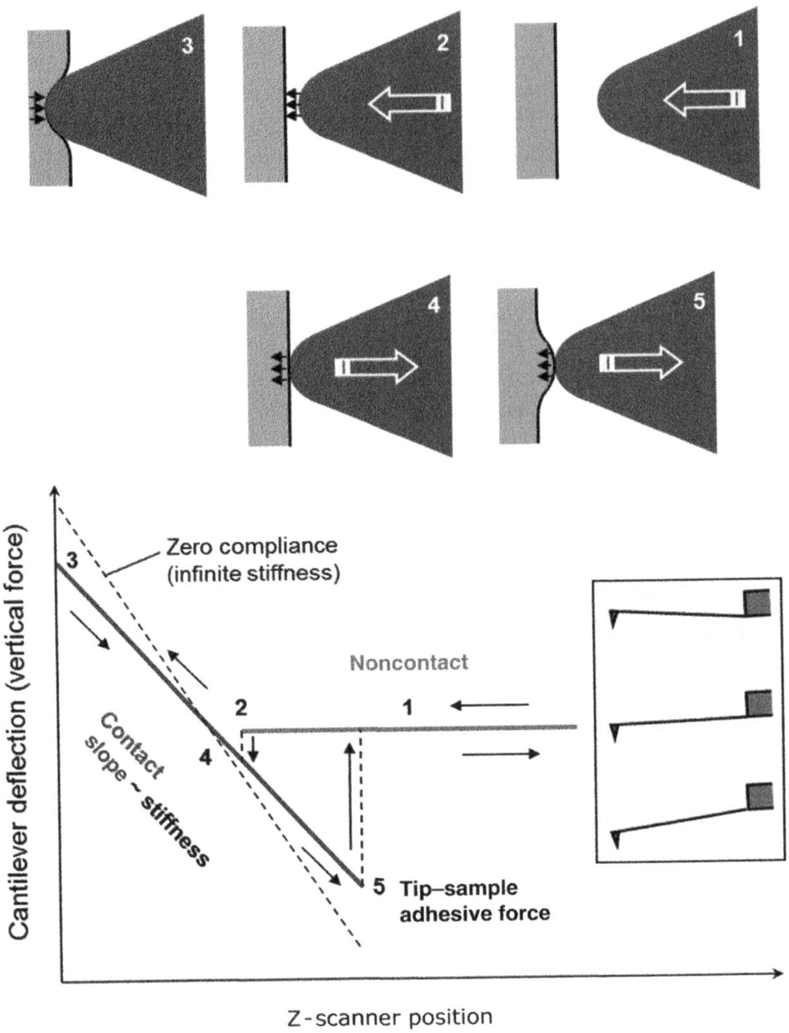

FIGURE 1.5 Tip–sample illustrations corresponding to select locations in a schematic force-curve cycle. (1) Tip and sample far enough apart that the interaction force is zero. (2) Tip close enough to sample so that attractive forces are felt and cause the tip to jump to contact (overcoming the resistance of the cantilever). (3) Maximum approach point with significant indentation into soft sample and repulsive forces acting on tip due to the sample deformation. (4) Return to state of zero indentation during retraction. (5) State of final contact just prior to the tip's jump from contact as the maximum pulling force of the cantilever exceeds the tip–sample adhesion. Inset depicts the directions of cantilever bending relative to the unbent stage (exaggerated).

measurement of stiffness and adhesion as seen in a *force curve* with accompanying illustrations of tip and sample. In Section 1.5, we treat force curves in greater detail, but for now, we consider only in the context of stiffness and adhesion images collected in a mode known by at least two commercial names: *pulsed force mode* and *peak force tapping*. (This is described in greater detail in Section 6.5.) During approach or retraction of the *Z* scanner to bring tip and sample together and then move them back apart at a given pixel location, one can render the *contact slope* as a datum of qualitative material stiffness. (Quantitative stiffness requires comparing this slope to the zero-compliance slope as approximated on a very rigid sample, the dashed diagonal line in Figure 1.5.) One commonly measures tip–sample adhesion as the maximum pulling force sensed upon retracting the tip from the surface with the *Z* scanner [6]. These measurements can be readily calibrated; the *Z*-scanner movement is quantified by imaging known height changes atop calibration gratings and the vertical cantilever bending is calibrated to equal the *Z*-scanner movement on a rigid sample (Chapter 3). This is converted to cantilever spring force by multiplying by the cantilever spring constant. Height in this mode can be gauged from the *Z*-scanner position at the turnaround point at maximum force (an operator-specified signal from the split photodiode).

Friction during continuous sliding contact is semiquantified as the change of lateral force signal upon reversing the lateral sliding direction, as seen in a *friction loop*. This is depicted in Figure 1.6 for two cases: relatively low and high applied (loading) forces. The latter is controlled by the value of cantilever bending maintained during lateral scanning, as can be selected during force-curve viewing. The measurement of the height of the friction loop removes the difficulty of measuring the *true zero* of the lateral quad photodiode signal and further removes most topography-derived contributions to lateral force as well as other artifacts that are independent of lateral scanning direction, as discussed in Chapter 7 (wherein procedures for friction force calibration are also described). The heights of friction loops on different surface domains—that is, the relative amounts of hysteresis—provide ratios of friction force, meaning quantitative materials contrast.

In the following, we consider examples of stiffness and adhesion imaging (Figure 1.7) and friction imaging (Figure 1.8). These cases are chosen to demonstrate not only the differentiation of similar materials but also the identification of chemical changes and differences in crystalline defect concentrations. Thus, these nontrivial examples illustrate the sensitivity of AFM as an analytical tool.

The images in Figure 1.7a and b are simultaneously acquired topography and stiffness for a blend of two chemically similar polymers—poly(butyl methacrylate) (PBMA) and poly(lauryl methacrylate) (PLMA)—that nonetheless dramatically differ in stiffness, PLMA being soft and rubbery and PBMA being relatively rigid and *glassy* (an amorphous solid state) [10]. Moreover, the right side of each image contains the as-prepared material and the left side the same material *after exposure to a 2.0-MeV helium ion beam* (used in Rutherford backscattering spectrometry) that preferentially depletes hydrogen and oxygen, leaving a carbonized ("burnt") material. The topography contains a *reduced height* of about 600 nm from beam exposure at left due to the loss of atoms; the stiffness reveals a *lack of contrast* in

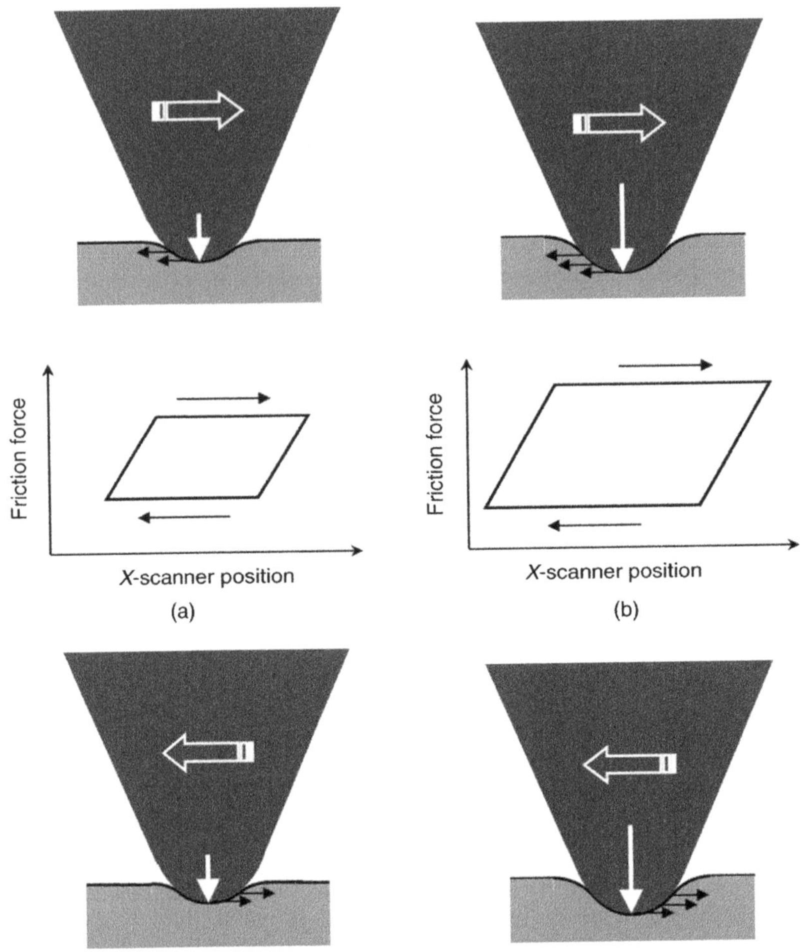

FIGURE 1.6 Friction loops and associated tip–sample illustrations for two cases of frictional imaging, (a) low and (b) high applied vertical force via different amounts of upward cantilever bending maintained as the tip slides over the surface.

the exposed region, whereas the as-prepared material at right contains soft (dark) and rigid (bright) domains, the phase-segregated polymer blend. The soft domains include a few large circles that correlate with circular dips or "craters" in topography; yet, many of the circular topographic features do not exhibit softness. There are also much smaller, soft circular domains.

But touching can be subtle indeed. The adhesion or pulling force needed to separate tip from sample is displayed in Figure 1.7c. Darker corresponds to lower adhesion. Here, we find a richer and subtler sensitivity to material differences at the surface. Most of the soft circular domains, but not all, exhibit lower adhesion—counterintuitively less "sticky", notably three large circular domains residing at the

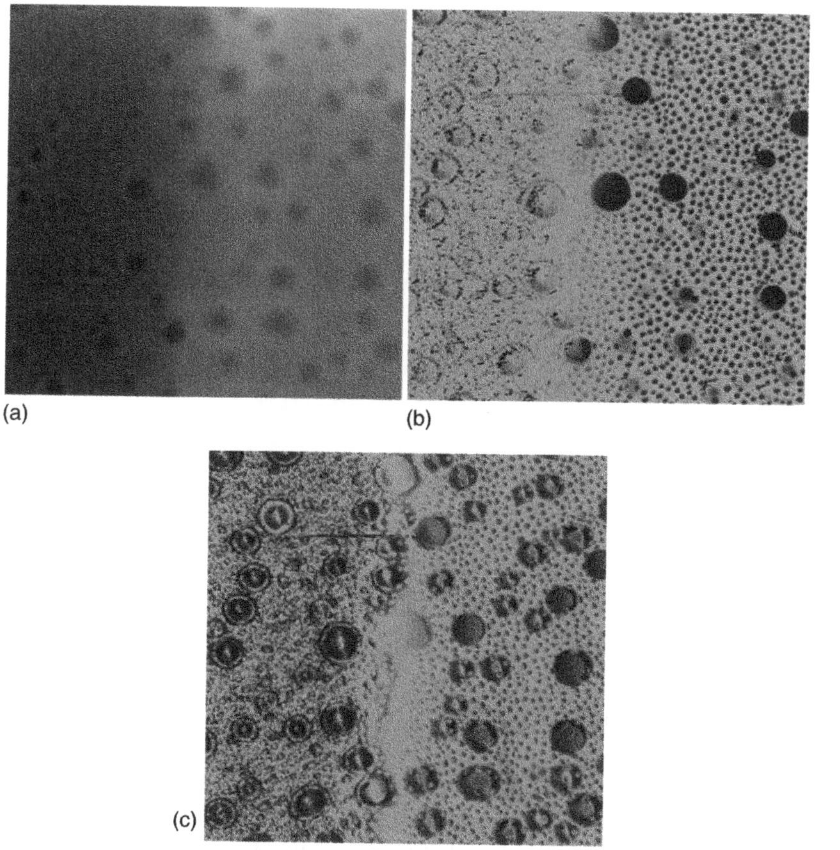

FIGURE 1.7 (a) Height, (b) stiffness, and (c) tip–sample adhesion images of a 75:25 blend film of PBMA and PLMA (spin coated onto a silicon wafer), 40 × 40 μm. The left portion of the imaged region had been modified by exposure to a 2-MeV beam of He ions.

boundary of the ion-beam-modified and unmodified regions. Moreover, there are many low-adhesion circular domains that do not seem to be soft. Even in the ion-beam-modified left side of the adhesion image, there are intriguing variations in tip–sample adhesion with little to no corresponding differences seen in the stiffness image.

All of these variations on materials contrast may seem bewildering for a seemingly simple, two-component system. Indeed, the complexity of Figure 1.7 is an example of what one often finds upon first viewing a property-sensitive image of a multicomponent sample: no shortage of contrast! In analytical science, a first goal is to measure differences. Then we have the potential to learn something. Sorting out what it all means, quantitatively and at a fundamental level, is always a remaining challenge. Some may balk at property-sensitive AFM imaging for this reason, while for many this challenge is the fun part! But our strongest motivation is the potential

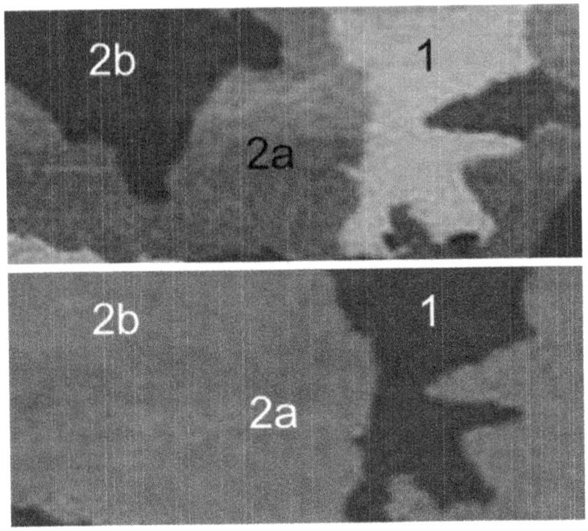

FIGURE 1.8 Ultrathin film (1.75 monolayers) of pentacene grown on an oxidized silicon surface, 7 μm across. Bottom image is topography of mostly second layer, partial first layer; top is simultaneously acquired friction force image.

payoff. From a utility standpoint, even qualitative and empirical findings that, say, correlate with material performance in technological applications can be very useful. In some cases, qualitative information obtained via material contrasting modes may be more important than quantitative topographic information.

Indeed, in some cases, topographic images tell us practically nothing, whereas the tip–sample interaction is astonishingly revealing. The magnitude of the sliding friction force can be exceedingly sensitive to *disorder* in crystalline organic systems as further discussed in Chapter 7. Our second example, in Figure 1.8, is a two-molecular layer film of pentacene, a molecule valued for its semiconductor properties and potential use in flexible electronic circuitry. The bottom topographic image, collected under continuous sliding contact, contains two shades corresponding to the surface heights of the first (dark) and second (light) layers, each about 2 nm thick. The top image displays the corresponding friction force and contains *three* shades, the brightest (highest friction force) measured atop the first layer, while both intermediate and low values are found *within the second layer*. The intermediate shade 2a, the higher friction within the second layer, corresponds to domains known to contain a higher amount of disorder in the form of *line dislocations*: flaws in the orderly packing of molecules into a 2D periodic array that result from stress, in turn, derived from a crystalline structure that is incommensurate with underlying crystal grains [11–13]. Understanding the fundamental, molecular-scale mechanisms of friction is a goal of the *nanotribology* research community [9]. But this example demonstrates how AFM can be highly useful even in the absence of first-principles understandings of *contrast mechanisms* (detailed identification of the kinds of molecular motion activated by the passing AFM tip).

1.3 MODIFYING A SURFACE WITH A TIP

Shear forces also can be used to "tear up" a material. A simple, practical use of this "abrasive" scanning is the analysis of multilayered films. Provided that the top layer is not too difficult to disrupt with the tip and the substrate or underlayer relatively impervious to this same scanning tip, the ability to expose the substrate or underlayer results [14]. One case is a polyvinyl alcohol (PVA) [15] film that can contain a discontinuous skin of highly crystalline and brittle polymer. It is quite easy to fracture or disrupt the skin and expose a more amorphous underlayer. An example is shown in Figure 1.9, where a subregion previously had been cleared down to the underlayer by scanning at an elevated applied force (i.e., by maintaining a greater upward bend of the cantilever). The larger region in Figure 1.9 was then imaged at a light force where further tearing did not result. The altered box is evident not only in the topography image at left but also in the corresponding friction force image at right. The friction also suggests that some ill-defined surface mixture of the two components has not resulted; the level of friction within the cleared region is equal to the level of friction found in the initial exposed underlayer at left. (Intermediate values are indeed found within the lip of material piled at the periphery of the cleared region).

One may wonder how well these scanning conditions, abrasive versus nonabrasive, can be controlled via the applied loading force. We have already mentioned in Section 1.2 that the magnitude and sign of force can be measured in force curves because the zero of force is measurable. The operator may thereby specify the value of force to be *maintained during imaging*, what we call the *setpoint*. Indeed, the operator may vary this setpoint and, thus, the applied force through different values and measure how the friction force changes. Even negative forces can be applied, which means pulling on the tip, with the cantilever bent down like a

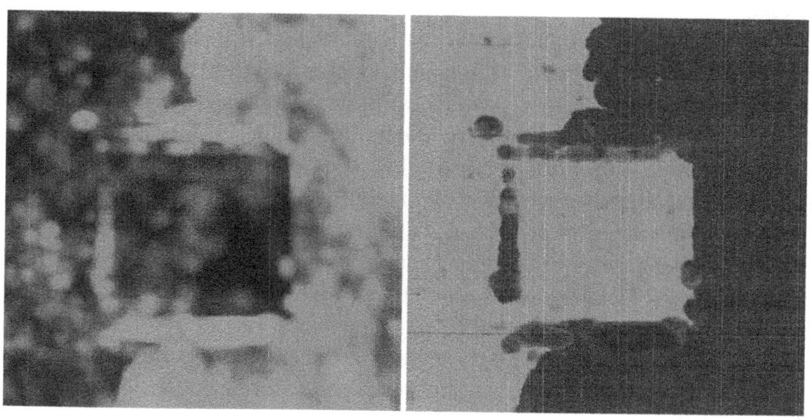

FIGURE 1.9 Topography (left) and friction force (right) images (1×1 μm) of PVA following abrasive scanning of a 500×500-nm subregion. The low-friction, highly crystalline topmost component is selectively disrupted.

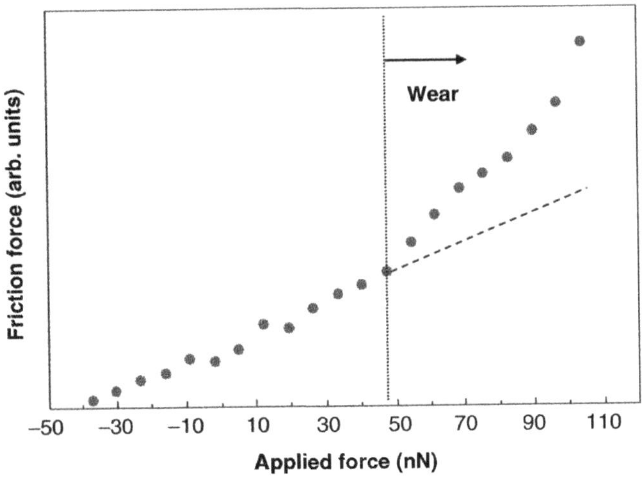

FIGURE 1.10 Friction force versus applied loading force atop a highly crystalline skin layer on a dry gelatin film.

stretched spring. In this case, contact is maintained by an even stronger adhesion force that pulls the tip in the opposite direction. We will discuss the analysis of quantitative friction–load data in Chapter 7. For the purposes here, one wishes to identify the onset of abrasion. This is typically seen as an increase of the *slope* of friction versus applied force, as shown in Figure 1.10 for the case of a (dry) gelatin film very similar to the PVA film examined in Figure 1.9, in which it contains a highly crystalline skin layer [14]. (Gelatin is a polypeptide derived from the protein collagen.) Thus, one can assign the initial low slope as intrinsic to friction in the absence of wear and friction forces above this extrapolated trend as due to wear processes.

This methodology has found utility in the biological as well as synthetic material realms. One example is a method to quantify cohesive strength of biofilms, specifically the extracellular polymer substances (proteins, polysaccharides) that serve as a "glue" to bind together a matrix containing bacterial cells, in the case of wastewater-treatment biofilms (Figure 1.11) [4]. *Cohesion* in and *adhesion* of biofilms is of great significance to many technological applications, whether this mechanical coherence is desirable, in the case of wastewater treatment, or undesirable, in the case of biofouling of surfaces that are preferred to remain clean. With successive AFM raster scans at relatively high loading forces, a gradual excavation of a hydrated biofilm matrix can take place (at 90% relative humidity), whereby chain molecules are disentangled and displaced by shear forces. During the course of this multi-raster scan treatment, one can reduce the loading (vertical) force to avoid abrasion, zoom out and acquire topographic images to assess the previous excavation process as done in Figure 1.9 (right image). Comparisons with an initial image of the pristine surface (Figure 1.9, left image) can be used to quantify the abraded surface. In particular, one can compute the volume of material

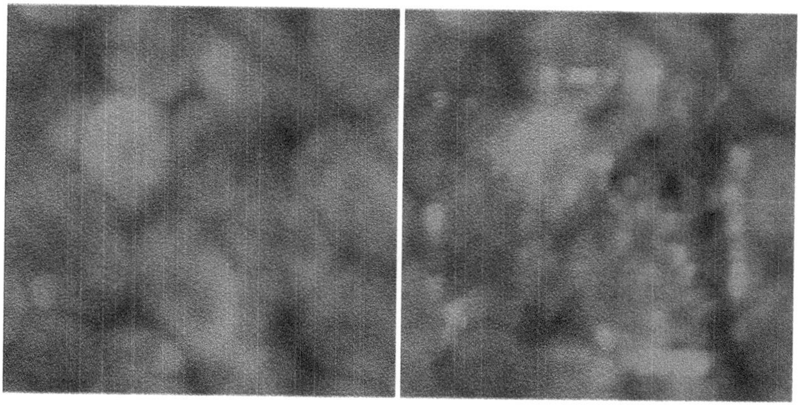

FIGURE 1.11 Topographic images (3.7 × 3.7 μm) of a wastewater treatment biofilm before (left) and after (right) AFM scanning at a destructive force within a square subregion (2.5 × 2.5 μm).

displaced by abrasive scanning. It is also possible to analyze the total friction force versus load to identify the fraction of frictional energy transfer that is responsible for abrading the biofilm, by extrapolating and subtracting the low-slope friction force that is unrelated to wear as suggested in Figure 1.10. By integrating the extra friction force due to wear over multiple raster scans, an aggregate frictional energy of wear can be measured. This energy, divided by the volume of film displaced, is then a measure of cohesive energy density [4], an intrinsic and exceedingly difficult property to determine by any method.

In addition to wear, many kinds of phenomena may be induced or catalyzed by tip–sample interaction. For example, it is well established that in air, *capillary transport* can take place, whereby molecules are transferred from tip to sample or sample to tip through a capillary *nanomeniscus* that forms at the tip–sample contact zone (Chapter 6) [16]. Even local oxidation of the surface can be carefully produced by an applied voltage bias between tip and sample under controlled humidity (to control the size of the meniscus). Many of these processes happen very rapidly. Indeed, just making a first contact of tip to sample can produce dramatic effects. In Figure 1.12, the initial touch of a freshly prepared gelatin film induced outward deformation ("doming") that extended many micrometers radially from the touch point (top-left image), together with a dramatic change in properties of this deformed region such as frictional response (top-right image). The presence and extent of this phenomenon is strongly dependent on film *age*. For gelatin, this means the *extent of a physical cross-linking network* as driven by collagen *renaturation*. Collagen is a connective protein (here extracted from mammal bone) that forms *triple-helical conformations* (shapes or structural arrangements). When chemically processed to form gelatin, these triple helices are "unwound" to isolate the individual molecules, but with passing time, triple helices reform as driven by the intrinsic biochemistry (hydrogen bonding of units in regular locations along the

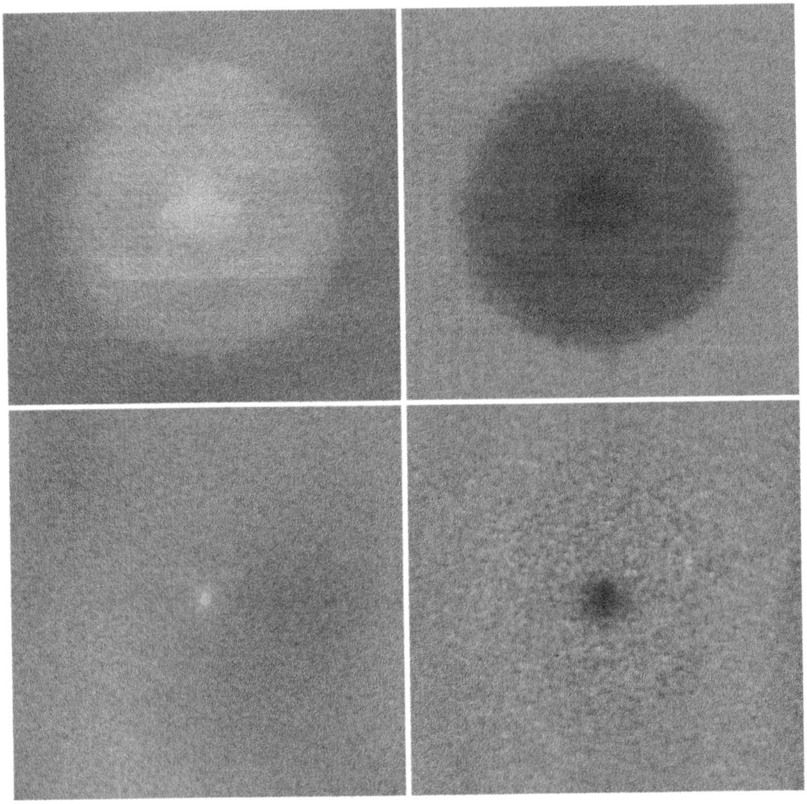

FIGURE 1.12 Height (left) and friction force (right) images (10 × 10 μm) of a (dry) gelatin film following approach and contact of tip to the center of the circular regions. Freshly prepared (top) and 1-week-old (bottom) films.

polypeptide chains). This produces *physical cross-links* that hold the polymer matrix together [17]. As the number of these cross-links increases, so does the stiffness of the matrix. Thus, an attractive force produces a doming of much smaller spatial extent as the film stiffens with age. After a week of aging, the touching of the same tip to surface produces an effect that is an order of magnitude smaller in spatial extent as seen in the bottom images of Figure 1.12 [18].

1.4 DYNAMIC (OR "AC" OR "TAPPING") MODES: DELICATE IMAGING WITH PROPERTY SENSITIVITY

When working with soft synthetic or biomaterials, or structures that weakly cohere or weakly adhere to substrate, one often discovers that the abrasion demonstrated in the previous section *cannot be avoided*, no matter how lightly one contacts the surface. A sliding tip, in the presence of tip–sample adhesion, may generate stress that exceeds a *yield point*, damaging the material under study. Multiple strokes by the tip, given a particular stroking direction, result in displacements that do not "relax away." Repeated scans show an additive effect [19]. In many cases, there is no gross tearing of material, rather, a subtler transformation of surface topography. In some cases, *repeated raster scanning is required to unambiguously identify these effects* and, thereby, assess whether the observed topography is "true." In the process, one can sometimes discover fascinating behavior. For example, on glassy polymers, one can observe the formation of *periodic ripples* perpendicular to the fast-scan axis of the raster pattern during repeated scanning. Systematic studies have unveiled quantitative relationships between geometric characteristics of the ripples—such as periodicity and angle of orientation—and fundamental thermokinetic responses of the polymer (a subtopic of Chapter 7) [19–21]. Indeed, this suggests the *usefulness* of scan-induced patterning of thin polymer films to gauge molecular mobility, especially those motions pertinent to tribology.

Returning to the goal of obtaining a "nonperturbative" image, of avoiding the preceding phenomena, what should one do? Empirically, it was learned in the first years of AFM that the biggest problem is indeed shear forces. (It is not too difficult to imagine that the sliding tip depicted in Figure 1.6b might tear the highly stressed material!) Thus, a very brief and purely vertical touch of tip to surface, with the tip remaining *off the surface most of the time* while scanning the surface, is key to avoiding or minimizing many of the above problems. The general term for such an imaging scheme is *intermittent contact*. There is more than one way to implement intermittent contact, different modes of operation. In Section 1.2, we showed images acquired with high-speed force-curve mapping [22], which uses the Z scanner to approach and touch tip to surface then retract, as few as one touch per pixel in the image. This is actually a rather less-known method and will be discussed in more detail in Chapter 6.

A much more common implementation of intermittent contact is called "tapping mode" by most users (originally, a vendor's trademark for a particular implementation) that also goes by the names dynamic, AC, and vibrating mode, or in some cases, noncontact AFM as will be discussed [23]. This scheme indeed *vibrates the cantilever* at or near its fundamental flexural *resonance frequency* such that many cycles of approach and retract occur per pixel location. Thus, a time-averaged *dynamic* interaction results. But this vertical cycle is not produced by the Z scanner; the vibrating cantilever does all the work. The *amplitude* of vertical tip oscillation must be sufficiently large to overcome adhesion between tip and surface. The amplitude also is commonly used to enable the tracking of surface topography, but with a number of caveats and potential pitfalls, as described in detail in Chapters 4 and 5. Tracking means that the Z scanner reactively displaces the distance between oscillating tip and sample to *keep the tip amplitude constant*, reduced from its amplitude when free of the surface. This is depicted in exaggerated form in Figure 1.13, for the case where the Z scanner is displacing the cantilever chip to accommodate changes in surface elevation (some instruments instead displace the sample vertically).

In Chapter 5, we also describe a property-sensitive imaging mode, known as "phase" imaging, which proceeds in parallel with topographic imaging in dynamic

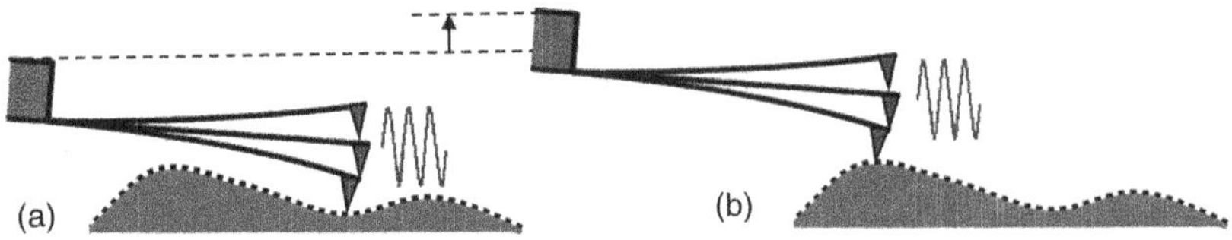

FIGURE 1.13 Exaggerated illustrations of topographic imaging via a vertically oscillating AFM cantilever/tip. An upward displacement of the chip to which the cantilever is attached allows the tip to maintain a constant oscillation amplitude whether (a) in a valley or (b) atop a hill. In actual operation, the oscillation amplitude is typically on the order of tens of nanometers, whereas the height of the tip is $\sim 10\ \mu m$.

AFM. We will see that this quantity can be more difficult to interpret than the adhesion, stiffness, and friction force images in the preceding sections, yet has proven exceedingly valuable to both fundamental and applied science and engineering. The "phase" is the *time shift between a sinusoidal driving signal that vibrates the base of the cantilever and the approximately sinusoidal motion of the tip end of the cantilever,* as the tip oscillates near and far from the sample surface. At first sight, it is not at all obvious why this measurement should provide materials contrast! Indeed of all AFM imaging modes, phase imaging is perhaps the most esoteric and almost certainly the most misinterpreted in published studies, conference presentations, and internal analytical reports. This is not because of a lack of published understandings. The core principles for understanding phase data were identified and published by the late 1990s in journals such as *Applied Physics Letters*, *Physical Review B*, and *Ultramicroscopy*. But with the massive growth in the presence of AFM's in laboratories of every sort, at institutions of every sort, coupled with the lack of formal training for most users of these systems, erroneous interpretations and poorly conceived instrument settings may often result.

There is firstly an intrinsic phase lag between the driving sine wave and the tip motion—which can be measured when the tip is oscillating far from the surface, out of *engagement* with sample—and secondly a phase shift resulting from tip–sample interaction. The latter shift results from the *modified resonant behavior of the cantilever* as discussed in Chapter 2 and later chapters. This phase shift provides materials contrast that may derive from *any and all portions of each approach–retract cycle*: whether the tip is sensing attractive forces far from contact (say due to a charged surface), or pushing into the surface, or breaking away, etc. Why use phase imaging if it *convolves* all of these different interactions? Well, it turns out that the extremely rapid, dynamic vertical oscillation can be controlled to provide an exquisitely *delicate* tip–sample interaction, even more so than other intermittent contact modes such as force-curve mapping. The most delicate case is a *noncontacting* oscillation; only attractive forces are sensed as the tip nears the surface because it does not get close enough to actually impact the surface, which would generate a repulsive force. This attractive interaction modifies the resonant behavior, as diagnosed in the phase shift; thus, the tip's oscillation amplitude decreases and enables the tracking of topography. Why this happens will become clearer in Chapter 2 and

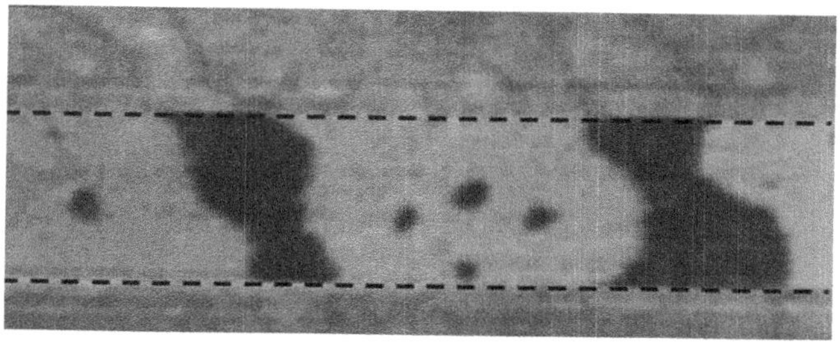

FIGURE 1.14 Topography image (1.4 μm wide) of an ultrathin polymer–surfactant complex film collected under two different parameter settings (inside and outside of dashed lines) in dynamic mode.

in further detail in later chapters. The very delicate interaction in this *attractive regime* allows difficult samples such as gels, nanoparticles weakly adhered to substrate, or even liquidy films to be imaged without tearing, plowing, puncturing, or other deleterious effects. Even some of the more robust, multicomponent materials—which can be imaged just fine with other modes—may be contrasted with *better resolution* in phase imaging because of the exceedingly brief (often $<1\ \mu s$) and delicate interaction.

By varying imaging parameters, one can easily toggle from a "true topography" imaging regime to a regime with penetration of tip into sample, and often selective to material components as discussed in Chapters 4 and 5. An example of this phenomenon is the topography image in Figure 1.14, an ultrathin complex film (<10 nm) containing silicone oil (polydimethyl siloxane), the short-chain (~ 2 nm) *amphiphilic* molecule cetyl trimethyl ammonium chloride (i.e., one end being cationic and polar and thus water loving, the other oily and thus water hating), and probably bound water [24]. The horizontal dashed lines mark the point during raster scanning at which *instrument settings were altered* so as to switch from "true" topography (top and bottom) to "false" topography (middle), where the tip selectively penetrates nanoscale-thick liquidy domains to reach near the solid mica substrate a few nanometers below. The operational strategies for exploring such behaviors are discussed in Chapter 5.

So far we have not emphasized nanometer-scale lateral resolution. But AFM tips are sharp enough, and the brief tip–sample interaction is delicate enough, to enable a touching zone that is only ~ 1 nm across. This takes us into a regime of resolution that light microscopy cannot reach, far below the wavelength of visible light. One common example is indeed phase imaging, as described above, to discern *phase segregation* in *block copolymer* systems. Notice we are making use of two completely different meanings of the word "phase!" This of course adds to the confusion. Phase segregation refers to the tendency of certain substances to separate from one another, such as oil and water, provided thermal motion and time. Thus, one obtains domains rich in one type of molecule or another. Block copolymers are long-chain molecules containing long "blocks" of one homopolymer or another. That is, each long block contains repeated "mers" of the same chemistry, one polymer.

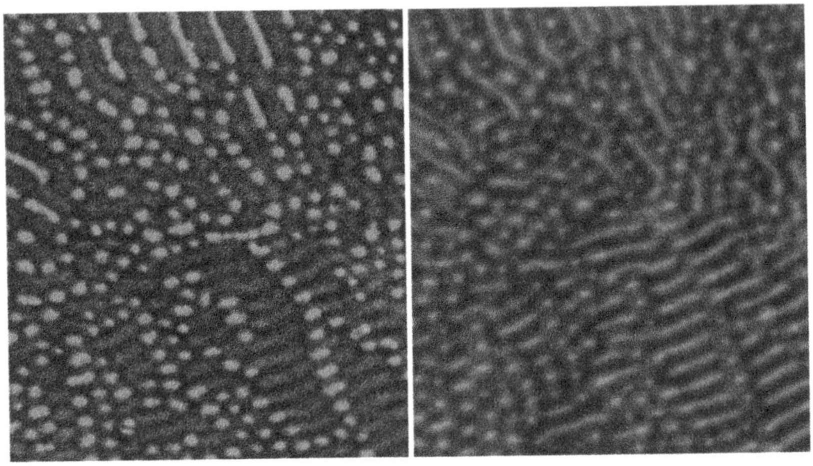

FIGURE 1.15 Dynamic-mode phase images (1 μm across) of a thick poly(styrene–isobutylene) triblock copolymer coating acquired using different imaging parameters to achieve different depth sensitivities.

Our example is an ABA block copolymer, meaning each long-chain molecule has an A block of polymer constituting the end chains, here polystyrene (PS), whereas the middle of each molecule is a B block of a different polymer, here polyisobutylene (PIB). Solid PS by itself is a rigid and glassy polymer, and PIB is soft and rubbery. The total length of the two PS blocks is about one-fourth of each chain molecule. The two phase images in Figure 1.15—here referring to the measured shift between two sine waves, driving and response—reveal phase-segregated PS and PIB domains, bright and dark, respectively—here referring to segregated materials on the ~50 nm scale. This means that like blocks from different chain molecules tend to bunch together to create PS-rich or PIB-rich domains. The lengths of the block chains result in 50-nm domains. Thus, longer molecules with longer blocks would produce larger than 50-nm domains, and shorter would produce smaller domains. The shape of the domains relates to the relative size of A and B; for our case, the tendency is for cylinders of PS to form in a matrix of PIB. An annealing step, exposing the polymer to solvent vapor for many hours, results in the orientation of cylinders either perpendicular or parallel to the surface, a more thermodynamically favored state [25].

The images in Figure 1.15 look similar because they correspond to the same surface region; yet they are different. The left image was acquired with a "light touch" and is thus "sharper." Moreover, it contains three levels of contrast as compared to two in the right image acquired under "pounding" conditions. This example illustrates how the exquisite sensitivity enables not only nanoscale lateral resolution (*X–Y*) but also vertical resolution (*Z*). If imaging parameters are selected to produce an extremely delicate touch—the case of the left image—then one can perceive the difference between PS at the surface (bright) or 1–2 nm below a "skin" of PIB (intermediate tint). The latter tends to segregate to the surface–air interface due to its lower surface energy [26]. Selecting imaging parameters to produce a stronger push into the surface—the right image—makes the tip "feel" the stiffer PS equally, whether at the surface or under 1–2 nm of soft PIB. The result is a single level of response for PS (brighter) instead of two levels. (We will discuss these operational parameters in Chapter 5.) Also, the boundaries between regions are more blurred in the right image; by pushing deeper into the polymer, a larger *contact area* between tip and sample is formed, meaning coarser image resolution.

1.5 FORCE CURVES PLUS MAPPING IN LIQUID

It is not only imaging that is enabled by nanoscale force probing. The *distance dependence of force* between tip and sample, acquired in a force curve,[8] can elucidate many physical traits of materials and biomolecules. Note that this differs from the *time-integrated* sensitivity of dynamic AFM, the preceding section, to these same distance-dependent forces. (In that case a variety of forces—including attractive and repulsive, short and long range—are convolved into one number, the phase shift.) Examples include the strength of interfacial attraction or repulsion, molecular binding or unbinding, rigidity and viscous resistance to tip motion, as well as fundamental studies of these phenomena via rate and/or temperature dependences [7]. Although typically acquired at a single-point location within the *X–Y* range of the scanner, force curves can also be collected using *X–Y-mapping routines*, meaning a 3D domain or "data cube." The detailed algorithms for doing this differ among commercial AFM systems and go by commercial names such as *Force Volume* and *SPS Imaging* in addition to the already-mentioned *Pulsed Force Mode* and/or *Peak Force Tapping*. Moreover, the data can be analyzed within subregions of the explored domain, although commercial data-analysis routines remain somewhat limited at this writing. (Third-party software and public-domain routines exist for this purpose.)

The details of force-curve acquisition and data interpretation are discussed in Chapters 2, 3 and 6, and force-curve mapping also is discussed in Chapter 6. Here, we preview two examples in processed form. The three full-cycle (approach-retract) force curves in Figure 1.16a were acquired under deionized water immersion from three surface locations on a thin-film system: (1) a silanized glass substrate, cleared by abrasive AFM scanning (at elevated force), (2) surrounding lip of disrupted polyacrylamide film, and (3) undisturbed polymer (swelling ~300%). This sample region was prepared for the purpose of measuring the water-swollen film thickness and analyzing the behavior of chain molecules in different conformational states. Conventional (sliding) contact-mode height images in water are shown in Figure 1.16b and c, as (a) color-mapped height to examine *Z* values and (b) differential height ($\delta Z/\delta X$) to enhance texture.

[8] Also called force–distance curves, force-displacement plots, force calibration plots, force spectroscopy, approach-retract curves, or force profiles.

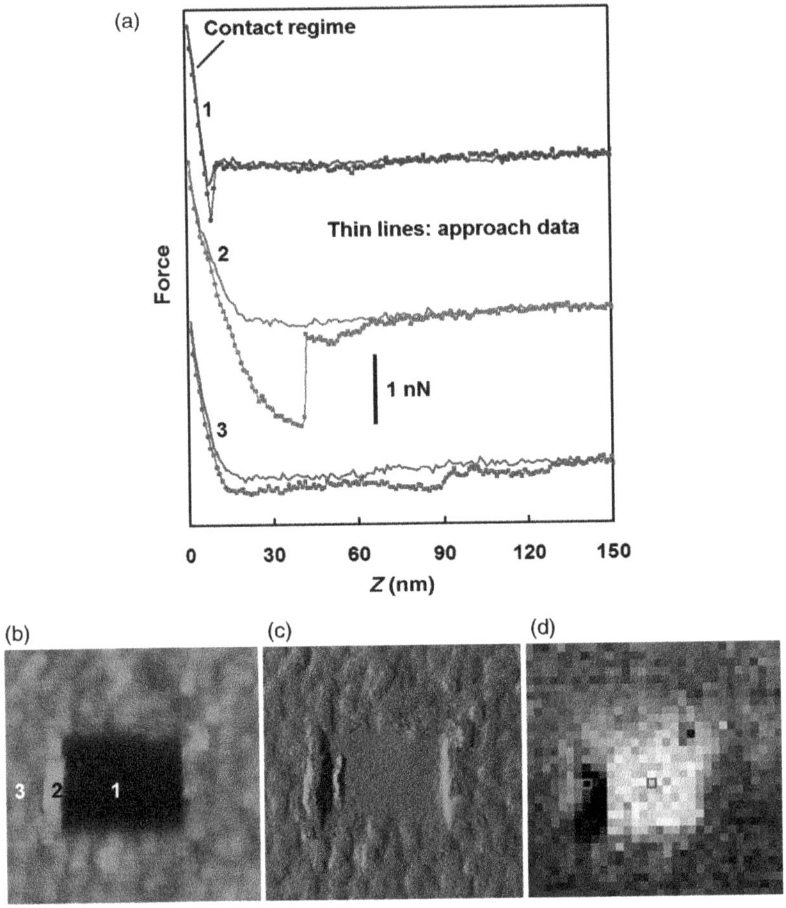

FIGURE 1.16 (a) Raw force curves obtained in deionized water at three characteristically different locations on a scan-abraded polyacrylamide film in water as imaged over 1.5 × 1.5 μm in (b)–(d). (b) Contact-mode height image of film and exposed glass substrate collected in conventional (sliding) contact mode. (c) Differential height image (slope of surface along horizontal) to highlight texture. (d) Stiffness mapping, the steepness of each force curve in the contact regime near the turnaround point. Open squares denote the locations of the force curves in (a).

There are several interesting features contained in the three force curves. During approach from right to left, short-range attractive forces (a steep downward dip) are felt above the bare glass within a 5-nm distance, but not either case of polymer. Upon "contact," the steepness of growing positive force (resistance) reflects mechanical stiffness that differs among locations. Figure 1.16d contains a qualitative stiffness mapping generated with a custom software algorithm [27], brighter being higher stiffness. The sites of the three force curves are denoted by corresponding open squares. The stiffest contact is obtained on the exposed glass substrate, whereas the softest contact is on the disrupted polymer to the immediate left of the exposed substrate. During retraction, dramatic differences in tip–sample adhesion are visible, indeed very large variations in the size of the *hysteresis* or *irreversibility*. The hysteresis loop is particularly large at location 2 because of the polymer disrupted by abrasive scanning. Weak attractive forces are sensed to distances of ~100 nm on region 3, the undisturbed polymer, meaning long chain molecules remain attached to the tip, spanning the gap between the tip and the film.

Such rich and characteristic behaviors (further examined in Chapter 6) can be qualitatively or semiquantitatively probed by even novice AFM operators in a matter of minutes including setup time. Even a qualitative examination may provide sufficient answers to one's analytical questions. As with topographic imaging, however, AFM can tell us much more—with more effort—because of the *quantitative* nature of the measurements. In later chapters, we will delve into systematics as well as some important realities and caveats. But for the moment, let us look at one example in a little more detail: relatively short-range attraction between SiO_2 tip and a *silanized* (i.e., made *hydrophobic*) glass substrate. We see in Appendix 2 that the result of summing atomic or molecular dipole–dipole interactions between a spherical-ended model tip and a flat surface is in an *inverse-squared* distance dependence of attractive force, which becomes negligible beyond a few nanometers. We will also see that immersing in a liquid medium can add further complexities that are nevertheless open to quantitative and fundamental analysis. Figure 1.17 contains a force–distance relationship for a SiO_2 tip and a silanized

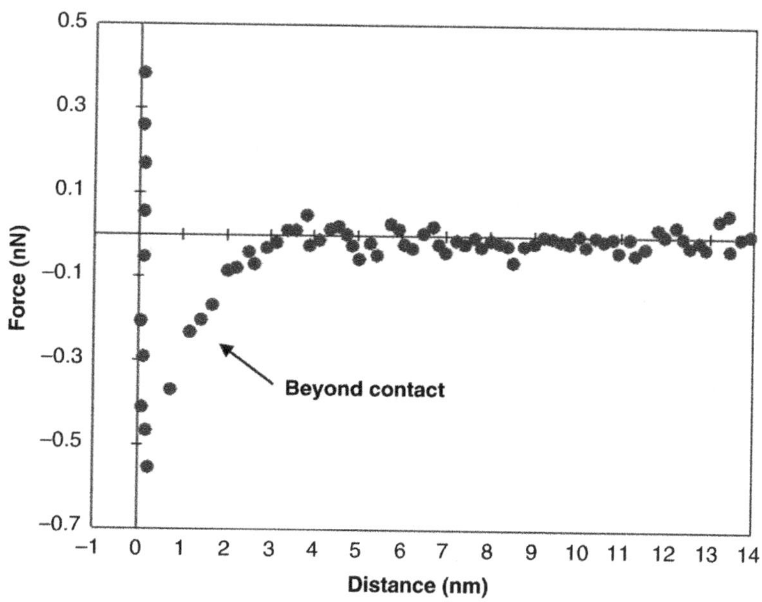

FIGURE 1.17 Force–distance relationship in water between a SiO_2 tip and a silanized SiO_2 surface.

SiO_2 surface immersed in deionized water, acquired at much higher data density along the distance scale than the top force curve in Figure 1. Moreover, this scale has been calibrated to account for both the *controlled* distance displacement of the *Z* scanner and the *uncontrolled* vertical bending of the cantilever (Chapter 3).

An important parameter in the theoretical expression describing the attraction between tip and sample beyond contact is the *Hamaker* constant, which relates to the mutual *polarizability* of pairs of interacting atoms. The dipole–dipole nature of forces responsible for this behavior on uncharged surfaces ensures a very short-range force compared to the *electrostatic* force between charged surfaces, or for that matter between a charged and uncharged surface. In Chapter 2, we will explore the world of intersurface forces in greater detail, including cases where the presence of more than one type of force results in both short- and long-range forces, even producing *nonmonotonic* (e.g., oscillatory) force–distance relationships, where the net force alternates between attractive and repulsive as a function of distance.

In-liquid surface force characterization can be made even more rigorous with control of *tip chemistry*. Most commonly, hydrophilic versus hydrophobic terminal chemistries have been prepared, primarily using alkyl thiolate self-assembled monolayers attached to gold-coated tips. Using ionizable terminal groups, tip charge states can be selected by choice of pH or ionic concentration in solution, allowing the analyst to toggle between attractive and repulsive tip–sample interactions indicative of the local surface charge state. Thus, force–distance measurements have become a useful tool in the hands of analytical and surface chemists, revealing nanoscale variations. Indeed, biomolecules or biochemical functionalities also can be attached to the tip or, in some cases, placed at the end of a polymeric "tether" (e.g., polyethylene glycol (PEG)), to provide *stereochemical* freedom: the ability of the functional groups at the end of the tether to rapidly explore orientations and thereby bond to complementary stereochemical groups on a biological surface. This "fly fishing" mode then can be used to search for sites of *specific* biological adhesion such as those operative on cell surfaces (e.g., in ligand-receptor bonding). These sites usually require much a stronger pulling force by the AFM cantilever in order to separate the tip-attached group from the biofunctional group on the sample surface—probably involving relatively strong hydrogen bonding—and are thus differentiated from *non-specific*, weaker interactions of purely dipole–dipole character (Chapter 2).

1.6 RATE, TEMPERATURE, AND HUMIDITY-DEPENDENT CHARACTERIZATION

From a scientific standpoint, AFM has matured into more than nanoviewer, nanomanipulator, or nanosensor of interfacial phenomena, fascinating and useful as these capabilities may be. The systematic variation of measurement rate and/or sample temperature can provide rigorous analyses of activated processes ranging from molecular motion (e.g., as manifest in the frictional properties of polymers [28,29]) to electronic transport (e.g., electronic conduction mechanisms in organic semiconductors) [30]. Thus, AFM can be used as a nanoscale dynamic mechanical

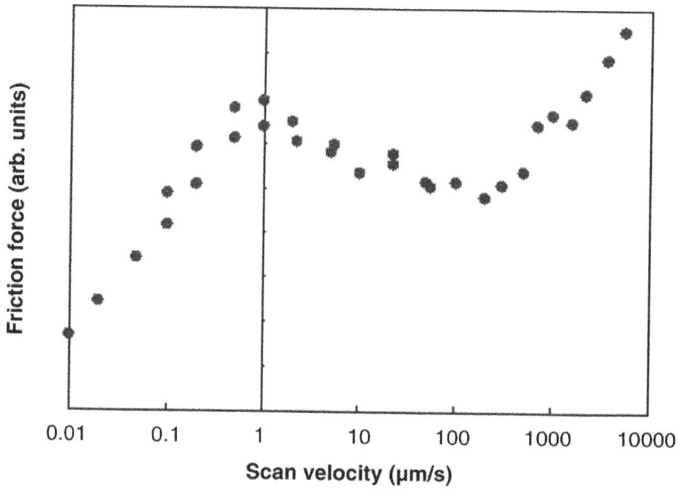

FIGURE 1.18 Scanning velocity dependence of friction force for a Si_3N_4 tip on a (dry) gelatin surface.

or thermal analyzer (DMTA) or rheometer [31], among other things, and ultimately as a tool for fundamental research in condensed-matter physics or physical chemistry. Here. we give three examples involving rate-dependent sliding friction, temperature-dependent stiffness and adhesion, and humidity-dependent phase imaging.

Figure 1.18 contains a graph of friction force versus tip scanning velocity, in micrometers per second ranging nearly *six decades of rate*—varied via a combination of scan size and scan frequency (lines per second)—on a gelatin film. A peak around 1 μm/s represents a dominant activated molecular motion—typically, rotational isomeric [32] or "turnstile" type—that tends to be in synchronization with, and thus excited by, the passing tip at a particular scanning velocity [33]. Another way to think of it is the characteristic time during which the passing tip interacts with the motional group. Different kinds of motion occur over very different time scales, as is well known from conventional *viscoelasticity* characterization [10]. Indeed, above 1000 μm/s, the friction force steeply increases again because another, much faster characteristic motion is preferentially excited. If our rate measurement window extended one or two decades higher, this peak would be resolved (friction force would again decrease). Such dissipative peaks, separated by several decades of rate, embody molecular motions of very different spatial extent, with longer range and *cooperative* motions (multiple synchronous activations) requiring more time, thus producing peaks at lower rates [10]. Thus, the scanning rate can be explored to bring out contrast in complex materials. For example, a polymeric material may contain local domains of heightened water content, increasing the activation of cooperative molecular motion, and thus the imaged friction force, at certain scanning rates [34]. We shall discuss such processes in greater detail in Chapter 7. This rate-dependent spectrum of activated motion can be relevant to all

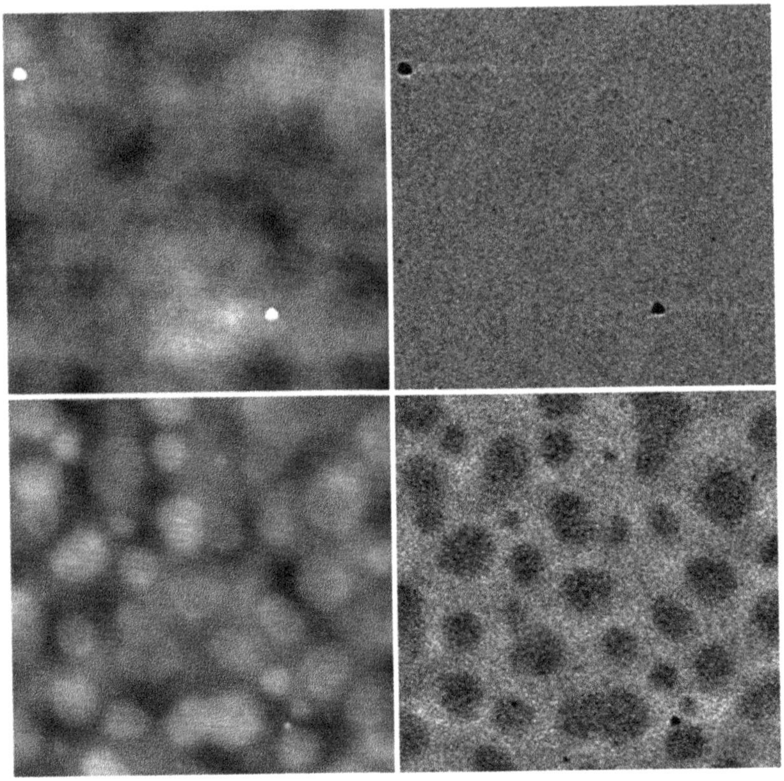

FIGURE 1.19 Rapid force-curve mapping AFM topography (left) and tip–sample adhesion (right) images 2.5 × 2.5 μm of a 50:50 blend of PMMA and PEMA at sample temperatures of 70°C (top) and 95°C (bottom).

sorts of polymeric technologies as well as the processing stages during manufacturing [35]. The characteristic rates also may be *modified by confinement to molecular-scale dimensions*, say in thin films or nanocomposites. Thus, one needs a tool such as AFM that can interrogate these behaviors down to the nanometer scale.

Our second example of this section, in Figure 1.19, is a blend of two methacrylate polymers that are glassy (rigid) at room temperature: polymethyl methacrylate (PMMA) and polyethyl methacrylate (PEMA). We compare topographic and pull-off force ("adhesion") images (left/right) collected at two sample temperatures, 70°C and 95°C (top/bottom). At 70°C, there are indistinct hills and valleys seen in height contrast (apart from two particulate contaminants) and essentially no adhesion contrast. At 95°C, there appear to be additional islands in height with distinct edges, and correspondingly dark adhesion domains (less sticky). The reason, at 95°C, the PEMA has softened because it is now rubbery instead of glassy—we have exceeded its *glass-to-rubber transition temperature*, T_g—whereas the PMMA has remained rigid (its glass–rubber transition being well above 100°C). Upon softening, the tip pushes farther into the PEMA by about 3 nm to achieve the force setpoint that is being maintained constant via feedback. The additional Z displacement needed to reach this point is contained in the topography image—being rendered as a lower elevation—even though it is not really topography (as discussed in Chapter 4). The onset of adhesion contrast results from the phenomenon known as *adhesion hysteresis* (Chapter 6) that includes a *memory* of the tip–sample contact area. Upon reversal, the pulling force required to break contact relates back to the tip–sample contact area reached at the maximum pushing force, greater on the softer polymer. A more continuous measurement of penetration and/or adhesion versus slowly ramped temperature can be used to precisely measure the glass-to-rubber transition temperature [36]. As with rate dependence, the T_g's of many polymeric materials are important to myriad applications as well as behavior during processing, and can be strongly affected by nanoscale confinement.

Our third example of this section, in Figure 1.20, is a phase image of a block copolymer coating of interest for its biodegradability. It is imaged in dynamic AFM

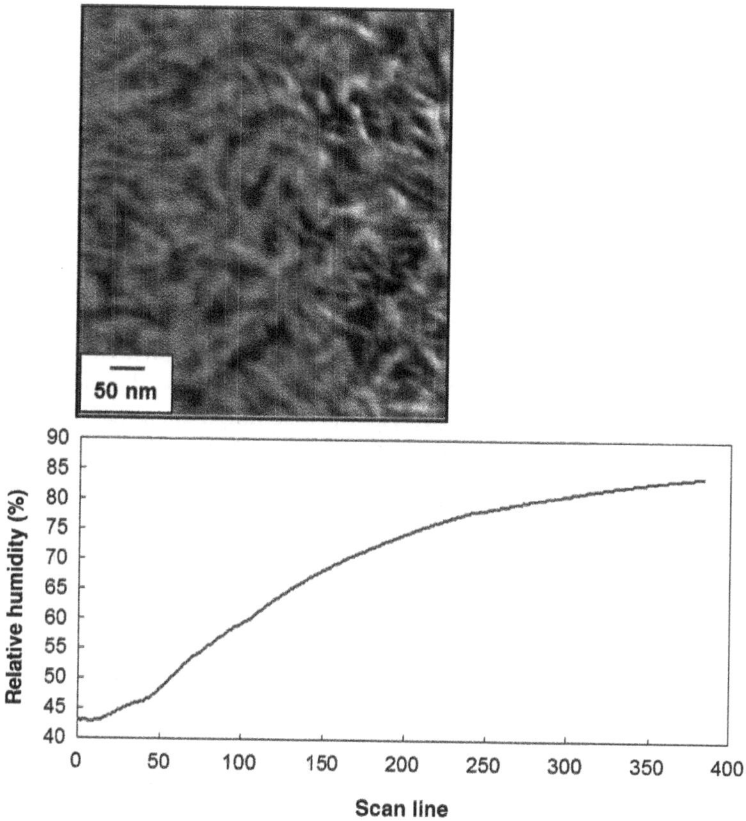

FIGURE 1.20 Phase image of PolyActive, a block copolymer of PBT and PEG, during the ramping of relative humidity from 43% at the beginning of the image at the left end to 77% at the end of the image at right.

under a continuous ramping of humidity from low to high as the image was collected from left to right (i.e., rotated by 90° from the usual orientation). This polymer contains both polar and thus hydrophilic (water absorbing) and oily and thus hydrophobic (water repelling) blocks; the latter tend to crystallize, further lessening the potential for water molecules to penetrate. Specifically, the hydrophilic blocks are (glassy) PEG and the hydrophobic blocks are polybutylene terephthalate (PBT). At low humidity, at which comparably little water has been absorbed into the hydrophilic blocks, we find only minor phase contrast. As the humidity reading rises above ~65%, strong phase contrast results (an effect that is reversible with humidity decrease). This image was acquired under settings that produce strong intermittent contact and thus the phase contrast derives primarily from mechanical differences. The water-swollen PEG comprises a soft matrix within which rod-like domains of stiffer, unswollen PBT produce brighter phase. A more detailed and quantitative discussion of phase is the principal subject of Chapter 5.

1.7 LONG-RANGE FORCE IMAGING MODES

In several chapters, we will consider attractive forces sensed from distances ranging from nanometers to micrometers, meaning not only dipole–dipole forces but also electrostatic forces. These forces may be very weak, requiring AC (dynamic) modes to detect and quantify with good signal-to-noise ratio. We will see that the same type of "phase" measurement discussed in Section 1.4, and via this phase, the resonance frequency of the cantilever, can allow one to measure the *gradient* of force versus distance (as discussed in Chapters 2 and 9), which, in turn, reflects the strength of attraction between tip and sample. But how do you image phase or cantilever resonance frequency at well-defined distances of tens or hundreds of nanometers, yet be sure that a variation in surface height, and thereby tip–sample distance, is not the source of contrast? The most common approach, discussed further in Chapter 9, is a tracking procedure called *interleave*, illustrated in exaggerated form in Figure 1.21. This works by first collecting a single line of height data $Z(X)$ in dynamic mode, then repeating this same scan trajectory but at a distance of typically nanometers or tens of nanometers, relative to the topographic trace measured in the first pass, by offsetting $Z(X)$ by a fixed amount, $Z(X) + Z_0$. Certain

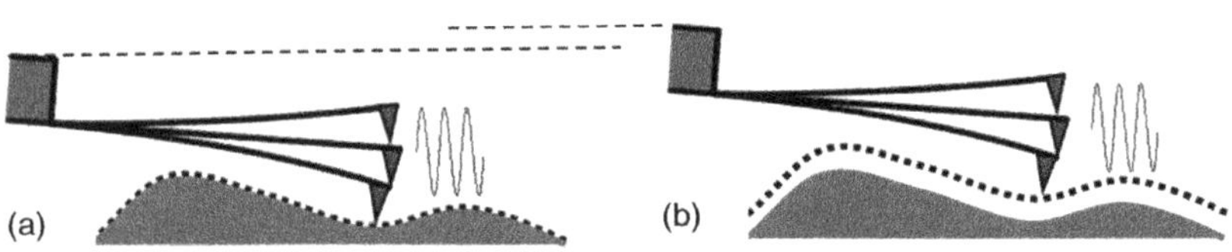

FIGURE 1.21 Exaggerated illustration of (a) a first scan to acquire topography and (b) a second interleave scan at a fixed mean displacement from the surface while phase or frequency shift are measured. The oscillation amplitude and vertical displacement are typically on the order of tens of nanometers, whereas the height of the tip is ~10 μm.

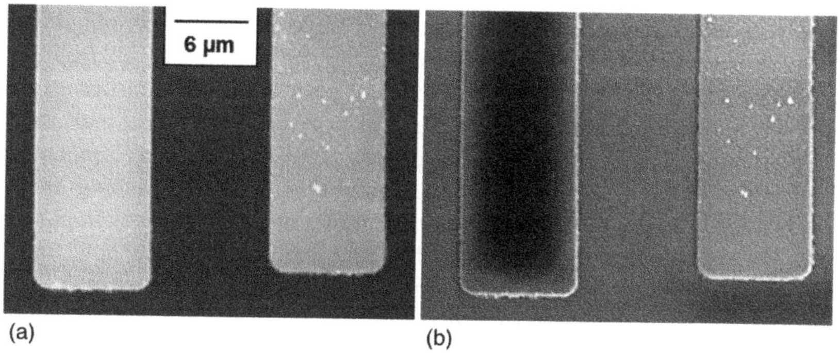

FIGURE 1.22 (a) Height and (b) phase images of strips of ~50-nm tall aluminum (left) and gold (right) on a silicon substrate. The phase image was collected during the interleave scan where the oscillating tip was pulled an extra 50-nm away from the surface and the sample biased by 3 V. Darker phase corresponds to stronger attractive force.

parameters may be altered for the second pass, including the application of a voltage bias between tip and sample to detect electrostatic differences (e.g., charging). Even magnetic differences can be probed at this relatively large distance by employing a tip with a magnetized coating. These modes are usually called electrostatic (or electric) force microscopy (EFM) and magnetic force microscopy.

Figure 1.22 exemplifies the sensitivity of interleave-based phase imaging in EFM, with tip and sample displaced an extra 50 nm apart while the sample is biased by 3 V. The high (bright) domains in the height image (a) are metal strips, aluminum and gold, which were lithographically created on silicon, as is common in microelectronics. In the phase image (b), we see that the strength of attractive forces sensed at such a "large" distance (compared to the range of dipole–dipole forces) is nevertheless different for each of the three materials. Darker phase corresponds to stronger long-range attraction as discussed in detail in Chapter 9. Variability within each metal region derives in part from surface contamination left over from the lithographic growth process.

Even under zero bias there can be differences in the strength of long-range forces sensed by the tip due to intrinsically different *surface potentials*. On metals, this corresponds to different *work functions*, the energy needed to take an electron from the *Fermi level*—the highest-energy occupied electronic state (which can be excited slightly higher to produced electrical conduction)—to the *vacuum level*, just enough energy to escape the solid. A related interleave-based imaging mode actually maps surface potential by adding a variable voltage bias to tip or sample so as to negate the intrinsic surface potential difference of tip and sample, thereby "nulling" the force. The variation of applied bias with position across the surface, as needed to null the force, provides an image of surface potential variations. This mode is most commonly called Kelvin-probe or just Kelvin force microscopy (also, scanning surface potential microscopy) and is further detailed in Chapter 9.

In summary, we see that electromagnetic phenomena can be imaged with AFM. One approach uses interleave to enable the measurement of "long-range" forces in parallel with topographic imaging (the latter obtained via short-range forces). Figure 1.22 demonstrates that tip–sample distances can be carefully controlled down to the nanometer scale to enable the probing of both long- and short-range forces, even though the imaged region may be tens of micrometers in size, and using just one device, the *X–Y–Z* scanner under electronic feedback. The development of the first commercial AFMs *and* the further equipping with advanced scanning modalities such as interleave occurred within just a fraction of a decade. This was in part due to the already advanced scientific understanding and engineering know-how for *piezoelectric materials* and scanners made from them, as well as fast electronic feedback control. We will discuss these subsystems in greater detail in Chapters 3, 4 and 9.

1.8 PEDAGOGY OF CHAPTERS

The overview of Sections 1.1–1.7 is partly intended to help point the reader to the chapter or chapters of greatest interest or urgency. Overall, the chapters are pedagogically arranged by grounding the technique in the concepts of distance-dependent forces in Chapters 2 and 3. Chapter 2 explores the topic from the standpoint of physical phenomena, divorced from instrumentation realities. Chapter 3 then drills down into the methods by which the AFM instrumentation actually measures these distance-dependent forces, whether in their own right or as a means to enable topographic imaging. Given these capabilities, Chapter 4 covers the methodology of topographic imaging in both quasistatic (contact) and dynamic ("AC"/"tapping") modes, including numerous realities and caveats that the AFM user will want to understand so that artifacts are avoided, accuracies are not overstated, etc. Issues affecting the choice of tip or cantilever and imaging mode will be discussed in the process. Chapter 4 also clarifies the nature of topographic data as manipulated during postprocessing. Chapter 5 focuses on phase measurement and imaging—for both diagnostic and materials contrast reasons—in dynamic AFM, including several difficulties, caveats, and calibration issues as well as examples of the kinds of information obtained. Chapter 6 covers tip–sample adhesion including both solid–solid (tip–sample) and capillary adhesion forces. It further describes mapping methods that examine force–distance measurements (force-curve mapping) in the quasistatic sense (maintaining a balance between tip–sample force and cantilever-applied force), as well as the distance dependence of amplitude and phase within dynamic AFM. Chapter 7 covers lateral-force-derived methods within (continuous) contact modes. Chapter 8 describes common data postprocessing and analysis methods. Chapter 9 treats more advanced dynamic methods for probing long-range electromagnetic forces via interleave methods, as well as newer dynamic methods that examine higher vibrational modes of cantilevers, for probing both long- and short-range force responses.

The appendices are intended to augment and in some cases provide greater mathematical detail on topics treated in the chapters. Appendix 1 describes three spectral methods for cantilever spring constant calibration, an augmentation to the conceptually simpler calibration method of Section 3.7. Appendix 2 uses integral calculus to derive van der Waals force–distance expressions for various tip geometries as discussed in Section 2.2. Appendix 3 uses power balance concepts and calculus to derive energy dissipation expressions useful to phase data interpretation, as discussed in Section 5.3. Appendix 4 provides an explicit geometric description of a capillary meniscus in the circular approximation, as discussed in Section 6.3.

REFERENCES

[1] Binnig, G., C.F. Quate, and C. Gerber, Atomic force microscope. *Phys. Rev. Lett.*, 1986, **56**(9): 930.

[2] Sarid, D., *Scanning Force Microscopy*. Revised ed., 1994, New York: Oxford University Press, p. 263.

[3] Haugstad, G., et al., Atomic force microscopy of AgBr crystals and adsorbed gelatin films. *Langmuir*, 1993, **9**(6): 1594–1600.

[4] Ahimou, F., et al., Biofilm cohesiveness measurement using a novel atomic force microscopy methodology. *Appl. Environ. Microbiol.*, 2007, **73**: 2897–2904.

[5] Zangwill, A., *Physics at Surfaces*. 1988, Cambridge: Cambridge University Press, p. 454.

[6] Drelich, J. and K.L. Mittal, eds. *Atomic Force Microscopy in Adhesion Studies*. 2005, Leiden-Boston: VSP.

[7] Bhushan, B., ed. *Springer Handbook of Nanotechnology*. 2nd ed., 2007, Berlin: Springer.

[8] Bhushan, B. and H. Fuchs, eds. *Applied scanning probe methods XII: Characterization*. Nanoscience and Technology, 2009, Berlin/Heidelberg: Springer.

[9] Gnecco, E. and E. Meyer, eds. *Fundamentals of friction and wear on the nanoscale*. Nanoscience and Technology, ed. P. Avouris, et al. 2007, Berlin: Springer-Verlag, p. 714.

[10] McCrum, N.G., B.E. Read, and G. Williams, *Anelastic and Dielectric Effects in Polymeric Solids*. 1967, London: Wiley.

[11] Puntambekar, K., et al., Structural and electrostatic complexity at a pentacene/insulator interface. *Adv. Funct. Mater.*, 2006, **16**: 879–884.

[12] Schwoerer, M. and H.C. Wolf, *Organic Molecular Solids*. 2007, Weinheim: Wiley-VCH Verlag, p. 438.

[13] Kalihari, V., et al., Observation of unusual homoepitaxy in ultrathin pentacene films and correlation with surface electrostatic potential. *Adv. Mater.*, 2009, **21**: 1–7.

[14] Haugstad, G., et al., Probing biopolymers with scanning force methods: Adsorption, structure, properties, and transformation of gelatin on mica. *Langmuir*, 1994, **10**: 4295–4306.

[15] Finch, C.A., *Polyvinyl Alcohol: Properties and Applications*. 1973, London: Wiley.

[16] Piner, R.D., et al., Dip pen nanolithography. *Science*, 1999, **283**: 661–663.

[17] Djabourov, M., Architecture of gelatin gels. *Contemp. Phys.*, 1988, **29**(3): 273–297.

[18] Haugstad, G., W.L. Gladfelter, and R.R. Jones, Scanning force microscopy characterization of viscoelastic deformations induced by precontact attraction in a low cross-link density gelatin film. *Langmuir*, 1998, **14**: 3944–3953.

[19] Schmidt, R.H., G. Haugstad, and W. Gladfelter, Scan-induced patterning in glassy polymer films: Using scanning force microscopy to study plastic deformation at the nanometer length scale. *Langmuir*, 2003, **19**: 898–909.

[20] Schmidt, R.H., G. Haugstad, and W.L. Gladfelter, Scan-induced patterning and the glass transition in polymer films: Temperature and rate dependence of plastic deformation at the nanometer scale. *Langmuir*, 2003, **19**: 10390.

[21] Schmidt, R.H., G. Haugstad, and W.L. Gladfelter, Correlation of nanowear patterns to viscoelastic response in a thin polystyrene melt. *Langmuir*, 1999, **15**(2): 317–321.

[22] Krotil, H.U., et al., Pulsed force mode: A new method for the investigation of surface properties. *Surf. Interface Anal.*, 1999, **27**(5–6): 336–340.

[23] Garcia, R. and R. Perez, Dynamic atomic force microscopy methods. *Surf. Sci. Rep.*, 2002, **47**: 197–301.

[24] Haugstad, G. and A. Avery. Probing the morphology and tribo-response of nanostructured fluid films for personal care applications, in Technical Proceedings of the 2005 NSTI Nanotechnology Conference and Trade Show. 2005, Anaheim, CA: Nano Science and Technology Institute.

[25] Lazzari, M., G. Lui, and S. Lecommandoux, eds. *Block Copolymers in Nanoscience*. 2006, Weinheim: Wiley-VCH.

[26] Karim, A. and S. Kumar, eds. *Polymer Surfaces, Interfaces and Thin Films*. 2000, Singapore: World Scientific Publishing, p. 294.

[27] Haugstad, G. Probing swelling and molecular conformation on polymeric coatings for biocompatibility, in 225th ACS National Meeting. 2003, New Orleans, LA: American Chemical Society.

[28] Hammerschmidt, J.A., G. Haugstad, and W.L. Gladfelter, Probing polymer viscoelastic relaxations with temperature controlled friction force microscopy. *Macromolecules*, 1999, **32**: 3360.

[29] Overney, R.M., et al., Glass and structural transitions measured at polymer surfaces on the nanoscale. *J. Therm. Anal. Cal.*, 2000, **59**: 205–225.

[30] Choi, S.H., B. Kim, and C.D. Frisbie, Electrical resistance of long conjugated molecular wires. *Science*, 2008, **320**(5882): 1482–1486.

[31] Aklonis, J.J. and W.J. MacKnight, *Introduction to Polymer Viscoelasticity*. 2nd ed., 1983, New York: Wiley, p. 295.

[32] Wade, L.G., *Organic Chemistry*. 2nd ed., 1991, Englewood Cliffs: Prentice-Hall, p. 1233.

[33] Hammerschmidt, J.A., et al., Polymer viscoelastic properties measured by friction force microscopy. *Macromolecules*, 1996, **29**(27): 8996–8998.

[34] Haugstad, G., et al., Probing biopolymer films with scanning force methods. *Mat. Res. Soc. Symp. Proc.*, 1995, **355**: 253–258.

[35] Ferry, J.D., *Viscoelastic Properties of Polymers*. 1980, New York: Wiley.

[36] Tsui, O.K.C., et al., Studying surface glass-to-rubber transition using atomic force microscopic adhesion measurements. *Macromolecules*, 2000, **33**: 4198–4204.

2

DISTANCE-DEPENDENT INTERACTIONS

2.1 GENERAL ANALOGIES AND TYPES OF FORCES

Why be concerned about the distance dependence of force? First, this is how we are going to image surfaces. We need a means of getting vertical information—*surface height*—and we do this by being sensitive to force versus distance. Second, as previewed in Sections 1.5 and 1.6, we hope to obtain *analytical information* about materials, interfaces, and even individual molecules, information that can be extracted from the distance dependence of measured forces, perhaps after some analytical curve fitting. Third, we hope that by measuring distance-dependent forces in an array of locations across a surface, we will be able to *map the spatial distribution* of materials, even quantify *physical changes* within a given material due to the state of, say, its liquid environment (e.g., pH, ionic concentration), the relative humidity (presence of adsorbed or absorbed water, perhaps even inducing crystallographic changes), the sample temperature (e.g., local phase transformations, viscoplasticity), previous interactions with the tip (e.g., tribological disentangling of a polymer matrix), the presence or lack of biomolecular adhesion (e.g., possible denaturing), and potentially other changes in physical response or biological function.

Measuring a phenomenon that changes with distance, in order to gauge distance, is not a new concept. A pressure altimeter measures the elevation of an airplane via

Atomic Force Microscopy: Understanding Basic Modes and Advanced Applications, First Edition.
Greg Haugstad.

the decrease of air pressure as a function of vertical distance (above sea level). An exponential relationship between air pressure P and elevation Z

$$P(Z) = P_0 \exp(-mgZ/RT) \tag{2.1}$$

(with m = average molecular mass, T = average temperature at sea level, and R = gas law constant) can be plotted as linear on a semilog scale, giving utility over a wide range of distances provided reasonable signal/noise in the measurements at large distance. There are caveats related to the calibration of such an altimeter, for example, related to geographic differences in the relationship between pressure and altitude. Similarly, we will find that the distance dependence of forces sensed by AFM can vary at different locations over a heterogeneous surface, requiring some corrective calibration procedures in order to accurately measure topography while at the same time revealing material properties.

More advanced altimeters are aboard satellites and use electromagnetic (EM) waves—a radar pulse reflected from the sea surface—to gauge altitude. Here, the travel time of pulse from satellite to sea surface and return echo back can be converted into distance, thanks to the (relatively) constant speed of light. Thus, this measurement does not suffer from the sorts of calibration issues that air pressure does, related to geographic variance. A version of the same general technology has also been used to map the earth's surface topography (by the European Remote-Sensing Satellite). But here, given the complexities of reflecting EM waves from very different kinds of land masses, for example, forested versus rocky desert, special calibration procedures must be invoked. Again this is somewhat analogous to AFM: a method that to first approximation provides topographic information, but at finer detail is sensitive to the "material properties" (here terrain type).

Speaking of EM waves, rather than time how long it takes to travel to and from a surface, can we use interferometry? We know from basic physics that a difference in optical path length much less than one wavelength can produce significant measurable differences. Cannot we reflect a focused, monochromatic beam of light from a surface and use the degree of constructive and destructive interference to gauge distance, thus height, with precision well below the wavelength of light? Yes! This is how "phase-shifting interferometry" surface profilometers work to generate a 3D reproduction of surface topography. So why bother with AFM and the complexities of touching? First, there is the problem of lateral resolution. The ability to spatially resolve topography on the lateral scale is poor with optical interferometry, limited by diffraction to something on the order of the wavelength of visible light, meaning hundreds of nanometers. Second, there is the fact that optical interferometry is a *periodic function of distance*, rather than a *monotonic* (increasing or decreasing) function. So changes of height across a surface need to occur in a gradual way such that one can count the cycles of interference. Then, of course, there is the importance of the *other* information AFM provides via force measurements, under the headings of materials contrast and quantitative property assessment. This is not to say that optical methods cannot reveal more than topography; certainly they can, as seen in special contrast modes of light microscopy such as differential interference contrast (sensitive to index of refraction, i.e., optical density, and, thereby, crystallinity), polarization contrast (sensitive to anisotropy, i.e., crystalline orientation), and others. Here again, it seems that any physical phenomenon or probe that can be used to gauge topography is sensitive to additional material attributes. But "touching," in its multifarious force manifestations, is a uniquely powerful probe as previewed in Chapter 1 and detailed in the coming chapters.

Speaking of force, what of gravity? Can a low-flying satellite (e.g., the GOCE at 260 km) sense a large object, such as an isolated mountain like Kilimanjaro in Tanzania, by the change of gravitational force, and thus be able to "image topography?" Suppose you model the earth as a perfectly smooth sphere (of radius 6400 km) and treat Kilimanjaro as a 30° cone (mass #1) with center of mass 5.9/4 km above the sphere (mass #2). If the satellite is orbiting 260 km above the smooth sphere, what is the relative gravitational force due to Kilimanjaro? Approximating the masses as 6.0×10^{24} and 2.3×10^{15} kg, respectively (the latter 8.6×10^{17} cm^3 times 0.027 kg/cm^3), the relative strength of gravitational pull on the satellite is

$$F_1/F_2 = (Gm_1 m_{\text{satellite}}/r_2^1)/(Gm_2 m_{\text{satellite}}/r_2^2) = (m_1/m_2) \cdot (r_2/r_1)^2) \qquad (2.2)$$

where r_2 and r_1 are the distances from satellite to the earth's center and from satellite to Kilimanjaro point mass, respectively; thus

$$F_1/F_2 = 2.3 \times 10^{15}/6 \times 1024) \times [(6400 + 260)/(260 - 1.5)]^2 = 2.5 \times 10^{-7}.$$

Although the gravitational tug of Kilimanjaro is only a tiny portion of the total, in measurement science, "seven orders of magnitude smaller" does not necessarily mean unmeasurable. (Indeed, some satellite-born gravity sensors detect very weak gravitational variations such as due to tides.)

Direct force measurements with typical AFM microcantilevers and the usual laser-beam deflection scheme can range over $10^{-11} - 10^{-6}$ N. But extremely weak, distance-dependent *force gradients* can be measured with the same technology via *resonance frequency shifts*, as will be discussed. Small additional force gradients can be measureable on top of the much larger force gradient of a Hookean spring-mass system (AFM cantilever with tip), especially in vacuum due to a high *quality factor* (Section 2.4.2). In this case, a 7-decade range of force-derived signal is perhaps achievable.

When you begin to count up the powers of ten, it would seem that gravity cannot play a role in AFM. But what of small distances? Gravity's inverse-square distance dependence is the same as that of electrostatic forces between point charges. Consider two singly ionized *atoms*, say xenon (unreactive). If one separates these ion centers by 1 nm (say by adsorbing to a ~10 K surface in vacuum [1]) what is the relative strength of gravitational (attractive) and Coulomb (electrostatically repulsive) forces between the two ions? The gravitational force is

$$F = Gm_1 m_2/r^2 = 3.2 \times 10^{-33}\ \text{nN}$$

whereas the electrostatic force is

$$\begin{aligned} F &= (1/4\pi\varepsilon_0)(q_1 q_2/r^2) \\ &= 2.3 \times 10^{-1}\,\text{nN} \end{aligned} \tag{2.3}$$

using

$m_1 = m_2 =$ xenon atomic mass $= (0.1313\,\text{kg/mol})/(6.023 \times 10^{23}\,\text{atoms/mol}) = 2.18 \times 10^{-25}\,\text{kg/atom}$,
$r =$ interparticle distance $= 1 \times 10^{-9}\,\text{m}$, and
$G = 6.67 \times 10^{-11}\,\text{m}^3\,\text{kg}^{-1}\,\text{s}^{-2}$, $\varepsilon_0 = 8.85 \times 10^{-12}\,\text{C}^2\,\text{N}^{-1}\,\text{m}^{-2}$, and $q_1 = q_2 = 1.60 \times 10^{-19}\,\text{C}$.

The gravitational force between two ions 1 nm apart is 32 orders of magnitude weaker than the electrostatic force! Even if one attaches a 10-μm radius tungsten sphere to an AFM cantilever and brings vanishingly close to another such sphere, the gravitational attraction of spheres is a still meager 10^{-12} nN. In AFM, gravity is utterly insignificant compared to electrostatic forces.

What forces are present between *neutral* atoms and molecules, which (presumably) still dwarf gravitational forces? Of the four fundamental forces in nature (the other two being the strong and weak forces important to the nuclear scale), the EM force manifests itself in the most diverse ways. Most of the forces we unavoidably encounter in the material domain—and, necessarily, in materials characterization—relate to *dipoles*. Each fundamental material element, be it atom or molecule, bears a net dipole moment—an internal displacement of negative and positive charge (i.e., polarization). But what of the xenon atoms considered above? The inert gases—helium, neon, argon, krypton, xenon, and radon—are spherical atoms with no net dipole moment *in the absence of interaction with other atoms*. The interaction of electronic charge distributions is the key. Upon bringing together two perfectly inert, spherical atoms, *fluctuating* electronic charge distributions (i.e., instantaneous dipole moments) begin to exhibit *correlations in their orientations*. The electrons of one atom tend to repel the nearest electrons of the other atom such that at any instant, a net dipole on one atom induces a net dipole on the other atom and vice versa. Averaged over time, there is a nonvanishing correlation in the orientations of these dipoles, even though the *time-averaged* charge displacement on each is zero. These correlating dipoles tend to attract each other [1], a force that falls off as distance to the seventh power [2].[1] This decrease of force with distance is many decades steeper than the Coulomb attraction or repulsion of (unlike- or like-charged) ions. Given such a steep force–distance relationship, one

[1] This is assuming nonretarded interactions, that one dipole senses the state of the other instantaneously; in truth, there is a finite time for the interactions to propagate at the speed of light. At larger distances, the correlation of dipoles degrades more steeply with increasing distance as their polarizations becomes more "out of phase." In crossed cylinder experiments, the retarded regime has been identified at distances greater than roughly 20 nm. See Stewart, A. M., V. V. Yaminsky, and S. Ohnishi, Langmuir, 2002, **18**: 1453.

begins to perceive why AFM is able to track nanoscale variations of topography, of surface height.

What forces derive from contact between solid probe and solid sample? *What IS contact at the atomic level?* We know that atoms are not hard, opaque spheres such as billiard balls. Rather, they consist of an exceedingly small, almost point-like nucleus (diameter $\sim 10^{-6}$ nm) and a wispy electron cloud that comprises almost the entire atomic volume. Yet we also know that atoms do not ordinarily interpenetrate. There must be something that resists interpenetration, and we know that the explanation lies in the domain of quantum physics. One must consider the fact that the nucleus is positively charged and essentially stationary (i.e., beyond the domain of nuclear physics), whereas the electrons are negatively charged and, in a more dynamic state, including wavelike behavior. As two atoms come together, the outermost electrons of one atom encounter their like-charged counterparts on the other atom and exert Coulomb repulsion. But the electrons on atom A also are now closer to the nucleus of atom B and begin to feel attraction to it because the electrons of atom B do not completely *screen out* the nucleus of atom B. Depending on the types of atoms, the result may be a stable configuration whereby electrons are effectively shared between the two nuclei, forming a *covalent bond* and thus a two-atom molecule. This means an attractive force is present between the atoms; if you could attempt to pull them apart, you would experience a force of tension, the covalent bond. But in many other cases, again depending on the kinds of atoms, there is simple repulsion: no quantum mechanical benefit of the atoms "getting together." And, in all cases, if you *continued* to push the atoms together, quantum mechanics ultimately would resist the process as no new electronic bonding states become available and the Pauli exclusion principle requires that the various electrons of atoms A and B cannot occupy the same space and energy states.

Some analytical models describe this repulsion by using exponents even greater than the (inverse) power of seven mentioned above for dipole–dipole attraction, meaning a *steeper* growth of repulsive forces than attractive forces (apart from sign) with decreasing distance. In truth, the repulsion is best modeled with an exponential form, from purely quantum physics considerations (as described by a Morse potential [3] which, upon solving Schrodinger's equation, well predicts vibrational energy states in diatomic molecules [4]). But a more common and practical representation, from the standpoint of speeding computational algorithms, is the Lennard–Jones (L–J) [5] "6–12" potential—with potential energy of attraction growing as inverse distance to the sixth power (thus, force, the gradient of potential, growing as the seventh power as stated above) and potential energy of repulsion growing as the *12th* power (a convenient square of the sixth power that an algorithm would calculate first). On the plotted nanoscale in Figure 2.1, this 12th power form approaches a nearly vertical trend (that of an infinitely rigid "wall"). The figure shows how these two distance-dependent functions add together to produce a *potential well,* centered at a distance that is the energetically favored configuration, a weak dipole–dipole "bonded" state (a physical, not chemical, e.g., covalent, bond). This distance then becomes the natural separation of atoms or molecules upon packing together into "van der Waals" solids, where there is no quantum mechanical

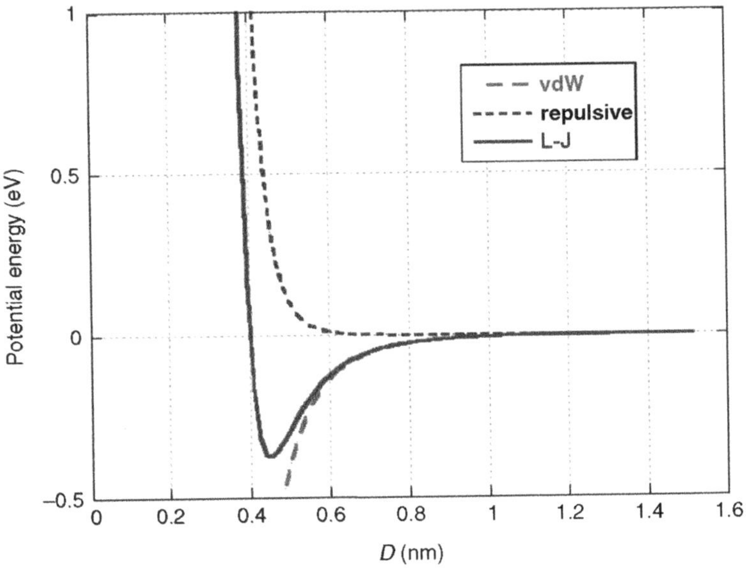

FIGURE 2.1 Distance dependence of L–J potential overlaid on attractive (r^{-6} van der Waals) and repulsive (r^{-12}) potentials.

(e.g., covalent) or ionic bonding between atoms or molecules. The term van der Waals, incidentally, applies to *any* dipole–dipole interaction [6]; this includes not only the fluctuating dipoles discussed above, known more specifically as *dispersion* or *London* forces [7], but also interactions between permanent dipoles (such as the molecule CO) sometimes called *Keesom* forces, as well as dipoles induced by others atoms or molecules that are charged or permanent dipoles, known as *Debye* forces [2,6].

2.2 VAN DER WAALS AND ELECTROSTATIC FORCES IN A TIP–SAMPLE SYSTEM

2.2.1 Dipole–Dipole Forces

Of course, in AFM, our tip and sample each are not a single atom or a molecule! Rather, they are extended solids or, more generally, *condensed matter* (which includes things such as gels, cells, or liquidy films), comprised of many, many atoms. Dipole–dipole interaction energies $-C/r^6$ can be mathematically integrated as point interactions (under simplifying assumptions) to derive a force–distance expression for extended bodies. A common model used for tip–sample van der Waals interactions is that of a sphere or paraboloid attracted to an infinite half space (Appendix 2). The assumption is that the end of the tip is rounded off and thus essentially a paraboloid in shape. Because the force falls off so steeply with distance, the actual shape of the tip above the rounded end may not matter.

It is straightforward (Appendix 2) to first integrate the nonretarded point-dipole/point-dipole potential energy

$$U(r) = -C/r^6 \tag{2.4}$$

over the half-space medium (a flat sample), yielding a $1/D^3$ dependence of potential energy with D being the distance between point dipole and surface (of the half-space medium). Then this expression can be integrated over point dipoles (atoms) in the tip [2], say if endmost portion is modeled as a paraboloid of revolution with a radius of curvature R. With the additional assumption that the distance between the nearest point of tip to sample is much smaller than the height of the paraboloidal part of the tip, one obtains a $1/D$ dependence of potential energy, where D is now this distance between tip end and sample surface. Force is given by the gradient of potential energy as

$$F(D) = -\frac{A_H R}{6D^2}. \tag{2.5}$$

The parameter

$$A_{\mathrm{H}} = \pi^2 C n_1 n_2, \tag{2.6}$$

where $n_{1,2}$ are the number densities of point dipoles (atoms) in tip and sample, has units of energy, and is known as the *Hamaker* constant for the two-material system [8].

One is immediately faced with the meaning of zero distance. Obviously, when the two solids are in contact, the van der Waals force does not become infinite. The distance D must include a small positive offset such that distance remains finite upon contact. Physically, this offset arises from the fact that the distance scale used to describe a van der Waals attractive potential between atomic dipoles is *zeroed at the center of each atom*. The bottommost tip atom and its nearest counterpart atom on the surface do not interpenetrate such that their centers coincide, of course. As a first approximation, the smallest distance that these centers can approach should be roughly two atomic radii, a number on the order of 0.3–0.4 nm. However, when a solid sphere or paraboloid is brought up to a solid surface, the total van der Waals potential as integrated in Appendix 2 is not zeroed precisely at a distance that is the sum of two atomic radii. Comparisons of the theoretical relationship between van der Waals force and surface energy to experimental measurements have (remarkably) indicated that for most solids the effective zero of distance can be empirically taken as $D_0 = 0.16\,\mathrm{nm}$ [2].

Is the simplifying assumption of *pairwise additivity* of induced dipole interactions really valid? As we integrate over sample and tip, the polarization of one differential element is assumed to depend on one other differential element as captured by the $1/r^6$ interaction. But these two dipole charge distributions are subject to polarization by the third, fourth, and so on, dipoles in the vicinity. How can dipole 1, induced by its interaction with dipole 2, also be induced by its interaction with dipoles 3, 4, and so on, in such a way as to produce a nonzero effect? One might suspect that all these directional charge displacements would somehow average out to zero, that the induced-dipole/induced-dipole interaction would go away once atoms assemble into condensed matter. Yet we have one key fact that AFM tip and sample are separated (on opposite sides of the *XY* plane, for integration) suggesting some net polarizing effect. This is to be expected just as the adsorption of spherical atoms on a surface, and a second layer of such atoms on the first, result from a net polarizing effect (e.g., the case of xenon [1]). But these intuitive speculations are just that. A rigorous physical treatment of all dipoles is a *many-body problem*. The same essential problem was solved by the Russian physicist Evgeny Lifshitz using quantum field theory (and later with simpler approaches by other authors), where each medium is instead treated as a *continuum of EM fluctuations* [9]. The final result is that the functional dependence on distance D is the same as obtained above, but the expression for Hamaker constant A_H is much more complicated. To calculate A_H one must integrate over the frequency dependence of a lengthy expression containing the *complex dielectric function* $\varepsilon(\omega)$ of up to *three* media, which for AFM means tip, sample, and, in general, the immersion medium. These frequency-dependent fluctuations include not only electronic displacements within neutral atoms but also mobile electrons in metals and ions in solution, as well as vibrating and rotating permanent dipoles within molecular media (whether solid or liquid); in short, any motion in response to an imposed EM field (as generated by everything else). The result is a theoretical relationship between A_H and the *dielectric functions* ε_i and *indices of refraction* n_i (the symbol n denotes number density elsewhere in this chapter and in Appendix 2), *thermal energy* $k_B T$, and the principal *electronic absorption frequency* ν_e (taken as similar for all three media, $\sim 3 \times 10^{15}$ Hz):

$$A_H \approx \frac{3k_BT}{4}\left(\frac{\varepsilon_1-\varepsilon_3}{\varepsilon_1+\varepsilon_3}\right)\left(\frac{\varepsilon_2-\varepsilon_3}{\varepsilon_2+\varepsilon_3}\right) + \frac{3h_P\nu_e}{8\sqrt{2}}\left[\frac{(n_1^2-n_3^2)(n_2^2-n_3^2)}{(n_1^2+n_3^2)^{1/2}(n_2^2+n_3^2)^{1/2}\left[(n_1^2+n_3^2)^{1/2}+(n_2^2+n_3^2)^{1/2}\right]}\right] \tag{2.7}$$

where h_P is Planck's constant and the numerical subscripts apply to the case of solid media 1 and 2 interacting across medium 3. The first term derives from "zero-frequency" interactions (i.e., where one or both dipoles are permanent), whereas the second term derives from fluctuating electronic charge distributions (London dispersion forces, i.e., entirely induced dipoles). The magnitude of A_H is calculated to be in the $\sim 10^{-19}$–10^{-20} J range for most material systems (compare to $k_BT = 4 \times 10^{-21}$ J at room temperature); measured values are in reasonable agreement [2]. Experimental Hamaker constants are shown in Table 2.1 for amorphous SiO_2 (i.e., oxidized silicon tip) interacting across vacuum (or air) with a few materials commonly examined with AFM.

TABLE 2.1 Experimental Hamaker constants for amorphous SiO_2 interacting across vacuum with other materials.

Material	ε	n	ν_e (10^{15} Hz)	A_H (10^{-20} J)
Amorphous SiO_2	3.82	1.46	3.2	6.5
Polystyrene	2.55	1.59	2.3	6.9
Polymethyl methacrylate	3.12	1.50	2.7	6.4
Polydimethyl siloxane	2.7	1.4	2.8	5.4
Mica—$KAl_3Si_3O_{10}(OH)_2$	7.0	1.58	3.1	7.8

Equation 2.5, for $R = 10$ nm, implies that at closest approach (contact), the van der Waals attractive force F ($D = D_0 = 0.16$ nm) $\approx -4 \pm 1$ nN in air, using the Hamaker constants in Table 2.1. Employing values of ε and n in Table 2.1, Equation 2.7 predicts that for SiO_2–SiO_2 interaction, this force should drop by a factor of 10 upon immersion in water.

The Hamaker expression also makes two important predictions regarding attraction versus repulsion: (i) for identical tip and sample materials, the interaction is always attractive (positive A_H), whereas for different tip and sample materials, the interaction can be attractive or repulsive depending on the properties of the *immersion medium* relative to the tip and sample, and (ii) in vacuum, the interaction is always attractive irrespective of the relative compositions of tip and sample. Given the huge difference in dielectric constant and index of refraction of air compared to any condensed medium, prediction (ii) holds for air as well as vacuum.

If a tip is so sharp that the contribution of interactions beyond the modeled spherical end cap (i.e., the "shank" of the tip) are appreciable, then modeling the tip as a cone (Appendix 2) can lend insight into how the van der Waals force differs: a $1/D$ dependence. At the other extreme, if a tip end has been dulled (say after a "crash"), then it may approach the model of a flat-ended cone, in which case the force includes a contribution that goes as $1/D^3$ (Appendix 2). Real tips are expected to be intermediate to cone and paraboloid, or paraboloid and truncated cone, in which case the distance dependence of the van der Waals force is expected to include terms $1/D^{\zeta}$ with $1 < \zeta < 3$. One example model is a hyperboloid. Figure 2.2 compares log–log plots of the magnitude of $F(D)$ for four tip models (see Eqs. A2.14–17 and associated narrative). In addition to the very different exponents (slopes) as indicated above, the magnitude of force varies dramatically. In particular, at ~1 nm separation, a truncated cone exhibits roughly one order of magnitude larger force compared to the case of a rounded tip end, whereas an ideally sharp cone exhibits two orders of magnitude smaller force compared to a rounded end. Clearly, the details of tip shape must play a significant role in the distance dependence of force sensed in AFM.

2.2.2 Electrostatic Forces

Notwithstanding a presumption of purely van der Waals forces in many analyses of AFM force–distance behavior, it is not uncommon for real tip–sample forces to

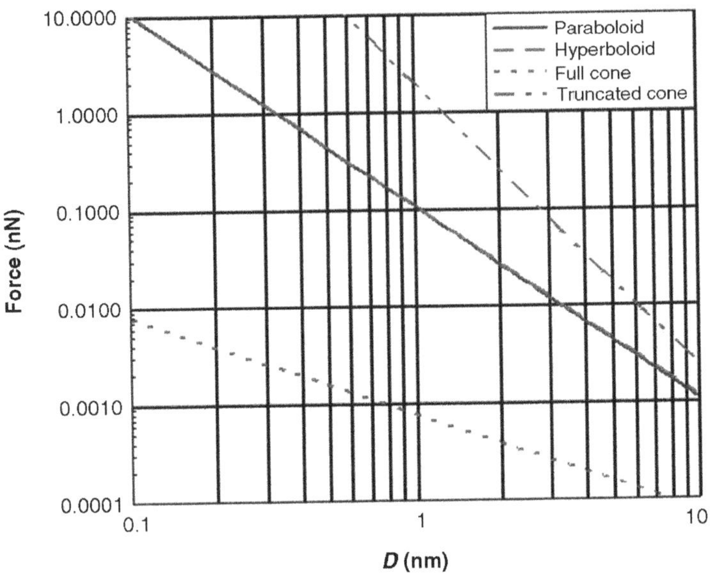

FIGURE 2.2 Comparison of the magnitude of distance-dependent van der Waals forces for four models tip geometries interacting with an infinite half-space sample, as derived in Appendix 2.

include *electrostatic* contributions. The presence of electrostatic forces may arise from the nonconductive nature of many samples, such that a net charge is present (e.g., on plastics). This generally results in longer-range *charge–dipole* interactions ($1/r^3$ potential), with the (polarizable) atoms in the tip, or potentially *charge–charge* interactions ($1/r$ potential), if the tip is nonconductive or conductive but voltage biased. Even if both tip and sample are idealized, grounded conductors, these will generally differ in *surface potential* because of different ionization potential or *work function*. (A more detailed discussion is in Section 9.1.) In this case, as tip and sample near one another, an electric field is unavoidably produced due to different surface potentials. The electric field can cause charge flow and/or dipole reorientations; thus, a longer range, attractive electrostatic force results, measureable to distances of at least tens of nanometers. Under a several-volt bias or given substantial sample charge, this distance can increase to at least hundreds of nanometers and even micrometers.

Modeling electrostatic interactions is more complicated than the much shorter-range dipole–dipole forces because the entire tip (and even the cantilever to which the tip is attached, Section 3.6.4) can contribute appreciably to the measured force. For the general case, a tip at potential V relative to sample—whether due to applied bias, surface potential differences, or trapped charge on tip or sample—will experience an attractive, distance-dependent force given by

$$F(D) = -\pi\varepsilon_0 V^2 g(D)$$

where ε_0 is the vacuum permittivity and the *geometric factor* $g(D)$ is proportional to dC/dD, the gradient of capacitance C with respect to tip–sample distance D [10,11]. If one models the tip simply as a sphere of radius R (i.e., treating only the rounded apex of the tip), then the following relation

$$g(D) \approx \frac{R^2}{D(D+R)}$$

provides a simple approximation [10]. But given that electrostatic forces normally can be sensed at distances of at least tens of nanometers, a sphere alone is usually not such a good model. Much more complicated forms of $g(D)$ have been derived to better describe the contributions of both the rounded tip apex and the tip *shank* [12,13].[2] Most commonly used is the model of a *spherically capped cone*,

$$g(D) = \frac{1}{\ln^2(\tan(\beta/2))}\left[\ln\left(\frac{H}{D+R(1-\sin\beta)}\right) - 1 + \frac{R\cos^2\beta/\sin\beta}{D+R(1-\sin\beta)}\right] + \frac{R^2(1-\sin\beta)}{D[D+R(1-\sin\beta)]},$$

and another model is the somewhat simpler *hyperboloid*,

$$g(D) = \frac{1}{\ln^2(\tan(\beta/2))}\left[\ln\left(1+\frac{H}{D}\right) - \frac{\left(D-\dfrac{R}{\tan^2\beta}\right)H}{D(H+D)}\right],$$

where β is the half-opening angle of the tip, H is the tip height, and R is the radius of curvature of the rounded tip apex.

Figure 2.3 compares the distance-dependent van der Waals force of Equation 2.5 with the electrostatic force for the case of a 250-mV potential difference between a flat surface and a hyperboloid tip (10 nm radius, 10 μm height, 35° full opening angle, Hamaker constant 6×10^{-20} J), and further shows the composite curve summing both force types. Such a potential difference can easily result from dissimilar intrinsic surface potentials (as discussed in Section 9.1). One might consider the presence of forces greater than 0.01 nN in magnitude in the 5–10 nm distance range to be an indicator of electrostatic interactions. Given that force goes as V^2, this force would be 16 times stronger for a 1 V potential difference and 400 times stronger for a 5 V difference. Thus, one begins to perceive that AFM can be a sensitive tool for electrical property characterization under applied bias, as discussed in Chapter 9. But even if such applications are not of direct interest, it is important to recognize the implications for degraded topographic imaging in the presence of large potential differences, given that forces are less localized (i.e., summed over a broader distance range and

[2] Also, the body of the cantilever to which the tip is attached, as discussed in Section 3.6.4.

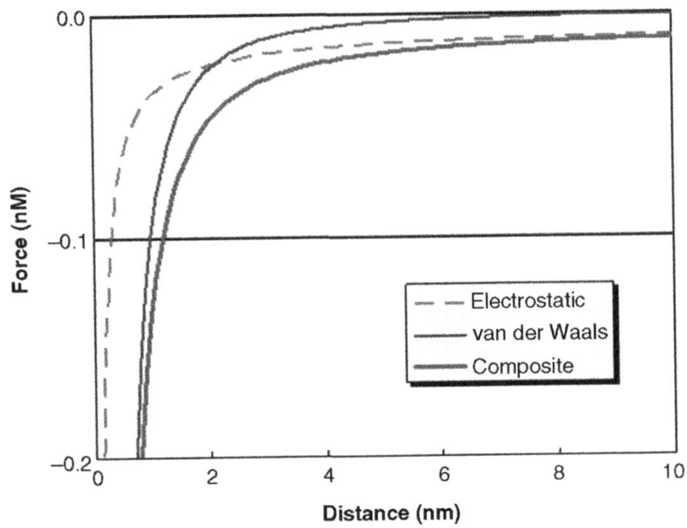

FIGURE 2.3 Electrostatic and van der Waals force–distance relationships for a hyperboloid tip of radius 10 nm, height 10 μm, full opening angle 35°, and Hamaker constant 6×10^{-20} J.

thus laterally "smeared") compared to the van der Waals case. Thus, unintended *sample charging*—which can produce large potential differences (i.e., many volts)—can significantly degrade topographic imaging resolution. Less obvious is the possibility that unavoidable differences in *surface potential* between tip and sample can produce long-range forces and potentially degrade topographic imaging, such as produce height artifacts. We will see that this especially is true in dynamic AFM and particularly when performed in the net attractive force regime (Chapters 4, 5 and 9).

2.3 CONTACT FORCES AND MECHANICAL COMPLIANCE

How should we describe repulsive forces due to *contact* between tip and sample? Should we integrate the repulsive r^{-12} term from the L–J potential over the interactive zone, in analogy to integrating the attractive r^{-6} term in deriving the expression for dipole–dipole interactions? The answer is no. The L–J expression for repulsive forces assumes that two atoms or molecules are *rigidly* pushed together, that this contact causes interpenetration of the electron clouds at the touching point, but that nothing deforms "behind each atom." This electronic interpenetration would increase very rapidly as tip and sample are brought together, resulting in a very steep (e.g., $1/D^8$) force–distance relationship. But in an extended solid, where atoms or molecules are packed together, touching another solid does not simply distort the electron cloud of those atoms *in direct contact at the interface*. It also results in *mechanical compliance* or strain of the solid below the surface; one can think of springs connecting atoms in the 3D solid and these springs compressing by gradually decreasing amounts as function of increasing distance away from the contact interface (i.e., into each solid). Thus, the *steepness of increase* of energy or force, as contacting materials are pushed further together, is *reduced* relative to that obtained if one integrated the r^{12} interatomic potential. On a scale of distance like that of Figure 2.1, the repulsion between tip and sample would grow quite gradually rather than the nearly vertical trend displayed for an L–J interaction. Indeed, on soft materials such as polymers, the atoms usually are not in a *closed packed* state with the outermost electron clouds touching. Instead, there can be substantial *free volume*: empty space or voids into which atoms and molecules can be pushed via the forces of contact. (These voids may be subnanometer in size [14], but there are many voids for many different atoms to move into.) Thus, the potential energy or force can be a much more gradually increasing function of displacement of AFM tip into sample on such a (soft) material. And if the AFM tip itself is covered by soft material (e.g., contaminated from previous scanning of soft materials), the energy or force can increase even more gradually as function of displacement of tip into sample.

Whether rigid tip or sample, or soft tip or sample, the question of how contact force or energy increases as a function of tip displacement into sample is central to most AFM imaging and measurement modes, as well as to other instrumentation such as nanoindentors [15–18]. It has therefore received extraordinary attention by experimentalists as well as theorists and computational researchers. Analytical expressions have been derived to describe continuum material systems under various simplifying assumptions [19]. (An awareness of these assumptions is sometimes lacking when analysts apply the expressions.) The formalism considers surface *tractions*—force per unit area—operating at points along the contact interface, both normal and tangential to this interface [19], to derive relationships between force, displacement, and contact geometry, and usually under the assumption of *linear elasticity*, which means stress σ and strain ε are related by Hooke's Law,

$$\sigma = E \cdot \varepsilon$$

where the proportionality constant E is Young's modulus of elasticity. (More complicated tensor relationships of this general type apply to the actual 3D interfacial contact problem [20], and include both compressive and shear forces.) For linearity to hold requires that the size of the dimensions of the contact area must be small compared to (a) the characteristic dimensions of each contacting body and (b) the (underformed) radii of curvature of the surfaces in contact. In AFM, condition (b) often may not strictly hold, given the small radius of curvature of the tip. A second assumption of the Hertz model is the absence of *viscoelasticity*: creep or viscous flow of material under force over some characteristic time. Finally, the contact is assumed to be "frictionless," here meaning a lack of tangential tractions at the interface (i.e., shear stress).[3] It should

[3] The meaning of "frictionless" can depend on the technical community using the term. The contact mechanics community uses friction force to mean an elastically resistive tangential force that is reversible (what some would call static friction), whereas the physical science community often uses friction force to mean a resistive drag force during translational motion derived from energy dissipation, that is, irreversibility. The latter is generally adopted in this book.

be apparent to the reader that one or more of the above assumptions may not hold for some classes of materials problems, including certain polymers (dependent on temperature and rate).

We discuss more general adhesive contact mechanics models in Section 6.2 because it is a substantial subtopic in AFM. Here, we consider the much simpler case of the Hertzian model, which ignores attractive forces between tip and sample. The general Hertzian model treats two spheres in contact (a flat surface being the special case of zero curvature). For tip–sample systems, this allows one to consider rounded contacting surfaces, not only the shape of the tip but also the hill or valley locations on the sample, each with a characteristic radius of curvature. Three useful relationships are

$$a = (RF/K)^{1/3} \tag{2.8}$$

$$F = KR^{1/2}h^{3/2} \tag{2.9}$$

$$hR = a^2 \tag{2.10}$$

where a is the radius of the contact circle for the two spheres, F is the applied force (load), h is the total displacement (deformation depth), and

$$R = (1/R_1 + 1/R_2)^{-1} \tag{2.11}$$

$$\frac{1}{K} = \frac{3}{4}\left[\frac{(1-\nu_1^2)}{E_1} + \frac{(1-\nu_2^2)}{E_2}\right] \tag{2.12}$$

where $R_{1,2}$ are the radii of curvature of the two spheres (for a flat, horizontal surface, one value R_i value becomes infinite, so the term drops out) and K is a function of the elastic constants of each solid; the Young's moduli E_i (N/m^2 or Pa) and the dimensionless Poisson's ratio ν_i quantifies the strain perpendicular to loading force divided by strain parallel to loading force. The reciprocal of the expression in brackets is sometimes called the *reduced elastic modulus* E^*.

For an essentially nondeformable tip (i.e., with much higher modulus than sample) pushing into a flat compliant sample of elastic modulus E and Poisson's ratio ν, R is simply the tip end radius of curvature. With elastic response then given by

$$K \approx \frac{4}{3}\left[\frac{E}{(1-\nu^2)}\right] \approx 1.5\,E,$$

the situation is described in Figure 2.4. Intuitively, the diameter $2a$ of the horizontal *contact circle* determines the approximate lateral resolution of the technique. If one assumes a tip of radius 10 nm and a sample modulus of 2.0 GPa (say that of a glassy polymer or plastic) and Poisson's ratio of 0.33, together with an applied load of 1.0 nN, the Hertzian expression predicts an indentation depth h of 0.35 nm and a

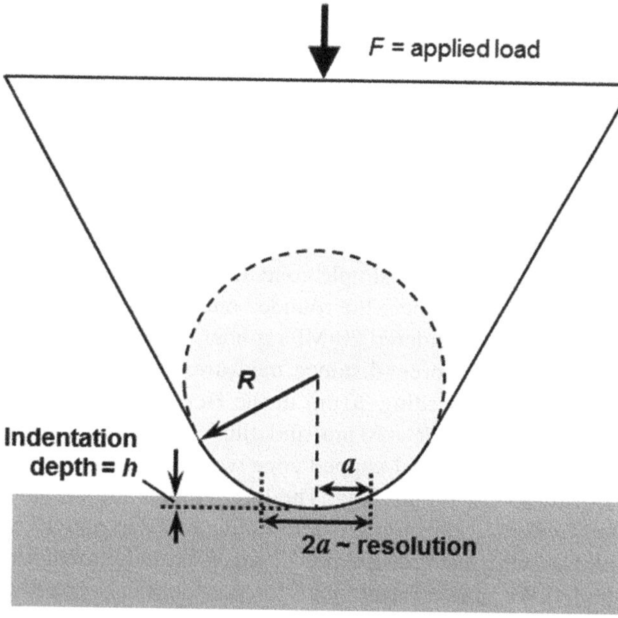

FIGURE 2.4 Illustration of a spherically ended AFM tip indenting into a flat, compliant half-space material by an amount small enough to be described by the classic Hertzian model.

contact radius a of 1.9 nm, meaning a lateral resolution of approximately 4 nm or 40% of R, about half of that illustrated in Figure 2.4. If the same tip pushes into a hundredfold softer polymer (a rubber) of modulus 20 MPa, however, then the Hertzian equations would predict a much larger indentation depth and contact radius, 4.8 and 6.9 nm, respectively. It is important to reiterate, however, that the classic Hertzian model derivation requires a contact radius $a \ll R$ and thus, by Equation 2.10, $h \ll R$. Sizes of contact geometries up to that depicted in Figure 2.4 (i.e., $a \approx 40\%$ of R) are actually described reasonably well by Equations 2.8–2.10.

Because intrinsic *yield points* of materials are expressed as force per unit area and can be exceeded in AFM (often on soft materials) resulting in plastic deformation and wear processes, it is worth considering the vertical *contact pressure* in the Hertz model. This increases from zero at the edge of the contact circle to a maximal value at the center of the contact zone, according to

$$p(\rho) = \frac{1}{\pi}\left(\frac{6FE^{*}2}{R^2}\right)^{1/3}\left(1-\frac{\rho^2}{a^2}\right)^{1/2} \equiv p_0\left(1-\frac{\rho^2}{a^2}\right)^{1/2}$$

where ρ is the radial coordinate and p_0 is defined as the maximal pressure at the center of the contact circle [19]. This expression predicts that the maximal contact pressure (i.e., at $\rho = 0$) exerted by a rigid tip of radius $R = 10$ nm on a material of reduced elastic modulus $E^* = 2.2\,\text{GPa}$ is $p_0 \approx 200\,\text{MPa}$ for $F = 1\,\text{nN}$. This pressure can exceed the compressive yield strength of many materials of this approximate modulus, such as glassy polymers. (We will see in Chapter 6 that tip–sample adhesion will act to increase the contact area and thereby decrease the contact pressure.)

Another complication is that on many soft materials within the analytical domain of AFM—such as soft rubbers, gels, or cells—penetration into the material can go beyond the rounded end of the tip. Thus, there are tip geometries other than spherical that better model the contact mechanics. These include paraboloid, hyperboloid, spherical-capped cone, and even a simple cone, the latter being reasonably valid for very large indentation depths where the rounded end of the tip is of minimal importance. Even on the above-considered 20-MPa rubber, a larger force of $F = 10\,\text{nN}$, as is easily reached in typical force–distance measurement modes, results in a predicted indentation depth exceeding 20 nm in the Hertzian model, well beyond the hemispherical end cap with $R = 10\,\text{nm}$ and thus an invalid use of this model. Instead, a model such as spherical-capped cone would be more appropriate, as suggested by the illustration of Figure 2.5a. The force–indentation relationship for this model is plotted in Figure 2.5b for a 10-nm radius end cap and 17.5° conical half angle, in comparison with a simple 3/2 power curve, the functional form of the classic Hertzian model. We note a lower slope for the spherical-capped conical mode, meaning a softer contact. The mathematical relationships needed to model a spherical-capped cone (Figure 2.5), as well as a spherical-capped four-sided pyramid, are

$$F = \frac{2E}{1-\nu^2}\left\{ah - \zeta\frac{a^2}{\tan(\beta)}\left[\frac{\pi}{2} - \arcsin\left(\frac{b}{a}\right)\right] - \frac{a^3}{3R} + (a^2-b^2)^{1/2}\left[\zeta\frac{b}{\tan(\beta)} + \frac{a^2-b^2}{3R}\right]\right\} \quad (2.13)$$

$$h = 2\zeta\frac{a}{\tan(\beta)}\left[\frac{\pi}{2} - \arcsin\left(\frac{b}{a}\right)\right] - \frac{a}{R}\left[(a^2-b^2)^{1/2} - a\right] \quad (2.14)$$

where β is the cone or pyramid half angle, b is the radius at which the tapered sides of the tip transition into the spherical end cap (of radius R), and ζ takes the value of $\frac{1}{2}$ for cone and $2^{1/2}/\pi$ for pyramid. Force F versus indentation h, as plotted in Figure 2.5b, can be numerically calculated by simultaneous solution of Equations 2.13 and 2.14 using a common spreadsheet program.

We note in passing that mechanical indentation of tip into sample means *as a function of negative distance* $D = -h$ between the end of the tip and the original sample surface plane. Thus, the Hertzian force increases with distance as $1/(-D)^{1.5}$, to be compared with the $1/D^8$ dependence obtained by integrating the 6–12 L–J relationship, a much steeper trend as distance approaches zero. Clearly, the total distance of solid compliance (elastic deformation) is huge compared to the aggregate amount of electronic overlap at the tip–sample interface (which would produce the steep resistance due to Pauli repulsion). Indeed, contact mechanics models assume that the repulsion felt during penetration of tip into sample is entirely due to

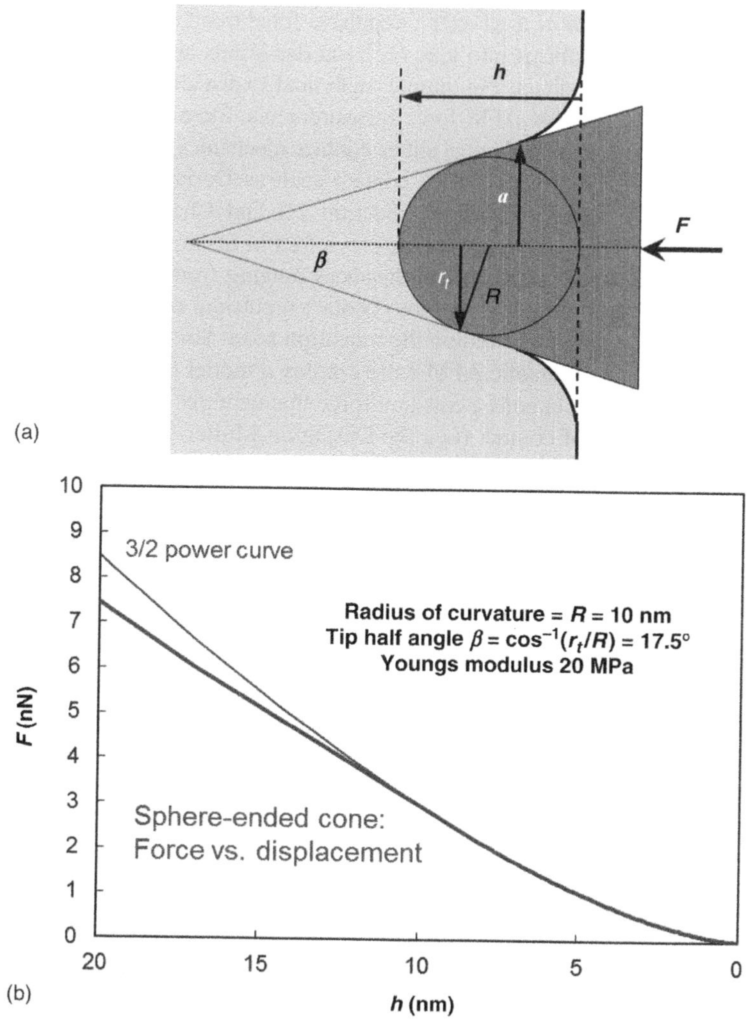

FIGURE 2.5 (a) Illustration of an AFM tip indenting into a flat, compliant half-space material beyond the spherical end cap of the tip. (b) Hertzian-type load–displacement curve for a sphere-ended cone indented into a flat half-space material, in comparison to a 3/2 power curve (classic Hertzian sphere).

3D mechanical resistance, not just Pauli repulsion at the interface. Mechanical resistance means that the conceptual "springs" connecting atoms are compressed. This resistive spring force is, of course, due to increasing electronic overlap, that is, Pauli forces. The point, in mechanical analysis, is that this overlap is not just at the tip–sample interface; rather, it extends well into the solids in three dimensions (by a variable extent).

The next question is how a $\propto(-D)^{3/2}$ repulsive force trend, for example, in contact (negative distance) transforms into a $\propto D^{-2}$ van der Waals attractive force trend, for example, just beyond contact. No simple analytical expression has been derived to contain both trends. In most AFM force measurements, there is no need for such an expression because one is analyzing either contact mechanics data (further discussed in Chapters 3 and 6) or forces out of contact such as Derjaguin–Landau–Verwey–Overbeek (DLVO), steric, and others (Section 2.5 and Chapter 6). But to model *dynamic AFM or force-curve mapping*, as previewed in Sections 1.4 and 1.5, wherein the tip is exploring forces over vertical distances ranging from beyond interaction up through van der Waals attraction and into contact repulsion, one needs to understand the entire range of behavior including the transition zone. Some computational methods (used in analyzing dynamic AFM data) employ a model that includes a Hertzian (3/2-power) force term but adds a constant force that captures the total van der Waals attraction at the onset of contact (i.e., the Derjaguin-Muller-Toporov (DMT) model covered in Section 6.2.2) and attaches this force trend to the -2 power distance dependence beyond "contact" (which is where the tip–sample distance is essentially two atomic radii, the distance between the centers of touching atoms) [21]. Others numerically connect a 2–8 force law derived from the L–J interaction with a contact mechanics form [22]. It is worth reiterating, however, that the use of a $-1/D^2$ form beyond contact *may only be strictly valid for distances* $D \ll R$, whereas in both dynamic AFM and force-curve measurements, D may range to distances greater than R.

In Figure 2.6, we qualitatively plot what such a force–distance relationship looks like on a soft material. Note that the repulsive force gradient (above dashed line), although relatively steep, is not extremely so compared to the strongest attractive force gradient just beyond the force minimum. (Compare this to the nearly vertical repulsive gradient in the L–J potential plot in Figure 2.1 that is intended to capture purely electronic overlap effects at the interface and no underlying 3D deformation.) We will see in later chapters that it is this *measurable repulsive force gradient*

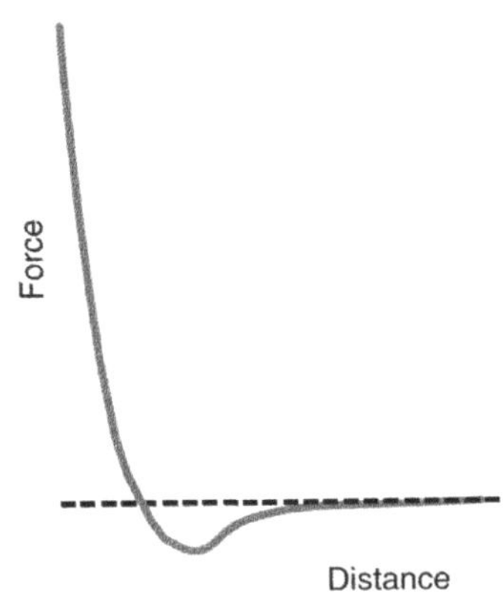

FIGURE 2.6 Schematic illustration of a force–distance relationship including attraction and "soft" repulsion.

that not only makes AFM enabling—as a quantitative nanomechanical measurement tool—but also makes the technique complicated from the standpoint of imaging surface topography. Why? In some cases, this mechanical compliance is effectively *contained in the topographic image*; apparent nanoscale "hills" and "valleys" can be present in the image due to variations in the repulsive force gradient from point to point across the surface, *rather than being real hills and valleys.*

2.4 DYNAMIC PROBING OF DISTANCE-DEPENDENT FORCES

We now consider the fundamental picture of dynamic AFM (previewed in Section 1.4 and commonly called "tapping mode"), where the tip is oscillated vertically and the intermittent interaction of tip and sample enables surface imaging, both topography and material properties. In Sections 2.2 and 2.3, we discussed the physical origins of force–distance relationships in AFM—variations in both magnitude and sign of force as well as its vertical gradient—between tip and sample. It seems intuitively obvious that if we consider a tip *continuously oscillating over a range of distances*, then the force–distance relationship should affect the overall behavior of the oscillating tip and the spring-like cantilever to which it is attached. But how?

Consider the following thought experiment; if we "ping" a tuning fork, causing it to ring with its characteristic pitch, then to bring the fork down to a solid surface until it interacts with the surface, we expect a change. What sort of change? Certainly, if we press hard to the surface we expect the ringing to quickly damp down to inaudible, extinguishing the vibration of the tuning fork. But we also can imagine controlling this touch with fine precision (say, using a Z displacement actuator as employed by an AFM) such that the ringing decreases gradually as a function of a very finely controlled approach distance, instead of suddenly. During this incremental reduction of the *amplitude* of tuning fork ringing (loudness) during approach, might we detect a change of *frequency* (pitch) of the tuning fork? One would be surprised if this were not the case. Accepting for the moment this bit of speculative intuition, . . . *why* would the pitch change? We will see in this and in later sections that these tuning fork concepts lie at the heart of dynamic AFM.

Understanding dynamic AFM requires examining the spring-like cantilever to which the tip is attached. Further details on cantilever, the laser/photodetector scheme for quantifying cantilever motion, artifacts in this system, and other instrumentation issues will be considered in Chapter 3. Here, we focus on the core principles needed to understand the technique in the context of distance-dependent forces.

2.4.1 Importance of Force Gradient

We treat the tip/cantilever system as an equivalent point mass m^* attached to a spring of stiffness k_0, a simple harmonic oscillator swinging toward and away from sample at its resonance frequency

$$\omega_0 = (k_0/m^*)^{1/2} \tag{2.15}$$

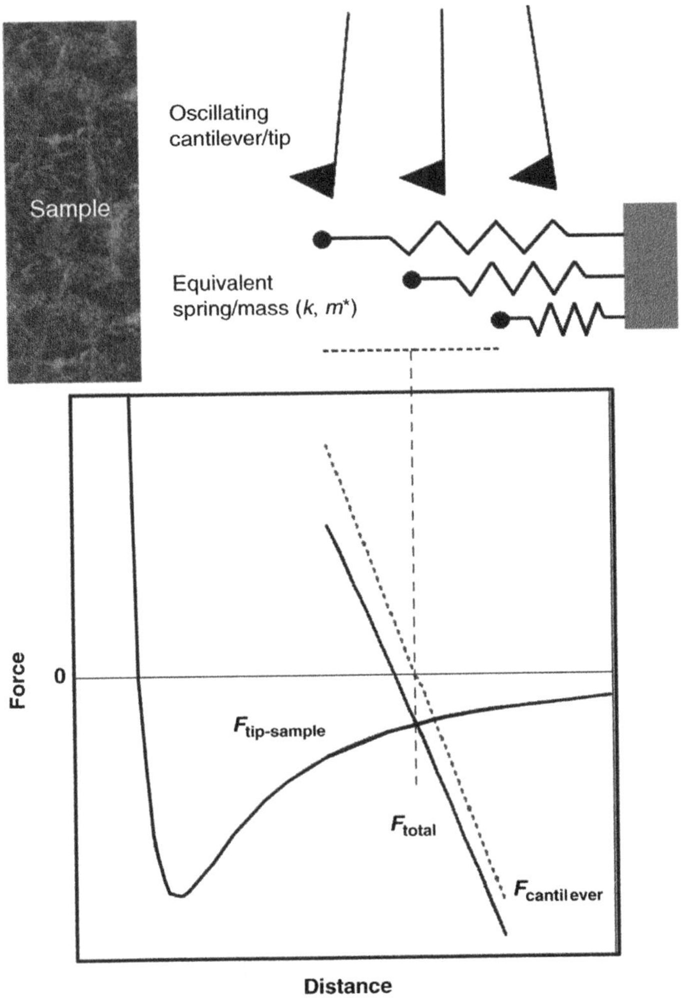

FIGURE 2.7 (Top) Illustration of oscillating tip/cantilever beyond a sample surface, and equivalent spring/point mass. (Bottom) Linear spring restoring force (dashed line) overlaid on attractive portion of tip–sample interaction, and approximately linear sum of reduced slope.

as known from introductory physics. This equivalence is depicted in the top of Figure 2.7 (albeit with exaggerated range of tip oscillation compared to tip size; in reality, the former is measured in nanometers whereas the latter has dimensions measured in micrometers!). As one case of interest, we can imagine the tip exploring distances in which only *attractive* forces are felt and with a *range of probed distance that is small relative to the full range of attractive forces*, as depicted in the bottom of Figure 2.7. The forces on the tip or equivalent point mass m^* are the sum of the cantilever's restoring force, $F_{\text{cantilever}}$ (dashed diagonal line of negative slope) and the tip–sample interaction force $F_{\text{tip-sample}}$ (light gray curve). Note that $F_{\text{cantilever}}$ is positive in value when pulling the tip from the surface (i.e., left to right), opposing the attraction of $F_{\text{tip-sample}}$. The cantilever force is that of a Hookean (linear) spring

$$F_{\text{cantilever}} = -k_0(s_Z - s_{Z0}), \tag{2.16}$$

where $s_Z - s_{Z0}$ is the tip displacement from the equilibrium position s_{Z0} (undeflected cantilever state, zero force). For convenience, we take the zero of the s_Z scale to be the point at which $F_{\text{tip–sample}}$ is zero (i.e., transitions from net attractive to net repulsive). The tip–sample interaction force is similar to that considered in Sections 2.2 and 2.3, a combination of relatively weak attraction at longer distances and steep, strong repulsion at short distances.

Because: (1) the range of distances explored by the tip is fairly small compared to the total range of $F_{\text{tip-sample}}$ and (2) the slope of the cantilever force is large in magnitude compared to that of $F_{\text{tip-sample}}$, the total force

$$F_{\text{total}} = F_{\text{cantilever}} + F_{\text{tip-sample}} \tag{2.17}$$

has the appearance of another diagonal line shifted slightly downward from the $F_{\text{cantilever}}$ dashed line, and of slightly *shallower slope*. This implies that the *tip/cantilever in the presence of tip–sample forces still behaves like a Hookean spring*, and thus like a harmonic oscillator. But a slightly reduced slope of $F_{\text{total}}(Z)$ reflects an effectively softer spring constant of this modified oscillator,

$$k = k_0 - \delta k = k_0 = \delta F_{\text{tip-sample}}/\delta s_Z \tag{2.18}$$

where the final differential term expresses the change of tip–sample force over the range δs_Z explored by the tip. Thus, the resonance frequency of the oscillator is modified from ω_0,

$$\omega = (k/m^*)^{1/2} = [(k_0 - \delta F\text{tip-sample}/\delta s_Z)/m^*]^{1/2}$$

or

$$\omega \approx \omega_0\left[1 - (1/2k_0)\delta F_{\text{tip-sample}}/\delta s_Z\right] \tag{2.19}$$

where the final approximate form is valid provided that the gradient of the tip–sample interaction force; $\delta F_{\text{tip-sample}}/\delta s_Z$ is small compared to the spring constant of the cantilever k_0. Revisiting the expression for van der Waals forces from Section 2.2 in terms of the Hamaker constant $A_H = 6.5 \times 10^{-20}$ J for a SiO_2 tip interacting with

SiO_2 sample in air, we calculate a slope (first derivative) of $A_H R/3D^3 \approx 0.2$ N/m (J/m^2) using $R = 10$ nm and $D = 1$ nm. A common cantilever spring constant for dynamic AFM operation in air is $k_0 \approx 40$ N/m. Thus, our value of $\delta F_{\text{tip-sample}}/\delta s_Z$ is comparatively small, a factor of 200 less than k_0, consistent with the requirement for the above approximation to be valid.

This result also means that $\omega/\omega_0 \approx 1 - (1/2)(1/200) \approx 0.9975$. Thus, the resonance frequency of the oscillator is reduced, albeit only slightly, by the presence of a weak attractive force gradient. So indeed the pitch of our resonant "tuning fork" has been modified by bringing it close to a material surface! Suppose we could monitor this shift as we raster scan a SiO_2 tip over a two-component material (at ~1 nm "fly height") varying in Hamaker constant. For example, let us consider a phase-segregated blend of polystyrene and poly(methyl methacrylate), Hamaker constants of $A_H = 6.9 \times 10^{-20}$ J and 6.4×10^{-20} J, respectively, meaning a difference of just under 10% in the tip–sample force gradient $A_H R/3D^3$. Thus, the two values of the modified resonant frequency, as we raster scan over the surface, *differ* by roughly a hundredth of a percent of the unmodified resonant frequency. Any variation of a quantity that is four orders of magnitude smaller than the quantity itself may seem insignificant. But again, in the world of science, a few powers of 10 are not necessarily too much.

How do we ensure a "fly height" that is only 1 nm from the surface, assuming we are dealing with nontrivial surface topography? We shall see in Chapter 4 that we can feed off the amplitude of an oscillating cantilever so as to *track this topography* at essentially constant fly height. At the same time, we can monitor another quantity, the phase of tip motion relative to the driving signal that is shaking the cantilever; this phase shift is in turn *sensitive to the resonance frequency of the cantilever*. Can a 10^{-4} variation in resonance frequency produce a measurable change of this phase shift as we scan over the surface? We will see below that the answer is a resounding "yes," to the point of producing rich and beautiful materials contrast, such as that previewed in Section 1.4.

But all is not as simple as suggested above. It is in practice difficult, essentially impossible in most cases, to stably operate with the small oscillation amplitude depicted in Figure 2.7, one that is much smaller than the range of attractive forces while centered just a nanometer or so off the surface. Part of the difficulty can be due to force gradients that exceed the cantilever spring constant, producing a nonequilibrium force condition that causes the tip to accelerate and snap into contact with the surface. (This is commonly stated in introductory treatments of AFM.) In reality, the stiff (e.g., 40 N/m) cantilevers that are typically used may not encounter such intrinsically steep force gradients. The greater, real-world difficulty is the presence of mobile, liquidy contaminants on both tip and sample surfaces; under van der Waals attraction, these may neck together to form a capillary bridge (Section 6.3), which in turn aggressively pulls the tip/cantilever down to the sample surface (an extrinsic and large effective force gradient). In either case, the stable operation of dynamic AFMs in ambient conditions requires a strategy just the opposite of that described above; one must choose an oscillation amplitude that is *very large compared to the range of van der Waals interactions*. Then the restoring spring force of

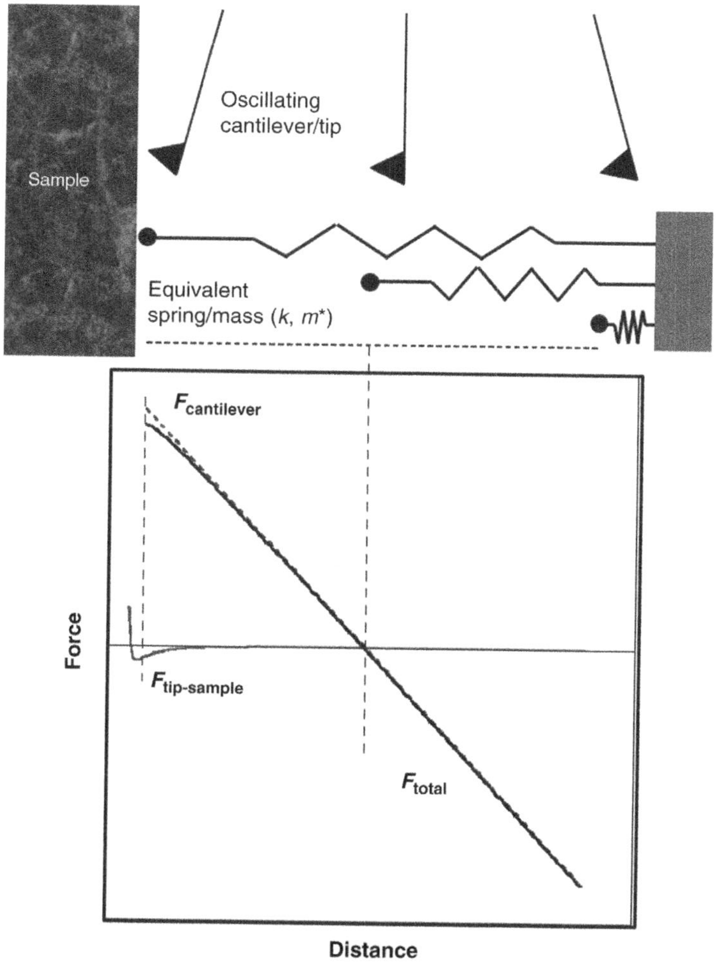

FIGURE 2.8 (Top) Illustration of oscillating tip/cantilever up to and far away from a sample surface, and equivalent spring/point mass. (Bottom) Linear spring restoring force (dashed line) overlaid on tip–sample interaction curve, and approximately linear sum with reduced slope near surface.

the cantilever is large when the tip encounters strong attractive forces, such that the tip can "pull out" of this interaction. But even if using a large oscillation amplitude, the principles outlined above still hold, such that measurable frequency shifts, and thereby materials contrast, remain possible. How can this be?

Figure 2.8 is a depiction of a large-amplitude tip oscillation above a material surface, analogous to that of Figure 2.7 for small-amplitude oscillation. The amplitude and center of oscillation are chosen to approach the tip close enough to the surface to yield attractive forces, but not so close as to reach into the repulsive part of the tip–sample interaction curve. We see in the graph that the addition of the $F_{\text{tip-sample}}$ curve to the $F_{\text{cantilever}}$ line to obtain F_{total} has only a small effect, because $F_{\text{cantilever}}$ becomes large as the tip nears the sample. If we mathematically fit a linear function to F_{total}, we find that the slope of this function is only slightly modified relative to $F_{\text{cantilever}}$, even less so than was the case in Figure 2.7 for small-amplitude oscillation. Nevertheless, this distance-localized interaction perturbs the resonant behavior of the cantilever in an equivalent fashion as in Figure 2.8, resulting in an effective reduction of the oscillator stiffness and thus resonance frequency. If instead the oscillator amplitude is chosen to produce an excursion of tip into the repulsive part of the tip–sample interaction curve at extreme left, one can imagine a resulting rise in F_{total} at the left end, such that the trend of F_{total} departs first below then above that of $F_{\text{cantilever}}$ as the distance approaches zero and becomes negative. Again, a mathematical line fit through this (nonmonotonic) trend may only slightly deviate from the linear behavior of $F_{\text{cantilever}}$, and indeed, this departure could be of either sign: a line of slightly lesser or greater slope, depending on the relative strength of attractive and repulsive forces encountered. This is the general rule in dynamic AFM; some combination of attractive and repulsive forces that effectively stiffens (steeper equivalent slope) or softens the oscillator and in some cases, is very close to doing neither because attraction and repulsion effectively offset each other. (So far, we are ignoring changing in damping, a later topic.)

Experimentally one can demonstrate that the cantilever motion remains highly harmonic (sinusoidal) but slightly shifted in frequency $\omega_0 = (k/m^*)^{1/2}$, as if the distance dependence of F_{total} is *effectively linear* of slope k slightly different from k_0. Thus, the concept of a shifted resonance frequency, due to the net effect of $F_{\text{tip-sample}}$ on the slope of the restoring force, remains valid to first approximation. If the tip reaches far enough into the $F_{\text{tip-sample}}$ curve such that the oscillator is net stiffened by the interaction, we anticipate that the *amount* of frequency increase could be a gauge of material stiffness, meaning modulus. How can we measure and image this frequency to distinguish materials based on modulus? It was mentioned earlier that the *phase lag* between the tip oscillation and the driving oscillation at the base of the cantilever is sensitive to frequency shift. To understand why, we must quantitatively examine this relationship. This process is in fact a practical first step to setting up an AFM for operation. But for now we will discuss the basic concepts divorced from instrumentation details.

2.4.2 Damped, Driven Oscillator: Concepts and Mathematics

In dynamic AFM, the cantilever behaves as a *driven harmonic oscillator with damping* as schematically illustrated in Figure 2.9. The driving signal is a sinusoidal vibration applied to the base of the cantilever, equivalent to a shaken beam from which a spring of stiffness k_0 and effective point mass m^* are attached. The damping of a freely oscillating cantilever (i.e., frictional resistance to its motion) is primarily due to the fluid medium, air or liquid, equivalent to a dashpot of damping coefficient b (i.e., a damping force proportional to the velocity of the mass, $F = -bv$) in parallel with the spring. (For operation in vacuum, the

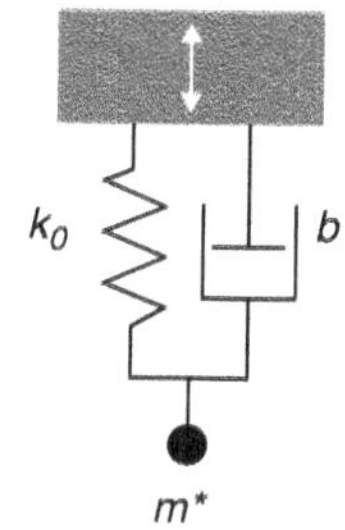

FIGURE 2.9 Schematic illustration of a damped, driven harmonic oscillator.

remaining damping is due to intrinsic anelasticity in the cantilever itself and other system components such as material interfaces between excitation signal and cantilever.)

The mathematical description of a damped, driven harmonic oscillator is a classic undergraduate physics problem, solving Newton's second law of motion (i.e., force = mass × acceleration). Here, we provide the *steady-state* ("particular") solution of this differential equation for the case of *underdamped* motion ($2m^*\omega_0 > b$, the relevant case for AFM). This solution describes the *frequency dependence* of the amplitude and phase lag of the sinusoidal motion of the mass relative to the driving sine wave, Figure 2.10 (note inverted phase scale, a common convention; instead some authors scale from 0 down to −180) [23]. There is a peak in amplitude at a frequency, the resonance frequency of the damped, driven oscillator (DDO) given by

$$\omega_0^{\mathrm{DDO}} = \sqrt{\omega_0^2 - \frac{1}{2}\left(\frac{b}{m^*}\right)^2} \tag{2.20}$$

where again $\omega_0 = (k/m^*)^{1/2}$, the resonance frequency of the undamped oscillator. A useful dimensionless parameter is the *quality factor* Q given by

$$Q = \frac{m^* \omega_0^{\mathrm{DDO}}}{b} \tag{2.21}$$

which quantifies the sharpness of the resonance and, importantly, is inversely related to the strength of damping. The value of Q in Figure 2.10 is 500 and the resonance frequency 300 kHz ($=\omega_0/2\pi$), which is fairly typical for common cantilevers used in dynamic AFM. This value can be approximately determined from the ratio of resonance frequency divided by the full width of the amplitude peak at ≈70% ($1/\sqrt{2}$) of its peak value. Importantly, the phase lag, our primary interest in this discussion, also changes steeply in the vicinity of this resonance, as detailed below.

The frequency dependences of amplitude and phase are given by Lorentzian and arctangent functions, respectively, as

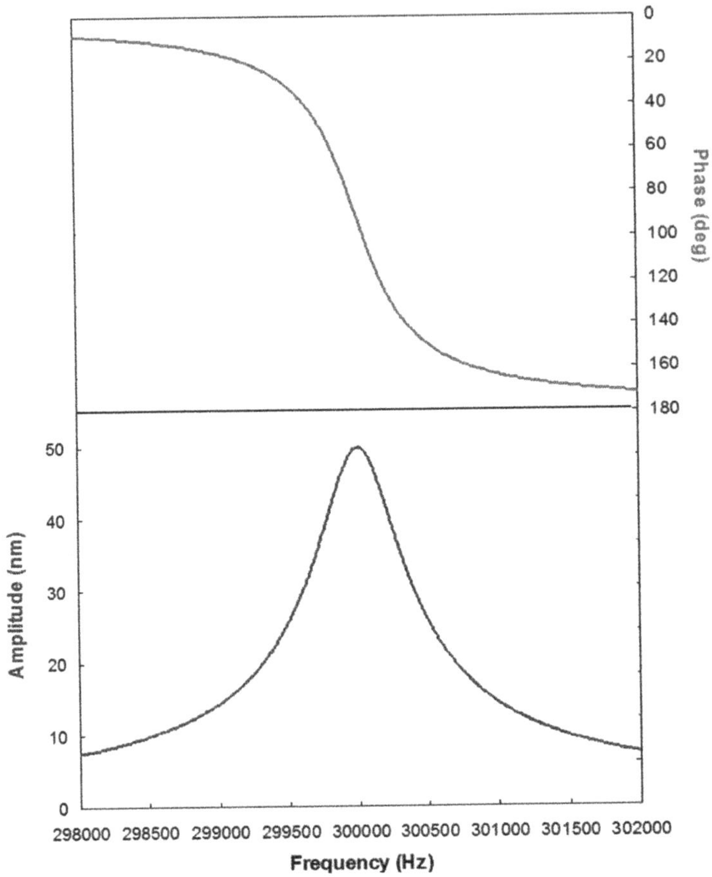

FIGURE 2.10 Frequency response of the amplitude and phase of a damped, driven harmonic oscillator.

$$A = \frac{A_{\text{drive}} Q}{\sqrt{\left(\omega/\omega_0\right)^2 + Q^2\left(1 - \left(\omega/\omega_0\right)^2\right)^2}} \tag{2.22}$$

$$\varphi = \arctan\left[\frac{\omega/\omega_0}{Q\left(1 - \left(\omega/\omega_0\right)^2\right)}\right] \tag{2.23}$$

where A_{drive} is the amplitude of the drive signal.

The time dependence of this resonant motion is illustrated in Figure 2.11. For AFM, the beam motion corresponds to the drive signal applied to the base of the

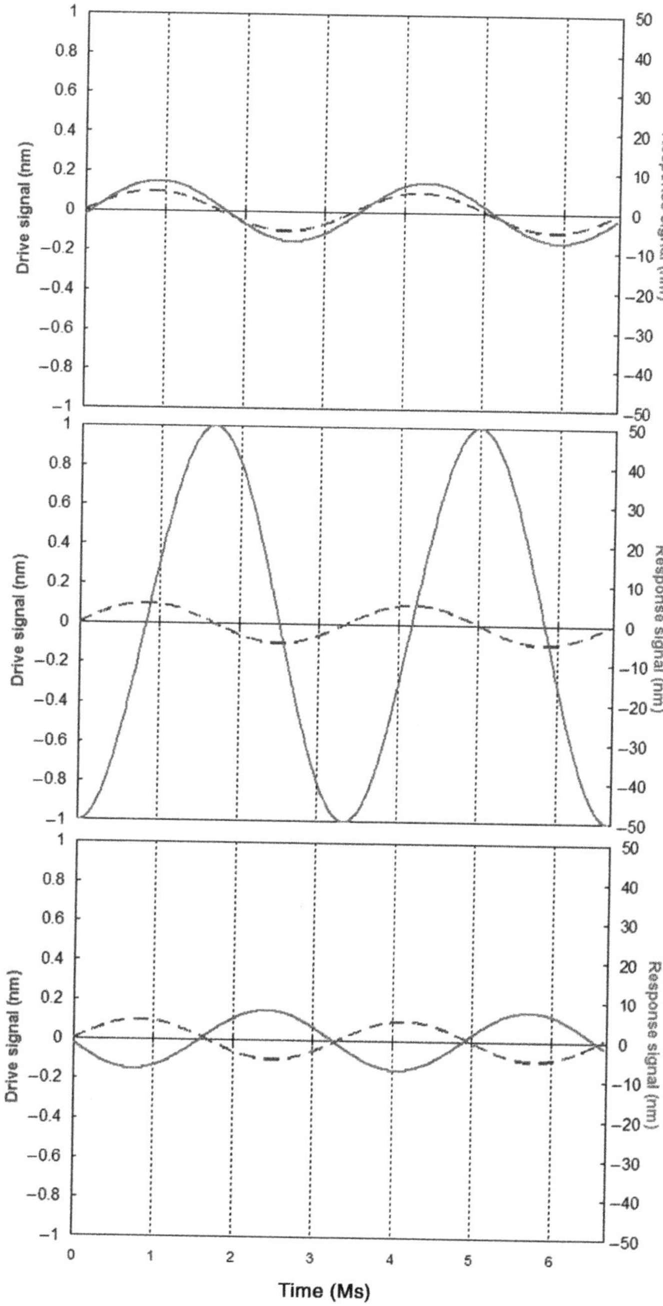

FIGURE 2.11 Time dependence of driving signal (dashed curves) and response (solid curves) of a damped, driven harmonic oscillator for driving frequencies below (top), at (middle), and above the intrinsic resonance frequency of the oscillator.

cantilever (dashed curve) and the point mass response signal corresponds to the motion of the tip (solid curve). At frequencies well below the resonance $\omega = \omega_0$, the phase lag approaches 0°, meaning the beam and point mass (driving and response signal) are moving up and down in near synchronization (top graph in Figure 2.11). At frequencies well above resonance, the phase lag approaches 180°, meaning the beam and point mass are almost exactly out of phase (mass moves down while beam moves up and vice versa, bottom graph in Figure 2.11). Precisely, at resonance, the beam and point mass are 90° out of phase (middle graph in Figure 2.11), in addition to the strong amplification of the response relative to drive signal. This means that the beam has maximum upward velocity when the point mass become stationary at the bottom of its motion and, subsequently, the beam has reached its highest point when the point mass achieves its maximum upward velocity, etc.

2.4.3 Effect of Tip–Sample Interaction on Oscillator

Now, back to the inclusion of tip–sample interaction for an oscillating AFM tip-cantilever. As already stated, the resonance frequency of the oscillator will be shifted by this interaction. Thus, the frequency dependence of amplitude and phase will shift to the right or left depending on whether there is a net repulsive or attractive interaction. Because the phase curve changes monotonically (i.e., increasing from below to above 90°) as drive frequency increases from below to above resonance, the phase lag indeed relates simply to the magnitude and sign of the frequency shift. Importantly, the change of phase is particularly steep near resonance, meaning a small shift of resonance frequency results in a substantial change in phase. Consider a shift of resonance frequency that is only one part in 10^4, the case stated earlier. This would amount to a frequency shift of about 30 Hz, translating into a change of phase at resonance a bit more than 5°. Figure 2.12 exemplifies this shift in each direction (i.e., due to net attractive or net repulsive forces for left- and right-shifted curves). We will see that real data acquisition systems have sensitivities of ~100 mV per degree of phase shift, meaning a 5° shift produces about a half volt of change in raw signal, easily measurable and with a good signal-to-noise ratio.

The above treatment is an oversimplification in several respects, but one is particularly glaring. Indeed, it may result in unfounded conclusions when using phase imaging for materials contrast (Chapter 5). We have discussed how the tip–sample interaction effectively changes the oscillator spring constant and thus its resonance frequency, thereby shifting the phase signal. But in Figure 2.9 and Equations 2.20–2.23, we are reminded that *damping* (dissipative) forces also are important in the resonant behavior of the oscillator. Upon tip–sample interaction, it is not only the oscillator spring constant k, hence the resonance frequency ω_0, that effectively changes; the damping coefficient b may effectively change because of additional energy dissipation (i.e., cantilever into sample), as integrated over the tip–sample interaction cycle (e.g., due to sample viscoelasticity in the case of polymers). Thereby, *the effective Q will change*. Thus, the frequency dependence of amplitude and phase in Figure 2.10 is not only *shifted* by the "stiffness" of tip–sample

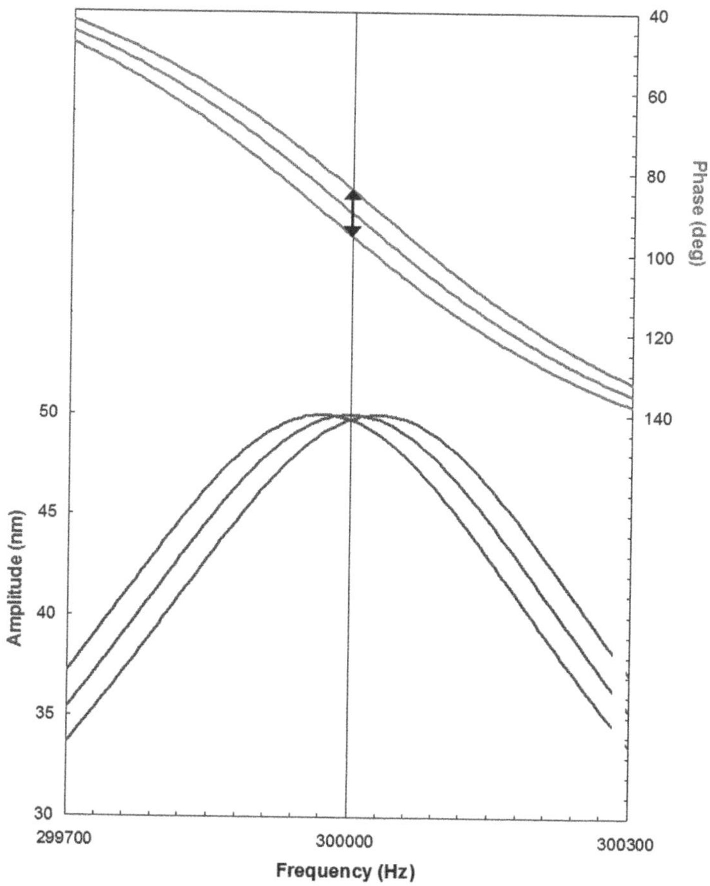

FIGURE 2.12 Illustration of the frequency-shifted amplitude and phase response of an AFM cantilever due to nondissipative tip–sample interaction that is net attractive (leftward) or net repulsive (rightward).

interaction (net positive or negative) but also *broadened* due to tip–sample interaction damping. (This additionally shifts the resonant frequency per Equation 2.18, i.e., both the *position* and the *shape* of $A(\omega)$ and $\varphi(\omega)$ are modified by tip–sample damping.) The result: materials contrast that can be obtained by imaging phase lag (as exemplified in Section 1.4) does not have a simple, one-to-one interpretation in terms of sample properties. For example, sample regions differing in elastic modulus will indeed give phase contrast, all other things being equal; but phase contrast does not necessarily imply variations of elastic modulus. It is important to consider other property differences (e.g., viscosity, surface potential, Hamaker constant) that could potentially produce the observed phase contrast. And, of course, "stiffness" in the most general sense has both positive (modulus) and negative (Hamaker constant) contributions related to different material characteristics (responses to indentation or dispersion forces).

Later, we will consider a more general and quantitative expression for interpreting phase lag, derived from energy conservation principles. We will see that phase can be directly and exclusively related to energy dissipation during tip–sample interaction. What then about stiffness? One must bear in mind that energy dissipation is an *extensive* quantity; its magnitude scales with the *size* of the dissipative tip–sample interaction zone, in addition to the *intrinsically* dissipative character of this interaction. A softer material will deform through a greater volume and form a larger contact area with the tip. Thus, viscoelastic dissipation internal to the deformed material, as well as dissipative processes taking place at the tip–sample interface (adhesion hysteresis due to conformational rearrangements, charge exchange, etc., during the making and breaking of contact) will be larger on a softer material, all other things being equal. For example, two polymers differing in elastic (*storage*) modulus but otherwise identical would produce phase contrast. But two polymers contrasted in phase are not necessarily different in elastic modulus.

In this section, we have assessed dynamic tip–sample interactions in terms of how distance-dependent forces affect oscillator behavior, ultimately modifying the phase lag between tip and drive signal. But the resonance shifting and damping that yields this change in phase lag must also change the amplitude, as both $A(\omega)$ and $\varphi(\omega)$ shift along the frequency scale. Obviously, the changes of both amplitude and phase should depend on the extent the AFM tip plunges into the tip–sample interactive relationship (e.g., Figure 2.8), as this determines the steepness (positive or negative) of the force–distance relationship and presumably the magnitude of damping. After bringing the vertically oscillating tip close enough to interact with the sample surface, one intuitively expects the amplitude of tip oscillation to decrease as the base of the cantilever and the sample surface are *further* approached (decreasing Z, Figure 2.13). Importantly, we have seen that this decrease of amplitude relates in part to the shift of oscillator resonance curve to higher or lower frequency.

It is instructive to contemplate in more detail how amplitude and phase change as a function of Z. If there were no dissipation, such that amplitude and phase change in strict accordance with the (simultaneous) shifting of $A(\omega)$ and $\varphi(\omega)$ along the frequency scale as Z decreases, what would be the relationship between phase and amplitude? This can be seen by numerically plotting φ versus A from Equations

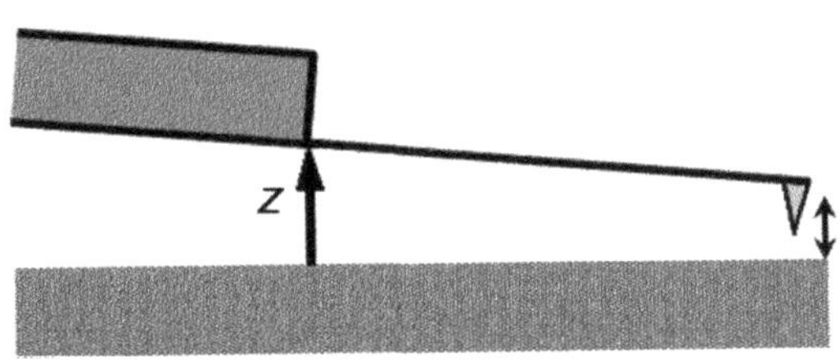

FIGURE 2.13 Depicted connection between the Z displacement of the cantilever chip and the amplitude of the oscillating tip.

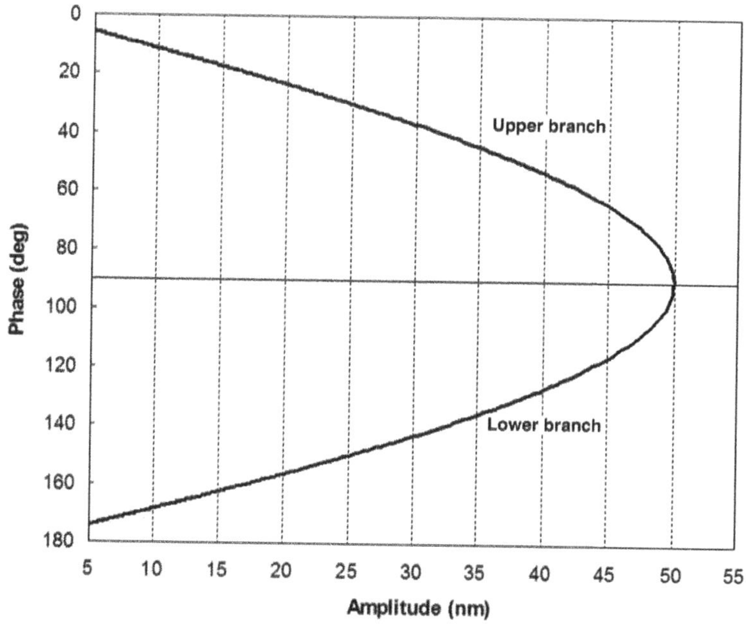

FIGURE 2.14 Relation between phase and amplitude of a damped, driven harmonic oscillator.

2.20 and 2.21 (eliminating the variable ω). Inspecting this relationship as plotted in Figure 2.14, the most remarkable observation is a *double-valued* phase versus amplitude curve. Thus, for any given reduction of amplitude (here from a maximum value of 50 nm at free resonance), there are *two* possible phase values. How can this be? The answer is the shifting of $A(\omega)$ and $\varphi(\omega)$ along the frequency scale can be due to either net attractive or net repulsive interactions. The upper branch of the curve in Figure 2.14 results from net repulsive interactions (phase lag less than 90°), whereas the lower branch results from net attractive. So then, as we bring oscillating tip and sample together and the amplitude decreases from 50 nm, which branch is followed? We know that upon approach, attractive forces are felt first. Thus, we expect that the lower branch is followed. But approaching sufficiently far will completely stop the cantilever oscillation and result in a static, repulsive contact. In between, as oscillating tip and sample are brought together, we anticipate that at some stage the oscillating tip will switch from a net attractive to net repulsive interaction where the tip is intermittently hitting the surface. In Figure 2.14, this could only happen if there were a large change of phase, a jump from the lower branch to the upper (e.g., at an amplitude of 40 nm). We will see in Chapter 5 that this qualitative behavior is indeed experimentally observed and, practically speaking, very important to the operation of an AFM. It, moreover, can be reproduced in numerical simulations assuming a force–distance relationship akin to that in Figure 2.6 [21,22].

This is a profound difference between dynamic ("tapping mode") AFM and quasistatic (contact mode) AFM; in the former case, *more than one possible state of tip–sample interaction* for a given value of Z (at the jump discontinuity). And, that these states differ in the distance to which the tip reaches into the tip–surface interaction, *the very distance relationship that is being used to identify the vertical position of the surface*, ultimately to image topography. So, is this two-state situation a problem for imaging topography? Absolutely! To fully understand the implications of this concept, we will examine real measurements of amplitude and phase lag as a function of Z (Chapter 5). We will see that jumps between two dynamic states can be endemic to dynamic ("tapping mode") AFM imaging. Indeed, steps to avoid such jumps often are the *principal optimization procedure* the operator must execute to obtain good instrument performance.

2.4.4 Energy Dissipation in Tip–Sample Interaction

The preceding discussion assumed no damping in tip–sample interaction, rather, the reduction of oscillation amplitude via resonance frequency shift only. How would the phase-amplitude relationship in Figure 2.14 change if including energy dissipation? Consider an effective change in the Q of a damped driven oscillator, in turn due to an *effective change in the damping force* coefficient b. Although b quantifies a *continuous* damping throughout the oscillation cycle, instead of damping during the brief tip–sample interaction, it is the only dissipative parameter available within this simple conceptual model. This treatment is exactly analogous to the effective change of the continuous oscillator stiffness k (and thereby resonance frequency ω_0) used to account for tip–sample interaction stiffness; thus, we are at least internally consistent in our crude model. In Figure 2.15, we overlay the phase-amplitude curves for three values of Q: the freely oscillating value of 500 (solid curve) plus values of 450 (dashed curve) and 400 (dotted curve) representing relatively light and heavy additional damping, respectively. Upon approach to a particular reduced amplitude, 35 nm, the predicted shift of phase is indicated with arrows for each case of additional damping, assuming that this results from being in the repulsive regime (upper branch) due to tip–sample contact dissipation. We see that only 10% and 20% decreases in the effective value of Q (i.e., increases in b) cause phase shifts of nearly 10 and 20°, respectively. Is this realistic? We recall that the damping force is $-bv$, where v is the velocity of the tip. The dissipated energy is given by the time integral of the quantity force × velocity, or bv^2. Given this quadratic dependence on velocity and the fact that velocity should be small in the vicinity of the tip's turnaround point (i.e., contact), compared to the high velocities attained near the midpoint of each up and down swing of the cantilever, can the additional dissipated energy during contact be equivalent to a 10–20% change of b during the entire oscillation cycle? At this stage, we have no way to assess this question. We need a more detailed analysis.

It turns out that one can derive a mathematical expression *relating the power dissipated in tip–sample interaction to the measured value of phase*. A more detailed treatment is provided in Appendix 3. Here, we describe the basic concepts and state the final result. By energy conservation, the energy per time imparted by the driving

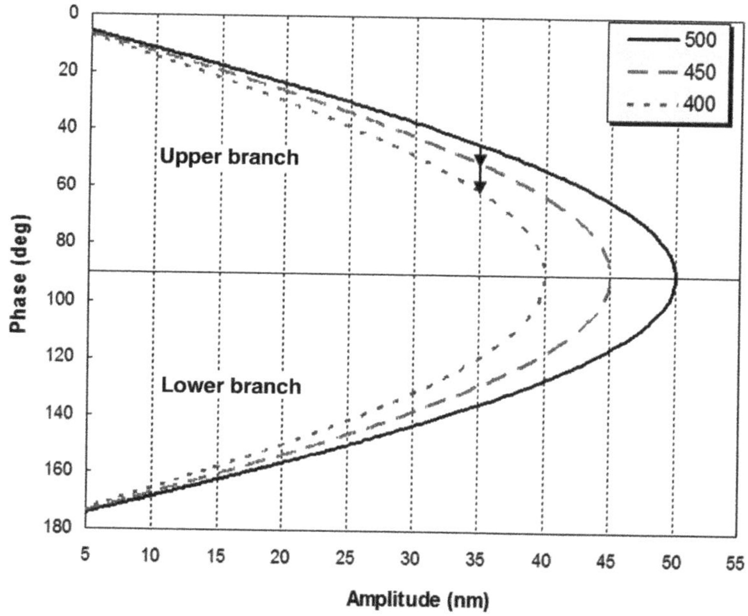

FIGURE 2.15 Phase–amplitude relationship for damped, driven harmonic oscillators at three values of Q.

force at the base of cantilever (force times velocity, i.e. power, integrated over one period of oscillation), must be equal the energy dissipated per unit time by the motion of the cantilever and the tip–sample interaction [24,25]:

$$P_{\text{driver}} = P_{\text{cantilever}} + P_{\text{tip-sample}}. \tag{2.24}$$

Integrating the driving force expression and subtracting the cantilever damping expression, along with the (typically excellent) approximation $\omega \approx \omega_0$ (in order to achieve a significant excitation of the cantilever as seen in Figure 2.10) results in the following relationship between tip–sample energy dissipation per encounter and phase lag φ:

$$E_{\text{tip-sample}} = \frac{\pi k A^2}{Q}\left(\frac{A_0 \sin\varphi}{A} - 1\right) \tag{2.25}$$

where A is the cantilever amplitude at frequency ω, reduced from a free amplitude A_0 at resonance frequency ω_0, and Q remains the quality factor of the *freely oscillating* cantilever (i.e., not an effective Q that includes tip–sample dissipative effects). To better understand this expression, we rewrite as

$$E_{\text{tip-sample}} \approx \frac{2\pi}{Q}\left(\frac{1}{2}kA_0^2\right)\frac{A}{A_0}\left(\sin\varphi - \frac{A}{A_0}\right). \tag{2.26}$$

This shows the importance of the ratio A/A_0 and the prefactor $\frac{1}{2}kA_0^2$, the kinetic energy of the freely oscillating cantilever at resonance, as well as the Q of the (freely oscillating) cantilever, in determining the magnitude of tip–sample energy dissipation.

It is worthwhile to check the relationship between φ and A predicted by this expression for the special case of zero tip–sample dissipation, to compare with Figure 2.14. The nontrivial solution to $E_{\text{tip-sample}} = 0$ is

$$\varphi = \sin^{-1}(A/A_0); \tag{2.27}$$

a good spreadsheet exercise is to show that this relationship reproduces the two-branched curve in Figure 2.14 given $A_0 = 50$ nm. Second, it is instructive to consider the φ–A relationship for the special case of a nonzero dissipation that is independent of A and A_0. This implies that bringing the tip and sample together and then apart dissipates a well-defined energy *independent of the kinetic energy of the oscillator* (probably not accurate if viscous damping, i.e., a velocity-dependent damping force, is present.) In Figure 2.16, we plot this relationship for three cases of tip–sample dissipation: zero (solid curve), relatively light (dashed curve), and relatively heavy (dotted curve), to compare with Figure 2.15. The light and heavy dissipative curves correspond to ratios of $E_{\text{tip-sample}}$ to $\frac{1}{2}kA_0^2$ (free oscillator energy) of 0.001 and 0.002. Using $A_0 = 50$ nm, $Q = 500$, and $k = 40$ N/m, these two tip–sample dissipations are a bit over 300 and 600 eV per cycle, respectively. Upon approach to a

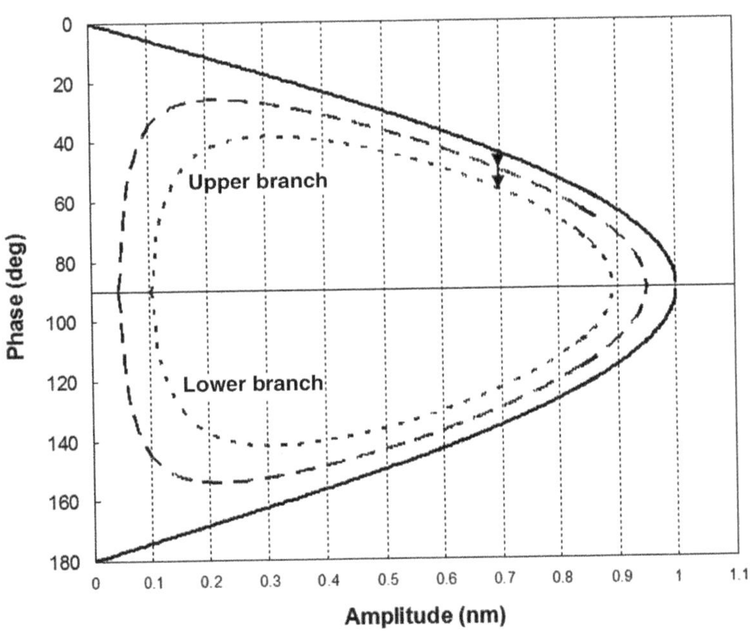

FIGURE 2.16 Phase–amplitude relationships for dynamic AFM for cases of zero tip–sample dissipation (solid curve) as well as finite tip–sample dissipation, cases of relatively weak (dashed) and strong (dotted).

reduced amplitude $A/A_0 = 35$ nm, the shift of phase is indicated with arrows for each case, assuming repulsive regime (upper branch) and presumably dominant contact dissipation. These phase shifts are quite similar in magnitude to those in Figure 2.15, which treated 10% and 20% increases in the damping coefficient b throughout the oscillation cycle. Thus, we find that fractionally small dissipations per tip–sample interaction can be manifest as very substantial phase shifts. Of course, hundreds of eV may not sound like a small number, given that chemical bond energies are on the order of a few eV. We shall see, however, that even sharp tips can have tip–sample interaction zones containing hundreds or thousands of atoms, and thus the dissipated energy presumably excites a large number of bond vibrations, meaning a fraction of an eV per bond, and this energy propagates away from the interface rapidly (i.e., does not "accumulate" from one tip cycle to the next). For perspective, bear in mind that electron microscopes can impart energies to a sample, per incident electron in the exciting beam, that are several powers of 10 greater than the above energy per tip–sample encounter; moreover, the number of electrons impinging per unit time is also much higher than the number of tip–sample encounters per unit time in dynamic AFM.

2.5 OTHER DISTANCE-DEPENDENT ATTRACTION AND REPULSION: ELECTROSTATIC AND MOLECULAR FORCES IN AIR AND LIQUIDS

In Sections 2.2 and 2.3, our treatment of tip–sample interactions was a picture of relatively weak attraction, sensed within just a few nanometers of the surface, and much steeper, mechanical repulsive forces sensed in contact. One need not spend a lot of time using AFM, however, before one encounters other kinds of interactions that may be much longer range as in the case of electrostatic forces, for example, a strong attractive force near a cut piece of plastic (forces that can also be much larger in magnitude than dipole forces), or long-range repulsive forces that are much weaker in both gradient and magnitude, for example, on polymer surfaces immersed in a favorable liquid medium. The science of distance-dependent interfacial forces in liquids, in particular, is very rich. It is beyond the scope of this book to cover all aspects. The reader may wish to consult the review article on AFM-based surface force characterization by Butt et al. [10] as well as monographs that treat the general subject [2,26–28]. In this section, we will conceptually introduce the major types of in-liquid surface force, with emphasis on the theoretical forces that the AFM user may encounter whether or not the intended goal is to study them. In other words, our emphasis in this section is diagnostic, not analytical. Practical issues of measurement calibration and basic data processing will be covered in Chapter 3.

2.5.1 Electrostatic Forces in Liquids: Superimposed on van der Waals Forces

We know from Section 2.1 that the Coulomb force between two point charges falls off as distance squared, whereas the force between point dipoles falls off much steeper, distance to the seventh power. In summing together the contributions of many electrostatically charged atoms or molecules, we therefore anticipate a distance-dependent force between AFM tip and sample surface that falls off much more gradually than the inverse distance squared dependence (Eq. 2.5) determined by summing over point dipoles in a sphere and half space. Because of this much more gradual fall-off with distance, we cannot approximate the tip as a sphere (i.e., more than the spherical end cap of the tip matters). Instead, even an approximate treatment must consider the overall shape of the AFM tip and possibly the contribution of the cantilever, although positioned a distance equal to the height of the tip (micrometers), or more, from the sample surface. But because of such a long range and thus gradually changing force with distance, it is quite feasible to oscillate the tip at a distance in dynamic mode such that the force varies essentially linearly over the amplitude explored by the tip. Thus, the picture of Figure 2.7 can be quite valid, such that the negative shift of cantilever resonance frequency can be simply related to attractive force gradient. This enables a whole plethora of analytical methods for both imaging and quantitatively assessing technologically important EM phenomena and materials. We will discuss in greater detail in Chapter 9 two of these methods known as electrostatic force microscopy and Kelvin probe force microscopy.

Measurable electrostatic attraction felt by an AFM tip can extend easily to 100 nm, even micrometers in the case of charged objects in air. An overall static charge, however, is *not required* for electrostatic forces to be present; the fact that tip and sample are generally not the same material means their surface potentials will differ, which produces an electric field—and thus an electrostatic force—as tip approaches sample (assuming both are grounded, thus have different vacuum levels [29,30]). Such a force may be only a few tens of nanometers in range and comparable in magnitude to van der Waals forces. We will further discuss this force, and how to probe it, in Chapter 9. In Chapter 3, we will learn that the baseline or "zero" of force measurements can contain weak distance dependencies due to nonidealities and artifacts, making the identification of the surface-potential-derived force even less obvious.

At the solid–liquid interface, given the fact that many solids ionize in polar liquids such as water, one finds that both electrostatic and van der Waals forces are important. The creation of a charged surface means that oppositely charged *counterions* have gone into solution. Some of these ions transiently bind to the surface within the so-called *Stern* or *Helmholtz* layer, while others form a diffuse, counterbalancing atmosphere of ions known as the *electric double layer*. Ions in this atmosphere are in continuous thermal motion but tend to be concentrated nearer to the charged surface, with a probability for occupation falling off exponentially from the surface, just like gases in the earth's atmosphere (Eq. 2.1). (The steepness of this exponential fall-off is quantified by the Debye length, the distance from the surface at which the argument of the exponential is -1.) Figure 2.17 depicts such a probabilistic scenario by using a commonly available spreadsheet random number generator to place particles parallel to a surface and with number density that falls off exponentially from the surface.

Often the liquid medium is chosen to contain ionic charges of the same sign as the surface and thus termed co-ions. The term double layer refers to the situation in applications where like particles bear like charges, such that the diffuse counterion

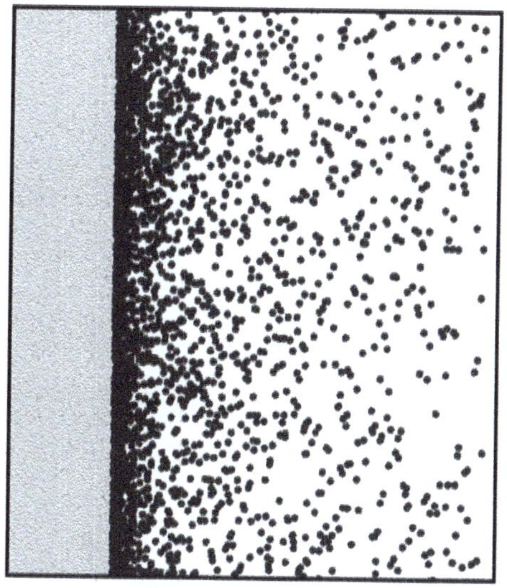

FIGURE 2.17 Simulated location of ions near an oppositely charged surface.

atmospheres near each surface comprise like-charged "layers" that repel as the particles near one another. With decreasing interparticle distance, this repulsive double-layer force will be superimposed on what is usually an attractive van der Waals force to produce a *nonmonotonic* force–distance relationship. This may include long-range exponential repulsion followed by dominant van der Waals attraction upon approach to short distances, followed by steep repulsion upon contact. The exact force profile, including the question of whether the longest range forces are repulsive or attractive, depends on the pH and/or ionic concentration of the medium. In aggregate, this complex assembly of distance-dependent interactions is described by the so-called *DLVO* theory.

In Figure 2.18, we plot two fictitious profiles of interaction energy versus distance up to but not including contact. The two curves result from the superposition of a repulsive-electrostatic term that falls off exponentially with distance and an inverse-distance van der Waals attraction, producing a peak in repulsion at 3–4 nm distance, below which there is a steep drop as van der Waals attraction becomes dominant. But if the electrostatic charge force is relatively small, as in the case of the lower curve, then there is also a local minimum around 10 nm distance because the inverse distance-squared dependence of van der Waals forces does not approach zero as rapidly as the exponential electrostatic repulsion does with increasing distance. This local minimum is the thermodynamic driving force for some slower particle coagulation processes.

2.5.2 Molecular-Structure Forces in Liquids

Variations between attraction and repulsion at very small distances can further result from *solvation* forces due to the molecules in the liquid medium and their

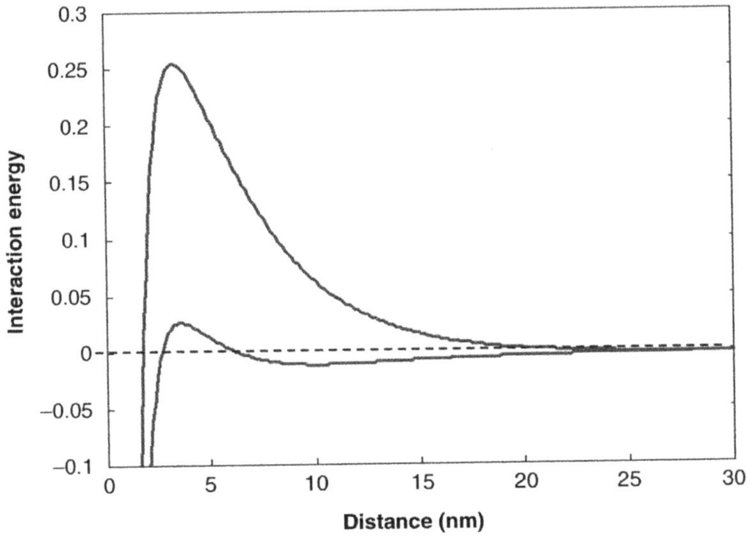

FIGURE 2.18 Distance dependence of DLVO interactions, for cases of relatively large (top curve) and small (bottom curve) electrostatic repulsions.

interaction with solid surfaces separated by nanoscale distances. These are non-DLVO forces due to the breakdown of continuum concepts on the order of molecular dimensions. Discrete, solid-like ordering of liquid molecular layers between flat surfaces can occur as depicted in Figure 2.19, producing oscillatory forces as a function of molecular-scale surface separations (i.e., on the order of at most a few molecular diameters) [2]. Forces alternate between repulsive and attractive with

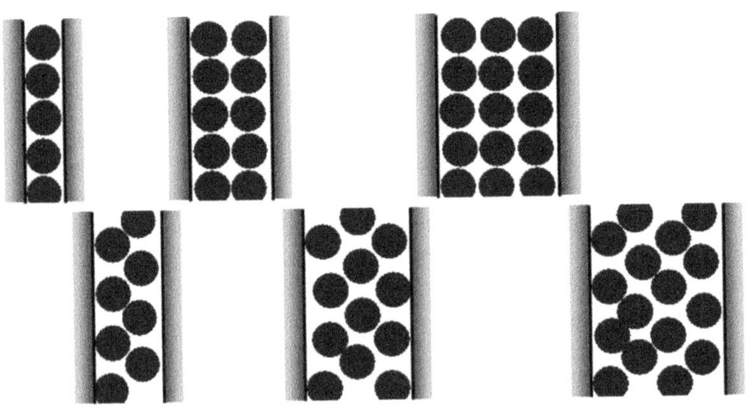

FIGURE 2.19 Depiction of relatively ordered and disordered liquids as a function of spacing of confining surfaces, higher density at spacings that are integral multiples of molecular diameters.

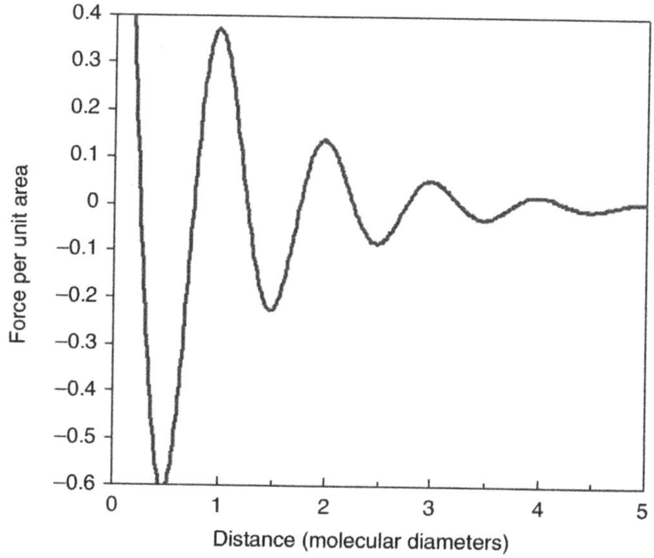

FIGURE 2.20 Numerical form describing the dependence of force versus distance due to solvation.

decreasing distance (up until reaching contact), with peaks spaced by distances of approximately one molecular diameter. The magnitude of these forces generally increases as distance decreases, sometimes described with an analytical form consisting of a sinusoidal function multiplied by an inverse exponential with decay length of one molecular diameter, as shown in Figure 2.20 [2]. Geometric characteristics of individual molecules thus are important because they affect molecular packing or "networking" (in the case of *hydrogen bonding*) and ion binding/correlation, as well as molecular orientation relative to solid surface.

Much of the experimental evidence of solvation forces has been achieved with the surface forces apparatus [2], which brings together atomically flat mica surfaces glued to macroscale crossed cylinders (radius of curvature ~1 cm). At separations of nanometers, these surfaces are effectively ideal planar surfaces and thus confer ordering to the liquid in between. One might expect that an AFM tip, carefully approaching an atomically flat surface, also could induce this kind of molecular ordering, and that the exquisite sensitivity of AFM force measurement is high enough to profile oscillatory forces. In practice, this capability has been demonstrated in only a handful of published studies, perhaps because the end of the tip does not provide a nearly planar surface even on a sub-10-nm lateral scale. The most dramatic examples probing oscillatory forces have employed *dynamic* (vertically oscillating tip) AFM because of its stability in the presence of both attractive and repulsive force gradients, as discussed in Section 2.4. An excellent example is a study by Han and Lindsay [31], where cantilever oscillation amplitude was measured versus *Z* position as AFM tip and graphite surface were brought together and apart, immersed in mesitylene. Local amplitude minima provided points at which a solid-like modulus could be calculated, ranging from ~1 MPa to 50 MPa for shallowest to deepest minimum with decreasing distance. In contrast, quasistatic AFM force curves suffer from jump instabilities during approach when the force gradient transitions from negative to positive upon reaching a local force maximum, such that the force minima are not navigated versus distance. Thus, a more staircase-like force–distance profile results [32].

In aqueous solutions, surfaces of silica and the aluminosilicate mica (common substrates for biomolecules and cells), certain clays, and small amphiphiles (short chain molecules with hydrophilic end groups and hydrophobic chains) including lipid bilayers and biological membranes exert repulsive forces between each other, even though the concepts of DLVO and solvation forces predict adhesion or coagulation [2]. The reason is an additional short-range, exponentially repulsive force that can be quite steep (compared to DLVO). Between hydrophilic surfaces or molecules, this is commonly known as a *hydration* or *structural* force. The underlying mechanism is the strong binding of water molecules to hydrophilic groups on the approaching surfaces or molecules. The steep character of this force relates to the substantial energy needed to disrupt a hydrogen-bonded network or dehydrate the two approaching surfaces.

2.5.3 Macromolecular Steric Forces in Liquids

At the surface of macromolecular materials in liquid media as well as biopolymeric systems, a rich variety of attractive and repulsive long-range forces result from the long-chain molecular nature. Some special methods to map these forces will be covered in Chapter 6. The emphasis in this section is to describe interfacial-force phenomena that any AFM user operating in liquids needs to understand, whether the intent is simple imaging, nanomechanical measurements, macromolecular interfacial science, or rigorous biophysics/biochemistry.

In many liquid environments, a polymeric material or film will swell as small-molecule liquid penetrates the matrix of chain molecules. Correspondingly, individual polymer chain coils will expand into the liquid medium to explore numerous configurations, thermodynamically driven by *entropy* (the number of available conformational states) and *enthalpy* (how energetically favorable these states are, dependent on the strength of chemical attraction to solvent molecules relative to intrachain attraction). Upon approaching such a surface with another object, even a sharp AFM tip with ~10-nm radius of curvature, this molecular freedom is reduced as the space available is confined to a smaller volume and the liquid medium is displaced. This spatial constraint decreases the entropy of the system, generating a repulsive *steric* force. The distance at which steric forces are sensed depends on: (i) the molecular weight as it determines the chain length Nl, where N is the number of statistical segments per chain (ratio of molecular weights of full chain to that of one segment, which may or may not be the same as one monomer, to be discussed below) and l is the effective segment length, (ii) the degree to which the particular polymer is enthalpically attracted to the particular liquid medium (the *quality* of solvent that the liquid medium serves as, for that polymer), (iii) temperature, as it affects the entropic driving force for chains to reach into the liquid, and (iv) the lateral density of molecular packing in the polymeric material, as it affects the number of chains reaching into the liquid per unit area.

To better understand the role of entropy in distance-dependent forces at polymeric surfaces (both the preceding steric repulsion and molecular extension forces of Section 2.5.5), it is worthwhile to consider in some greater detail the statistics of long-chain molecules [33–35]. In the simplest picture, a polymer is treated as a chain of *freely jointed segments* that can rotate in any direction. (Generally, each statistical segment is on the order of a monomer in size, but typically is larger although sometimes smaller.) Thus, one is not concerned about the space occupied by a given segment when contemplating the position of any other segment. This is the 3D random walk or "random flight" problem, where the polymer is treated as a "phantom chain" (i.e., it may penetrate itself).

We exemplify by starting with the segment at one end of the chain and taking randomly oriented steps of unit length ($l = 1$) until reaching the other end of the chain. We construct a 3D random walk using an everyday spreadsheet program, for $N = 1000$ segments. To do this, we use the program's random number generator to select polar and azimuthal angles (to an integral number of degrees) for each step relative to the previous location. A 2D projection of one such randomly generated 3D chain is shown in Figure 2.21, beginning at the origin denoted with a large dot

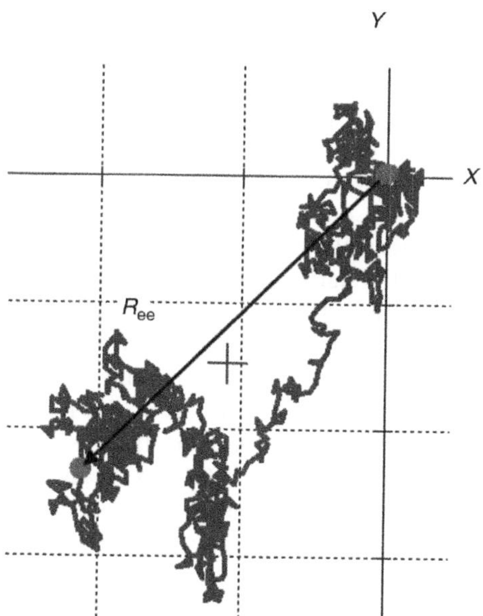

FIGURE 2.21 Two-dimensional view of a 1000-step 3D random walk simulating a single-polymer-chain molecule conformation. Dots locate the end points of the chain separated by R_{ee}, and the cross indicates its center of mass.

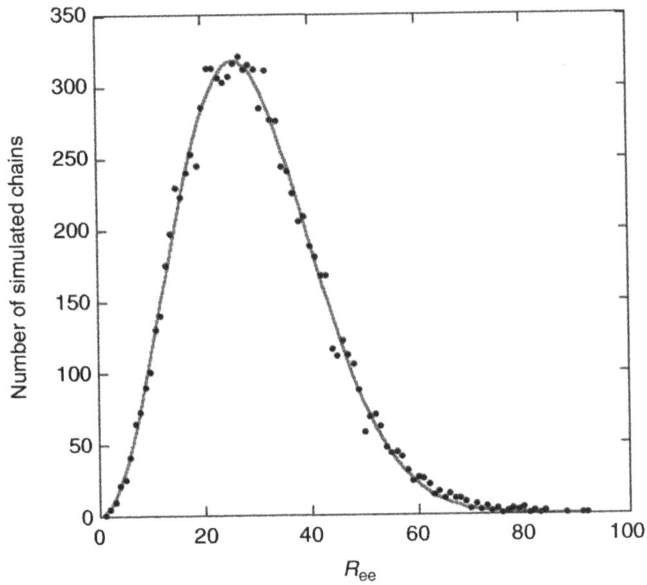

FIGURE 2.22 Distribution of values of chain end-to-end distance R_{ee} resulting from 10,000 simulations of 1000-step 3D random walk. Overlaid curve is a fit to the form $c_1x^2 \cdot \exp(-c_2x^2)$.

and ending at the other large dot. The vector R_{ee} denotes the 2D projection of the 3D displacement between these two end points, about 26 units in this case (tick marks being 10 units). Comparing many such 1000-step random walks, we find that R_{ee} ranges broadly; the start and end points in a few cases are very near each other (a small number of equivalent unit steps in a single direction) and in a few other cases very far apart (many tens of equivalent steps), but the bulk of simulated walks terminate 20–40 units from the starting point. To appreciate this distribution, 10,000 random walks, each containing 1000 random steps, were performed (using the same spreadsheet program with scripting for automation) and the 10,000 values of R_{ee} compiled into a histogram with unity bin size, Figure 2.22. This distribution of R_{ee} values was mathematically fit with a function of the form $c_1x^2 \cdot \exp(-c_2x^2)$, the appropriate analytical expression as derived from statistical principles [33]. The peak location is centered around $R_{ee} = 27$ and the median value of R_{ee} is 27.8. These are very small numbers considering that each chain consists of 1000 segments (i.e., a fully extended, linear chain would have $R_{ee} = 1000$). Notably, only 2 of the 10,000 chains had end-to-end distances longer than 90, and none over 100, out of 10,000 simulated chains!

Another important parameter is $\overline{R_{ee}^2}$, the average being taken over all possible whole-chain conformations. For our 10,000 simulated chains, this value was calculated to be 980. The statistics of a freely jointed chain predicts the well-known relationship $\overline{R_{ee}^2} = Nl^2$[33] meaning 1000 in our case ($l = 1$), within 2% of the simulation result. An even better known, indeed *experimentally accessible*, statistical metric describing a polymer chain is the *radius of gyration*, R_g. The definition of R_g is the standard (root-mean square) deviation of the distances of all chain segments from the center of mass of the chain (e.g., marked as a cross in Figure 2.21), *averaged over the ensemble* of (all possible) whole-chain conformations. A well-known relationship derived from statistical analysis [33] is

$$R_g^2 = Nl^2/6, \tag{2.28}$$

thus

$$R_g^2 = \overline{R_{ee}^2}/6. \tag{2.29}$$

The above expressions only describe a polymer coil in an "ideal" or *theta* solvent, where inter- and intrachain interactions, be they attractive, repulsive, or due only to excluded volume (the space occupied by the actual chain segments) as well as chain–solvent interactions, can be treated as negligible. This case can occur when inter- and intrachain attractive forces are just strong enough to precisely offset the opposing attraction of segments to solvent as well as the repulsive forces of excluded volume (at the theta temperature, i.e., dependent on entropy). In general, a polymer coil in a real solvent will be described by a radius that is collapsed or expanded relative to R_g, sometimes called the Flory radius $R_F = c_F R_g$, where c_F is dimensionless and may be less than or greater than 1. Mathematically, the statistical model of a polymer chain as a large number of independently oriented segments still applies, but with a simple increase or decrease of the effective size of the segments by the factor c_F.

Moreover, to account for the real restrictions of angles between adjacent backbone bonds along the chain (e.g., C—C—C being 109.5°, with turnstile azimuthal motion at this constant polar angle), one can model using a smaller number N of larger segments l that effectively rotate independently with respect to the adjacent segments (i.e., each segment is defined to contain a large enough number of constrained bond angles in series such that the end can be freely positioned with respect to the end of the adjacently defined segment). Thus, the basic statistical physics of the freely jointed chain model can be applied to many real systems. (Note that the functional relationship between R_g and the number of segments N in the chain will indeed depend on the above details. Different characteristic power law relationships $R_g \sim N^\nu$ have been derived and confirmed experimentally. For a theta solvent $\nu = 1/2$, whereas for good and bad solvents $\nu = 3/5$ and $1/3$, respectively, the latter reflecting a collapsed globule whose volume is simply proportional to N [33].)

Steric repulsive forces can be experimentally detected at distances of several times the radius of gyration. The preceding simulation results reveal why this is the case: we have $R_g = (980/6)^{1/2} \cdot l = 12.8$ and $\left(\overline{R_{ee}^2}\right)^{1/2} = 31.3$ for unity segment length l. It is easy for a polymer chain (in an ideal to good solvent) to extend several times farther into the solvent than 12.8 segmental units, indeed much farther than 31.3 units, as per the distribution of R_{ee} values in Figure 2.22.

There are some successful theoretical treatments to describe certain simple cases of steric repulsive forces. One example is end-attached but otherwise inert polymer brushes that are sparsely distributed in a theta solvent, where $R_F = R_g$. (This state is dependent on both the chemical structure and temperature, i.e., both enthalpy and entropy.) In this case, interchain interactions are negligible and each chain undergoes simple entropic confinement as the opposing surface (e.g., AFM tip) approaches. The theory predicts an approximately *exponential* distance dependence of force in the distance range $2R_g$ to $8R_g$. Another special case is two opposing, densely packed brush layers at separations below two times the fully extended chain length. Here, the Alexander-de Gennes theory predicts a two-term, *fractional power law* expression (9/4 power for repulsive term, 3/4 power for attractive term) for the force exerted per unit area of interaction between brush layers.

As another extreme case, if a bulk polymer or thick film is physically or chemically crosslinked (chains stereochemically interlocked or covalently attached to one another) but with low crosslink number density, then a zone of increasingly diffuse chains can range hundreds of nanometers out from the material "surface" in a good solvent. One then measures a very gradual increase of repulsive force as another surface (i.e., AFM tip) pushes into this zone, ultimately displacing solvent and densifying the polymer matrix. One should note that it may be difficult to conceptualize the material stiffness within simple mechanical models (e.g., Hertzian) that assume a characteristic material modulus; the "material" includes a penetrating solvent that may be variably displaced by the tip.

2.5.4 Derjaguin Approximation: Colloid Probe AFM

Whether one is exploring fundamental intersurface forces as an end goal or simply using such measurements to diagnose the nature of the surface on sample and/or tip, it can be helpful to convert the final numbers from measured forces to *interfacial energies*. Values of the interfacial energy have been tabulated for many systems. Clearly, the sharpness of the probe will affect the magnitude of force and thus the energy of interaction, these being *extensive* quantities. We wish to find a simple way to convert to the *intensive* quantity interfacial energy, defined as energy per unit area [2]. Often used for essentially this purpose is a general relationship between the interaction force between two nondeformable curved bodies and the interaction energy per unit area between *two equivalent parallel half spaces*, whether the surfaces are in contact or separated by a small distance D. The relationship is called the *Derjaguin approximation*. It is valid when the radii of curvature of the surfaces are much larger than the separation distance D and the range of the interaction. The Derjaguin approximation is often stated in somewhat restricted form; here, we state in its most general form as described by White [36]. It applies to any force law whether attractive, repulsive, or oscillatory, and to any quadratically curved surfaces, not just spheres as is common in various contact mechanics models such as the Hertzian model of Section 2.3. Moreover, each quadratic surface need not be axisymmetric, that is, may be described by two principal radii of curvature, and the principal axes of one surface are in general arbitrarily oriented with respect to the principal axes of the other surface. The Derjaguin approximation states that the interaction force F between curved bodies 1 and 2 with closest approach distance D is proportional to the interaction energy per unit area u of the parallel surfaces of half spaces separated by distance D:

$$F(D) = 2\pi(\lambda_1\lambda_2)^{-1/2}u(D) \tag{2.30}$$

with

$$\lambda_1\lambda_2 = \left(\frac{1}{R_1^a} + \frac{1}{R_1^b}\right)\left(\frac{1}{R_2^c} + \frac{1}{R_2^d}\right) + \sin^2\phi\left(\frac{1}{R_1^a} - \frac{1}{R_1^b}\right)\left(\frac{1}{R_2^c} - \frac{1}{R_2^d}\right) \tag{2.31}$$

where the superscripts refer to the different *principal axes* of surfaces 1 and 2, and ϕ is the rotational angle between their coordinate systems. An interesting (but usually unstated) "corollary" is that a torque exists ($dF/d\phi$ nonzero) between the bodies, which, if unopposed, will align the principal axes of the two bodies. (One might consider twisting forces in AFM due to this weak effect.) If an attractive interaction, the axes with smaller principal radii for each surface will align; if repulsive, the axis with smaller principal radius for one surface will align with the axis with larger principal radius for the other surface. In the case of an axisymmetric tip and flat sample, the Derjaguin approximation simplifies to the more well-known expression

$$F(D) = 2\pi Ru(D) \tag{2.32}$$

meaning a simple normalization of the measured force by $2\pi R$ is equivalent to a measurement of the interfacial energy of two half spaces of the tip and sample materials separated by distance D (again, units of energy per unit area).

Equation 2.32 is commonly used in measurements performed with the surface forces apparatus—with its crossed cylinders being mathematically identical to the sphere-flat geometry—as well as so-called colloidal probe AFM or simply colloidal probe microscopy (CPM), where the conventional AFM tip is replaced by a microsphere attached (usually glued) to the cantilever. (Commercial tipless cantilevers are often employed.) A measurement of R in CPM can be obtained with a light microscope, such that a precise measurement of $u(D)$ is obtained. CPM has become a well-established method within the AFM family [37], not surprisingly in research that focuses on quantitative interfacial and colloid science. It is also important to note that the use of a microsphere means that $R \gg D$ (an assumption in the derivation of the D^{-2} van der Waals force-versus-distance relationship) out to quite large distances D, tens or hundreds of nanometers. It is interesting to note, however, that at distances of order 100 nm or greater, *retardation* effects become significant. This results from the finite speed of light, the speed at which force interactions are communicated by virtual photons between atoms [38]. Thus, the ability of one fluctuating charge dipole to correlate with another is reduced. This effect of retardation on the van der Waals force–distance relationship has been measured with the surface forces apparatus [2].

For AFM, one could say that an identification of the D^{-2} force law at shorter distances as well as possible retardation effects at larger distances *requires* the colloid probe method, given that many commonly available AFM tips have $R \sim 10$ nm or smaller (meaning the D^{-2} dependence holds only for $D \sim 1$ nm or smaller; moreover, attractive forces are far too weak with such tips to measure out to retardation distances). Of course, the presence of nanoscale roughness on colloid probe and/or sample surface may produce a departure from a strictly $1/D^2$ dependence, provided that the presence of additional physical forces (electrostatic, steric) have not already done so.

We finally note that another expression, originally due to Hamaker, can describe the distance dependence of van der Waals energy $U(D)$ out to distances at which the special case $D \ll R$ does not hold, provided that there are only small differences between the dielectric constants of spherical tip, half-space sample, and intervening medium, and not accounting for retardation [6]:

$$U(D) = -\frac{A}{6}\left(\frac{R}{D} + \frac{R}{2R+D} + \ln\frac{D}{2R+D}\right). \tag{2.33}$$

The force expression $F(D)$ is then obtained from the negative derivative with respect to D, which after some algebra simplifies to

$$F(D) = -\frac{AR}{6D^2(1+D/2R)^2}. \tag{2.34}$$

2.5.5 Macromolecular Extension Forces (Air and Liquid Media)

If both surfaces contain polymer or if the approaching probe surface is energetically favorable to polymer adsorption, then the extension of polymer chains bridging between the two surfaces may result. Once a given chain molecule has attached between two surfaces, its desire to explore conformational states can produce an *entropically attractive* force. The various chain segments explore the available space, but if the chain is extended well beyond $\left(\overline{R_{ee}^2}\right)^{1/2}$, where R_{ee} is now the end-to-end distance of the bridging part of a chain molecule (i.e., generally not the full chain), there obviously will be far fewer conformational states achievable (see Figure 2.22); this yields a force pointing in the direction of more available conformational states, meaning *decreasing* surface separation (such that the chain is less fully extended), and thus *attraction*. This phenomenon is more likely after bringing surfaces together and then displacing them apart because strong adsorption or entanglement is much more likely to occur when the surfaces are pressed together with some external force. Then upon retraction, one is more likely to retain extending chain molecules out to substantial distances (tens or hundreds of nanometers).

So far, the most common AFM-characterized cases of this phenomenon involve *single molecule force spectroscopy*, where a single-chain molecule becomes extended until it detaches from one surface or disentangles from the pinning matrix. (An example, polyacrylamide, is treated in Section 6.4.2.1.) If the end-to-end molecular extension distance R_{ee} of the chain portion that is bridging between AFM tip and sample surface does not get too near its fully elongated (all-trans) state and intramolecular interactions are negligible, then the resistive force is entropic and well described by the *freely jointed chain* model, sectioned into small, identical segments that independently rotate with respect to attachments to adjacently defined segments. In a statistical mechanical treatment [39,40], the time-averaged equilibrium force F applied (e.g., by AFM tip, the average being over thermal fluctuations) to the Nth segment of a chain of N noninteracting model segments of *Kuhn length* l_K, so as to hold the chain extended at an end-to-end distance R_{ee} with segment #1 fixed in position, is given by the expression

$$R_{ee} = k_B T \frac{d \ln \zeta}{dF} \tag{2.35}$$

where ζ is the *partition function*

$$\zeta = \int_1 \cdots \int_N e^{-(-\vec{F} \cdot \vec{R}_{ee})/k_B T} dl_1 \ldots dl_N = \left[\int_0^{\pi} e^{F l_K \sin\theta / k_B T} \sin\theta \, d\theta \right]^N = \left[\frac{2}{u} \sinh u \right]^N$$

and $u = Fl_K/k_BT$. Here, a local *spherical coordinate* system was set up for each segment l_i such that the preceding segment l_{i-1} terminated at the origin, and with the polar axis chosen positive in the direction of the force F. Taking the derivative with respect to force F, one obtains

$$R_{ee} = N l_K \cdot \left[\coth\left(\frac{F l_K}{k_B T} \right) - \frac{k_B T}{F l_K} \right]. \tag{2.36}$$

The Kuhn length l_K quantifies the minimum size of segments that are large enough to be considered independent (i.e., adjacent segments are uncorrelated in orientation). The prefactor $Nl_K \equiv L$ is the length of the fully extended chain in its *trans* conformation. Thus, fitting this model to experimental data involves two free fitting parameters, L and l_K. L is generally unknown *a priori*, because the full polymer chain is not usually attached at the endmost segments, and second, the length of the full polymer chain is not known exactly (at best it is known only within some distribution per the polydispersity of the polymer sample). The expression in brackets is known as the Langevin function $\Lambda(u) = \coth(u) - 1/u$, a form that recurs in statistical physics. For example, the Langevin function arises in considering freely rotating, noninteracting gas molecules of nonzero dipole moment perturbed in their rotation by an external electric field. Presently, we consider freely rotating, noninteracting units of a polymer chain perturbed in their rotation by a force pulling on the chain. Thus, the statistical similarity is not such a surprise [40].

The freely jointed chain model is valid to describe the behavior of many polymers in good solvents (generally, non-theta solvent situations) if the Kuhn length can be chosen to be large enough to produce uncorrelated behaviors between adjacent segments yet small enough that the total number of segments in the chain remains very large. The larger the Kuhn length is, the more rigid the polymer chain.

At the extreme, the Kuhn length of a perfectly rigid chain (polymer rod) is identical to the full chain length L. At some point between this extreme and the opposite extreme of C—C distances along the chain backbone (or a slightly larger monomer length), a different model is often used to describe polymers that have a significant rigidity, especially biopolymers such as DNA and many proteins: the *worm-like chain*. Here, the polymer is not modeled as a chain of independent and freely rotating segments; in fact, the discrete, segmented structure of the chain is entirely absent from the model. Instead, the chain is considered a continuous string with constant bending modulus of elasticity. Thus, the chain orientation (unit tangent vector) at a given point is correlated with the chain orientation nearby; the *persistence length* quantifies the statistical range of this correlation (the distance at which it falls to 1/e). An exact treatment is numerical (i.e., involves a power series expansion), but an interpolation formula has been used,

$$F = \frac{k_B T}{l_p} \cdot \left[\frac{1}{4(1 - R_{ee}/L_{\text{contour}})^2} - \frac{1}{4} + \frac{R_{ee}}{L_{\text{contour}}} \right] \tag{2.37}$$

where l_p is the persistence length and R_{ee} is again the distance between the ends of the portion of the chain molecule that is bridging between the AFM tip and sample surface.

Figure 2.23 overlays plots of the Fl/k_BT versus $R_{ee}/L_{contour}$ for the freely jointed and worm-like model formulas on ordinate scales differing by a factor of two as if the force scale is identical but $l_p = 2l_K$ (hypothetical case), and with a negative sign to indicate attractive force. We see that the overall trend is qualitatively similar but

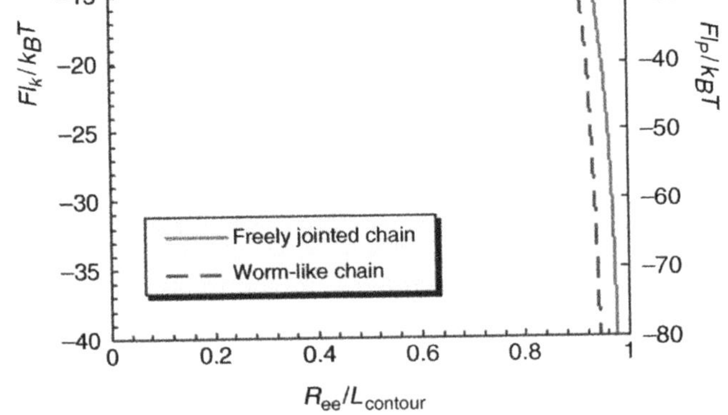

FIGURE 2.23 Force–extension relationships of freely jointed chain and worm-like chain models for single-chain stretching, in reduced coordinates.

quantitatively different, with the worm-like chain model more sharply increasing in force magnitude as $R_{ee}/L_{contour}$ approaches unity, a fully extended all-trans conformation. But what is most notable, in either model, is the steep rise in force magnitude. Just how large is this force? If we assume a Kuhn length of 0.3 nm in the freely jointed chain model, we calculate a force of 500 pN at an extension of $R_{ee}/L_{contour} = 0.97$. We will see in Chapter 6 that this magnitude of force is far above the noise.

The analysis of real single molecule force spectroscopy data out to the full extension length observed (which also is the highest signal/noise data) usually requires more complex expressions than those plotted in Figure 2.23. The discrepancy has suggested a need to include increases in enthalpy induced by chain extension. Why would this be? Upon nearing full chain extension, *bond bending* can become significant, an increase of the covalent bond angles between adjacent atoms along the main chain (e.g., greater than 109.5° in the case of alkanes). This confers an additional stiffness within the freely jointed chain model or a large increase in the bending modulus within the worm-like chain model. The preceding analytical expressions have been modified to incorporate these effects, and the resulting expressions have been aptly named the *extended* (or elastic or Hookean) [41,42] freely jointed and worm-like chain models. The extended freely jointed chain model adds a term to the length prefactor to allow for an effectively larger contour length,

$$R_{ee} = \left(L_{contour} + \frac{NF}{k_{segment}}\right) \cdot \left[\coth\left(\frac{Fl_K}{k_B T}\right) - \frac{k_B T}{Fl_K}\right] \tag{2.38}$$

where $k_{segment}$ is the effective extensional spring constant per segment and N is again the number of segments comprising the portion of the chain that is bridging between AFM tip and sample surface. Thus, the first and second terms in the prefactor derive from entropic and enthalpic considerations, respectively, with the latter describing a quantitative modification of *trans* conformation bond angles. The extended worm-like chain model includes the term F/Φ in two places,

$$F = \frac{k_B T}{l_p} \cdot \left[\frac{1}{4(1 - R_{ee}/L_{contour} + F/\Phi)^2} - \frac{1}{4} + \frac{R_{ee}}{L_{contour}} - \frac{F}{\Phi}\right] \tag{2.39}$$

where Φ denotes the specific stiffness of the polymer.

The result of curve fitting of real AFM data to a single-chain model is presented in Section 6.4.2.

Not all long-chain bridging interactions exhibit the characteristic force-extension profile of decreasing entropy, that is, one or more steep increases in attractive force as in Figure 2.23. Occasionally, one instead observes long-range, highly reproducible, constant-force "plateaus" (tens or even hundreds of nanometers along the distance scale), typically comprising a "staircase" if more than one plateau is present. In some cases, these plateaus are interspersed with steep entropic dips. We exemplify in Figure 2.24 for gelatin molecules (polypeptide derived from the protein collagen) bridging between a Si_3N_4 tip retracting from a graphite surface following contact [43]. Another such case will be considered in Section 3.6 (Figure 3.26,

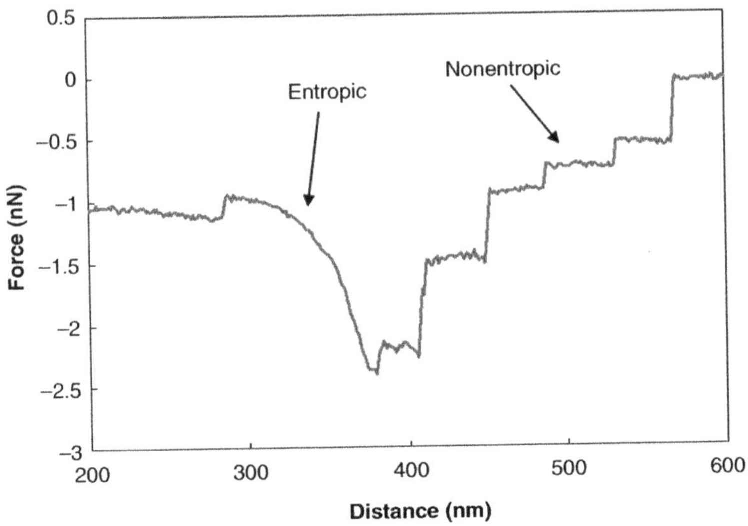

FIGURE 2.24 Distance dependence of force during retraction of a Si_3N_4 AFM tip from a graphite surface in air with a gelatin fibril bridging between tip and surface.

together with methodological aspects). Given the above discussion of the statistics of long-chain molecules, how can these molecular chains seemingly "avoid" entropic effects? Turning our viewpoint around, an immediate interpretation of this observation could be a lack of conformational freedom in the molecular entity bridging between two pinned points. This only can occur if there are long-chain molecules that are intrinsically rigid rods (i.e., not only short-chain molecules, such as liquid crystals and other small organics) or if multiple chains are bound together in such a way that conformational exploration is impossible. In this case, one envisions more of "peeling off" phenomenon. (As a thought experiment, note that a single-chain molecule gradually peeling from a surface so as to maintain a fully extended state would exhibit a strongly increasing entropic resistance: as the number of "liberated" segments N increases, the number of conformational states accessible increases. The tensional force required to constrain the chain to a single, fully extended, state *must* strongly increase.)

From a diagnostic perspective, it is important to note that on molecular materials in liquid environments, generally more than one class of force will be present, and even a given class such as molecular-chain bridging forces may not be simple, as illustrated above. This is true of chain bridging forces in even more respects because several conformationally free polymer chains may comprise a bridge, and this assembly may not behave as a simple superposition of several single-chain force-extension profiles, due to complex chain coupling and/or entanglement. Each chain under tension is attached or pinned at arbitrary points along the single-molecule chain, not the end points (barring some sort of functional end grafting). On polymers that bear a net charge (say, in water), steric repulsion and/or bridging forces may be

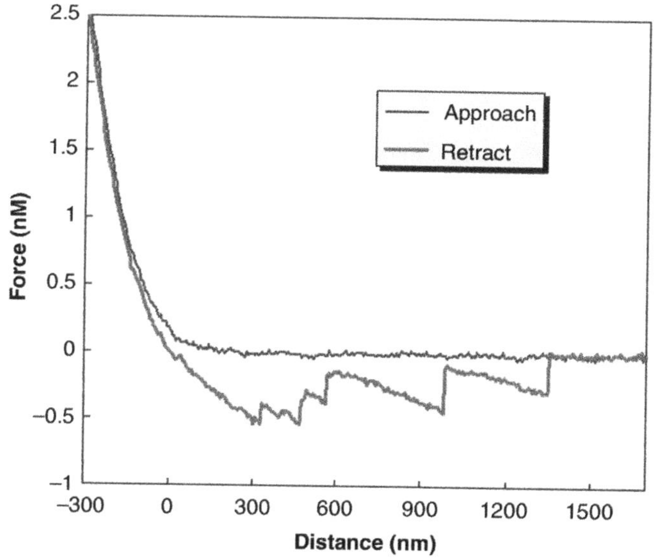

FIGURE 2.25 Force–distance cycle for a silicon nitride tip interacting with a gelatin gel in aqueous medium.

superimposed on DLVO forces and possibly hydration forces. Given that many of these forces fall off in magnitude with distance, and perhaps without characteristic mathematical profiles, it may be difficult to identify and separate the various contributions (e.g., steric from DLVO repulsion). Single-molecule stretching at large distances is easier to identify, however, even if superimposed on long-range repulsive or attractive forces of other types because these latter forces usually vary minimally at the larger distances at which single-molecule extension forces steeply grow in magnitude. Of course, rather than single-chain molecules extending within a given force–distance profile, several chains may extend in parallel and at differing degrees of extension for any nominal distance that the AFM tip has been retracted from a point of solid contact. "Sawtooth" force–distance profiles may result, with the vertical jumps occurring after different chains separate or within a given chain as steric restraints suddenly fail, liberating longer chain contours for stretching. This is illustrated in Figure 2.25, a complete force-versus-distance cycle (a methodological emphasis of Chapter 3) for the case of a silicon nitride tip interacting with a gelatin gel (covalently cross-linked with formaldehyde) in aqueous medium.

2.6 RATE/TIME EFFECTS

In the preceding sections of this chapter, we have treated the distance dependence of tip–sample interaction as being determined from the type of interaction, whether due to van der Waals attraction, contact deformation, molecular structure, entropic macromolecular effects, etc. But more generally, the *rate* or *timescale* over which the particular range of distance is explored can have a nonnegligible, even important, role. In this section, we describe two conceptual frameworks within which rate or timescale plays an intrinsic role. In Chapters 6 and 7, we detail analytical methodologies for exploring such behaviors in both distance-dependent (Chapter 6) and shear (Chapter 7) forces, that is, in order to characterize the properties of materials and molecular systems in a quantitative way.

2.6.1 Viscoelasticity

In the realm of soft materials and molecular systems, whether synthetic or biological, rate-dependent interactions are most commonly observed as *viscoelasticity*. An imparted force, or more precisely stress, does not uniquely determine a deformation or strain (i.e., per Hooke's Law, Section 2.3). Rather, the rate at which the stress is imparted, and the timescale over which the response is monitored, can determine the effect. Figure 1.18 previewed how the sliding friction force between an AFM tip and a polymer surface can vary substantially over a wide range of lateral scanning velocities. Similarly, if pushing an AFM tip into or pulling it out of its interaction with a viscoelastic material, the approach or retraction rate can be important. What is the source of this rate dependence? Phenomenologically, these materials are modeled using not only springs (restorative force components), as discussed in earlier sections and contained in Hooke's Law, but also *dashpots*. A dashpot is like a piston in a cylinder filled with some viscous fluid. The force of resistance of motion is proportional to *rate* of piston displacement, rather than the *magnitude* of displacement,

$$\sigma = \eta \frac{d\varepsilon}{dt} \tag{2.40}$$

where η is the viscosity of the material. Viscoelastic materials are usually modeled with a *combination* of springs and dashpots to capture the material's response to both the magnitude and rate of displacement. For example, a *Maxwell element* is a spring and dashpot in series. Thus, the spring force is also the force felt by the dashpot. A more physically accurate treatment of viscoelasticity is the *standard linear solid model*, which combines a spring in parallel with the Maxwell element, as depicted in Figure 2.26. Under a particular stress, the

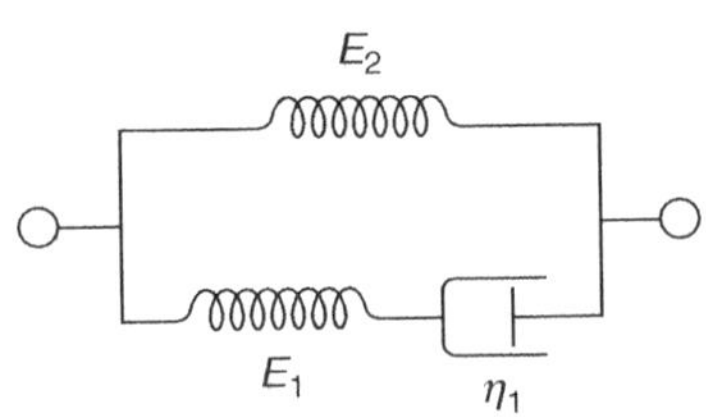

FIGURE 2.26 The standard linear solid model, which adds an elastic element of Young's modulus E_2 in parallel with a Maxwell element consisting of elastic and viscous subelements E_1 and η_1, respectively.

material will instantaneously deform to a particular degree of strain, then will continue to creep—deform in time—due to the behavior of the Maxwell element. Thus, in time, the material effectively softens, that is, an imparted force results in a deeper deformation than would result in the absence of creep. In the context of Figure 2.6, this would produce a less steep increase of force with decreasing distance (increasing indentation). In general the spring–dashpot type of modeled viscoelastic behavior can apply to compressive or tensile stresses during force–distance measurements, as well as shear stresses during sliding friction or shear modulation measurements (Chapter 7). In AFM, this means as the tip is pushed into, pulled out of, slid across, or laterally wiggled on the material under study.

2.6.2 Stress-Modified Thermal Activation

At finite temperature, seemingly equilibrium states can transition to other states due to thermal activation. An energetic barrier to this transition is exceeded via thermal fluctuations. Usually, room-temperature thermal energy $k_BT = 25\ \text{meV} = 4 \times 10^{21}$ J is small compared to the energy needed to disrupt a large assembly of attractive interactions at the tip–sample interface or to cause backbone conformational transformations along polymer chains embedded in a glassy polymer. But as stress increases while pulling away or pushing in, eventually such transitions occur, and with increasing probability with passing time. Conformational transitions, ultimately, on a scale as small as single isomeric bond rotations (i.e., *trans-gauche*) are at the root of viscoelastic behavior. So, in this regard, we are describing the same physics as in Section 2.6.1, but in a *molecular-scale* picture rather than *phenomenologically* as springs and dashpots.

But the breakage of an *interfacial bond* is a rather different concept. Such a bond or assembly of bonds could be between nonviscoelastic materials, both tip and sample. What is "soft" is the interfacial bond itself, even a large assembly of interfacial bonds. Thus, the force needed to pull out of a van der Waals attractive "bond" or a set of such bonds is not unique even though the energy may be unique in principle; rather, the force depends on the time allowed for the bond to rupture under stress and the random availability of thermal energy to enable this escape (i.e., k_BT). This time can be lengthy on a molecular scale; that is, if one compares the time under stress to the much shorter inverse vibrational frequencies of covalent bonds or phonons in the material ($\sim 10^{-13}$ s). Indeed, anytime there is a potential well describing the tip–sample interaction, the force exerted to escape this well is not "working by itself"; rather, it is aided by thermal fluctuations over very long periods of time on the scale of molecular and atomic vibrational motions. Failure to acknowledge this inescapable physics can lead to misconceptions; the AFM user may think that a measure of pull-off *force*, for example, is by itself a measure of a signature, characteristic bonding interaction. This is simply false. One cannot measure a material's "signature adhesive force" in the sense that one can, say, collect a signature vibrational spectrum using Raman or infrared spectroscopy. Even normalization to contact area (assuming one knows or can estimate it), thereby computing the more intrinsic quantity of stress, is still insufficient.

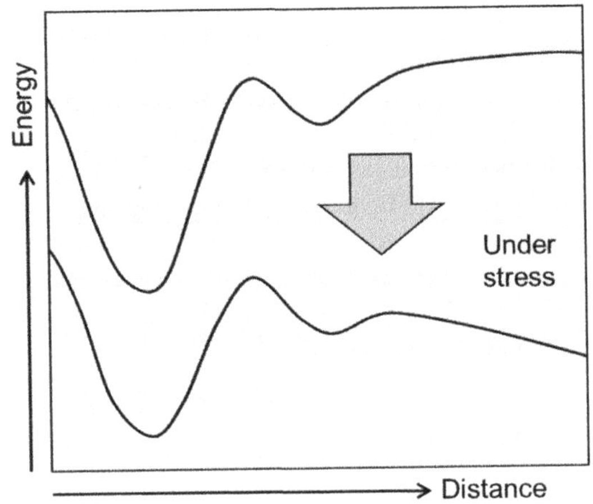

FIGURE 2.27 Illustration of the modification of an energy landscape by an imparted stress.

Fundamentally, one is instead measuring a force that produces stress that modifies the *energy landscape* in such a way that thermal activation becomes possible or easier. The stress actually lowers barriers to transitions between energy minima, as depicted schematically in Figure 2.27, making such transitions more probable. In some cases, such a transition will "liberate" the previously bound system, enabling the AFM tip to break free of the surface. In other cases, a transition will occur between different bonded states, whether at the tip–sample interface or within the sample (e.g., a transition between conformational states within an extending fibril of polymer). In any case, one must acquire many measurements and look at the *distribution of forces*, and, moreover, at different rates or temperatures, if seeking a fundamental understanding of the tip–sample interaction. This statement should not imply that single, at-a-point measurements of attractive interaction, or resistance to compression, are not useful. Often even single measurements are clearly different on different materials, thus diagnostically useful. But multiple measurements, generating a distribution of behavior (i.e., a histogram) are much better. Indeed, this is why *imaging* tip–sample interaction is so powerful; an ensemble of measurements is being acquired on each material. Toward this end, in Chapter 6, we describe in more detail the acquisition and processing of tip–sample interactive data sets in various imaging and mapping modes. The intent of the present section is to ensure that the AFM user does not begin with a flawed interpretive model of what AFM is probing in measurements involving the break of interfacial bonds.

2.6.3 Relevance to Other Topics of Chapter 2

In considering the topics of Section 2.5, one should be able to qualitatively predict those cases in which viscoelasticity or stress-modified thermal activation are more or less likely to be relevant. Electrostatic force profiles beyond contact should be set up by ionic or electronic displacements that are far faster than the inverse residence time at any particular distance. Same holds for van der Waals forces beyond contact, outside of a bonded state. Molecular structure forces in liquids also are expected to be set up very quickly, unless the liquid is highly viscous (or cooled to temperatures at which it becomes highly viscous). Macromolecular steric forces in liquids, due to expanded brush layers, are also expected to equilibrate very quickly. But if the polymer has low affinity for the solvent, and/or if it is compressed and solvent squeezed out by the indentation of the tip, the molecular motions may be slow enough that the system is better described as viscoelastic. Also, note that a microscale colloid probe approached very near a surface *may* produce strong molecular confinement, resulting in motions that are slow enough such that the force depends on measurement residence time, whereas a nanoscale-sharp AFM tip approached to the same distance from the surface *may not* produce such conditions, given that the confinement disappears only a short lateral distance away. Entropic macromolecular extension forces normally derive from the rapid exploration of conformational space, likely much faster than the rate at which these forces are measured. But the breakage of the tip–molecule or molecule–molecule bond is expected to be described within the stress-modified thermal-activation framework. Indeed, the rate dependence of this unbinding as well as conformational transitions along the stretched molecule (e.g., DNA) have been the subject of numerous investigations within the biophysics community including interpretive methods [44–46]. Even more so, the breakage of strong van der Waals and hydrogen bonding associated with ligand-receptor "lock and key" adhesion has been (and continues to be) exhaustively analyzed within the conceptual framework of stress-modified thermal-activation, in terms of models such as the Bell–Evans [45] (within which one can extract two parameters, the time for bond failure in the absence of stress and the distance of displacement for unbonding) and the more general Hummer–Szabo [46] (valid even if activation energy barriers are not sharp, which precludes using the Bell–Evans model).

How relevant are these rate/time issues to *dynamic* AFM? Given that the distance dependence of force is being explored down to submicrosecond timescales (i.e., a small fraction of the period of oscillation at frequencies up to several hundred kHz), a soft material like a rubber may behave more like a rigid, glassy polymer. Indeed, this is one advantage of dynamic AFM, the phenomenon of *dynamic stiffening*; a creep-susceptible material may not have time to do so during the brief time interval that an AFM tip is touching it. But this very short time frame is not the only one of relevance. There is also the time frame over which a given conformer in a soft material is *repeatedly* stressed by the interaction with the tip. If the material holds no memory of previous interactions, then this longer time frame is not relevant. But if stress accumulates from one encounter to the next, then the brief timescale of each interaction may not be the only important timescale. For example, during the course of dynamic approach-retract measurements, wherein amplitude and phase are monitored as Z is ramped to bring oscillating tip and sample together then apart (e.g., Section 5.1.2), it is entirely possible that the much longer time frame for Z cycling (often $\sim$1 s) may be important. Even during imaging, the relevant time over which the tip is repeatedly interacting with a given conformational unit (e.g., a monomer of a polymer) may be ~1 ms or more, rather than submicrosecond. In this case, the scanning rate (lines per second) as well as the scan line spacing (possible overlap of contact diameter from one line to the next) may be important to the material response. For example, one may observe that an image of a polymer is sharper when the grid of measurement points are more widely spaced (e.g., 512 points per 1000 nm), such that each pixel roughly corresponds to the diameter of tip–sample contact, rather than zooming down to produce a higher pixel density (e.g., 512 over 500 nm). In other words, a 500-nm *digital zoom from the larger image* may be sharper than an as-acquired 500-nm image.

REFERENCES

[1] Haugstad, G., et al., A new probe of metallization microstructure on semiconductor surfaces. *J. Vac. Sci. Tech. B*, 1991, **9**(4): 2416–2422.

[2] Israelachvili, J.N., *Intermolecular and Surface Forces*. 2nd ed., 1991, San Diego: Academic Press, p. 450.

[3] Morse, P.M., Diatomic molecules according to the wave mechanics. II. Vibrational levels. *Phys. Rev.*, 1929, **34**: 57–64.

[4] Schiff, L.I., *Quantum mechanics*. International Series in Pure and Applied Physics, ed., 1968, New York: McGraw-Hill.

[5] Lennard-Jones, J.E., Cohesion. *Proc. Phys. Soc.*, 1931, **43**: 461–482.

[6] Parsegian, V.A., *Van der Waals Forces*. 2006, New York: Cambridge University Press, p. 380.

[7] London, F., The general theory of molecular forces. *Trans. Faraday Soc.*, 1937, **33**: 8–26.

[8] Hamaker, H.C., van der Waals attraction between spherical particles. *Physica*, 1937, **4**(10): 1058–1072.

[9] Lifshitz, E.M., The theory of molecular attractive forces between solids. *Sov. Phys. JETP*, 1956, **2**: 73–83.

[10] Butt, H.-J., B. Cappella, and M. Kappl, Force measurements with the atomic force microscope: Technique, interpretation and applications. *Surf. Sci. Rep.*, 2005, **59**: 1–152.

[11] Jackson, J.D., *Classical Electrodynamics*. 2nd ed., 1975, New York: Wiley, p. 848.

[12] Hudlet, S., et al., Evaluation of the capacitive force between an atomic force microscopy tip and a metallic surface. *Eur. Phys. J. B*, 1998, **2**: 5–10.

[13] Colchero, J., A. Gil, and A.M. Baro, Resolution enhancement and improved data interpretation in electrostatic force microscopy. *Phys. Rev. B*, 2001, **64**: 245403-1–245403-11.

[14] Ferry, J.D., *Viscoelastic properties of polymers*. 1980, New York: Wiley.

[15] Fischer-Cripps, A.C., Critical review of analysis and interpretation of nanoindentation test data. *Surf. Coat. Technol.*, 2006, **200**(14–15): 4153–4165.

[16] Van Landingham, M.R., Review of instrumented indentation. *J. Res. Natl. Inst. Stand. Technol.*, 2003, **108**: 249–265.

[17] Warren, O.L. and T.J. Wyrobek, Nanomechanical property screening of combinatorial thin-film libraries by nanoindentation. *Meas. Sci. Technol.*, 2005, **16**: 100–110.

[18] Fischer-Cripps, A.C., *Nanoindentation*. Mechanical Engineering Series, 2nd ed., 2004, New York: Springer.

[19] Johnson, K.L., *Contact Mechanics*. 1985, New York: Cambridge University Press.

[20] Slaughter, W.S., *The Linearized Theory of Elasticity*. 2002, Basel, Switzerland: Birkhauser.

[21] Garcia, R. and R. Perez, Dynamic atomic force microscopy methods. *Surf. Sci. Rep.*, 2002. **47**: 197–301.

[22] Anczykowski, B., et al., Basic properties of dynamic force spectroscopy with the scanning force microscope in experiment and simulation. *Ultramicroscopy*, 1996, **66**: 251.

[23] Marion, J.B., *Classical Dynamics of Particles and Systems*. 2nd ed., 1970, New York: Academic Press.

[24] Cleveland, J.P., et al., Energy dissipation in tapping-mode atomic force microscopy. *Appl. Phys. Lett.*, 1998, **72**: 2613–2615.

[25] Garcia, R., et al., Phase contrast in tapping-mode scanning force microscopy. *Appl. Phys. A*, 1998, **66**: S309–S312.

[26] Butt, H.-J., K. Graf, and M. Kappl, *Physics and Chemistry of Interfaces*. 2nd ed., 2006, Weinheim: Wiley-VCH, p. 373.

[27] Adamson, A.W., *Physical Chemistry of Surfaces*. 1982, New York: Wiley.

[28] Erbil, H.Y., *Surface Chemistry of Solid and Liquid Interfaces*. 2006, Oxford: Blackwell.

[29] Kittel, C., *Introduction to Solid State Physics*. 6th ed., 1986, New York: Wiley.

[30] Zangwill, A., *Physics at Surfaces*. 1988, Cambridge: Cambridge University Press, p. 454.

[31] Han, W. and S.M. Lindsay, Probing molecular ordering at a liquid-solid interface with a magnetically oscillated atomic force microscope. *App. Phys. Lett.*, 1998, **72**(13): 1656–1658.

[32] O'Shea, S.J., M.E. Welland, and T. Rayment, Solvation forces near a graphite surface measured with an atomic force microscope. *App. Phys. Lett.*, 1992, **60**(19): 2356–2358.

[33] Tanford, C., *Physical Chemistry of Macromolecules*. 1961, New York: Wiley.

[34] Flory, P.J., *Principles of Polymer Chemistry*. 1953, New York: Cornell University Press.

[35] Flory, P.J., *Statistical Mechanics of Chain Molecules*. 1969, New York: Interscience Publishers.

[36] White, L.R., On the Deryaguin approximation for the interaction of macrobodies. *J. Coll. Interf. Sci.*, 1983, **95**(1): 286–288.

[37] Kappl, M. and H.-J. Butt, The colloidal probe technique and its application to adhesion force measurements. *Particle Particle Syst. Characteriz.*, 2002, **19**(3): 129–143.

[38] Feynman, R., *Quantum Electrodynamics*. 1961, Boulder, CO: Westview Press.

[39] Weiner, J.H., *Statistical Mechanics of Elasticity*. 1983, New York: Wiley.

[40] Hill, T.L., *An Introduction to Statistical Thermodynamics*. 1960, Reading, MA: Addison-Wesley.

[41] Janshoff, A., et al., Force spectroscopy of molecular systems-single molecule spectroscopy of polymers and biomolecules. *Angew. Chem. Int.*, 2000, **39**: 3212–3237.

[42] Hugel, T. and M. Seitz, The study of molecular interactions by AFM force spectroscopy. *Macromol. Rapid Commun.*, 2001, **22**: 989–1016.

[43] Haugstad, G., Probing mesoscale molecular domains at surfaces of polymeric condensed matter with scanning force microscopy. in 227th ACS National Meeting. 2004, Anaheim: American Chemical Society.

[44] Noy, A., et al., Chemical force microscopy: Exploiting chemically-modified tips to quantify adhesion, friction, and functional group distributions in molecular assemblies. *J. Am. Chem. Soc.*, 1995, **117**: 7943–7951.

[45] Evans, E., Probing the relation between force–lifetime–and chemistry in single molecular bonds. *Annu. Rev. Biophys. Biomol. Struct.*, 2001, **30**: 105–128.

[46] Dudko, O.K., G. Hummer, and A. Szabo, Theory, analysis, and interpretation of single-molecule force spectroscopy experiments. *Proc. Natl. Acad. Sci USA*, 2008, **105**(41): 15755–15760.

3

Z-DEPENDENT FORCE MEASUREMENTS WITH AFM

3.1 REVISIT IDEAL CONCEPT

In this chapter, we delve into the details of how an AFM actually measures the distance dependence of force, usually called a *force curve*. The distance dependence of force not only enables topographic imaging, the focus of Chapter 4, but also can be measured in its own right to probe nanoscale properties. Before considering real AFM force curves, however, let us list the anticipated components of an *idealized* measurement:

i. a tip whose end is a spherical cap or solid-of-revolution, with known radius of curvature (nanometers) and chemical composition;
ii. a force sensor that measures the magnitude and sign of tip–sample force normal to the surface ("vertical"), in actual force units with ~0.01 nN precision and ranging up to micronewtons, and whose vertical position is not affected by these same forces;
iii. a Z-displacement device to programmatically vary the distance between sample and force sensor with ~0.01 Å precision and a range of tens of micrometers (to handle surface locations atop tall hills or within deep valleys);

Atomic Force Microscopy: Understanding Basic Modes and Advanced Applications, First Edition.
Greg Haugstad.

iv. a device for accurately *measuring* this Z displacement, independent of devices in (ii) and (iii);
v. an unambiguous indicator of the point at which the tip reaches the surface (i.e., the zero of distance).

Given these components, the force–distance curve depicted in Figure 3.1 would be no challenge to measure, whether distances are measured in micrometers, nanometers, or Angstroms, or forces are measured in micronewtons, nanonewtons, or tens of piconewtons. Increments of distance from surface could be determined with enough accuracy and precision to navigate along the curve from a state of zero force far away (at right) to a growing but weak attractive force moving leftward, followed by a transition to a (usually) steeper repulsive force upon contacting the surface. We have seen in Chapters 1 and 2 how such force–distance curves can reveal important characteristics of materials and interfaces related to mechanical properties, polarizability, surface charge, and intermolecular forces, including those important to biology as well as synthetic materials and interfaces. Thinking ahead to imaging, one can imagine the repeated collection of these curves within a 2D grid of locations across a surface, provided a precise X–Y positioning device. Also, the measurement of the point at which the tip reaches the surface (component (v)) is such that the corresponding value of Z displacement provides a "height" measurement. An accurate reproduction of surface topography could result (Chapter 4). If the force–distance trend varied significantly in shape from point to point along the surface, careful analysis could provide a high-resolution mapping of material or interfacial heterogeneities (Chapters 5–7). This would include absolute, quantitative evaluation of properties such as modulus or surface energy given a known, simple shape and composition of the end of the tip (component (i)) as well as the highly controlled and quantitative force–displacement measurements (components (ii)–(iv)).

In conventional AFMs, none of the ideal components (i)–(v) is completely realized. In the next sections, we will examine some of the actual instrument components in detail, to understand how the measurements depart from ideality while

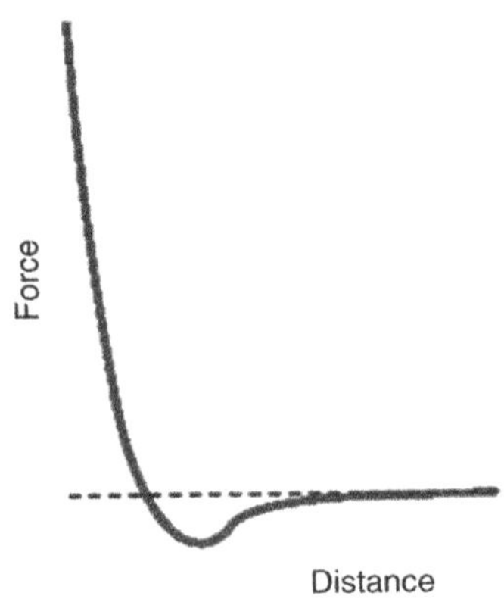

FIGURE 3.1 Schematic force–distance relationship broadly relevant to AFM.

quantifying their (nevertheless impressive) range and precision, with an eye toward practical applications. We will better understand the unavoidable coupling between these components, for example, how the force measurement scheme (ii) affects the tip-to-sample distance (iii). In the process, we will describe procedures that can help the user control the linearity, accuracy, and precision of measurements (i.e., within components (ii)–(iv)). We will also identify certain "second-generation" hardware components that can improve force–distance measurements.

3.2 FORCE-Z MEASUREMENT COMPONENTS: TIP/CANTILEVER/LASER/PHOTODETECTOR/Z SCANNER

3.2.1 Basic Concepts and Interrelationships

The force sensor is a sharp tip attached to a microcantilever that *displaces vertically*, that is, bends in the presence of force (Figure 3.2). The force F is measured as this vertical displacement s_z times the operative cantilever spring constant k

$$F = ks_z. \tag{3.1}$$

A nominal spring constant is provided by the manufacturer, but this represents the cantilever stiffness to forces applied normal to the tip, perpendicular to the cantilever. The *operative* stiffness k includes the effect of cantilever *inclination* angle α relative to the surface being probed, usually, $\alpha \approx 10$–$12°$, mainly determined by the

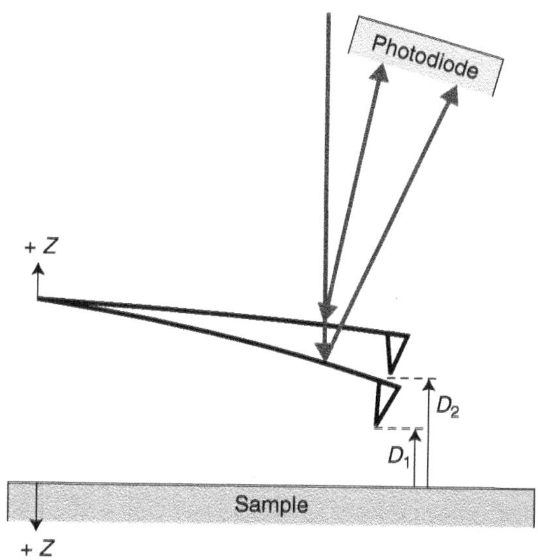

FIGURE 3.2 Depiction of reflected laser displacement for two values of tip–sample distance $D_{1,2}$ due to different degrees of cantilever bending. The operator controls Z, the position of either the cantilever base or the sample surface.

instrument's mounting of the chip to which the cantilever is attached, but generally dependent on surface tilt as well. Thus, $k = k^{\perp}/\cos^2\alpha$ [1],[1] where $k^{\perp}$ is the actual spring constant for a *force acting perpendicular to the long cantilever axis* (the value provided by the manufacturer). This correction for angle of inclination is typically 3–4%. Such a precise correction is small compared to grosser error in the nominal spring constant provided by the manufacturer. Better values can be determined by the end user adopting published procedures, discussed in Section 3.7 and Appendix 1.

The displacement s_z is indirectly measured by reflecting a laser off the cantilever top side and measuring the movement of the reflected laser spot on a position-sensitive detector, a split photodiode, at a distance on the order of centimeters. The displacement of the laser spot is strictly dependent on the *change of the angle of inclination* of the cantilever near the tip end, rather than the *vertical displacement* of the tip *per se*. In other words, the change of cantilever inclination serves as a proxy for the vertical tip displacement. This displacement s_z, otherwise called the cantilever *deflection*, usually is calibrated from a known displacement of the (precalibrated) Z scanner while in contact with a rigid sample that effectively does not deform under the pressing tip. The Z scanner of course is used to execute much larger displacements in a full cycle of force versus Z, such that force becomes negligible at the most extreme separation distance where tip and sample are far apart. The Z scanner can be used to vertically displace either the chip to which the cantilever is attached (the substrate from which it was grown) or the sample, depending on instrument design. *Positive* Z is usually defined to *increase* tip–sample distance (see Figure 3.2). The vertical displacement of the tip s_z due to changes of force, however, necessarily changes the tip–sample distance D, as depicted in Figure 3.2: two different forces, gauged from different amounts of cantilever bending via different laser spot positions on the photodiode, necessarily result in different tip–sample distances D_1 and D_2. In Section 3.2.2, we will demonstrate how s_z and Z measurements are combined to determine D.

One should be aware from the outset that if using liquid immersion (for biology, electrochemistry, or a host of other good reasons), then the laser trajectory is further dependent on optical mismatch of media at an interface (i.e., air–liquid) according to Snell's Law,

$$n_1 \sin\theta_1 = n_2 \sin\theta_2 \tag{3.2}$$

where n_1 and n_2 are the indices of refraction of the media and θ_1 and θ_2 are the angles of the laser beam relative to the normal of the interface, on either side of the interface (e.g., air and water). This is depicted in Figure 3.3. For a given vertical tip displacement δs_z, the reflected laser spot displacement at the photodiode, $\delta\theta_2$, will be *greater* for the immersed case than the in-air measurement (by ≈40% for water

[1] The square of cosine derives from the fact that the impact of tilt is imposed twice during the calculation: once to account for component of force perpendicular to the cantilever and again to account for the vertical component of tip displacement due to this force.

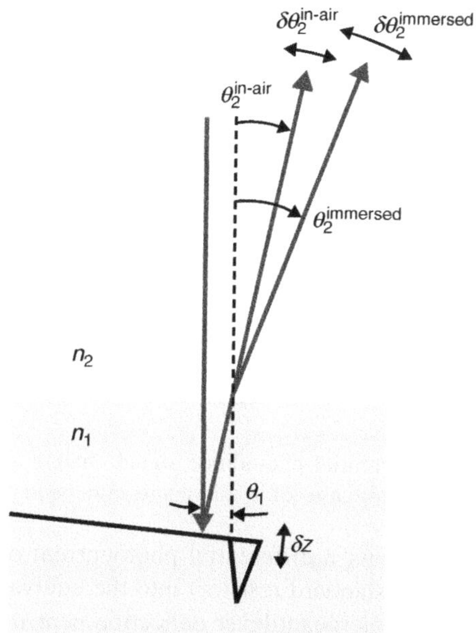

FIGURE 3.3 Differences in laser trajectories for in-air versus water-immersion environments.

immersion and 20° cantilever tilt). Apart from the obvious calibration issue, one should also note the implications for force sensitivity (immersed > in-air) as well as the accessible range of force measurements (immersed < in-air). (Note that some AFM liquid cells are *closed,* such that the laser passes from liquid to glass to air *en route* to the photodiode. Here, however, the liquid–glass and glass–air interfaces "cancel out" in terms of the ultimate angle of the laser beam exiting the cell, as described by Snell's Law for the two-interface system, $n_1\sin\theta_1 = n_2\sin\theta_2 = n_3\sin\theta_3$.)

Figure 3.4 illustrates schematically what a measurement of differential photodiode output signal versus Z displacement would look like in the absence of attractive forces or material compliance (deformation, i.e., penetration of tip) in air and water. The zeros of each scale are arbitrary, dependent on the exact vertical positioning of the photodiode as well as the location of Z within the full range of the device that displaces Z.

It is important to note that the laser spot position at the photodiode is not really measured as it might be using, say, a CCD array camera; rather, it is very simply gauged as the difference in the total intensity of the laser light impinging on two adjacent photodiodes.[2] (We will examine the *linearity* of this "knife edge" type of measurement in Section 3.2.3). This intensity differential is not, of course, a direct measurement of cantilever deflection. A calibration procedure is needed to convert

[2] The output signal is the difference of top minus bottom element outputs divided by the top plus bottom element outputs (to normalize out fluctuations in total laser intensity), times ~10 V.

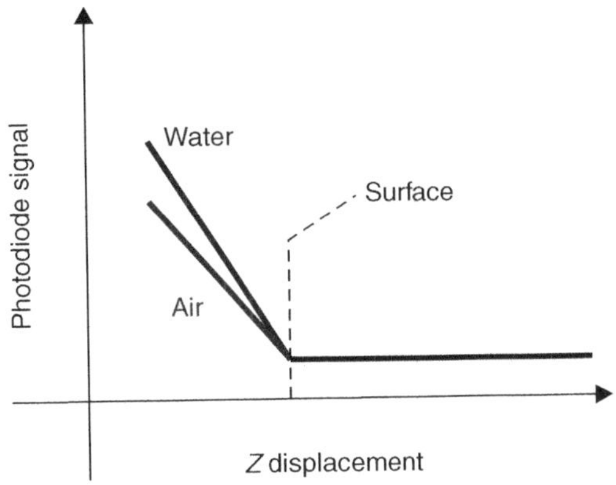

FIGURE 3.4 Schematic differential photodiode signal change at the onset of repulsive forces upon tip–sample contact for cases of in-air versus immersion in water.

the photodiode output (whether a differential photocurrent or differential voltage, i.e., a photocurrent across a standard resistor) into the equivalent vertical displacement of the tip, that is, the microcantilever deflection near its end, in length units. Then, as stated in Equation 3.1, this deflection is multiplied by the cantilever spring constant k (force per unit deflection) to determine force. *Changes* in cantilever deflection are thus converted to *changes* in force; the *zero* of force measurement is determined from the photodetector output signal for tip and sample "far apart," such that the force can be considered zero. There can be additional optical artifacts in the location of this zero that derive from the laser/photodetector measurement scheme, as well as thermal warm-up effects that can shift the zero. These phenomena are described in Section 3.4.

3.2.2 Tip–Sample Distance

As already stated, in Figure 3.2, it is obvious that the bending of the cantilever—which provides the force measurement—also changes the *distance* between tip and sample. Thus, to obtain a force-versus-distance measurement, the vertical displacement of the tip must be taken into account. The desired independent variable, distance, is primarily controlled using another device, the Z scanner (further described in Section 3.5). Then the vertical displacement of the tip *together with* the scanner Z displacement determines the tip–sample distance. Assuming the positive direction of Z-scanner displacement, whether the base of the cantilever or the sample, is defined so as to increase tip–sample distance, and tip displacement away from sample is also defined as positive, the measured tip and scanner displacement must be *added* to calculate tip–sample distance.

$$D = Z + s_z. \tag{3.3}$$

Some AFM software may invoke a different sign convention, such that subtraction is required rather than addition.

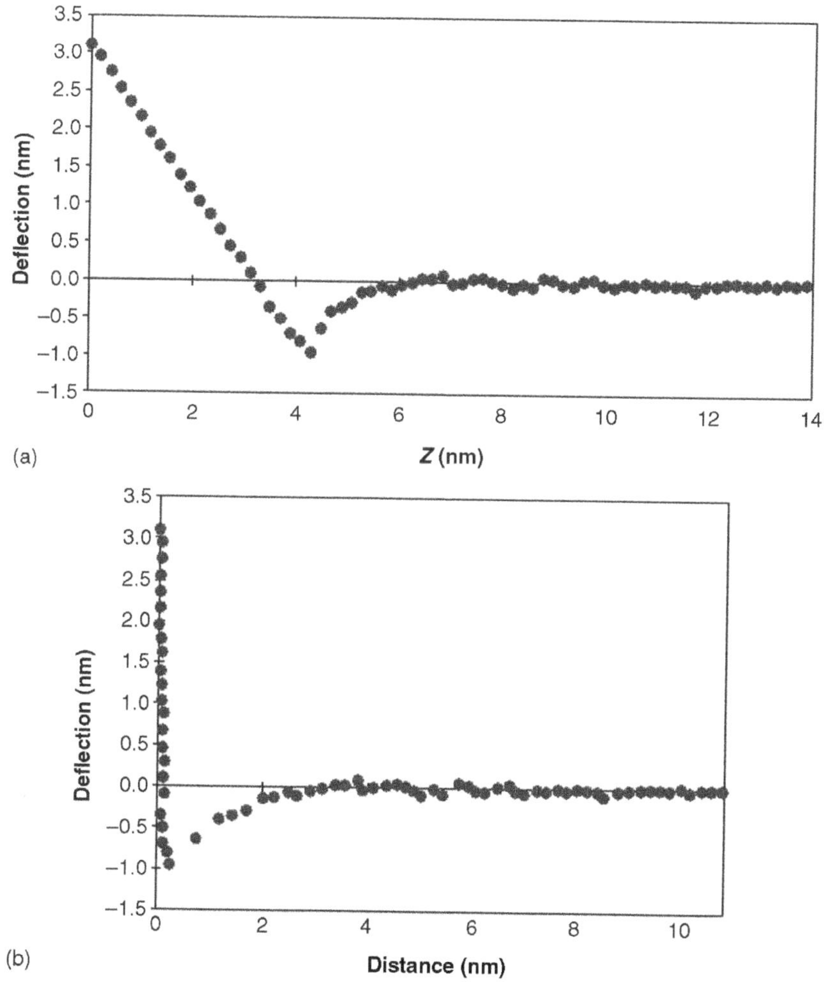

FIGURE 3.5 (a) Cantilever deflection versus *Z* during approach of SiO_2 tip to SiO_2 surface in water. (b) Same data after conversion of horizontal scale to distance.

To illustrate, consider the real AFM "force curve" in Figure 3.5 (previewed in Section 1.5), cantilever deflection data acquired in water while approaching a SiO_2 surface with a SiO_2 tip attached to a cantilever of stiffness ≈ 0.6 N/m. Note that the approximate scatter in the precontact deflection data is about $\sim$0.1 nm, meaning $\sim$0.1 nN in force. Softer cantilevers can drop this noise floor further. The top curve (a) is plotted versus the raw independent variable, displacement of the *Z* scanner (Section 3.5), zeroed at the left end point (farthest *Z* ramp to bring tip and sample together). The bottom curve (b) is plotted versus actual tip–sample distance, accounting for the vertical displacement of the tip by adding the deflection to *Z* in nanometers. One can start at the rightmost data point and, moving leftward, examine individual data points to see the effects of this mathematical operation. As the deflection becomes appreciable due to attractive van der Waals forces, the addition of these negative numbers to Z pushes the data points leftward in the distance plot. Upon transitioning to repulsive forces at contact, the addition of these positive numbers to Z pushes the data points rightward, such that the contact domain changes from a diagonally sloped trend in the top graph to a nearly vertical trend in the bottom graph.

In the preceding, one step was left out: the conversion of deflection numbers from the raw differential photodiode output, typically in volts (but in some instruments in photocurrent, rather than voltage drop across a standard resistor), to values in nanometers. To do this, we simply multiply by a coefficient called the *inverse sensitivity* S^{-1}, in nm/V units: the displacement of the Z scanner (precalibrated by imaging a grating containing features of known height, Chapter 4) per unit change of output of the differential photodiode.[3] This gives, as an end result, an essentially vertical deflection-versus-distance trend (Figure 3.5b). Why? We posit that because both tip and sample are quite rigid materials, the amount of *indentation*—tip penetrating into sample—is negligible on the scale shown in Figure 3.5b. In Section 2.3, we discussed the simple Hertzian model that predicts the amount of indentation of a spherically shaped tip (modeling a rounded off end) into a flat half space. (In Chapter 6, we will generalize to models that additionally account for attractive forces.) Applying this model to identical SiO_2 tip and sample assuming a Young's modulus of 73 GPa and Poisson's ratio of 0.165, a 10-nm radius Si tip and a force up to +1.9 nN (maximum 3.2-nm deflection times 0.6 N/m cantilever stiffness), the indentation is predicted to reach ≈ 0.05 nm. This small number is consistent with the tip–sample distance calibration in Figure 3.5b, which results in nearly zero change of tip–sample distance within the positive-deflection contact regime at left.

Quantitative nanomechanical analysis of force curves on softer materials can be performed by using a rigid material like silica to calibrate the value of S^{-1}, because the indentation of tip into silica or stiffer materials is essentially zero. To appreciate, we compare in Figure 3.6b two theoretical force curves comparing silica and polystyrene (Young's modulus 2.8 GPa, Poisson's ratio 0.34) within the simple Hertzian model, Figure 3.6a (i.e., neglecting attractive forces and treating polystyrene as an elastic rather than *viscoelastic* solid). Here, we performed the inverse of Figure 3.5 (the experimental case): we started with theoretical force versus indentation (i.e., negative distance D) in Figure 3.6a and calculated the Z-scanner displacement needed to produce this indentation, Figure 3.6b, if using a cantilever with a spring constant of 1 N/m and tip of radius 10 nm. We see that the difference in the resulting theoretical force-curve slopes is small but seemingly measurable (Figure 3.6b), suggesting that in a real experiment, a modulus for polystyrene might be determined. Second, we note that the force-curve slope on silica, by counting the shown divisions of rise over run, is indeed essentially unity (apart from a minus sign), meaning

[3] When we ask how sensitive (S) is our cantilever deflection measurement, we think of how much the measured signal changes (here, a voltage) for a given amount of deflection in length units. Thus, S is in units of V/nm. Also note that some use the term optical lever sensitivity, or OLS.

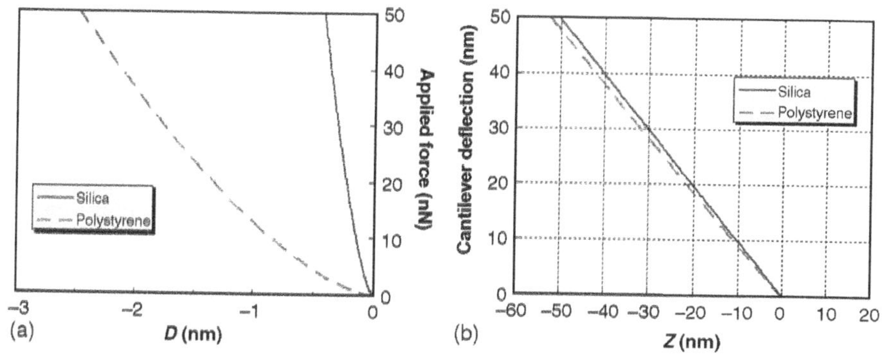

FIGURE 3.6 (a) Hertzian force–distance relationship (zero attractive forces) for indentation of a SiO_2 tip into a material of modulus equal to that of SiO_2 or polystyrene. (b) Corresponding prediction of deflection of cantilever versus *Z*-scanner displacement given a cantilever spring constant of 1 N/m.

it should serve our purpose as sufficiently rigid to provide a calibration of S^{-1}. That is, we assume that the raw force-curve slope on glass, in volts of differential photodiode output per nanometer of *Z*-scanner displacement, is effectively a *measurement of S*.

3.2.3 Finer Quantitative Issues in Force–Distance Measurements

It is important to examine the aspects of geometry in greater detail for this theoretical force curve on polystyrene: approximately 50 nm of upward tip movement to produce roughly 2.5 nm of vertical tip displacement into sample. Given that the tip is attached to a 3D object, the cantilever, how do we know it is displacing only in the *Z* direction? We do not; geometry, in fact, dictates otherwise. The cantilever cannot bend upward to displace the tip in *Z* without displacing it along *X*. This is illustrated in Figure 3.7, for an exaggerated vertical displacement of more than the tip height (~10 μm) and treating the cantilever bending as that of a hinged lever. On this scale, the lateral tip displacement ΔX for a vertical tip displacement ΔZ is obvious. In reality, we are dealing with much smaller displacements, nanometers

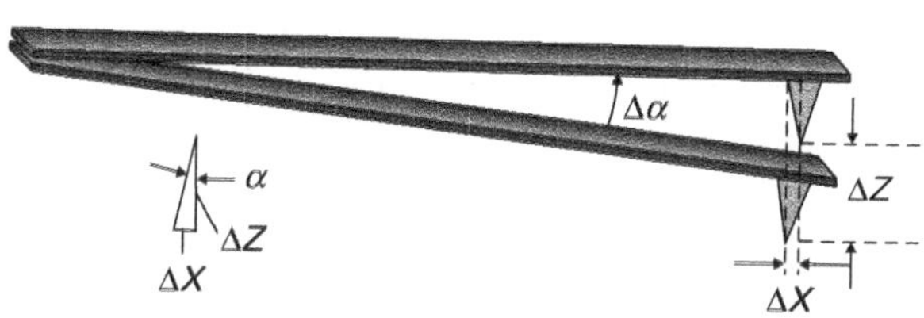

FIGURE 3.7 Depiction of geometric relationship between vertical and horizontal AFM tip displacement.

instead of micrometers, but the geometric constraints are essentially the same. For $\Delta Z = 50\,\text{nm}$, $\Delta\alpha$ is extremely small, such that we may consider the cantilever tilt angle α to be a constant (and typically $\approx 12°$). Then, $\Delta X = \Delta Z \tan(\alpha) = (50\,\text{nm}) \cdot \tan(12°) \approx 10\,\text{nm}$. Note that this lateral tip movement is much greater than the 2.5 nm of vertical indentation! Apart from this purely geometric consideration, there is also a more physical question: whether the tip actually moves 10 nm along X or whether static friction instead keeps it pinned at a point. We will return to this tribological issue in Chapter 7 in the context of lateral force measurements and imaging. We will see that a pinned tip is unlikely for 50 nm of Z displacement, but may be the case for, say, a 5-nm Z displacement. Thus, the overall force curve, from the first touch during approach to the highest force achieved at turnaround to the final tip–sample separation during retraction, may involve stick-to-slide and slide-to-stick transitions at points *other than* the initial touch and final break of contact. Even within the "stick" regime, there is necessarily a lateral component of force, due to cantilever geometry, in addition to the normal component (i.e., there are both normal and shear stresses in the contact). Knowing that we ultimately wish to quantify mechanical properties such as elastic modulus, we should recognize that the lateral stress must play a role in the most precise, quantitative measurements. As with all nonidealities and caveats covered in this chapter, it is good to register in one's mind early on, to foresee limitations and avoid overly precise interpretative statements that would require a more careful data treatment.

The preceding indicates that the tip may *slide across the surface while indenting into it*, during force-curve measurements ostensibly performed to probe vertical response (and thereby, say, determine Young's modulus). What about plastic response: material yield that produces permanent (or at least residual) deformation? The preceding discussion suggests that one might attempt to cross-check the quantitative measurement of yield, that is, a "roll-over" in force versus distance, with post-measurement topographic imaging. We consider an example of the latter in Figure 3.8a, a topographically imaged trench approximately 300 nm in length and up to 20 nm in width, created in an ultrathin film of the glassy polymer polyvinyl alcohol (PVA, deposited on mica) using a Si_3N_4 tip a few tens of nanometers in radius attached to a cantilever of stiffness $k \approx 0.6\,\text{N/m}$ [2]. The trench was produced while ramping the Z displacement by micrometers so as to reach forces of order $\sim 1\,\mu\text{N}$, which enabled the AFM tip to penetrate the PVA and apparently reach the mica substrate. Examining the data in cross section in Figure 3.8b, one can see that the tip gouged deeper and deeper into the material during the upward ramp of Z as load increased (right to left, top to bottom in Figure 3.8a). There is a relatively abrupt penetration from 0.27 nm to 0.76 nm depth, observed in the cross-sectional data trace in Figure 3.8b. The width of the trench also provides a glimpse of the contact diameter during the plastic deformation process. This width must correspond to the domain sufficiently stressed to induce yield. Of course, the contact is somewhat smaller at loading forces below yield, so the scratch width might be considered an upper bound on the size of the tip–sample contact zone under elastic conditions. In Figure 3.8a, however, we find that the width of the trench does not change much for a large range of increasing force in the yield portion of the force

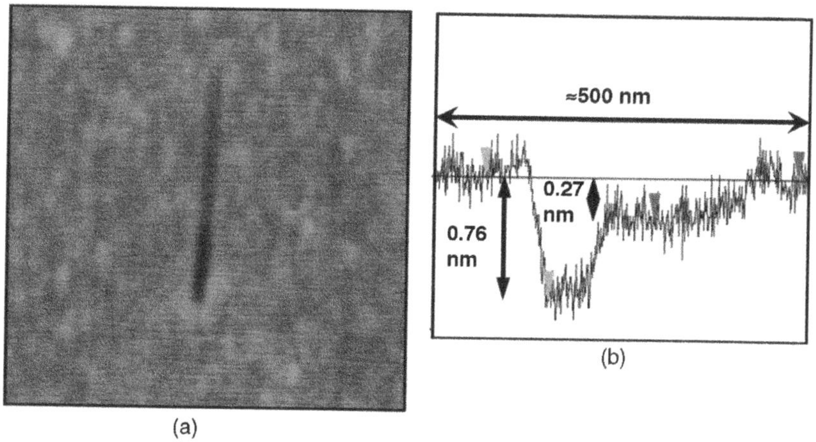

FIGURE 3.8 Topographic image (a) and cross section (b) of scratch produced during force–Z measurements due to lateral tip displacement as vertical force increases (top to bottom in (a), right to left in (b)).

curve, suggesting that the trench width is a pretty good estimate of the contact diameter at slightly lower load, just below the yield point. This value could be cross-checked for approximate consistency with the force-curve data, at forces just below the critical value at which yield initiated, analyzed within an elastic contact mechanics model. In addition to a force–indentation relationship as considered above, these models generally predict an analytical relationship between the radius a of the (elastic) tip–sample contact diameter and the (elastic) indentation depth h. In the Hertzian case, this relationship is $a^2 = hR$ (Section 2.3).

Figure 3.7 depicts the cantilever as a rigid rod hinged at one end. But the fact that we are using a *flexible* cantilever as force sensor implies that it bends internally; there is no hinge where the cantilever attaches to its support structure. So where, exactly, does the cantilever bend? Intuitively, we expect a continuous change of inclination all along the length of the cantilever. A laser spot of finite diameter, reflected near the end of the cantilever, samples some portion of this potentially variable inclination. We wish to know the mathematical shape of this bent cantilever, $z(x)$, and thus whether its inclination varies substantially along the length of the cantilever, potentially making the exact laser spot position an issue. (Lowercase symbols are used here to distinguish from vertical and lateral scanner positions Z and X discussed in the preceding paragraph.) Deformable body mechanics provides an analytical solution for a simple rectangular cantilever; this means ignoring the presence of the tip and other complications of shape, or reflective coatings, of real cantilevers. Here, we state the key points and final result; a more detailed treatment of the deflected cantilever shape is included in Section 9.2.1.

An external force produces a torque on the cantilever that induces bending. At any location x along the cantilever length (attached end defined as $x = 0$), and through any short cross-sectional element inside of the cantilever at x, there is a

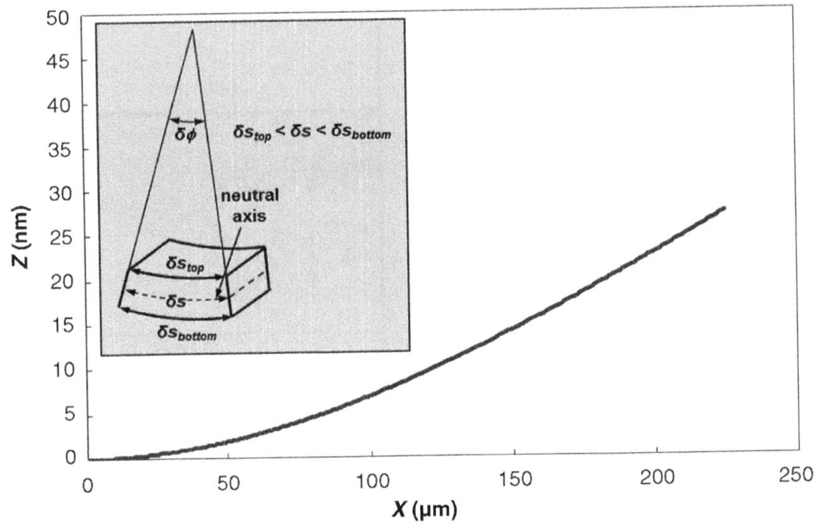

FIGURE 3.9 Edge view of vertical displacement Z of cantilever versus position X for an upward force located at the end of the cantilever. Inset: cross-sectional differential element of cantilever.

force balance at that element.[4] For a cantilever of length L, the torque at any location $x < L$ due to an external upward force F at $x = L$, perpendicular to the cantilever axis, is balanced by the cantilever's local bending moment M (Section 9.2.1), which derives from the differences in *curved deformation* in the cross-sectional element: compression in the top half and expansion in the bottom half, respectively, above and below the *neutral surface*, where material is neither compressed nor expanded, as shown in Figure 3.9 inset. The amount that δs_{top} is reduced and δs_{bottom} is extended (relative to δs at the neutral surface) is inversely proportional to the cantilever's Young's modulus E and cross-sectional area moment of inertia

$$I = wt^3/12, \tag{3.4}$$

where w and t are the width and thickness (height) of the cantilever cross section, respectively. (Thus, in cantilever fabrication, halving the cross-sectional width w or thickness t increases the amount of cantilever bending, and thus sensitivity to force, by factors of 2 and 8, respectively.) The relationship describing the equilibrium of torques and moments provides a differential equation that can be integrated twice to yield the result

$$z(x) = \frac{Fx^2}{6EI}(3L - x). \tag{3.5}$$

[4] Assuming static equilibrium, which means that the change of force must occur very slowly compared to the inverse of the cantilever's resonance frequency.

This functional form is graphed in Figure 3.9 for a rectangular beam with specifications of $L = 225\ \mu\text{m}$, $w = 35\ \mu\text{m}$, $t = 2.0\ \mu\text{m}$ constructed from silicon of Young's modulus $E = 179$ GPa, and an external upward force $F = 30$ nN localized at $x = L$. (Note the greatly expanded vertical relative to horizontal scale.) This produces a vertical displacement of 27 nm at $x = L$. By Hooke's Law, we calculate $k = F/s_z = 1.1$ N/m and produce the relationship

$$k = Ewt^3/4L^3. \tag{3.6}$$

Note that a cantilever half as thick but otherwise identical, under the same force, would result in 220 nm of deflection (to two significant digits), meaning $k = 0.14$ N/m. The precision of force measurement would at first sight be eight times greater as well; but this calculation does not take into account differences of *noise*. (Note that softening a cantilever increases its intrinsic thermal noise, Appendix 1; there is still a net gain of precision, but by less than the factor predicted in the preceding.)

Importantly, almost all of the bending (curvature) of the above model cantilever involves the interval $0 < x < 125\ \mu\text{m}$, such that the inclination is nearly constant for $125 < x < 225\ \mu\text{m}$. This suggests that the laser beam-deflection gauge of force should *not* be very sensitive to the positioning of the laser spot within this outer portion of the cantilever: the local inclination and thus the position of the reflected laser spot at the photodiode should be nearly independent of spot position. Some researchers have experimentally examined the bent shape of real cantilevers by using a highly focused, micropositioned laser beam and comparing with theoretical predictions (particularly to understand higher frequency vibrational modes in dynamic AFM, Section 9.2) [3]. Any AFM user, however, can easily assess the variable cantilever inclination by measuring the sensitivity S from the slope of force curves in contact as discussed in the context of Figure 3.5. Our model predicts that for a given Z displacement of the contacting tip via Z-scanner displacement, the movement of the reflected laser spot at the photodiode should be about the same provided that the incident laser is positioned within roughly the outer half of the cantilever. Let us compare this prediction with real experimental measurements.

Figure 3.10a contains six force curves acquired during approach to contact (in air on a silicon wafer), utilizing six laser spot positions on a ~400-μm-long rectangular cantilever as approximately depicted in Figure 3.10b. (The light microscope was focused on the rightmost tip end of the cantilever, which was closest to the underlying silicon surface, making the base of the rectangular cantilever appear to be wider because of a defocusing.) For positions A–C, where the laser spot is located within the right half of the cantilever, the slopes or sensitivities S are within 5% of 8.0 mV/nm. For positions D–F, where the laser spot is located within the left half of the cantilever, the slopes substantially decrease to 5.8, 3.6, and 2.0 mV/nm, respectively. Also note that the magnitude of the vertical *jump-to-cont*act varies in proportion to the measured contact slope. This jump is due to *attractive* forces that, in turn, mobilize adventitious contamination on tip and sample (here, mainly water, presumably) to produce a capillary liquid neck that bridges to the tip and strongly pulls it to the surface [4]. (We will discuss capillary forces, including the role of

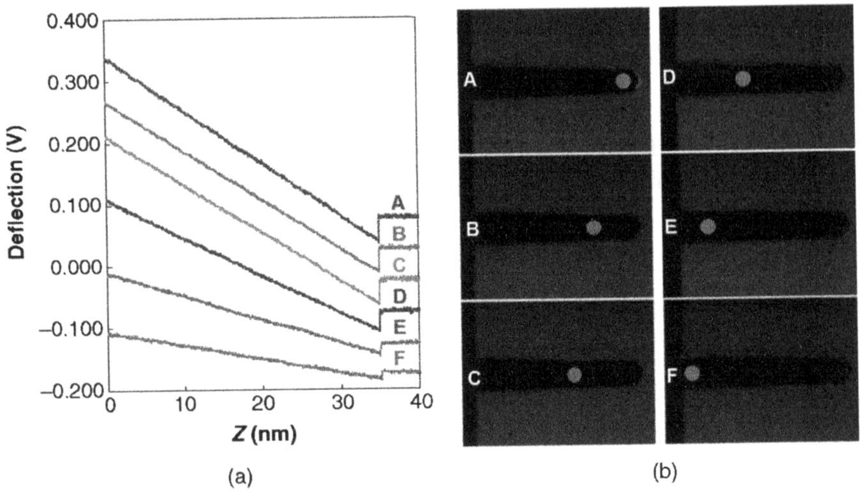

FIGURE 3.10 Approach force curves (a) for different laser spot positions on the cantilever (b).

humidity, in greater detail in Chapter 6). Clearly, accurate measurements of cantilever deflection, and thus force, require proper calibration to the slope of contact deflection versus Z displacement; the latter is calibrated by imaging the known heights of a grating (Chapter 4).

Another question is the *linearity* of the force measurement. The laser spot position at the photodiode depends on the inclination of the cantilever, which by taking the derivative of Equation 3.5 (for a rectangular cantilever) is

$$\frac{dz}{dx} = -\frac{Fx^2}{2EI}. \tag{3.7}$$

This equation, similar to Equation 3.3, is linear in force, as we hope. Second, the distance from cantilever to photodetector is in centimeters such that small variations of this distance, by at most micrometers (the maximal cantilever deflection for maximal Z-scanner displacement), are four or more orders of magnitude smaller, meaning no measurable effect on the laser spot position at the photodiode; such small variations cannot be the source of observed nonlinearity in the overall response of the measurement. (The force-derived variations of cantilever *inclination*, not variations in cantilever-to-photodiode *distance*, are the only significant variations). But a third question is the photodiode itself: as already stated, it does not measure spot position *per se*, rather it measures the difference of integrated spot intensity on either side of a split photodiode junction (each of which produce a photocurrent across a standard resistor to give voltages V_1 and V_2). The spot intensity profile is of course not uniform, as needed for purely linear response, rather some Gaussian-convolved Lorentzian form (the latter would be produced at the photodiode by a spherical wave emanating from a point source, the focused spot on cantilever) as

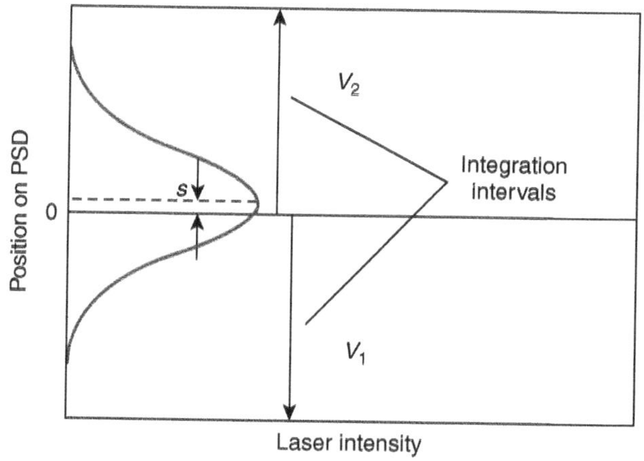

FIGURE 3.11 Depiction of laser spot intensity at the position sensitive detector and different photodiode element outputs V_1 and V_2 depending on the spot displacement relative to the photodiode junction.

depicted in Figure 3.11 [5]. Thus, we anticipate that as the laser spot moves across the photodiode junction (0) by some increasing amount s, the split photodiode differential output will change in an approximately linear fashion *only if this movement is very small.* Indeed, in Figures 3.5a and 3.10a, the measurement appears to be linear, for cantilever deflection changes measured to be a fraction of a volt of raw photodiode differential signal.

We see in Figure 3.12a, in contrast, that if the laser spot movement at the split photodiode corresponds to *several volts* of signal change, considerable nonlinearity indeed results. In detail, the nonuniform laser spot intensity profile produces a cubic term (second term in a power series expansion of Lorentzian or Gaussian), which is opposite in sign to the linear term and comprises as much as ~10% of the overall

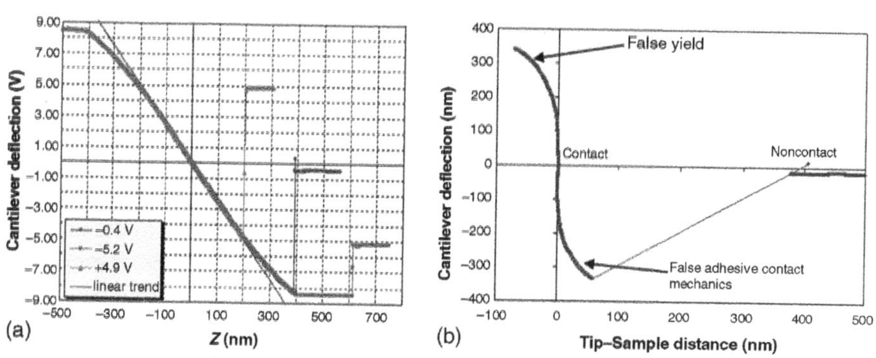

FIGURE 3.12 (a) Retraction force curves spanning the entire range of measurable cantilever deflection, for three photodiode positions corresponding to three deflection baselines. (b) Conversion to distance abscissa for middle force curve in (a).

signal [5]. In the overlaid, raw force curves in Figure 3.12a (acquired during retraction from a silicon surface including the *jump-from-contact* up to the zero force baseline upon break of tip–sample *adhesion*), one can see from inspection that the approximately diagonal, contact portion of the plot is not truly linear, rather exhibits a slight curvature. This is easy to perceive if comparing to a true diagonal line, as shown, or by viewing the page at a shallow angle. Three force curves are overlaid; the photodiode was offset to three different vertical positions, thus three noncontact baselines (zeroes of deflection) are shown near +5, 0, and −5 V. A consistent shape of the nonlinearity is measured irrespective of this position, indicating that it is independent of the magnitude of deflection of the cantilever (i.e., it is not derived from a nonlinear relationship between cantilever deflection and laser spot movement).

The magnitude and importance of the nonlinearity become more glaring upon conversion of the abscissa to tip–sample distance, shown in Figure 3.12b for a single force curve (here, actual cantilever deflection is zeroed slightly below the zero of differential photodiode output). In this example, zero indentation was assumed, so as to produce a vertical, zero-mechanical-compliance data trend in contact. Such a vertical trend is indeed observed for cantilever deflections measured within approximately ±100 nm of zero output from the differential photodiode. The operator might limit the ultimate loading such that the positive deflection data stay within the approximately linear regime. But upon retraction, a large tip–sample adhesion, such as that exemplified in Figure 3.12b (for a fairly blunt tip), will unavoidably force operation beyond the linear response window. Upon analyzing these data with adhesive contact mechanics models as discussed in Chapter 6, inconsistencies between data trend curvature and analytical model are inevitable at the extremes of negative deflection. Similarly, if raising the ultimate load to high positive cantilever deflections in order to cause material yield (e.g., Figure 3.8), the force–distance "roll-over" signature of yield may be difficult to unambiguously identify (i.e., differentiated from the nonlinear photodiode response). The most obvious solution is to use a much stiffer cantilever such that forces reach yield points at values of cantilever deflection that remain well within the linear photodiode output regime (i.e., well below 100 nm of cantilever deflection).

3.3 PHYSICAL HYSTERESIS

In describing the preceding optical issues, laser spot positioning and differential photodiode nonlinearity, we have also encountered two interesting physical phenomena: the jump-to-contact and jump-from-contact. These are due to the presence of attractive forces between tip and sample. Absent attractive forces, force curves would resemble those in Figure 3.4: a simple onset of increasing repulsive force during approach and subsequent reduction to zero repulsion during retraction, *at the same value of* Z. This means no hysteresis over a closed cycle of force versus displacement, meaning no net work done, and no energy dissipation. Even if AFM is new to a user, this must seem impossible. We commonly experience the generation of sound upon touching objects, which necessarily means a transfer of energy;

we know intuitively that it requires net work to touch and then break away from anything that is sticky. How are such physical realities manifested in force curves?

It turns out that even in the absence of interfacial mobile species (e.g., adsorbed water that produces a capillary meniscus between tip and sample, treated in detail in Section 6.3), hysteresis is generally expected in AFM when using *soft cantilevers* as employed in contact mode. Why? Briefly, the presence of an attractive force gradient (dF/dD) exceeding the spring constant of the cantilever results in a condition of *force imbalance*; thus, the tip accelerates toward ("jumps to") or away ("jumps from") the sample. In physics terminology, the system is not quasistatic at all points during the approach–retract cycle. To see how this occurs, we consider in Figure 3.13 (top) a "2–8" force versus distance (D) relationship modeling ideal tip–sample interaction (integration over 6–12 Lennard-Jones interactions between a spherical tip and a half-space sample, each SiO_2, determining the Hamaker constant, but treated as mechanically incompliant, see Section 2.2) and its idealized

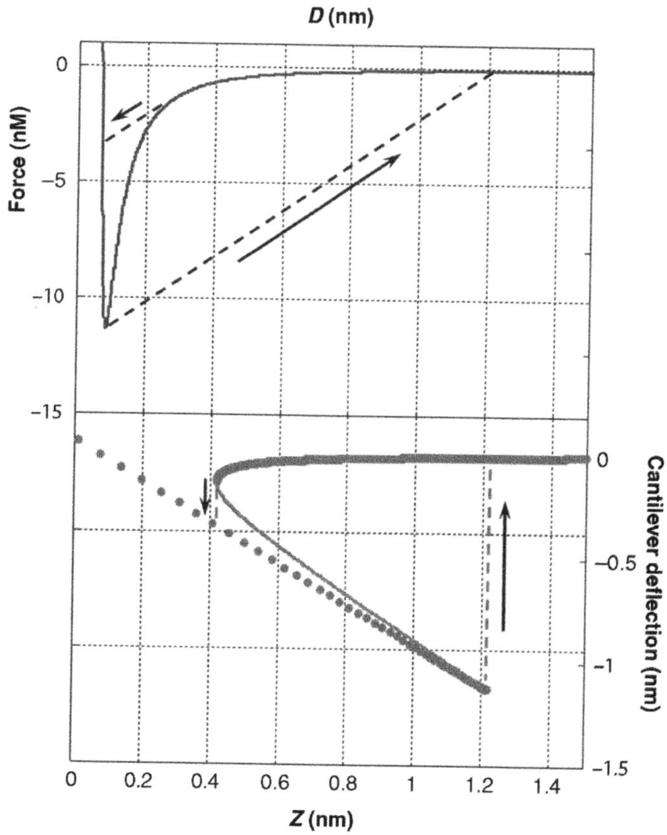

FIGURE 3.13 (Top) Theoretical 2–8 force–distance relationship. (Bottom) Correspondingly predicted deflection-Z force-curve measurement, circles representing measurable values and thin line representing bypassed values during jump-to-contact and jump-from-contact.

measurement, cantilever deflection versus Z displacement, as shown in Figure 3.13 (bottom) for the case of a $k = 10$ N/m cantilever. This idealized measurement was directly calculated from the force–distance relationship shown above it, in the same fashion that Figure 3.6b was calculated from Figure 3.6a (which conversely, you might say, treated the presence of mechanical compliance but no attractive forces). There are two locations where deflection versus Z reaches infinite slope reflecting jump instabilities; these locations result from the force gradient exceeding the value of k. Upon approach, the onset of attractive van der Waals forces within a nanometer from contact results in a downward deflecting cantilever. When the attractive force gradient is equal to 10 N/m (diagonal dashed line), the instability results in a jump-to-contact. Upon further approach of Z followed by retraction, the jump-from-contact occurs upon rounding the bottom of the force–distance curve. Both of these jumps bypass the part of the force–distance curve (top) that has positive slope greater than k. These points correspond to the vertical-line portion of the deflection–Z plot. Even in an idealized measurement, there is no way to "get onto" this part of the curve because it is bounded by locations of infinite slope at which jumps occur between other allowable points.

The short range of van der Waals forces[5] can make the role of force gradient more difficult to perceive than depicted in Figure 3.13 when using a typical soft contact mode cantilever of order $k = 1$ N/m or smaller; in most cases, upon close inspection, one will find that there is little or no measurable deflection of the cantilever just prior to the jump-to-contact, because it is a *capillary bridging* phenomenon causing the instability, as mentioned earlier, rather than the condition of force gradient exceeding k. Observing a dominant role for (noncapillary) force gradient usually requires the presence of long-range attractive *electrostatic* forces between charged sample and polarizable tip, (i.e., charge–dipole forces). Then a plot of deflection versus Z containing substantial jump-to-contact as well as jump-from-contact distances will result, purely due to long-range forces. An experimental example is shown in Figure 3.14a and b on a thick polypeptide film, dry gelatin, using a nominal $k = 0.6$ N/m cantilever (silicon including integrated tip) and calibrating S^{-1} (nm/V of raw deflection) to give a -1 contact slope. In Figure 3.14a, a bending down of the cantilever is observed over the course of several hundred nanometers of approach, terminated by a 60-nm jump-to-contact. In the deflection (s_Z) versus distance (D) plot of Figure 3.14b, this jump is seen to occur along a *slope of unity*, $ds_Z/dD = 1$, which is equivalent to $dF/dD = d(ks_Z)/dD = k$, as anticipated. (Two nonidealities are also observed in contact, the detector nonlinearity as in Figure 3.12 and some apparent contact hysteresis that is in fact Z-scanner hystersis, treated in Section 3.5.) Interestingly, one also observes an effect of charge transfer during contact (which brings tip and sample closer to the same electrostatic potential): the vanishing of attractive force immediately following the jump-from-contact and appearance of weak repulsion. Full retraction provides time for charge

[5] Note that van der Waals forces are often categorized as long range because the term short range is reserved for covalent and hydrogen bonding. But compared to electrostatic or magnetic forces, van der Waals forces are extremely of short range.

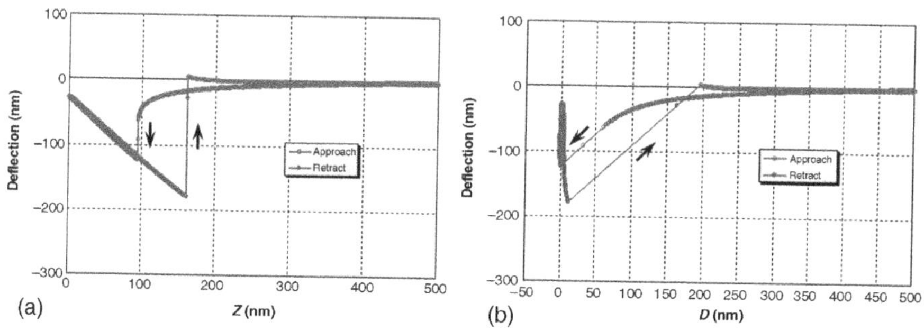

FIGURE 3.14 (a) Real force curve on a charged gelatin (dry) surface. (b) Corresponding data after converting from *Z*-scanner displacement to tip–sample distance abscissa.

dissipation from the tip such that upon approach, the strong charge–dipole attraction is again experienced in repeated force curves. (The shown force curve was acquired amidst several force-curve cycles before and after.)

As a final note, one should recognize that jump-to-contact or jump-from-contact also could involve the stretching of a very soft yet solid material, whether the sample itself or possible contamination of the AFM tip. It is possible for such a "nanofibril" to be softer than the cantilever; in the absence of liquidy contaminants, such a fibril could be the spring-like element (in series with the cantilever) whose spring constant is being exceeding by the attractive force gradient. (In Chapter 6, we discuss contact mechanics models appropriate to describe soft materials, including outward deformation, i.e., stretching, in the presence of attractive forces.) If there is a soft contaminant mass on the tip, one may find that immediately after the jump-to-contact, the force curve exhibits a near-horizontal data trend with further approach, indicating minimal resistance to the compression or collapse of this fibril. During retraction, one may observe that an otherwise normal-looking jump-from-contact, in fact, does not return to the zero force baseline; rather there remains a weak attractive force that gradually decreases in magnitude until the force signal continuously returns to zero at a significant distance (e.g., tens of nanometers) beyond the principal jump-from-contact. The more one uses AFM on soft and/or mobile materials, the more variations of these kinds of observations one may catalogue. One quickly learns that the "textbook" concept—van der Waals attractive force gradient exceeding spring constant of cantilever—is the exception rather than the rule.

3.4 OPTICAL ARTIFACTS

The nonidealities discussed in Section 3.2, recapped as follows, derive from simple geometric and device realities that are pretty obvious upon examination:

1. tip–sample distance is affected by both tip and *Z*-scanner displacements;
2. tip displacement cannot be purely vertical due to cantilever attachment;

3. the magnitude of laser spot displacment at the photodiode depends on the inclination of the cantilever, which varies along its length, and on changes in refractive index at media interfaces;
4. the intensity varies across the laser spot at the photodiode, resulting in nonlinear response that can be substantial if the spot is displaced to near the full measurable range.

Other nonidealities in force curves—again in the vertical scale, the measurement of force—derive from a more esoteric *optical interferometry* artifact, certainly found in first-generation instrument designs. It results first from the fact that the laser light directed at the cantilever is not completely reflected from the cantilever. The laser intensity does not abruptly drop to zero at the visual edge of the laser spot. Instead, it extends beyond the apparent edge of the spot such that this light "spills off" the cantilever and reflects off the sample surface. Second, the cantilever can be so thin (~1 μm) that a significant portion of the laser light travels right through the cantilever and reflects off the sample surface. (Indeed, in some cases, one can literally see sample surface features illuminated by the laser, through the cantilever, using an overhead light microscope.) In aggregate, this laser light reflected from the sample can partially interfere with the laser light incident upon and reflected from the cantilever. As a result, *the distribution of laser intensity at the photodiode can vary as the distance between cantilever and sample varies due to causes other than force*. Most commonly, this results in a wavy and tilted baseline in force curves in the noncontact regime, easily visible if ramping Z many hundreds of nanometers above a reflective sample, as in Figure 3.15 (sample silicon). Here, we examine raw deflection in volts versus Z displacement during approach, over a range of 800 nm, resulting in three "waves" in the noncontact regime separated by a distance that relates to the wavelength of the laser light and geometric parameters of the optical path. The inset depicts the essence of the problem, scattered laser light from sample surface. The amplitude of the wavy baseline is roughly 50 mV, translating to a variation of the "zero" of cantilever deflection of about 1.5 nm.

We will see in Chapter 4 that for ordinary, in-air imaging, the potential variation of contact force due to this artifact is usually small. But if analyzing force curves to extract quantitative information about surface forces as considered in Chapter 2, the baseline artifact can, in fact, be problematic. One strategy to reduce this effect is to coat cantilevers with reflective metals such as gold or aluminum. (This changes the mechanical properties of the cantilever and can be detrimental to performance in variable environments, temperature and humidity). The reflective coating reduces the amplitude of the "waves" in the noncontact regime. But in fact the data in Figure 3.15 were acquired with a gold-coated silicon nitride cantilever (i.e., the waves are already reduced, compared to uncoated cantilevers). Another strategy used in newer commercial AFM systems ("second generation") is a low-coherence laser; this greatly reduces the interference phenomenon. The effect will also be reduced on nonreflective samples compared to highly reflective ones. Another published strategy is to modulate the laser current at high frequency [6]. In Chapter 7,

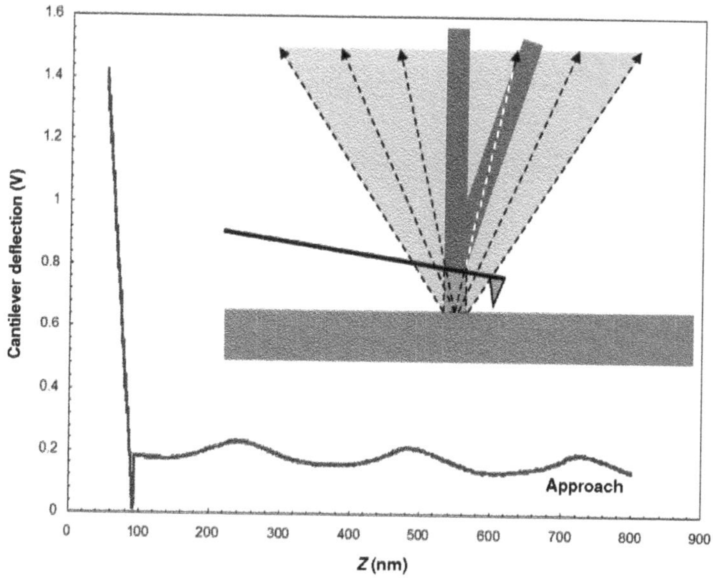

FIGURE 3.15 Force curve on approach including wavy baseline (zero) of cantilever deflection, due to interference of incident and reflected laser light as depicted in inset.

we draw attention to another manifestation of the interferometry artifact, in the background of lateral force images.

There are also practical, procedural strategies one can use to lessen the optical artifact described above. In the process, one gains better insight into the mechanism of the artifact. It is easy to verify that both the choice of laser position on the cantilever and the relative tilt between sample and cantilever affect the amplitude of interference fringes and the tilt of the baseline. In Figure 3.16, long-range force-curve baselines (acquired with a special closed-loop Z scanner, discussed in Section 3.5) for two laser spot positions are compared, one near the tip end of the cantilever (bottom) and the other roughly half way up the cantilever length (top). The curves are vertically displaced for clarity. Most notably, the baseline tilt is less severe for the halfway laser spot position. One also notes that the shape of the waves look different. Some of this is found to be spurious and related to the presence of two laser modes, meaning two "wavelets," which are out of phase and not exactly the same wavelength, and one of which evolves with passing time on the scale of minutes (possibly due to *mode competition*, the exchange of power between two longitudinal modes of a laser [7]). So a "pure," single sine-wave baseline is not always seen, and in some cases, two distinct waves can be seen superimposed and phase shifted, evolving in time. On average, the peak-to-valley amplitude tends to scale with the overall slope of the baseline. Together, this suggests that placing the laser spot nearly halfway back from the tip end of the cantilever can be a useful strategy for lessening the variable baseline effects. Recall in Section 3.2 that the sensitivity of the measurement of cantilever deflection was nearly as good for this

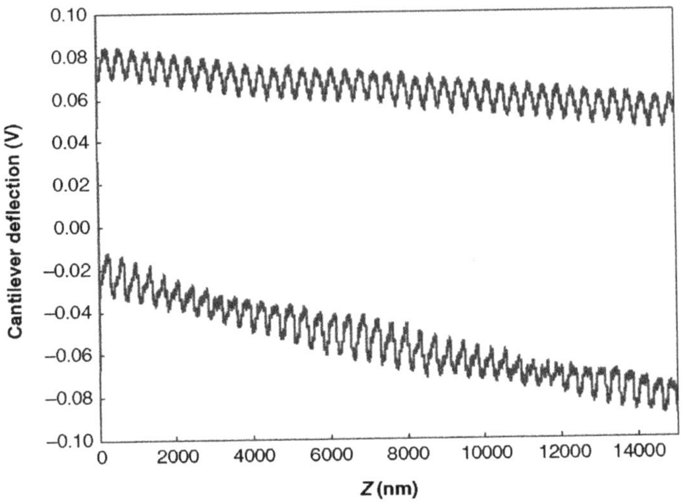

FIGURE 3.16 Effect of laser spot positioning on interference artifact within first 15 μm beyond contact, as seen in the baseline of deflection measurement.

laser spot position as compared to a position near the tip end. Indeed, the sensitivities S, determined from the contact slopes for the two experimental configurations in Figure 3.15 (not shown), differed by only 3%.

Figure 3.17 compares force-curve baselines (approach data) for different angles between the incident laser beam and the sample normal. In addition to parallel (as in Figure 3.2), we also examine angles ranging of approximately $\pm 10^\circ$. In all cases, the angle between incident laser beam and cantilever normal is fixed at $\approx 12^\circ$

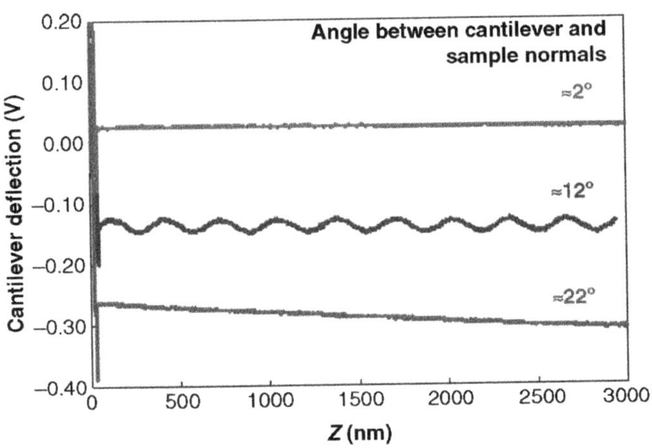

FIGURE 3.17 Effect on deflection baseline of variable cantilever and laser orientation with respect to sample surface.

(by instrument design). The corresponding angles between cantilever and sample were approximately 2°, 12°, and 22°, as indicated in the figure. This comparison was performed by tilting the head of the AFM via two independently adjustable threaded legs of a tri-leg assembly on which the head rests (a truly first-generation design, later scanner housings having replaced three degrees of freedom with one purely vertical degree of freedom, for convenience in coarse approach). Here, we see that the wavy optical interference artifact essentially vanishes for a significant tilt between incident laser beam and sample normal. Second, we see that tilting in such a way as to *reduce* the angle between cantilever and sample results in a quite horizontal baseline, whereas tilting in the opposite direction, such that the cantilever tilt relative to sample is *increased*, results in a severely tilted baseline. In addition to operational utility, these findings indicate mechanism: the dominant laser interference is between incident and sample-reflected laser light, rather than interference between cantilever- and sample-reflected laser light at the photodiode. By tilting the sample-reflected light away from the incident laser trajectory, the wavy interference baseline is precluded. Note, however, that this strategy pertains to mirror-like surfaces, whereas laser light scattered from very rough surfaces may yield much lesser interference with incident laser light, while scattering more light forward to the photodiode.

A final note of caution regarding the baseline (i.e., zero) of force curves is the possibility of substantial vertical drift due to thermal warm-up. A common observation with some commercial AFM systems is a gradual downward drift of force baseline with passing time, significant in the first hour since a cold start (at least tens of nanometers equivalent, in the measured cantilever deflection) but often not completely stable after even 2 h of warm-up. In force curves, this baseline is captured for each measurement cycle, thus thermal drift can be accounted during data processing. The larger concern is methods that are affected by the zero of force but for which the baseline is not routinely captured, such as *load-sensitive* contact-mode imaging—especially topography on moderately soft materials (such that contact mode is possible but tip measurably indents) or frictional imaging on any material, the subject of Chapters 4 and 7.

3.5 Z SCANNER/SENSOR HARDWARE: NONIDEALITIES

Next we consider the realities of components (ii) and (iii) listed in Section 3.1: the Z-scanner displacement actuator and the measurement of this displacement. The displacement typically utilizes the inverse piezoelectric effect:[6] an applied Z bias distorts a piezoelectric crystal, often in the shape of a tube, causing it to extend or retract. Given applied voltage ramp ranges of order ± 200 V, a Z piezoscanner will typically displace over a range on the order of ± 3000 nm or ~ 15 nm/V. Thus, millivolt-regime control of applied bias corresponds to *sub-Angstrom* control of Z displacement. The simplest and most commonly employed method, to measure Z

[6] Some scanners use other mechanisms but usually with coarser precision of Z displacement.

displacement, is to just invoke such a linear coefficient (i.e., 15 nm/V) to convert from the measured *set* voltage to a length scale, meaning an *assumption* of linear response (indeed a very important assumption). This coefficient is determined by measuring the Z-scanner displacement required to traverse a step of known height on a calibration grating. There usually is no independent *read* of the Z-scanner displacement, that is, no sensor gauging it, unless using a special type of "second-generation" scanner equipped with such a sensor (to be discussed).

Notwithstanding the preceding statements, piezoelectric scanners are in fact known to be *intrinsically nonlinear* devices and moreover exhibit a wide variety of problematic behaviors, namely, *hysteresis* (irreversibility), *creep* (time dependence), *cross coupling* (between axes), and *aging*. In Chapter 4, we discuss all of these nonidealities in the context of 3D scanning (topographic imaging). Creep and cross coupling are especially relevant to multidimensional imaging. Nonlinearity and hysteresis are highly relevant to several modes of operation and indeed are the most glaring issues in force-Z measurements, thus are the focus of this section (at a phenomenological level). Creep in Z during force–Z measurements is mainly noticed as a leftward or rightward drift of the entire data plot during the repeated collection of force curves (although this may also result from sample drift in Z, e.g., if adhesive tape is relaxing, as discussed further in Section 4.2.7). It can be minimized in by performing several cycles over a particular Z ramp size of interest and at a particular Z offset (just off the surface to avoid repeated tip–sample contact, if desired) prior to taking the measurements of interest. (Slight rate dependence in Z-scanner displacement, presumably due to creep, will be exemplified below.) We note that in most "first-generation" AFM designs, which comprise the majority of systems in use to this author's knowledge, there has been no attempt to engineer around these problems of Z nonlinearity, hysteresis, and creep, just as there has been no attempt to engineer around the force measurement nonidealities of Sections 3.2 and 3.4. Only the more recently available closed-loop Z scanners (here considered "second generation") can largely remove the effects of nonlinearity, hysteresis, and creep from force-Z measurements, but not without introducing other drawbacks (in resolution and feedback-circuit oscillations).

Figure 3.18 demonstrates nonlinearity and hysteresis in a plot of actual Z displacement versus a nominal Z-position ramp. Here, the actual Z displacement was measured using a special ("second-generation") scanner equipped with inductive sensors for measuring Z displacement (as can be employed in *closed-loop* operation). The hysteresis loop is for a 1500-nm nominal Z ramp size, which if linear would be along the diagonal dotted line. The nominal Z-position ramp values are *incorrect* because they convert a voltage ramp to a Z displacement in nanometers via some linear coefficient c_1 (nm/V), whereas at least a quadratic correction, meaning potentially a third term in a power series expansion

$$c_0 + c_1 V + c_2 V^2 \tag{3.8}$$

where c_0, c_1, and c_2 are constant coefficients, would be invoked to account for the nonlinear relationship between actual Z movement and Z ramping voltage;

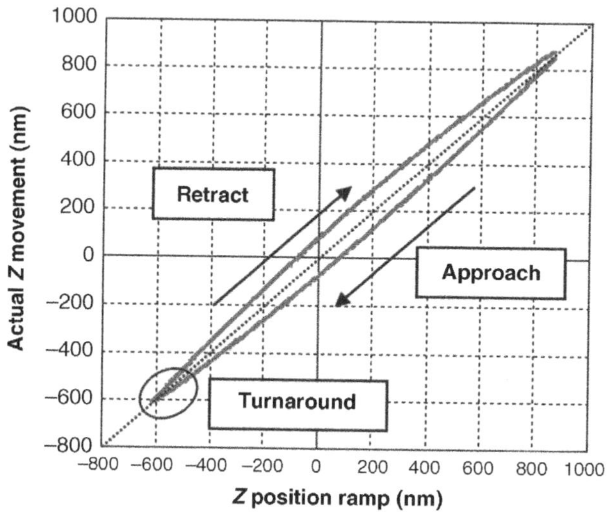

FIGURE 3.18 Nonlinearity and hysteresis of actual *Z*-scanner displacement for a nominal *Z* position ramp of 1.5 μm.

moreover, different corrections for ramp up and ramp down (at least opposite in sign) to account for the hysteresis of the scanner. Also note that the slope of actual movement is steeper upon initial retraction compared to the final approach just prior to turnaround. This means that the *Z* data points will be more greatly spaced immediately upon retraction than at the end of approach, with calibration implications. We will return to this particular issue in terms of its implications for real force curves in Section 3.6.

Most AFM force-curve analyses involve much smaller *Z* ranges than shown in Figure 3.18. In these cases, one finds that the hysteresis loop is considerably smaller, the phenomenon being highly path dependent. This is demonstrated in Figure 3.19a, an overlay of four *Z* ramping cases, three of the same 500-nm size but offset by constant increments of 500 nm and a fourth 1000-nm in size. The vertical offsets were not controlled by any setting made by the operator; they result from the preceding scanning history. Generally, these offsets are not a problem, as the user simply offsets *Z* by some arbitrary amount to produce a sought maximum force at the ultimate approach of tip and sample. What is measurable in Figure 3.19a is both a consistency of (average) slope for the 500-nm scans (i.e., good performance) and an inconsistency of slope for 500-nm compared to 1000-nm scans, specifically a discrepancy of 9%. Thus, quantitative analyses at different surface locations or between different samples, whether mechanical properties or long-range forces, are best compared for a given *Z* ramp size, not between data sets acquired at different *Z* ramp sizes.

Another variable one would prefer to keep constant for quantitative comparison is ramp rate. In most cases, one necessarily holds this variable constant for physical reasons, to avoid introducing different time responses. But in other cases, rate is

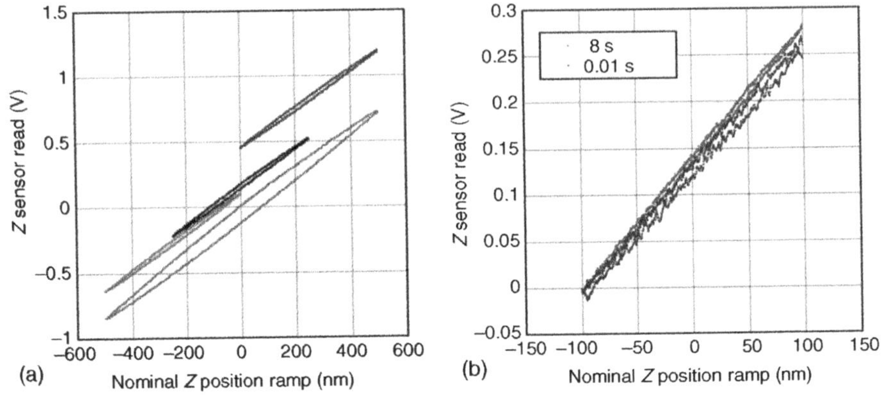

FIGURE 3.19 Dependence of Z-scan hysteresis and offset on (a) the size of the Z ramp and previous ramping history, and (b) Z-ramp rate.

intentionally varied to examine fundamentally dynamic material or molecular responses (i.e., kinetics), such as *viscoelastic* properties of polymers or *activation energies* for biological adhesion. One needs to examine, then, whether the Z-scanner response varies over the range of rates of interest, for a fixed Z ramp size. Figure 3.19b overlays 200-nm ramp responses of the Z scanner for ramp times spanning nearly three orders of magnitude: 0.01 s and 8 s. The magnitude of Z hysteresis is somewhat larger for the fast ramp cycle case (0.01 s), but perhaps more importantly the average slope of response is about 6% lower for this case. This is a pretty small variation in piezoscanner response over a wide range of ramp rates. (One also notices more noise in the faster measurements, as the integration time per data point is shorter and thus fluctuations less fully averaged out). Whether it is significant to an application will depend on exactly what is being quantitatively extracted from the data. This small difference could measurably affect, for example, fitting parameter values in single-chain molecule extension models (Section 2.5.5).

Lacking a special scanner equipped with a Z displacement sensor as used in Figures 3.18 and 3.19, one can exploit the optical interference artifact (Section 3.4) to reveal and indeed quantify the nonlinearity and hysteresis of the Z scanner provided that the Z ramp is large enough to observe interference fringes. In Figure 3.20a, equivalent force data (changes of deflection signal due to the optical artifact, above a highly reflective graphite surface) versus Z completely out of contact are plotted, overlaying the ramp-up or approach data with the ramp-down or retract data. It is obvious that the interferometric waves do not crest at the same values of Z during the approach and retraction strokes of the piezoscanner. In Figure 3.20b, the crest and trough positions are plotted versus the Z ramp. The quadratic deviation of these positions above or below the reference dashed line, the linear trend, is apparent, as is the opposite sign of this quadratic deviation for retraction compared to approach data. This strong hysteresis between approach and retract, as well as its quadratic nature, are more obvious in Figure 3.20c, where the intercrest and intertrough

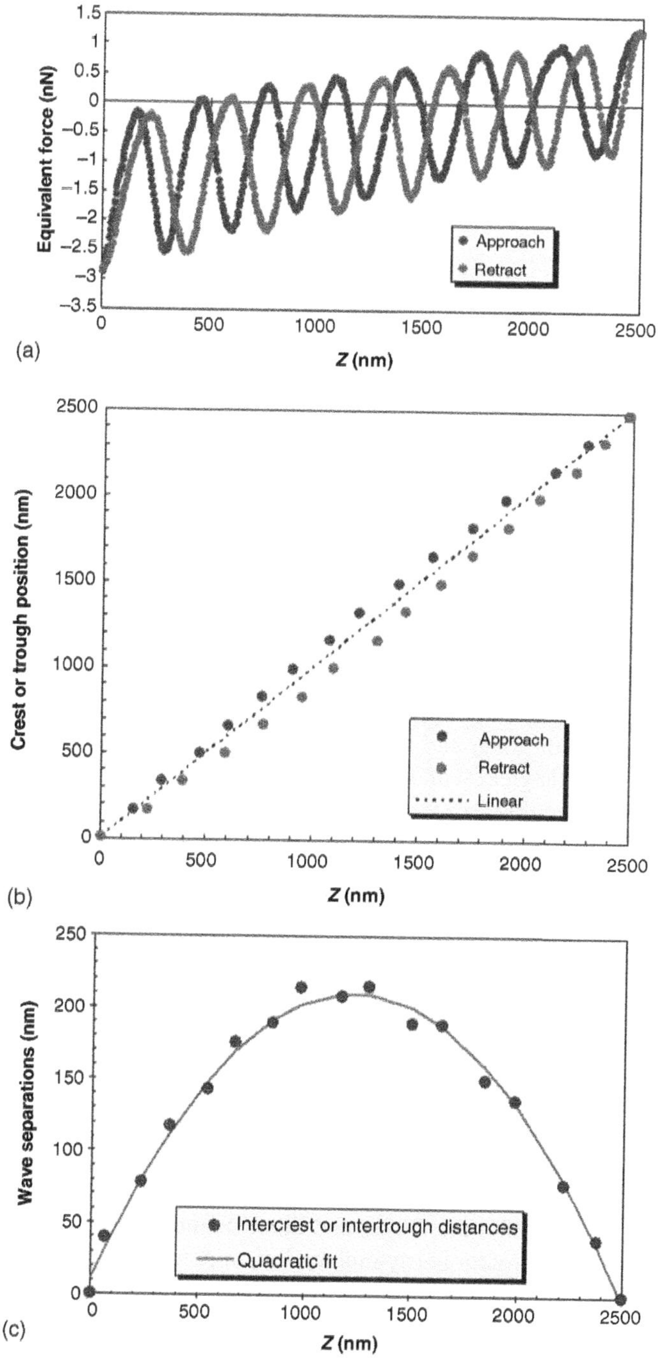

FIGURE 3.20 (a) Hysteresis of Z ramp seen in optical interferometric baseline artifact. (b) Baseline peak and valley positions as a function of nominal Z-ramp position. (c) Quadratic relationship between crest and trough positions in baseline.

distances are plotted versus Z (its value halfway between adjacent crests or troughs), removing the linear component. Here, the overlay of a mathematical curve fit of the form $y = y_0 + a(x - x_0)^2$ reveals the purely quadratic nature of this hysteretic nonlinearity, at least to within the noise of this analysis.

3.6 ADDITIONAL FORCE-CURVE ANALYSIS EXAMPLES

It is important to appreciate the extent to which the nonidealities and phenomena of Sections 3.2–3.5 impact real force-curve applications. To this end, we will consider in this section further examples of force-curve applications and in the process, discuss additional questions related to data processing as well as data interpretation. The particular examples also highlight some interesting behaviors in nanoscale interfacial science.

3.6.1 Glassy Polymer, Rigid Cantilever

Our first example involves quantitative analysis of elastic modulus of the polymer polystyrene (as considered theoretically in Figure 3.6). This glassy polymer is considerably softer than the calibration sample (for determining S^{-1}, here quartz) but not nearly as soft as, say, rubbery polymers or gels (or cells). To obtain useful mechanical responses on such systems—substantial indentation with minimal sliding of tip across the surface—one typically uses a rigid cantilever, here 40 N/m nominal (silicon with integrated silicon tip). At first sight, one may be concerned about reaching plastic yield with such a rigid cantilever. Actually, this is a common measurement methodology, to force a *plastic indent* during the upward force ramp then measure the force–distance signature of purely *elastic recovery* during retraction. One can subsequently image and measure the size of the permanent (or at least residual) indentation to help quantify the area of the contact at maximum force, together with the *slope* of force versus tip–sample distance upon retraction, S_{max}, to evaluate the modulus via the equation

$$K = \frac{\sqrt{\pi}}{2} \frac{S_{max}}{\sqrt{A}} \tag{3.9}$$

with K being the "reduced modulus" as given by Equation 2.12. This slope is greater in magnitude than during approach (wherein plastic yield results in greater displacement into the material) as depicted in Figure 3.21. This slope-of-elastic-recovery analysis is called the Oliver and Pharr method [8]. It formally applies to any tip with an axis of rotational symmetry, but, in practice, is used *even more generally* with apparent success [9]. The most rigorous application of this method involves making several indents of increasing depth to determine an "area function" of the tip (cross section versus distance from tip end), which is usually one of several "standard" shapes (e.g., including three- or four-sided pyramid). Knowing this shape provides a basis for numerically relating contact slope, area function, and

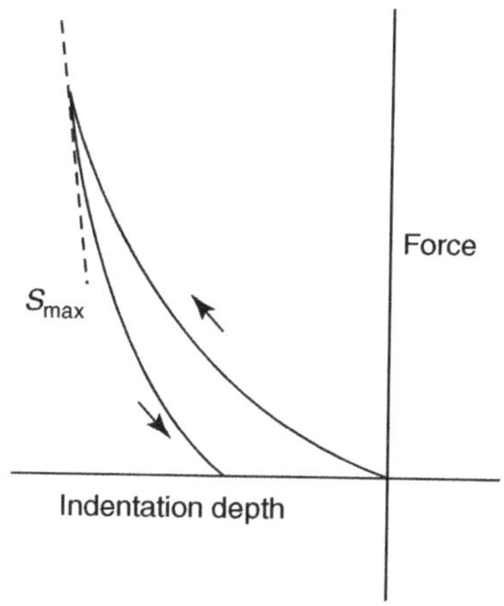

FIGURE 3.21 Schematic depiction of a mechanically hysteretic indentation cycle and designation of maximum slope upon retraction, S_{max}, used in the Oliver and Pharr method.

modulus. Thus, one need not have an analytical expression for the complete load–displacement relationship, such as the Hertzian expression for a spherical indenter.

As a first step in this analysis, we examine a calibration force curve on quartz, cognizant of the nonidealities and artifacts discussed in Sections 3.2, 3.4, and 3.5. In Figure 3.22a, we have calibrated the raw deflection signal from volts to nanometers by using a value of $S^{-1} = 10.9$ nm/V so as to produce a slope of -1 (dotted line) when plotted as deflection (nm) versus Z (nm). This of course assumes that we are calibrating the vertical scale to a precalibrated *linear* horizontal scale, which we know from Section 3.5 is not exactly the case. Most notably, we see appreciable hysteresis in the contact portion of the force curve. We do not expect that plastic indentation to elastic recovery is the source of hysteresis on this rigid calibration sample, and indeed we note that the sign of the hysteresis is the opposite of that in Figure 3.21. Upon plotting deflection versus tip–sample distance D (and re-zeroing both scales) in Figure 3.22b, we observe a hysteresis loop shape that indeed does not look like plastic–elastic hysteresis. Recalling from Section 3.5, the sense of Z-scanner hysteresis—greater displacement during retraction per unit change of Z voltage—explains the observation that the retraction data corresponds to *greater* apparent distance. We also note that the range of deflection exceeds 90 nm meaning $90/10.9 = 8.2$ V. This is well into the nonlinear response regime of the differential photodiode (Figure 3.12b). The quadratic scanner hysteresis and the cubic photodiode response together generate the measured trend of the data in contact in Figure 3.22b. To use these operational parameters for calibration is clearly difficult because the slope of the initial retraction data on our material of interest (polystyrene) presumably will be affected by the determination of S^{-1}, and, moreover, the overall shape of the data acquired on polystyrene will be affected by the nonlinearities in the scanner and the photodetector. One would prefer to reduce the Z stroke of the scanner, recalling the comparative hysteresis in Figure 3.19a, and also

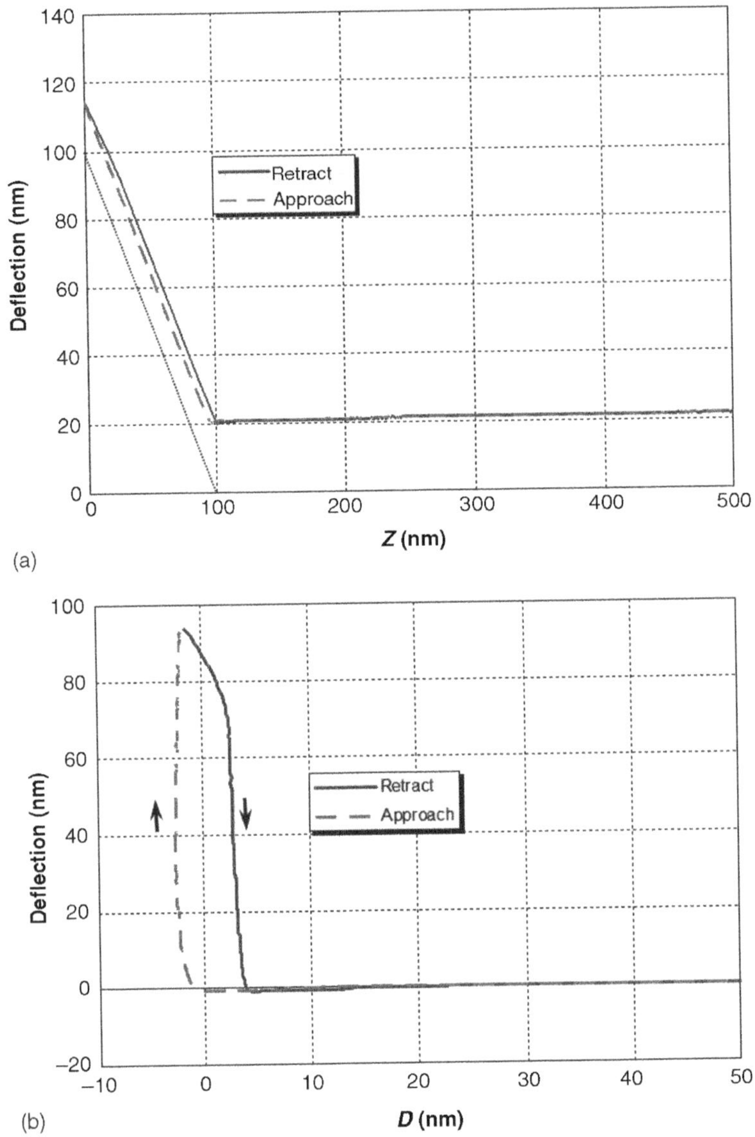

FIGURE 3.22 (a) Full force curve of silicon tip on quartz calibration sample, following calibration of the deflection scale (from volts to nanometers) to achieve a slope of unity magnitude in contact. (b) Same data as in (a) after conversion of the horizontal scale to tip–sample distance, zeroing the scale at the midpoint of the scanner hysteresis loop.

reduce the maximum cantilever deflection attained such that the photodetector nonlinearity does not manifest itself.

In Figure 3.23a are overlaid full-cycle force curves for the quartz calibration sample and polystyrene using a Z stroke of 100 nm instead of the 500 nm used in Figure 3.21, while keeping the maximum deflection below 1 V of differential photodiode output signal (i.e., by limiting the ultimate Z approach point or, equivalently, the Z scan start point for a given Z ramp). The deflection scale was again calibrated from volts photodiode output to nanometers (using $S^{-1} = 11.7$ nm/V) so as to produce unity magnitude slope on quartz. We immediately notice a lack of hysteresis in the quartz force curve and hysteresis in the polystyrene curve that is consistent with the plastic–elastic recovery behavior depicted in Figure 3.20. We also note that on approaching polystyrene, there is a sudden decrease of deflection indicating measurable attractive forces and a gradual upturn to net repulsive force upon further approach, whereas on quartz, we see no such behavior. Indeed, careful examination of the baseline on quartz over large Z strokes revealed that these variations in the baseline were primarily due to the optical artifact discussed in Section 3.4. Figure 3.23b contains the retraction data following conversion to force–distance form. The baseline of force (computed via deflection times $k = 40$ N/m) has been zeroed at the bottommost value in each curve. The calibration procedure results in a vertical trend for the data on quartz in contact, required by our basic assumption of zero indentation on the calibration sample. We note that given this calibration, the tip indents 5 nm into the polystyrene at 300 nN force.

For comparison, Figure 3.23c displays the predictions of the Hertzian contact mechanics model for a silicon tip with radius of curvature $R = 10$ nm. We find that at 300 nN force, this model predicts about 8 nm of indentation by the tip instead of 5 nm. But we also note that the model predicts finite penetration of the tip into quartz for this force range. Thus, a more precise calibration of S^{-1} (roughly 10.5 nm/V) would result in a bit over 1 nm of indentation into quartz at 300 nN force; thereby, the same calibration would produce a slightly larger indentation into polystyrene of about 6 nm. There remains a $\approx 25\%$ discrepancy between theory and experiment. But we also note that the *shape* of the data trend on polystyrene is not quite that of the Hertzian model: the increase of force with indentation is more gradual at first, then steeper, than the Hertzian prediction. We will see in Chapter 6 that a treatment of attractive forces can be provided by more complex contact mechanics models. The JKR model, in particular, predicts a force–distance trend closer to that in Figure 3.23b. We will also discuss the *viscoelastic* (time dependent) response of polymers that may in general modify the shape of the curve (and the fact that the modulus of a given polymer also depends on molecular weight as well as the preparation of the particular solid of interest, e.g., film casting from solvent or extrusion). Another issue is the shape of the tip; given a sufficiently large indentation, the tip cannot be regarded as a hemisphere, rather, the steepness of the side walls becomes important. (We look more quantitatively at this effect in example #3 of this section.) Some analysts invoke a conical tip to model its larger shape, for which Hertzian-type contact mechanics predicts a force increasing as indentation depth squared rather than the 1.5 power modeled in Figure 3.23c for a spherical indentor

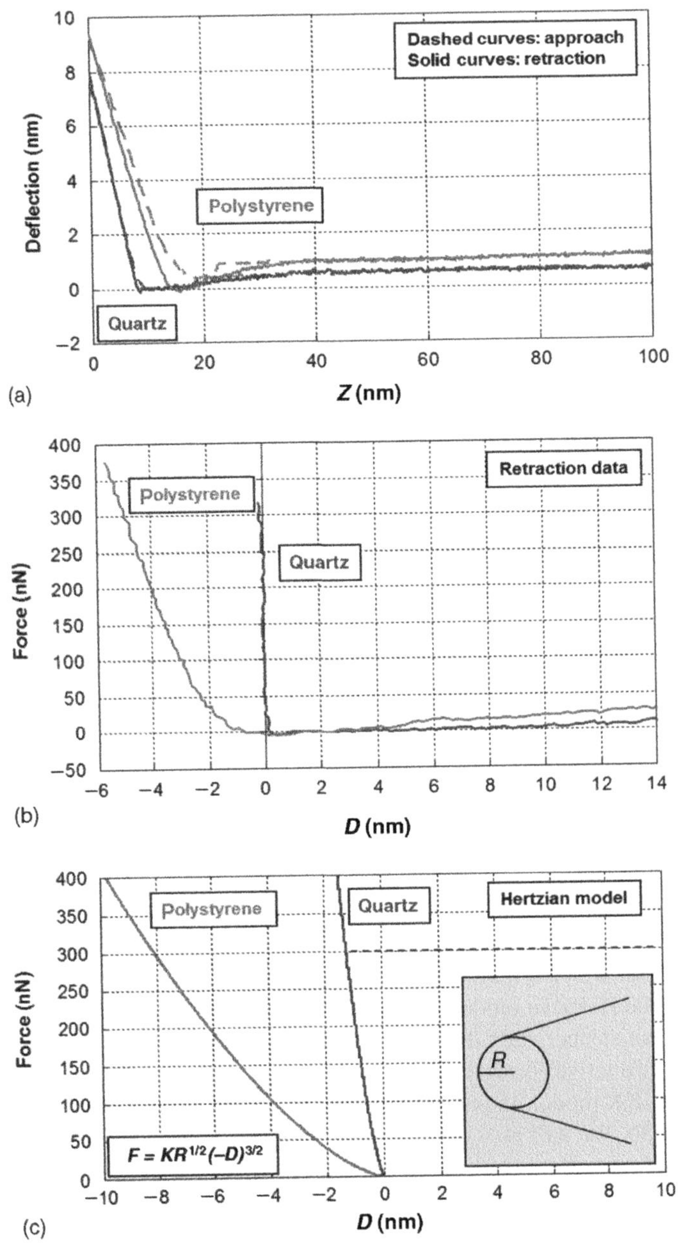

FIGURE 3.23 (a) Force curves on quartz and polystyrene employing rigid cantilever and calibrated to unity slope magnitude on quartz. (b) Same data plotted as force versus tip–sample distance, zeroing force at lowest point of each data set. (c) Theoretical force–distance relationship within Hertzian spherical contact mechanics model.

(see Eq. 2). This is consistent with the onset of a steeper slope at greater indentation depths, but in the case of Figure 3.23b, this indentation is perhaps not deep enough to require a conical analysis.

Remaining questions include the exact shape of the tip and thus the contact zone on the nanoscale, and uncertainties in the tip radius of curvature and cantilever spring constant as specified by the manufacture. In addition to imaging the shape of the residual indentation (as mentioned earlier in the context of the Oliver-Pharr method) or scratch (e.g., Figure 3.8a), other approaches to determine tip shape include imaging a sample containing vertical spikes of very small radius of curvature or terraces with vertical edges at least ~100 nm tall, or imaging the tip with scanning or transmission electron microscopes. The calibration of cantilever spring constant is a question that has spurred a vast literature. Many of the methods involve measuring phenomena such as the thermal noise of the cantilever or special modes of operation that we are not yet positioned to discuss within our introduction of physical and instrumentation concepts. Thus, we forego a broader treatment of these methods until Appendix A1. But we are positioned to discuss in Section 3.7 one very simple and intuitive method: acquiring force curves atop precalibrated cantilevers.

The preceding issues pertain to the absolute measurement of modulus. It is worth stating that *relative* measurements can be even more promising. The implicit error bars in the preceding discussion can be effectively reduced if only comparing the relative moduli of materials. Within the signal or noise of the data in Figure 3.23, one expects to be able to distinguish between moduli of glassy polymers that differ by only 10% even though the determination of *absolute* modulus to this level of precision may be difficult to impossible. Using the same tip/cantilever in a single experimental session, with a single setup to position the laser and photodiode, and consistent choice of the Z ramp size and ramp rate, leaves little to vary except the property of interest, modulus. (This assumes of course that other material properties such as surface energy are similar, which of course may not be true. In this case, more complex adhesive contact mechanics may be needed as discussed in Chapter 6.)

3.6.2 Gels, Soft Cantilever

Our second example, Figure 3.24, examines a case of extremely low modulus and partially ionizable material: the aqueous, biologically derived gelatin (under water immersion, i.e., in the gel phase), which as stated earlier is a polypeptide derived from the protein collagen. (This gelatin was treated with formaldehyde to produce a network of covalent cross-links in addition to the physical crosslinks that form due to the thermodynamic drive to "renature" collagen-fold triple helices, i.e., protein folding.) Using a soft Si_3N_4 cantilever, nominally $k = 0.6$ N/m, with integrated Si_3N_4 tip, is sufficient to produce substantial indentation totaling hundreds of nanometers at forces below 20 nN on the several-micrometer-thick gelatin coating (on mylar) of Figure 3.24a. Clearly, in this case, the tip has indented far beyond the rounded end of the tip ($R \approx 50$ nm) such that the common spherical-tip form of the Hertzian model cannot possibly apply to all of the data. Here, it is useful to examine

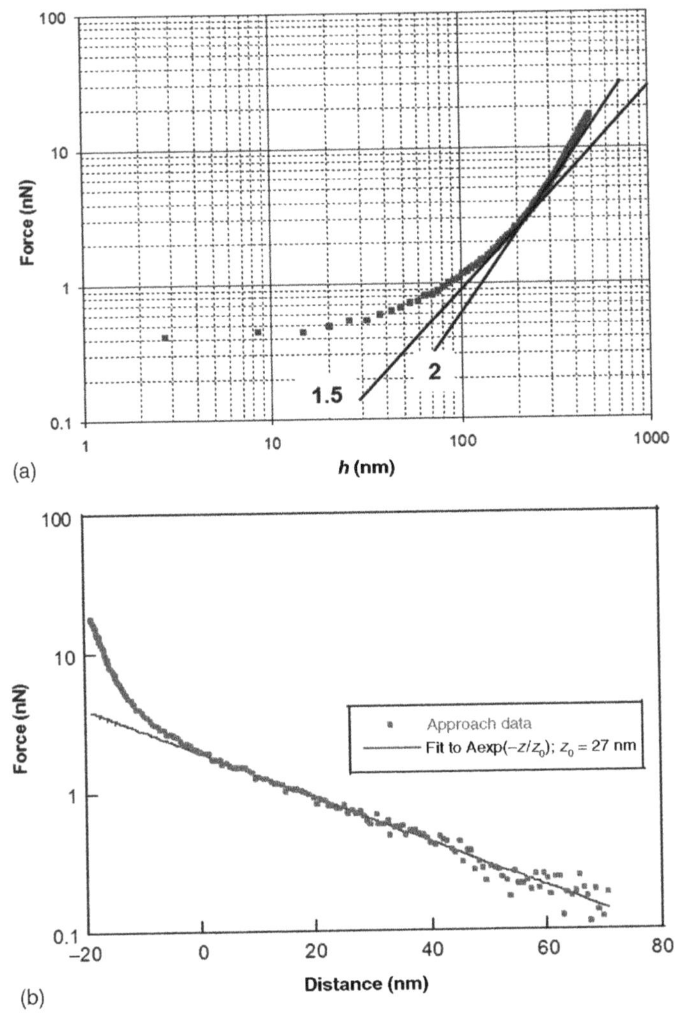

FIGURE 3.24 (a) Force versus indentation data on a thick cross-linked gelatin gel in water, presented in log–log form with overlays of power-law trends of exponent 1.5 and 2. (b) Force versus distance on thin gelatin film on mica in water, and overlaid exponential fit used to estimate the zero of distance.

the data on a log–log plot of force versus indentation h ($= -D$, i.e., reversing the usual sense of the horizontal scale, with a zero to be discussed) and overlay straight lines corresponding to relevant power laws. (The deflection calibration was made to a silicon nitride film surface using the same mounted tip in the same experimental session, also immersed in water.) We know that the spherical-tip form of the Hertzian model predicts force increasing as the 1.5 power of indentation. The other Hertzian form considered here is the conical tip [10], where force increases with indentation to the second power, according to

$$F = (2/\pi)\, Kh^2 \tan\beta \tag{3.10}$$

where β is the cone-included half-angle and K and h are defined as in Section 2.3. We note that at forces above roughly 4 nN, force increases steeper than the 1.5 power trend and is nearly consistent with the conical trend. This force corresponds to an indentation more than 200 nm deep and thus well beyond the spherical end of the tip, consistent with the steeper power law form. Utilizing Equation 3.10 and the approximate range of force and indentation consistent with Figure 3.24a, for a "cone" of half-angle 35°, gives $K \sim 100$ kPa. The tip is of course not a cone, rather, in this case, a four-sided pyramid. But this shape has been modeled with the conical form in the nanoindentation literature with reasonable success. (One mathematically transforms the actual pyramidal geometry into an equivalent cone, by finding a cone angle that gives the same area-to-depth ratio as the pyramidal shape [10].) Thus, we are not surprised to see near agreement with the power-2 conical scaling form in Figure 3.24a.

At the other extreme of Figure 3.24a, low force, we observe a growth of force with increasing indentation that is much shallower than that of the spherical tip power law (1.5). A polypeptide such as gelatin is well known to bear a net charge dependent on pH (here, expected to be a positive charge). This should result in an exponentially repulsive (DLVO) force versus distance trend with a neutral tip, as discussed in Section 2.5.1. In addition, gelatin chains with conformational freedom likely extend into the aqueous environment to produce a sterically repulsive force at the tip–sample interface at small h. Together, these long-range and short-range effects, and substantial tip penetration, produce a plethora of distance-dependent forces that meld into one another with increasing displacement of tip toward sample. In particular, this example illustrates how the identification of the *zero* of distance can be essentially impossible in such cases of gels (and other kPa-regime modulus, solvent-laden objects with interfacial charging, and/or steric forces). In our case, one can in principle make a semi-log plot of force versus distance and use the departure from the exponential (DLVO) form upon approach to estimate the onset of contact and thus the zero of distance. But this "contact" may represent the outermost chains of free polypeptide sterically interacting with the tip, rather than a "surface" of a medium with a well-defined modulus. In Figure 3.24b, we consider such a semi-log plot for the case of a *thin* gelatin film on a mica substrate. Here, a departure from the exponential trend is easily identified and used to define the "zero" of distance. This is possible because the contact mechanics of tip/gel is not that of a sphere against a very soft, infinite half space; rather the contact is stiffened by the presence of the underlying substrate a few nanometers below the gel surface. Thus, the DLVO interaction does not meld gradually into steric/contact types of interactions, rather is distinctly shallower in its distance dependence. Together, the examples of Figure 3.24 are intended to underscore that the zero of distance, the "touching point," is generally not easy to determine, especially on immersed polymeric systems.

3.6.3 Molecular-Chain Bridging Adhesion

Most quantitative force-curve analyses utilize *Z* ramp sizes that are small compared to the full range of the *Z* scanner. Exceptions include very long-range electrostatic forces in the case of highly charged surfaces in air, or molecular bridging forces for high-molecular weight chain molecules that may comprise a stretched fibril between tip and sample. In this latter case, the *magnitude* of force may be small though its *range* is very large. An example is Figure 3.25, which further demonstrates the effects of both *Z*-scanner hysteresis (Section 3.5) and the laser interferometry artifact (Section 3.4). We examine a raw force curve acquired during a spatially mapped indentation experiment with a rigid cantilever (which produced a grid of holes and surface rupturing, described in Section 6.4.2.2). The wavy baseline signal is apparent in both the noncontact data during approach and the lengthy adhesive bridge during retraction, where polylauryl methacrylate is clinging to the tip. Upon subtracting the data acquired during retraction from that acquired during approach (not shown), much of the wavy baseline is removed, with the caveat of the *Z* hysteresis discussed above, which means the approach and retract sets of data do not quite "line up." Here is where acquiring a same-sized "dummy" force curve, with the tip and sample not quite reaching each other, can characterize the baseline including *Z* hysteresis (as in Figure 3.20). In principle, one can mathematically fit this baseline with a multisinusoidal function of variable wavelength (as a function of *Z*) to account for nonlinearity, and using appropriate *Z* offsets, subtract this function from the data of interest such as that in Figure 3.25, to remove the wavy artifact from the data trend.

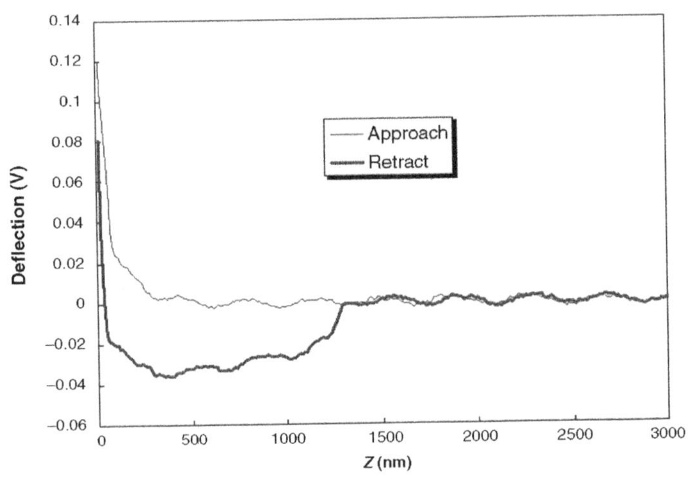

FIGURE 3.25 Raw force curve obtained on polylauryl methacryrlate illustrating the significance of the wavy optical baseline in both approach and retract data, and further its dependence on *Z*-scanner hysteresis.

The preceding data processing is of course complicated. One would prefer, if possible, to simply subtract the optical baseline of the approach data from that of the retraction data to examine details of the latter. This process can work well if the Z nonlinearity and hysteresis are not too severe (Z stroke not too large, or closed-loop scanner used) and, moreover, if measures are taken to minimize the optical artifact as per Section 3.4 (reflective cantilever coating, position of laser spot on cantilever, laser path tilt relative to sample normal, or use of low-coherence laser). Such data processing is presented in Figure 3.26 for the case of a fibril of the proteinaceous gelatin (tip-attached) bridging between the tip and a gelatin film surface, in ambient air. Figure 3.26a contains the raw force-curve cycle, in photodiode output (deflection in V) versus Z-scanner position in nanometers. Figure 3.26b contains a fully processed force-curve cycle (left scale) as well as the difference in force sensed during retraction with force sense during approach (right scale). The former contains the optical baseline as observed from 30 nm to 200 nm along the distance scale; this baseline varies by the equivalent of 0.3 nN of force over this range of distance. Although a small amount, on the scale of the physical phenomenon being probed, the optical artifact is substantial. The behavior of the nanoscale fibril of gelatin cannot be properly examined until the wavy optical baseline is subtracted out; whereupon one observes a prominent plateau of force, $F = -0.27$ nN, through most of the range of distance, but with some characteristic negative departures. One is observing two fundamentally different kinds of resistances to molecular extension:

1. the predominantly entropic resistance discussed in Section 2.5.5 in the context of the freely jointed chain model, here seen as negative excursions (arrows) from the plateau force, in which case each step of distance reduces the volume that a randomly fluctuating polymer chain may explore in conformational space;
2. a nonentropic force that must involve a peeling or unzipping phenomenon (here perhaps capillary mediated, although the phenomenon has been observed under immersion too), where each molecular-scale unit along the chain or fibril requires a fixed energy to unbind—force times a small change of distance—but does not affect the length of a randomly fluctuating polymer chain coil, because this would contribute to a changing entropic force.

Note that precise, quantitative analyses of behavior #1 would be impeded by the presence of the variable baseline due to the optical artifact: proper mathematical fits of data segments to the above chain models would not be possible. Second, the demarcation of distance intervals wherein constant or variable force is exhibited (i.e., the above two behaviors) would be difficult in the absence of the subtraction performed in Figure 3.26b.

One caveat about the subtraction strategy is that the analyst must consider the issues of noise: subtracting one data set from another necessarily increases the noise, as the random fluctuations add in quadrature,

$$<\Delta F>^2 = <\Delta F_1>^2 + <\Delta F_2>^2. \quad (3.11)$$

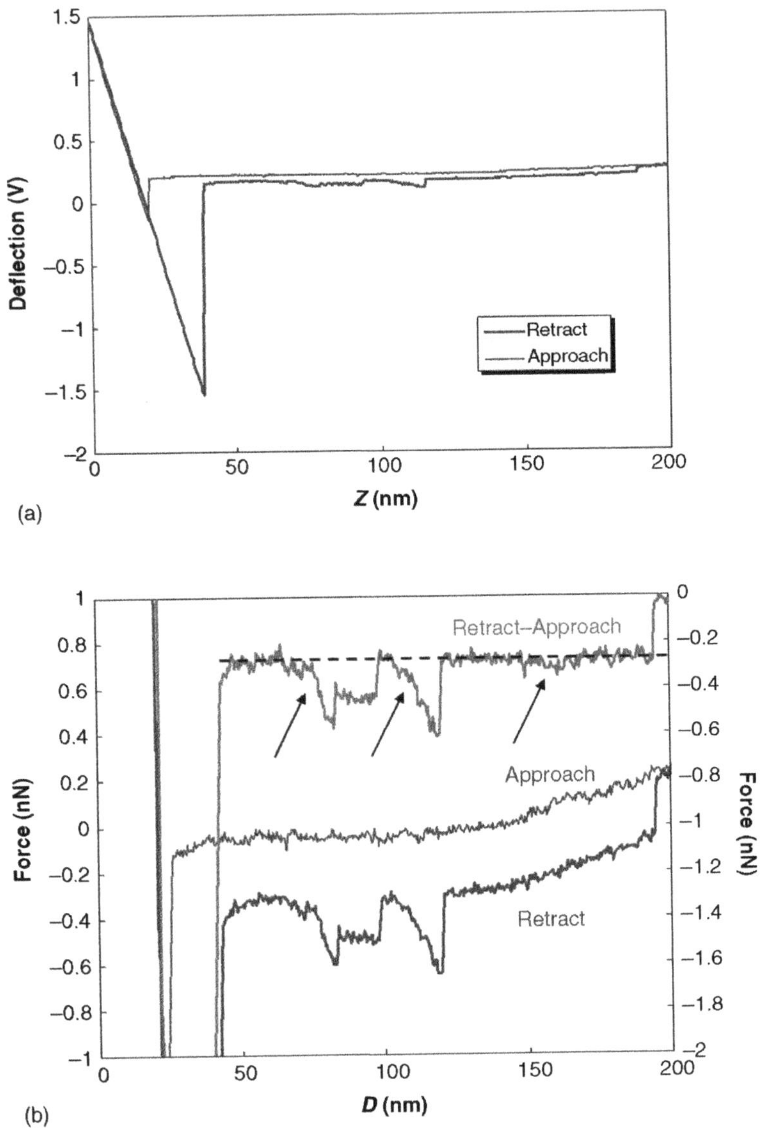

FIGURE 3.26 (a) Raw force curve with a gelatin fibril bridging between tip and graphite surface in air. (b) Force–distance plot of same data as in (a) and difference of retract minus approach data, plateau force marked with dashed line.

This fundamental relationship can be confirmed by subtracting a smooth polynomial curve fit to the variable force baseline in the approach data and calculating the standard deviation of force of the resulting data, then comparing with the standard deviation of the (retract minus approach) subtracted data in the plateau region.

Thus, to achieve the most precise analysis of molecular behavior via analytical modeling of the retraction data, a mathematical baseline subtraction is preferred over the subtraction of the approach data. But such a fitting procedure is usually too arduous to perform on every force curve obtained in even a single experimental session, easily numbering in the hundreds, whereas a simple script or custom data-processing program (e.g., Matlab, Visual Basic, LabView) that takes the difference of retract and approach data that can be implemented to enable rapid sampling or screening of data sets.

Another methodology for analyzing chain bridging forces but without issues of baseline is to *modulate distance* (i.e., vertically oscillate at a much higher frequency than the Z ramping cycle). This can be done in a number of ways, examples being (i) adding a second Z-piezoelectric device, independently controlled with a function generator, in series with the AFM's Z scanner; (ii) breaking out the Z signal (many commercial systems provide BNC access) and adding a time-domain sine wave by wiring an op-amp-based circuit (a method discussed in Section 7.4.2 in the context of setpoint ramping); or (iii) oscillating the cantilever, either at a vibrational resonance as done in dynamic AFM or at arbitrary frequencies with non-resonant methods (e.g., using a magnetic cantilever in a time-varying magnetic field, as discussed in Section 5.2.6). Both the amplitude and phase of the force response can be analyzed to provide *viscoelastic* characterization [11–13], meaning both stiffness and damping of the bridging molecule(s). Even single-chain modulation measurements have been demonstrated in both low-frequency (nonresonant) [14] and high-frequency (resonant) [15] implementations.

3.6.4 Bias-Dependent Electrostatic Forces in Air

The long range of electrostatic forces (Section 2.2) under tip or sample bias—with weak attraction measureable to distances of at least hundreds of nanometers—suggests that proper treatment of the deflection baseline signal is needed for quantitative analysis. An example case is shown in Figure 3.27. Part (a) contains four raw force curves for a grounded Pt–Ti-coated tip (40 nm in radius, full angle 40°, 20-μm tall) attached to rectangular n-doped Si cantilever, approaching an aluminum strip 6-μm wide and 50-nm tall (capped by native oxide, lithographically grown on doped silicon) and biased by 0–5 V. With the tip-scanned AFM employed here, the typical force-curve background is tilted in the opposite direction as that in Figures 3.15 and 3.16 (and waviness is suppressed using a low-coherence laser), rising with increasing tip–sample distance. This is clear from a nearly identical tilt in the force-curve baselines under no bias and 1 V bias. Upon subtracting a linear background as determined from the zero-bias case (shown approximately as dashed line), and following the usual calibration of deflection sensitivity, computation of distance from Z plus deflection, and multiplication of deflection in nanometers by the spring constant (3.5 N/m), the processed force-versus-distance plots in Figure 3.27b result.

Forces under +1 V external bias (shown) were nearly identical to those at −1 V bias (unshown), indicating that the magnitude of the surface potential difference ΔV_s between tip and sample under zero external bias was certainly less than 0.25 V.

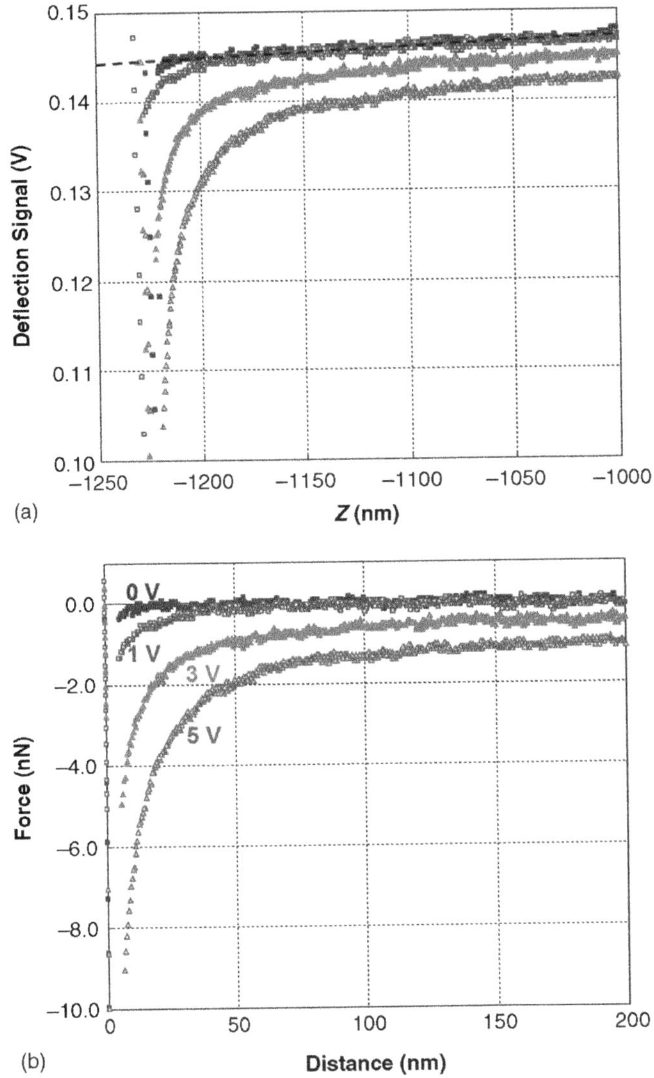

FIGURE 3.27 (a) Raw force curves for grounded tip approaching metallic sample at biases of 0–5 V. (b) Corresponding data converted to force–distance following linear background subtraction.

Nevertheless, a finite ΔV_s, whether intrinsic or due to trapped charge in the oxide, produces a weak attractive force measurable to 10–20 nm distance in the unbiased case, well beyond the range expected for purely van der Waals forces (see Figure 2.3). At applied biases, the strength of attractive force is proportional to $(V - \Delta V_s)^2$, as is treated further in Chapter 9.

We note that at 3 V and 5 V bias, there remains a measurable tilt in processed force–distance plots, and more notably a significant force at the largest distances, due to actual long-range electrostatic forces. Indeed, one can mathematically analyze such curves with Equation 2.9 for a spherically capped cone, a model that generates variable forces at hundreds-of-nanometer distances because it accounts for the entire tip, not just a spherical apex. What is still missing from this model, however, is the *cantilever body*, which is known to produce a force interaction with the sample that is measurable up to even larger distances (micrometers) [16–18]. Indeed, in Figure 3.27b, the significant and voltage-dependent force offsets at the largest distance, 200 nm, result not only from the contributions up the shank of the tip but also from the cantilever itself. A rectangular cantilever interaction with sample has been modeled with the relationship

$$g(D) = \frac{1}{3}\frac{1}{2\pi}\frac{wL}{(D+H)^2}\frac{1}{\left(1+\dfrac{L}{D+H}\right)\tan\alpha} \tag{3.12}$$

where w and L are the width and length of the cantilever, H is tip height, α is cantilever tilt angle, and D is distance; one inserts this expression into Equation 2.8, $F(D) = -\pi\varepsilon_0 V^2 g(D)$, to obtain force for voltage difference V. The 1/3 prefactor is a calibration coefficient needed to account for an *effectively different spring constant* for measuring forces distributed along the entire cantilever [17]. Up to now, we have only considered forces that are concentrated at the tip location, effectively localized at (or very near) the end of the cantilever. When summing forces exerted at all points along the cantilever, the resistive spring force due to bending of the cantilever is instead *continuously variable* from its tip end to its pinned base. (One also can think of the cantilever length continuously varying, with force applied to the end). In order to combine the cantilever–sample electrostatic force with the tip–sample electrostatic force into a single force measurement, which of course is the reality, one must invoke a correction factor of 1/3 for the distributed force. Thus, the effect of this latter force on the overall bending of the cantilever, as measured via cantilever inclination near the tip end, is reduced by distributing the force along the cantilever.

3.6.5 Screened Electrostatic Forces in Aqueous Medium

Our final example, Figure 3.28, involves DLVO (Section 2.5.1) interfacial forces between a silicon tip (with native oxide) and a glass surface in deionized water (mineral ions removed). Both surfaces *ionize* and bear a negative charge in neutral-pH water, which results in repulsion of tip from sample. The strength of this repulsion *decays exponentially with distance* due to Debye screening of the surface charge by positive *counterions* in solution (here H +, liberated from the silicon oxide surfaces) as well as ions due to the autodissociation of water (H_3O^+). The range of this force depends on the magnitude of surface charge density and thereby the concentration of counterions available for screening (Section 2.5.1).

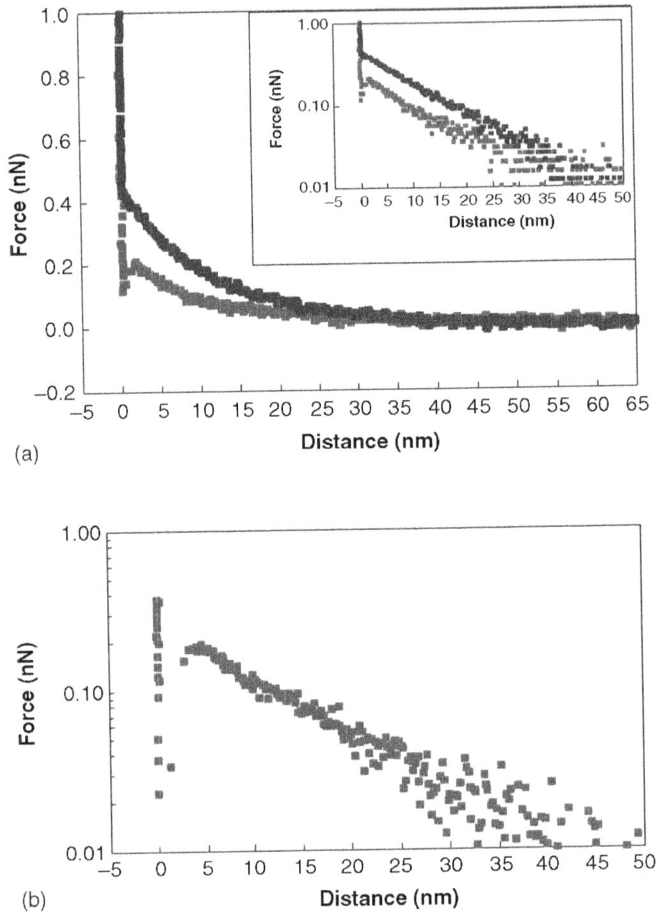

FIGURE 3.28 (a) Force versus distance between silica tip and clean glass surface in water, overlaying measurements using a sharp (lower data) and blunted (upper data) version of the same tip. Inset: semi-log plot of same data. (b) Similar results using a sharp tip but invoking finer digitization of the acquired data.

(An ionic solution optionally can be used, typically to increase the screening and thus decrease the force range.) As the surfaces approach to nanometer-scale distances, one may or may not observe a transition to dominant attractive van der Waals interaction, dependent on the magnitude of charging and on the geometry of the interaction zone, in turn a function of tip shape/bluntness. Examples of the latter issue are included in Figure 3.28a; the overlaid force–distance plots compare data obtained with a fresh commercial tip of ~10-nm nominal radius of curvature with that of the same tip blunted by high-force scanning (on glass). The semi-log plot of the same data in the inset reveals the exponential character of the force–distance relationship.

In AFM, semi-log plots often reveal the *digitization* of force signal, as is apparent at relatively large distances (e.g., 30 nm) in Figure 3.28a inset, where the force signal is small, due to the limits of 16-bit data acquisition (i.e., 65,536 increments of measureable force, upon analog-to-digital data conversion of photodiode array voltage to binary numbers). Most commercial AFM data-acquisition systems give the operator the *option to reduce* the range of measureable deflection signal (sensible if forces are very weak) and thereby reduce the size of digitization to below that of the *noise limit*. Figure 3.28b demonstrates a force–distance semi-log plot of data acquired at a 10-fold finer digitization than used to acquire the data in Figure 3.28a, such that noise (certainly, thermal fluctuational noise of the cantilever, perhaps laser noise) is dominant at the larger distances and lacking discrete plateaus due to digitization.

Higher signal/noise AFM measurements of DLVO forces often employ the colloidal probe method, as well as controlled variation of pH and/or concentration of added ions [1].

3.7 CANTILEVER SPRING CONSTANT CALIBRATION

Thus far in this chapter, we have converted cantilever deflection data to force by multiplying by the manufacturer-specified normal spring constant k (after converting the deflection data from differential photodiode output signal in volts to vertical tip displacement in nanometers, such that the converted force curve has unity slope in contact). But manufacturer-specified values of k should not be used for the most precise work (although closer to reality than values provided a decade or two ago). Variation in cantilever dimensions, particular thickness t in Equation 3.6, result in large variation in spring constant (although variations are somewhat reduced within a given wafer batch). In the case of silicon nitride cantilevers, significant variations in Si_xN_y stoichiometry and thus Young's modulus E can add to this problem. To obtain a better value of k, it remains incumbent on the end user to perform calibration measurements.

The most conceptually simple, as well as broadly applicable method (i.e., for a *wide range of k values and both rectangular and V-shaped cantilevers*), and requiring no additional electronics or spectral analysis (the latter is discussed in Appendix 1), involves acquiring a force curve against a special precalibrated reference cantilever. These are commercially available and typically include several rectangular, tipless, single-material (e.g., silicon of reasonably known density and modulus) beams per chip, with precisely measured resonance frequencies and plan-view dimensions (length and width), perhaps to three significant digits or better. There are other calibration methods that relate resonance frequency to cantilever dimensions such that the above enable a precise determination of reference cantilever thickness; then k is calculated (by the manufacturer) from the dimensions and modulus. (These measurements of length, width, and resonant frequency of course can be repeated by the user as a check on manufacturer numbers). Typically, the different reference cantilevers per chip span two orders of magnitude in k values,

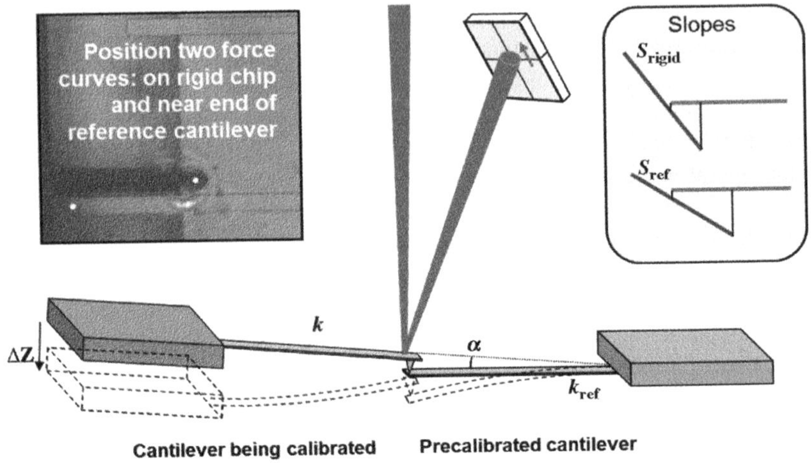

FIGURE 3.29 Illustration of a cantilever calibration method that requires pushing the tip against a precalibrated, tipless cantilever and comparing the force-curve contact slope to that acquired on a rigid surface, such as the chip to which the precalibrated cantilever is attached.

enabling calibration of a wide range of AFM cantilevers such as used in the preceding section.

Figure 3.29 illustrates the concepts of the method. Provided that the tip is positioned very close to the end of the reference cantilever and centered on its principal axis (white dot in lower left of left inset), with the two cantilevers antiparallel, and a force curve obtained with minimal tip sliding both at this location and on another, rigid surface (typically, the chip to which the reference cantilever is attached, as shown at center of left inset), one can calculate the operative spring constant k of the unknown cantilever via the relationship

$$k = k_{\text{ref}}\left(\frac{S_{\text{rigid}}}{S_{\text{ref}}} - 1\right) \tag{3.13}$$

where k_{ref} is the spring constant of the (presumed horizontal) reference cantilever and S_{rigid} and S_{ref} are the slopes of the raw force curves acquired on the rigid surface (e.g., chip of reference cantilever) and the reference cantilever, respectively [19]. In the latter case, obviously both cantilevers will bend, such that for a given change of Z, the magnitude of the deflection of the cantilever of interest is *reduced* by the magnitude of the deflection of the reference cantilever. This correspondingly reduces the slope of the force curve S_{ref} relative to S_{rigid}, as depicted in the right inset of Figure 3.29.

Normally, the positioning of the tip is performed using a light microscope in an overhead view. With some instruments, this positioning may be the precision-limiting measurement, as opposed to the above-mentioned determination (usually by the manufacturer) of reference cantilever dimensions, modulus and resonance frequency. To account for the precise tip position at some measurable distance from the end of the reference cantilever, a correction can be invoked that accounts for the fact that k_{ref} is inversely proportional to reference cantilever length cubed (Eq. 3.6), and the latter effectively reduced by the placement of the tip of the cantilever to be calibrated. A corrected value k of the cantilever being calibrated can be determined from a value k_{off} obtained using Equation 3.13 but with a finite distance of the tip-touch, ΔL, from the end of the reference cantilever of length L:

$$k = k_{\text{off}} \left(\frac{L}{L - \Delta L} \right)^3. \tag{3.14}$$

This expression is exact for rectangular cantilevers and within 4% for V-shaped cantilevers [20].

It is worth reiterating that the *operative* value of spring constant k in Equations 3.13 and 3.14 includes the role of the angle of cantilever inclination, that is, the measurements of interest are performed by a tilted cantilever, and thus the operative k includes the effects of tilt. As discussed in Section 3.2.1, the spring constant $k^{\perp}$ for forces perpendicular to the cantilever is slightly reduced, according to $k^{\perp} = k \cdot \cos^2\alpha$. The square cosine factor for most AFM designs is ≈ 0.96. But some other calibration methods (Appendix 1) in fact determine $k^{\perp}$ and thus must be converted to k by dividing by $\cos^2\alpha$.

REFERENCES

[1] Butt, H.-J., B. Cappella, and M. Kappl, Force measurements with the atomic force microscope: Technique, interpretation and applications. *Surf. Sci. Rep.*, 2005, **59**: 1–152.

[2] Haugstad, G., Contrasting static-to-kinetic friction transitions on layers of an autophobically dewetted polymer film using Fourier-analyzed shear modulation force microscopy. *Trib. Lett.*, 2005, **19**: 49–58.

[3] Rabe, U., K. Janser, and W. Arnold, Vibrations of free and surface-coupled atomic force microscope cantilevers: Theory and experiment. *Rev. Sci. Instrum.*, 1996, **67**(9): 3281–3293.

[4] Sacha, G.M., A. Verdaguer, and M. Salmeron, Induced water condensation and bridge formation by electric fields in atomic force microscopy. *J. Phys. Chem. B*, 2006, **110**(30): 14870–14873.

[5] Haugstad, G. and W.L. Gladfelter, Force-displacement measurements in a beam-reflection scanning force microscope: Calibration issues. *Ultramicroscopy*, 1994, **54**: 31–40.

[6] Kassies, R., et al., Removing interference and optical feedback artifacts in atomic force microscopy measurements by application of high frequency laser current modulation. *Rev. Sci. Instrum.*, 2004, **75**: 689–693.

[7] Siegman, A.E., *Lasers*1986, Mill Valley, CA: University Sciences Books.

[8] Oliver, W.C. and G.M. Pharr, An improved technique for determining hardness and elastic modulus using load and displacement sensing indentation experiments. *J. Mater. Res.*, 1992, **7**(6): 1564.

[9] Oliver, W.C. and G.M. Pharr, Measurement of hardness and elastic modulus by instrumented indentation: Advances in understanding and refinements to methodology. *J. Mater. Res.*, 2004, **19**(1): 3–20.

[10] Fischer-Cripps, A.C., Critical review of analysis and interpretation of nanoindentation test data. *Surf. Coat. Technol.*, 2006, **200**(14–15): 4153–4165.

[11] Aklonis, J.J. and W.J. MacKnight, *Introduction to Polymer Viscoelasticity*. 2nd ed., 1983, New York: Wiley, 295.

[12] Ferry, J.D., *Viscoelastic Properties of Polymers*1980, New York: Wiley.

[13] McCrum, N.G., B.E. Read, and G. Williams, *Anelastic and Dielectric Effects in Polymeric Solids*. 1967, London: Wiley.

[14] Sakai, Y., et al., Nanorheology measurement on a single polymer chain. *Appl. Phys. Lett.*, 2002, **81**: 724–726.

[15] Watabe, H., et al., Dynamic force spectroscopy on a single polymer chain. *Macromolecules*, 2006, **39**: 5921–5925.

[16] Belaidi, S., P. Girard, and G. Leveque, Electrostatic forces acting on the tip in atomic force microscopy: Modelization and comparison with analytic expressions. *J. App. Phys.*, 1997, **81**(3): 1023–1030.

[17] Bonaccurso, E., H.-J. Butt, and F. Schonfeld, Electrostatic forces acting on tip and cantilever in atomic force microscopy. *Phys. Rev. B*, **74**.(8): 1–8

[18] Guriyanova, S., D.S. Golovko, and E. Bonaccurso, Cantilever contribution to the total electrostatic force measured with the atomic force microscope. *Meas. Sci. Technol.*, 2010, **21**: 025502.

[19] Torii, A., et al., A method for determining the spring constant of cantilevers for atomic force microscopy. *Meas. Sci. Technol.*, 1996, **7**: 179–184.

[20] Sader, J.E., Parallel beam approximation for V-shaped atomic force microscope cantilevers. *Rev. Sci. Instrum.*, 1995, **66**(9): 4583–4587.

4

TOPOGRAPHIC IMAGING

In this chapter, we first introduce the idealized principles of topographic imaging with AFM (Section 4.1). "Idealized" means in the absence of *nonuniform force response* across the image, which, in general, may complicate image interpretation and, in some cases, dominate the contrast generated (i.e., topography may not be the only attribute imaged). Idealized also means divorced from the *geometric* details of the tip–sample interaction zone as well as *instrumentation issues* pertaining to calibration, precision, speed, etc. The idealized case is intuitive and simple to grasp, in both quasistatic and dynamic modes. In Section 4.2, we proceed to a much lengthier discussion of real-world nonidealities. These include issues due to the actual hardware employed for scanning, measuring tip–sample interaction, and actuating the feedback control system to track surface topography. In addition to such realities, we discuss a number of limitations, artifacts, caveats, and interpretive issues. These include: (i) mechanical *compliance* of the imaged materials, which on heterogeneous systems "leaks into" the so-called height image; (ii) the finite response time (speed) of the control-loop feedback system used to track topography; (iii) *X–Y–Z* scanner nonidealities; (iv) complications due to the *shapes* of the interacting tip and local surface; and (v) drift due to thermal warm-up and sample shape/preparation as well as complications external to the instrument itself. Although we will examine real data throughout Section 4.2, finer treatments of data analysis are delayed until Chapter 8, as these pertain to not only topographic but also property-sensitive data (Chapters 5–7). One can jump to Chapter 8 at any point, however, to delve further into topographic data processing.

Atomic Force Microscopy: Understanding Basic Modes and Advanced Applications, First Edition.
Greg Haugstad.

4.1 IDEALIZED CONCEPTS

In an idealized laboratory, one can conceive a scheme to map the topography of a material surface via *isoforce contours*, depicted in cross section in Figure 4.1a. One first imagines a force–distance relationship that *localizes* the interaction between the surface and a vanishingly small (point) vertical force sensor. This ideally would be a force that diminishes very steeply with distance. Long-range forces would instead add contributions from many locations along and below the material's surface, necessarily delocalizing the interaction. Based on the concepts in Chapter 2, we also know that the typical force–distance relationship usually involves a transition from relatively weak attractive noncontact forces, necessarily including van der Waals forces (but potentially also electrostatic), to a repulsive contact force that steeply increases upon approach. (More complicated cases, in liquid environments, were introduced in Section 2.5.) Ideally, this transition would be very abrupt, such that the location of the surface—the *zero* of distance—can be determined with sub-Angstrom-level precision. Second, one imagines a scanner that precisely moves the force sensor in an *X*–*Y* plane approximately parallel to the sample surface plane. (A scanner instead could be used to move the *sample* in the *X*–*Y* plane while the force sensor is fixed, as is indeed the case in some real AFMs.) Third, one imagines an extremely fast *Z*-displacing device that can almost instantaneously move the force sensor toward or away from the surface (or surface toward or away from sensor),

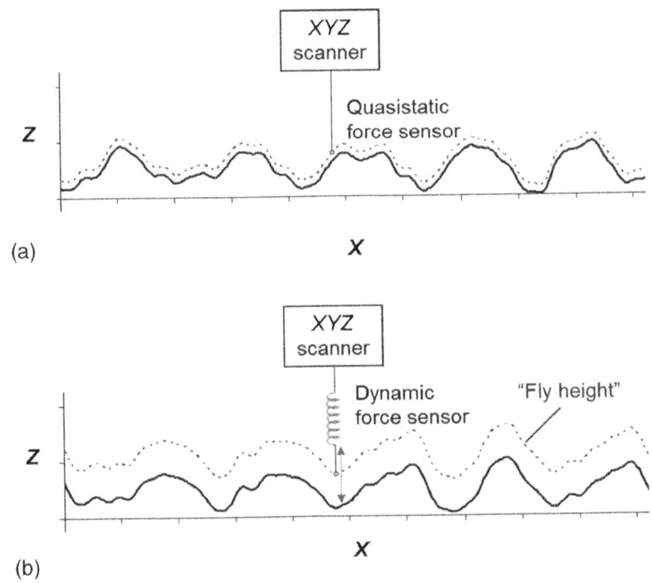

FIGURE 4.1 Cross-sectional depiction of an idealized point-like force sensor tracking a material surface topography as a contour of constant (a) quasistatic force or (b) dynamic amplitude during vertical oscillation with respect to some "fly height."

via feedback, such that the sensor read maintains an essentially constant value at the operator-selected *setpoint*. Thus, the force read is the *process variable*, in the language of *control systems* [1,2], which enables the Z-displacing device to maintain a constant distance between sensor and surface while traveling in X and Y. This distance could be selected to be zero, the point of contact; or some other distance *relative to* this zero. Fourth, one imagines a Z-displacement sensor that precisely *measures* these reactive changes of Z, which mirror the actual changes of surface elevation. Given these components, one can conceive of a programmatic X–Y raster scanning pattern together with reactive Z displacement such that the point sensor maintains a constant distance value, with one possible case being zero distance (contact). These Z displacements thus reproduce the topography of the surface, a *height image*. This is the idealized conceptual framework of quasistatic or "contact-mode" AFM depicted in Figure 4.1a. "Idealization" in the sense of instrumentation does not, however, preclude shear forces as the sensor tangentially scans the surface. These unavoidable friction forces are the principal subject of Chapter 7.

The preceding describes a *continuous and tangential* tracking of surface topography via isoforce contours. To construct an image containing a finite amount of data, one must sample the value of Z at *discrete* points within a grid of locations in the X–Y plane, as suggested by the dotted curve in Figure 4.1a. But within the same idealized world, one could just as easily approach and retract Z at each discrete X–Y location rather than tangentially scan along a given surface height contour. The ultimate approach value of Z would be that at which the sensed force reaches some particular value, whether positive or negative (repulsive or attractive). These values of Z over a discrete array of X–Y locations would construct a topographic image consisting of isoforce contours, as in the preceding case. But additionally, the entire force-Z trend could conceivably be probed at each X–Y location. Second, a purely vertical approach and retract at each X–Y location would *remove forces derived from shear motion.* This is the idealized framework of force-curve mapping modes such as force volume (Bruker), SPS spectroscopy (Agilent), digital pulsed force mode (Witec), and peak force tapping (Bruker). Although intermittent contact occurs, these modes are still quasistatic in nature; at each point within each approach–retract cycle, a well-defined force is measured.

Dynamic modes of AFM are more complicated. With a bit more imagination one can conceive of a similarly ideal, point-like force sensor attached to a *vertically driven, oscillating spring*, which does not itself directly interact with the surface. The point-like force sensor would thus remain a finite distance from the surface most or all of the time. Most importantly, the *force gradient* dF/dZ felt by the sensor alters the composite oscillator's spring constant (i.e., spring plus dF/dZ) and thereby changes the resonance frequency of the oscillator. Because the force gradient changes with Z, the *average value of force gradient over a complete oscillation cycle* produces a net positive or negative shift in the resonance frequency. In principle, a feedback system could vertically displace the midpoint of the oscillator (its DC offset, so to speak) and track contours of constant resonance frequency shift, in order to track surface topography. This is the scheme for *frequency-modulation* (FM) AFM, a variant of dynamic AFM used primarily in ultrahigh vacuum imaging environments and not the subject of our treatment. Alternatively, one could utilize the fact that a changing resonance frequency *necessarily produces a changing oscillator amplitude* at a fixed excitation frequency, as discussed in Section 2.4.2 with regard to Figure 2.9. Thus, a feedback system could, in principle, vertically displace the midpoint of the oscillator and track contours of constant oscillator amplitude, in order to track surface topography. This is the framework for amplitude modulation (AM) AFM, the variant of dynamic AFM that is predominantly used in air and in liquids. In the case of either FM or AM AFM, the midpoint of the sensor's oscillation range, its "fly height" so to speak, is what is effectively being moved in *X–Y–Z* to generate a topographic image, shown in cross section in Figure 4.1b. (Some AFMs actually move the sample, producing the same effect.)

One should note that a given value of force does not necessarily translate into one unique value of distance, as seen in the *nonmonotonic* relationship between force and distance discussed in Chapter 2 and depicted again in Figure 4.2. Here, a given attractive force can be found at *two* distances, for instance, the circled values at distances D_1 and D_2. Such "ambiguity" of course can be avoided by operating at positive quasistatic force. But even in principle, the (quasistatic) force-feedback tracking system cannot operate at the outer distance value D_2. The reason is that the force–distance slope must be negative in order to properly drive the tracking system; the feedback must be hard-wired to displace the sensor away from or toward the surface dependent on the *sign* of the deviation of the read signal from the setpoint. If attempting to operate at D_2, the feedback would displace the sensor

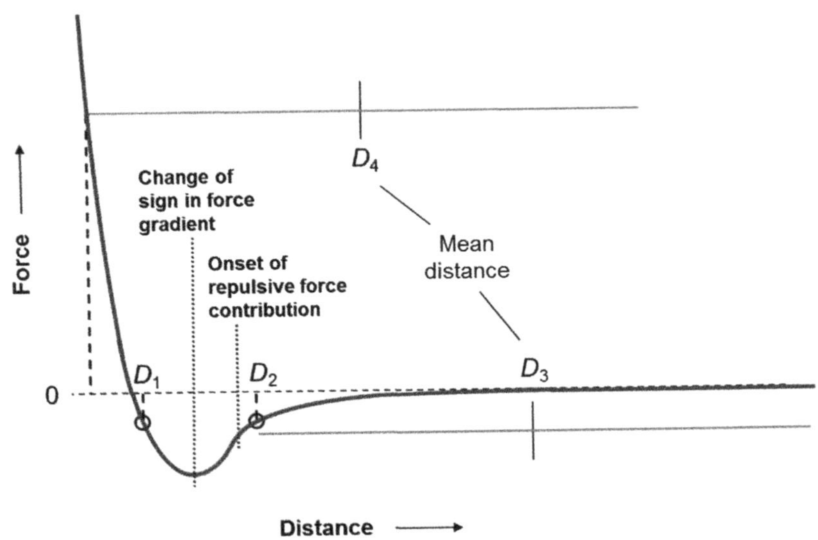

FIGURE 4.2 Depiction of prospective quasistatic operating points corresponding to a particular net attractive force at distances D_1 and D_2, and prospective dynamic operating points with net attractive versus net repulsive integrated force gradients at corresponding mean distances D_3 and D_4.

distance in the wrong direction for a given deviation of force relative to setpoint, preventing surface topography imaging. (We need not even consider the issue of jump-from-contact discussed in Chapter 3 for a real force sensor.)

In dynamic mode, there is a similar ambiguity, but it has much more significant ramifications. For a given value of integrated force gradient (i.e., over the complete oscillation cycle), the mean value of distance probed by the oscillating sensor could have two different values, *yet drive the feedback properly in either case*. Under AM, a deviation of amplitude below the setpoint can be due to: (1) an increasingly negative shift of resonance frequency under a net negative integrated force gradient or (2) an increasingly positive shift of resonance frequency under a net positive integrated force gradient. In either case, the feedback would react to increase the sensor's mean distance. Thus, the relationship between amplitude and distance is the *same at two different values of mean distance*, such as depicted by D_3 and D_4 in Figure 4.2. (Note that the onset of repulsion occurs at the inflection point in the force–distance curve, where the slope of increasing attraction with decreasing distance is a maximum [3,4].)

To more graphically perceive how an AFM feedback system works, one considers the Z-dependence of the quantity being measured: either the cantilever's quasi-static deflection or its amplitude of dynamic oscillation. We have already seen examples in Chapters 1–3 of force curves: measurements of cantilever deflection versus Z-scanner position. In dynamic mode, there is a precise analog, measurements of cantilever amplitude versus Z-scanner position. In both idealized cases, there is, to first approximation, a simple linear relationship as depicted in Figure 4.3a and b; these sketches are indeed very similar to commonly observed data because they include the effects of attractive forces. (Recall that increasing Z, from left to right, corresponds to tip displacement away from sample or sample from tip.) In contact mode (a) these attractive forces produce the jumps to and from contact and resultant hysteresis loop over an approach–retract cycle. In dynamic mode (b), for a choice of drive frequency slightly below cantilever resonance frequency (as is commonly recommended in instrument manuals and discussed further in Chapter 5), this may similarly result in jumps of amplitude, and resultant hysteresis loop over an approach–retract cycle. Attraction causes the resonant peak to shift to lower frequency as depicted in Figure 2.9. This results in an amplitude increase if driving at a frequency below resonance (i.e., the operating point "rides up the peak"), which causes the sensor to get *closer* to the surface, further increasing the attractive force gradient and thereby causing a further shift, etc. This positive *physical* feedback of the resonant response (not to be confused with the feedback used by the *instrumentation*, or *servo*, to track topography during *X–Y* scanning) produces the abrupt increase depicted in Figure 4.3b during approach [5]. The highest amplitude is reached once the sensor achieves (intermittent) contact, whereupon the steep repulsive force gradient acts to halt the negative shift of the resonance response; a repulsion-induced shift in the opposite direction reduces the cantilever amplitude (again, see Figure 2.9), which causes the sensor to penetrate less far into the repulsive part of the tip–sample interactive relationship; this negative physical feedback of the resonant response locks the oscillator in place in the net repulsive regime.

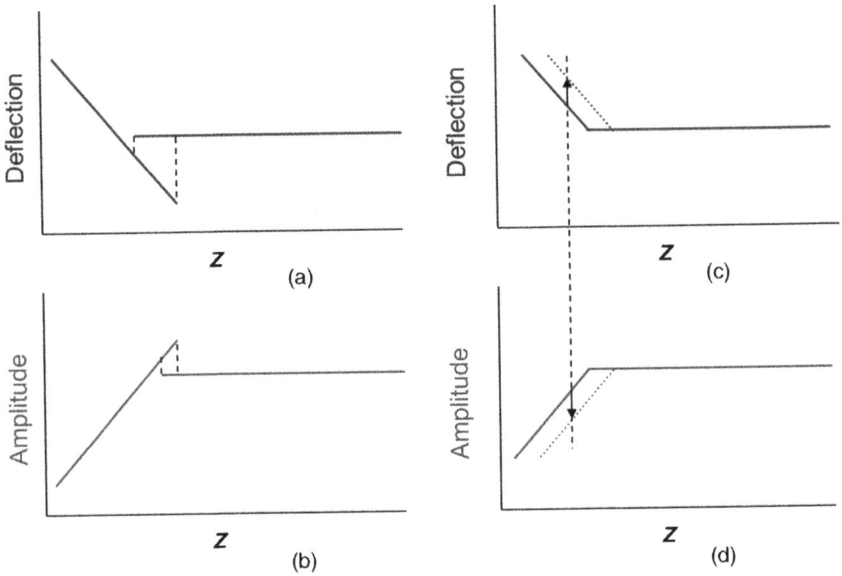

FIGURE 4.3 Schematic dependence of a cantilever's quasistatic deflection (a) and (c) or amplitude of dynamic oscillation (b) and (d) on Z scanner position, in the presence (a), (b) or absence (c), (d) of attractive forces. Dotted lines in (c) and (d) depict a shifted Z dependence at a higher surface location.

During Z retraction, the oscillator remains locked in this net repulsive dynamic state until the explored repulsive force gradient has weakened to the point that the resonant response switches back to net attractive, with the resonant response again shifted substantially to lower frequency or perhaps unshifted as the nearest sensor approach may be beyond measurable attractive forces.

In the absence of net attractive forces (as is conceivable for certain combinations of tip/surface/liquid medium), the jumps and hysteresis loops of Figure 4.3a and b would be absent as depicted in Figure 4.3c and d. But in any of cases a–d, one can imagine a choice of imaging setpoint (deflection or amplitude) such that a change of surface height (when moving from one *X*–*Y* pixel location to the next) would necessarily result in a *deviation from the specified setpoint* deflection or amplitude. A higher surface elevation, for example, would mean that the diagonal data trend would lie farther to the right as shown with dotted lines in Figure 4.3c and d. Thus, for a given operating point—deflection or amplitude at some Z along the diagonal portion—a lateral movement to a location of higher surface elevation would cause a deviation from the setpoint: up in deflection or down in amplitude. The tracking mechanism would react to this change and displace Z in an attempt to null (precisely offset) the deviation and return to the setpoint. (The sign of the Z displacement, of course, will differ for contact and dynamic modes of operation.) And the opposite reactions would apply to a lower, instead of higher, surface location.

4.2 THE REAL WORLD

A real AFM system consists of hardware components that of course do not react instantaneously, do not provide infinitesimal spatial resolution of force, cannot be positioned with arbitrarily fine precision relative to the surface of interest, etc. This section is divided into a number of subsections treating the realities of AFM instrumentation and operational concepts and procedures. Section 4.2.1 describes, in block form, the hardware components used to actuate the control system introduced in concept in Section 4.1 (in both contact and dynamic modes). This is followed in Section 4.2.2 with examples of real data collected in contact mode and the basics of rendering data as images and histograms. Section 4.2.3 treats some aspects of force control and sample compliance, topics expanded in later chapters under special methods. Section 4.2.4 covers the performance of, and user adjustments to, the Z feedback control system. Section 4.2.5 drills down into the details of the scanning hardware including the realities of piezoelectric scanners: creep, nonlinearity (broadening the topic of Section 3.5), hysteresis, and cross coupling, as well as some calibration and noise issues. Section 4.2.6 addresses difficulties derived from tip shape and state. Section 4.2.7 completes the treatment of practical realities by covering fine points related to instrumental drift and complications derived from experimental geometry. Thus, in addition to understanding in basic terms how AFM is able to image topography, a goal in Section 4.2 is to drill down into enough detail to understand how to better operate the instrument and properly interpret the so-called height or topography image; in the process, we will encounter some interesting, even fascinating, examples. Some issues important to height data will be discussed in greater depth in Chapter 5, armed with complementary phase imaging.

4.2.1 The Basics: Block Descriptions of AFM Hardware

Figure 4.4a describes the feedback-actuated system for tracking surface topography in contact mode (i.e., quasistatic operation). It expands on Figure 1.4 to schematically include

- further details on the differential measurement of cantilever deflection at the quad photodiode, the so-called "*Top-bottom*" signal (in reality, this differential signal is normalized to total output in order to remove fluctuations, then multiplied to provide a measurement range of around ± 10 V, typically corresponding to hundreds of nanometers);
- the comparison of this differential read signal with the operator-specified *Setpoint*;
- the use of the deviation or *Error Signal* (of the former read relative to the latter set), which may be positive or negative, to force the *Z servo*, via feedback, to increase or decrease the *Z signal*, respectively (positive Z being defined as rightward in Figure 4.3), typically a bias applied to a piezoelectric device;
- the Z-signal-driven displacement of the *Z scanner* (to which either the *cantilever chip* or the *sample* is attached, depending on instrument design), which

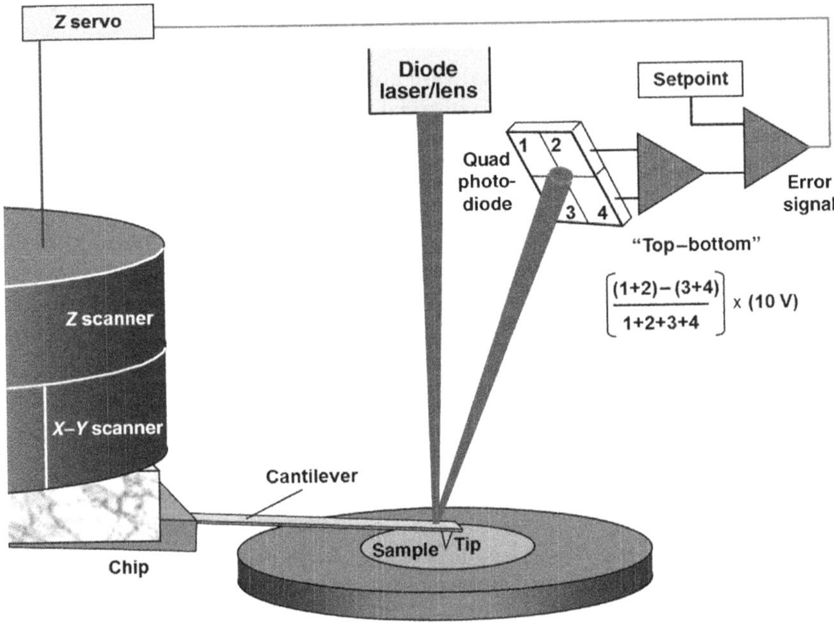

FIGURE 4.4 Schematic representation of the feedback-loop control system in contact-mode AFM.

necessarily changes the cantilever inclination (due to tip–sample contact) and thus moves the laser spot at the quad photodiode, so as to null the error signal (i.e., keeping the Top–Bottom read signal equal to the Setpoint);

- the *X–Y* scanner that programmatically raster scans the tip (i.e., point-by-point and line-by-line) while the *Z* control circuit is operative.

The *Z* displacements that provide approximate height data are, at least by design, independent of the *X–Y* raster scanning algorithm. Some piezoscanners use separate crystals for the *Z* and *X–Y* movement, although both are commonly (but not always) contained in a single device. At this point, the reader may wish to contemplate the fact that the very same hardware that is programmatically displacing *Z* as the *independent variable* in force curves (Chapter 3) is instead *displaced by feedback control* during topographic imaging. But as discussed in Chapter 3 and restated above, most *Z*-scanning systems actually use the *voltage applied* to the *Z* scanner (times a calibration coefficient, i.e., in nm/V) as a *proxy* for a measurement of *Z* position. Only a type of special scanner equipped with a separate, highly linear *Z* sensor (usually, an inductive or a capacitive device) directly *measure* physical *Z* displacement. We will consider these concepts in greater detail in Section 4.2.5.

The hardware system for the typical implementation of dynamic AFM requires a few more components (Figure 4.5) than for quasistatic operation (Figure 4.4). As discussed in Section 4.1, it is the *amplitude* of an oscillating tip that is used to drive

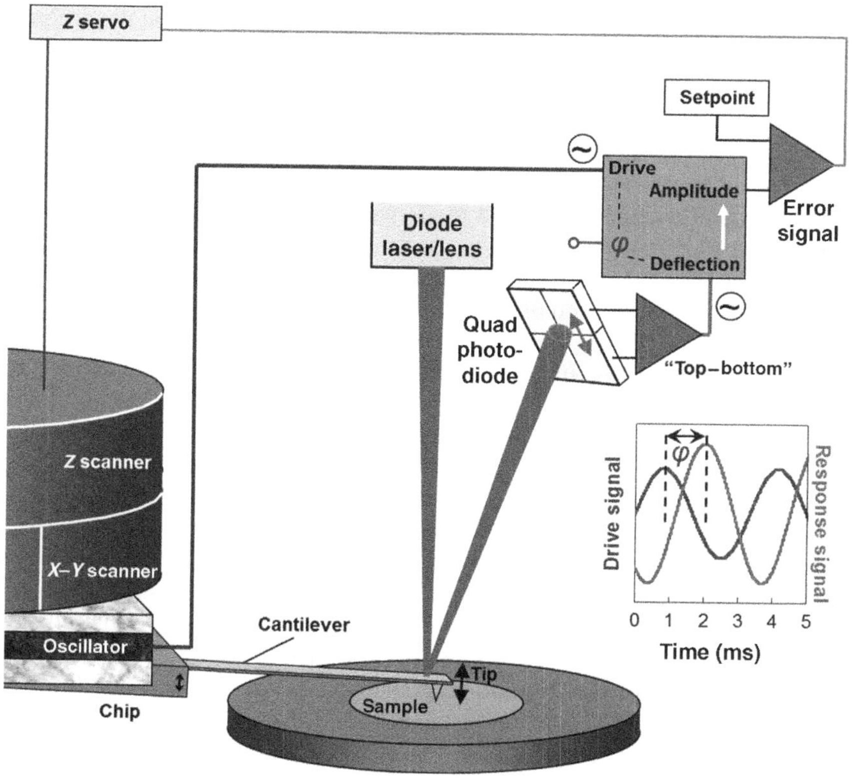

FIGURE 4.5 Schematic representation of the feedback-loop control system in dynamic (AC/"tapping")-mode AFM. Additionally, the phase shift measurement between driving oscillator signal and tip's oscillating response.

the feedback control system in order to track topography (at amplitudes *reduced* relative to the free-oscillation amplitude). The oscillation is generated by imparting a high-frequency vibration, near cantilever resonance, to the chip to which the cantilever is attached. Some instruments use a *piezoelectric bimorph*—a thin sandwich of two piezoelectric crystal slabs—to generate the vibration; some instruments employ other kinds of vibrating elements (or use a magnetic field to directly drive a magnetic cantilever, a very special case). In all conventional cases, a physical vibration transmits to the cantilever chip, but *after propagating through other materials and interfaces* along the path from source of vibration to chip. Generally, the amplitude of the physical vibration against the chip is not directly measured. The operator only controls the amplitude (and frequency) of the *AC drive signal* that electrically actuates the vibrating device. This AC drive signal is typically generated by the same piece of electronics (large gray box in Figure 4.5, absent in Figure 4.4) that is, used to measure the response, the time-varying "Top-Bottom" signal at the output of the quad photodiode. The electronics measures the amplitude of this AC response and outputs a DC signal *proportional to the AC amplitude*; this DC signal is compared to the setpoint to actuate the feedback (Z servo). Note that the difference, the error signal, must undergo a sign change to drive the electronics to displace Z in the correct direction; if the surface slope is positive, meaning height increasing as the tip scans laterally, then the amplitude will tend to deviate below the setpoint value (as per Figure 4.3d), whereas if the surface slope is negative, then the amplitude will tend to deviate above the setpoint. This is in the opposite sense as the deflection signal changes in contact mode when traversing upward versus downward surface topography. Because the same electronic circuit handles the Z displacement for both contact and AC modes, the sign of the difference between DC amplitude signal and the setpoint is flipped (in hardware) to produce the correct direction of Z displacement in AC mode.

The same electronics compares the relative *phase* (φ) of the drive and response AC signals (the graph in Figure 4.5). We will see that this phase measurement is important to identifying and stabilizing the regime of operation, net attractive or net repulsive, and, moreover, is often useful to provide materials contrast because of the way the phase is affected by subtleties of tip–sample interaction (e.g., Section 2.4.4). But, in practice, given interfaces between *instrument* components (e.g., from vibrating device to glue to steel to clip to chip) as well as potential capacitance in the circuitry, there may be several contributions to the overall phase shift between the response and drive signals. The operator needs to subtract out these miscellaneous instrumental contributions to the phase shift and thereby leave only the *physically meaningful* phase shift between the base and tip ends of the cantilever, the phase lag that we know should be 90° at resonance for a driven oscillator with damping prior to tip–sample engagement, as in Figure 2.6. Commonly, this step is done by driving at resonance and precisely "zeroing the phase"; but, in truth, one is forcing a 90° phase shift at resonance. Typically, the software will call this number zero, given that a 0–180° phase scale is linearly mapped into a $+10$ to -10 V signal in the data acquisition system (as normally used for input signals); thus, $90° \rightarrow 0$ V, and $+10$ to -10 V is thereby converted to $+90°$ to $-90°$, rather than the physically correct 0–180° (or equivalently 0 to $-180°$). There can be additional caveats about the nature of the phase signal output from the electronics, dependent on details that can vary among commercial designs, such as the use of a lock-in amplifier, which provides *true phase*, as opposed to some other circuitry that does not [6]; this phase outputting electronics can, internal to itself, introduce a phase shift that becomes part of the *total apparatus phase shift*, that is, all items introducing phase shifts other than the physically meaningful phase shift between the base and tip ends of the cantilever.

The block diagrams of Figures 4.4 and 4.5 assume that the tip is already *engaged* to the sample surface, under feedback control. But how do we reach this state of engagement? Most AFM systems employ a motor drive with much coarser Z control to bring the tip and sample together, from the initial distances of tens or hundreds of micrometers apart after initial instrument configuration. Typically, the motor simply turns a threaded linear actuator such that a rigid piece of hardware is displaced along Z. The series relationship between this coarse Z motor, the fine Z scanner, and the cantilever spring/tip are schematically depicted in Figure 4.6 for

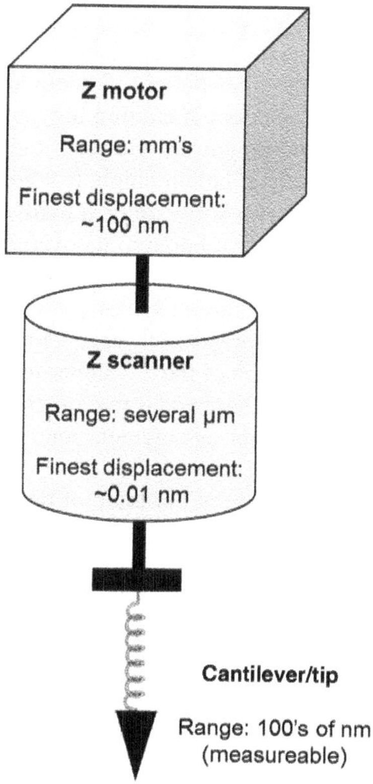

FIGURE 4.6 Schematic depiction of the series relationship between *Z*-displacing elements: *Z* motor, *Z* scanner, and cantilever spring/tip.

the case of a tip-scanned AFM. Also shown are the *Z* ranges typical for each component. The feedback control circuit is active while the motor is approaching, such that upon tip–sample engagement (upward change in cantilever deflection in contact mode or downward change in cantilever amplitude in dynamic mode), the feedback system can react in *Z* so as to keep either deflection or amplitude near the initial setpoint value. Typically, the engagement algorithm leaves the *Z* scanner approximately half extended, such that the *Z* scanner can displace either up or down during imaging by roughly half its full range.

But this limitation is misleading. Note that the relationship between the *Z* motor and *Z* scanner is analogous to the relationship between the *Z* scanner and the cantilever/tip. Just as the *Z* scanner has a greater range than the measureable *Z* displacement of the tip, the *Z* motor has a much greater range than the *Z* scanner. Once engaged, the operator retains the ability to *offset the absolute position of the Z scanner* by actuating the *Z* motor, usually in increments of hundreds of nanometers. This can be enabling; if engaging the tip to a microdevice, for example, the tip may land at a very high or low point relative to the overall range of heights one wishes to explore on the device. (The ability to control this landing point is of course dependent on the nature and quality of *video access*, usually viewing with an overhead light microscope while varying its focal plane to infer surface height; some instruments afford more than one microscopic viewing angle. Sample reflectivity also affects the quality of video access.) If landing low, one may wish to offset the absolute position of the Z scanner downward via the Z motor to enable imaging up to the highest locations (conversely, if landing high, to offset this position upward to access the lowest locations). Either a software or a hardware indicator will depict the engaged position of the Z scanner relative to its overall Z range; this Z position will vary as the operator adjusts the Z motor. A second usage pertains to a surface that has a large tilt, such that the range of heights exceeds that of the Z scanner when acquiring large images (i.e., spread over large lateral distances). In this case, one can displace the absolute position of the Z scanner to accommodate the sample tilt without having to repeatedly withdraw and re-engage tip to sample. In sum, the ability to increment the Z motor after engagement lends convenience as well as a fuller actualization of the Z-scanning range.

Apart from the question of Z range is the issue of data precision, the finest displacement of Z that can be recorded. Here, the Z-scanner data-acquisition system can be the limitation. Many AFMs have 16-bit data acquisition, meaning the recorded Z-scanner position can be any of $2^{16} = 65{,}536$ levels. If the Z-scanner range is, say, 6.5 μm, then the finest recorded displacement of Z is 0.1 nm. On a very smooth surface, this is not good enough; digitization will be apparent upon inspection of a height profile. Figure 4.7 compares cross-sectional data extracted from 500 × 500-nm height images of a silicon wafer. The upper data were acquired by applying the 16 bits over the full 5.3-μm Z range of a scanner. The lower data

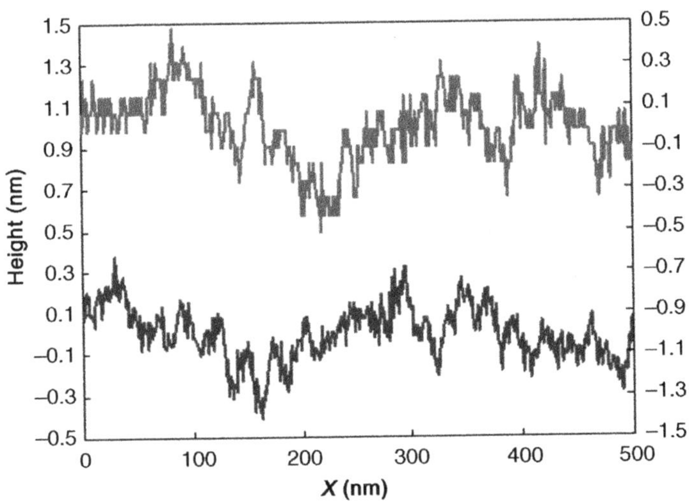

FIGURE 4.7 Comparison of height cross sections acquired for two levels of Z digitization, coarse (upper) and fine (lower). These corresponded to retaining the full range of Z (upper data) and reducing the limit of Z by a factor of 8 (lower data).

were acquired confining the 16 bits to 1/8 of this range, meaning eight times finer digitization of the signal. At this latter scale, the digitization is within the composite noise floor of the *Z* scanner and associated electronics ($\sim$1 mV), whereas digitization is obviously visible in the upper data as discrete vertical changes. Among other things, this measurably affects quantification of roughness; in evaluating the entire image from which the data in Figure 4.7 were extracted, the root-mean-square deviation from the mean (Section 8.2) is 10% *greater at the coarser digitization settings.*

Additionally, as mentioned above and discussed in the next section, the simple fact that any height image is tilted (sample surface not exactly parallel to *X*–*Y* scanner plane) means that under coarse digitization, the surface will be imaged as a staircase of steps (e.g., of 0.1 nm) rather than a smoothly inclined surface. But switching to, say, eight times finer digitization correspondingly reduces the *Z* range by the same factor (e.g., from 5300 nm to 662 nm), thereby reducing the ability to explore a large range of height or requiring more frequent use of *Z* motor steps. Another problem with reducing the *Z* range is drift (Section 4.2.7): a lesser ability to compensate for vertical movement of the sample (due to thermal warm-up, relaxation of the sample mount, etc.). One must pay much more attention to possible *Z* drift when operating with a reduced *Z* range.

4.2.2 The Nature of the Collected Data

Now let us consider an example of topographic imaging in contact mode. Figure 4.8a and b is contact-mode "Height" and "Deflection" image of a polycrystalline aluminum surface (with surface contamination, as we will see) captured over a 2 × 2-μm region during successive left-to-right excursions of the tip during an overall raster-scanning process. The scanning process is line-by-line downward at a rate of three lines per second. That is, the *fast-scan* axis is horizontal, the *slow-scan* vertical, to create a raster-scan pattern. This is in truth a zig–zag scanning pattern where the individual scan lines are not *exactly* parallel, also the case for the electron beam of a CRT television. (Some less-common AFM scanning systems may instead perform an array of precisely parallel line scans, where the displacement along the slow-scan

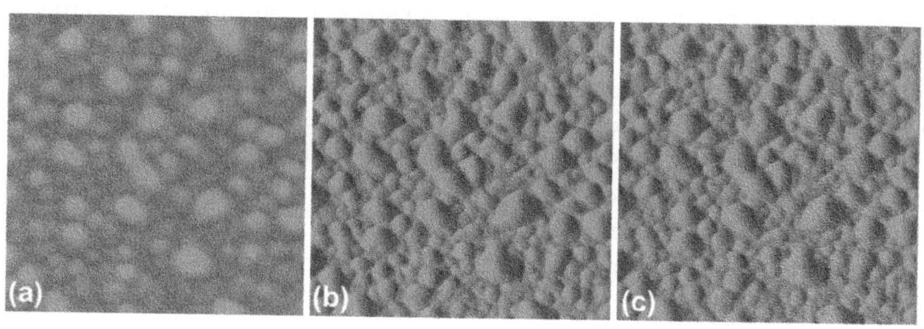

FIGURE 4.8 Images (2 × 2 μm) of (a) *Z* signal (height), (b) error signal (cantilever deflection relative to Setpoint), and (c) differential change of *Z* signal along the horizontal fast-scan direction, respectively.

axis is not continuous, rather in synchronization with the change of scan line.) Along the slow-scan axis, usually defined as Y, the scanner bias is simply ramped more slowly than that of the fast-scan axis (usually defined as X). The user may choose to collect data during either the left-to-right or the right-to-left excursions of tip relative to sample, often called "trace" and "retrace," and also while rastering up or down. Brighter signal denotes higher values of Z (a) and Error Signal (b). In (b), a more upward cantilever deflection, meaning a more positive deviation from the setpoint deflection value, is rendered brighter, whereas a more downward cantilever deflection or more negative deviation from setpoint is rendered darker. One result of this relationship is that the error signal varies essentially as the *first derivative* or *slope* of topography along the scan direction, $\delta Z/\delta X$. This relationship is empirically demonstrated if one creates a *differential height* image from the height data in (a) and compares to image (b). Image (c) is such a differential height image, calculated at each pixel location i along scan line j as the difference of height (Z) values at locations $(i+1, j)$ and $(i-1, j)$. Image (d) is essentially identical to image (b). Note that these images were created from entirely different kinds of voltage measurements (located at the left and right sides of Figures 4.4 and 4.5); in (c), the measurement is a change of the bias voltage applied to displace the Z scanner, whereas in (b), it is the difference between the output signal from the quad photodiode and the setpoint (the error signal).

It is worth underscoring that the data comprising an AFM "image" is a 2D matrix of numbers. We simply choose to render these numbers as colors or shades of brightness, typically using some gradient of grayscale or monochrome brightness, or variations in RGB-derived color. Thus, you might say that AFM images start out as numbers and are converted into pictures; whereas more traditional micrographs (e.g., light or electron microscopy) start out as pictures—acquired measurements of hue and/or intensity—then are digitized, say, for viewing on a computer screen in JPEG, TIFF, or other formats. AFM images also can be converted into these common graphical forms, but at this point, the images *are no longer AFM data*; there is no standard, simple conversion from colors and tints in a JPEG, for example, to heights in an AFM image.

Figure 4.9 compares the preceding image data in its more "native" graphical form rather than color or brightness, along one particular scan line (a single value of Y). In Figure 4.9a, the height or Z scale is shown at left (arbitrarily zeroed) and "stretched out" compared to the X scale as is common when viewing AFM topographic data. At right is the scale of the superimposed Error Signal: cantilever deflection (converted from volt to nanometer per calibration to a unity force-curve slope in contact) relative to setpoint. We see that the extremes of Error Signal are about ± 1 nm, whereas the Z displacements range over some 18 nm. (We will return to the issue of reducing this error via gain settings.) In Figure 4.9b, the Error Signal is directly compared to the calculated slope or first derivative of height $\delta Z/\delta X$. Note that the curves are nearly identical. This is the quantitative validation of what our eyes perceive in Figure 4.8b and c.

It is worth noting in passing that new users of AFM often "prefer" the deflection image to the height image. This is because we humans are not used to viewing surfaces that are color-coded or brightness-coded according to surface height. Instead,

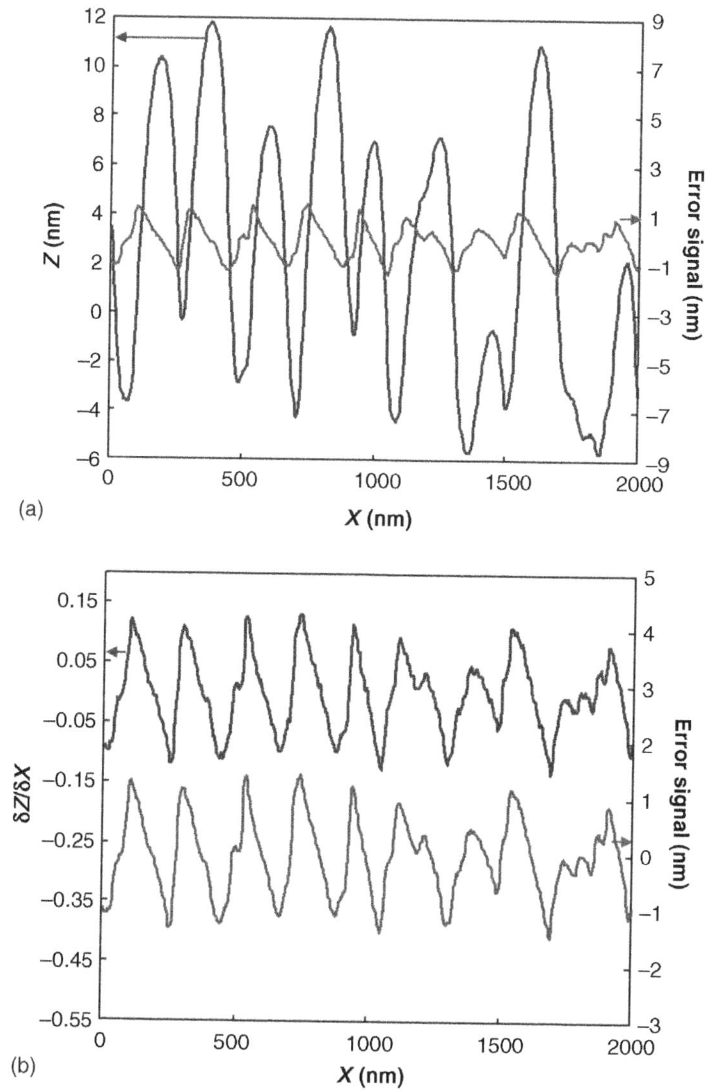

FIGURE 4.9 Graphical presentation of select single-line traces of data extracted from the images in Figure 4.8.

we are used to viewing surfaces that are relatively bright or dark due to the *reflecting and/or shadowing of incident light*. If it is inherently a multicolor surface, such as a tapestry, we do not perceive this as a multiheight tapestry. If we view a monochrome surface, such as painted concrete-block wall, via illumination, the reflections and shadows we perceive derive from the local *slope* of the surface with respect to the incident angle of the light source, relative to our viewing angle. We see different tints of a color depending on the variations of surface slope. An AFM deflection image generated during left-to-right tip scanning looks as if a surface illuminated by light incident from the left of the viewer (given the image rendering convention used in Figure 4.8b). Thus, one intuitively perceives texture and roughness in the deflection image but not in the height (Z) image, or at least not until the mind is trained to perceive bright and dark as high and low, such that a wide range of brightness is "seen" as a rough surface.

A bigger problem with conventional height image rendering is that many surfaces contain important features that are vertically small but superimposed on coarser topography spanning a large range of height. Thus, using a color or brightness gradient to map the entire height range does not always reveal the features one wishes to see, although the information is contained in the data. One can, however, adjust the range of color or tint assigned to a range of heights, as well as offset the color scale, to reveal generally high or low surfaces. This is demonstrated in Figure 4.10 for a solution-grown (macroscopic) AgBr crystal, the (111) face (previewed in Figure 1.1). AgBr microcrystals are the light-sensitive elements in traditional photographic film. Although imaging in a darkened room and with a cover over the AFM, the red laser light caused elemental silver grains or rods (Ag^0) to grow by photoreduction (presumably at atomic defect sites, because a perfect AgBr crystal should not be sensitive to red light), and along well-defined crystallographic directions [7]. Because the rods are ~10 nm tall, their nanoscale surface topography cannot be revealed in the same color scheme in which the AgBr surface topography is revealed (also nanoscale); instead the full range of colors (or brightness if using monochrome or grayscale) must be assigned to a narrower height range in either case. In Figure 4.10a, we see atomically flat terraces on AgBr separated by approximately 0.3-nm steps, the distance between adjacent atomic planes. The range of heights within a subregion of the image is quantified in the histogram of Figure 4.10c. Most of the counts (pixels) in the image correspond to a small range of low surface elevations, a complex of narrow peaks bracketed at left in Figure 4.10c. Conversely, a broad and very shallow signal under the bracket at right derives from the pixels representing the silver rods. These comprise a small fraction of the surface (thus low counts) and are quite variable in height (thus a broad signal). The color spectrum bars superimposed in Figure 4.10c depict the approximate range of colors employed to render visible each portion of the histogram in the image representations of Figure 4.10a and b. One can imagine defining a custom RGB color scale containing *two* steep color gradients, each centered at the AgBr or Ag^0 surfaces. This sort of customization of color gradients is available in some AFM post-processing software. But such a task may be a lot of work compared to simply viewing the error signal or differential height image, the latter shown in Figure 4.10d. Here, the "shadowing" effect is striking, again as if a light source is illuminating the surface from a shallow angle at left.

Histograms are broadly used in AFM to analyze data contained in both topographic and property-sensitive images. One might think of this as a quasispectroscopy. (We will see further examples in Chapter 5.) Thus, it is important to understand histogram representations. Although to some, the concept is trivial, many new users mistakenly interpret the vertical scale of the histogram as the image

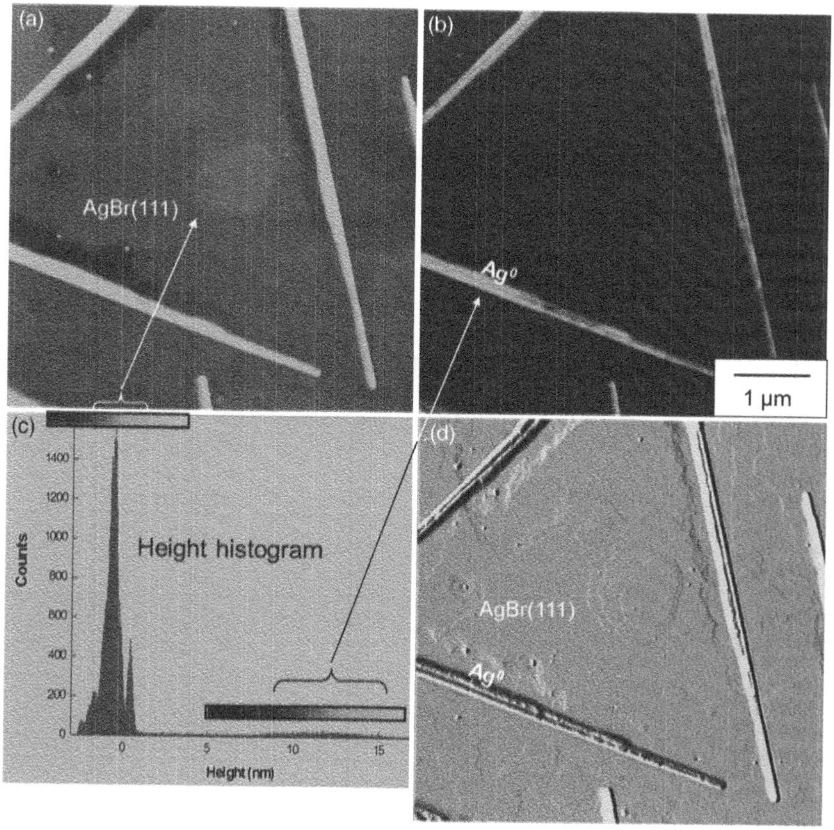

FIGURE 4.10 (a), (b), (d) Height images of the (111) crystal face of solution-grown (macroscopic) AgBr plus Ag^0 rods that form via photoreduction by AFM laser illumination. (a) Image data rendered as gradient of color, with the gradient centered on the AgBr crystal surface as depicted in the histogram of (c). (b) Same image data, with the color gradient spanning heights atop the Ag^0 rods as depicted in (c). (d) Same image data rendered as differential height.

data scale. Thus, to reiterate, it is the *horizontal* scale that represents the image data (height; in later chapters, friction force, phase, adhesion, . . .), whereas the vertical scale represents the pixel count or *frequency* of a given data value. To better understand a histogram representation of an image, consider a single line scan of height versus X, shown as the top graph in Figure 4.11, extracted from the height image in Figure 4.10. We subdivide the height scale (horizontal lines) into narrow intervals or *bins* and count the number of data points that lie within each bin. This number is then represented as a vertical bar in the bottom graph, the histogram plot. The tall surface feature in the height trace (Ag^0) consists of a small number of vertically spread data points, typically one count in any given bin and with plenty of blank bins in between. The two highest, circled data points must then produce the rightmost points in the histogram; a cluster of points around a particular height (inside the rectangle) yields a correspondingly concentrated histogram signal.

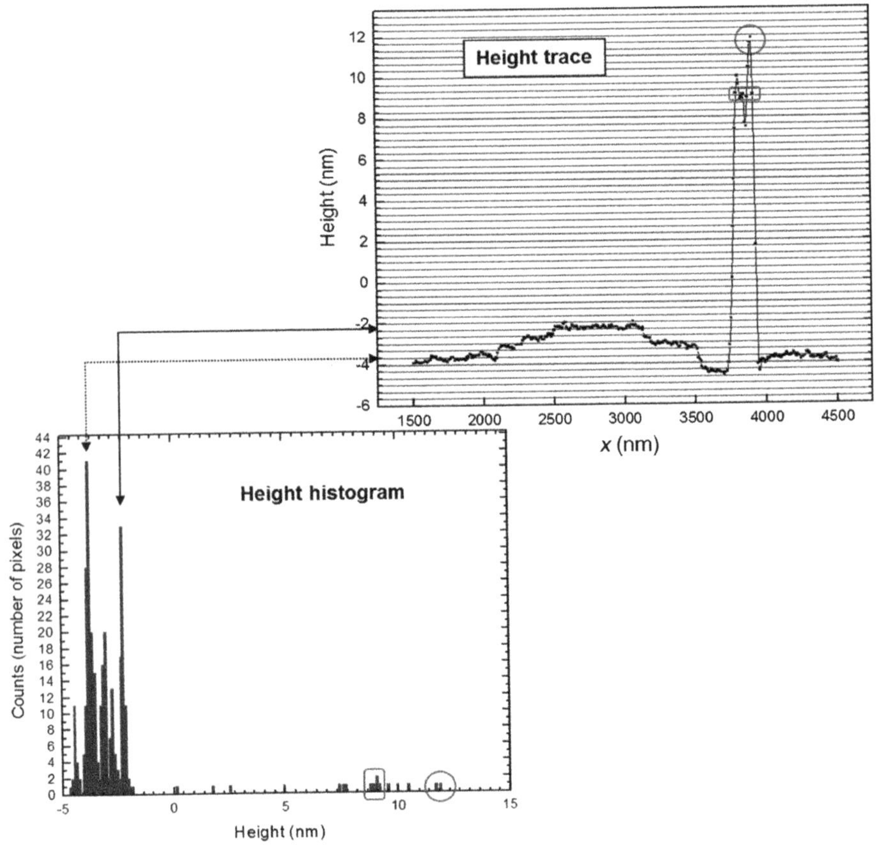

FIGURE 4.11 Relationship between a single trace of height data (top) and the same data in histogram form (bottom). Corresponding high points are encircled in each.

Note that the term *height* is being used in two distinct ways! On the one hand, there is the height of a surface being mapped within an image; on the other hand, there is the height of the histogram that represents a pixel count. This is part of the confusion. Again, it is a histogram *peak position* that is, used to quantify a characteristic height or other signal, just as it is in conventional radiative spectroscopies or mass spectrometries. In these latter techniques, the peak position is the characteristic feature (e.g., type of chemical bond, size of mass fragment), whereas the peak height quantifies the amount of each species. In AFM, the amount or height of the histogram is the number of pixels. That this signal is spatially resolved means that one can generate spectroscopic representations (histograms) *localized to certain regions*, even noncontiguous subregions of data dispersed across the overall imaged region. In Chapter 5, we will see examples of the usefulness of this "spatially dissected" quantification of image data.

The issue of small, potentially interesting height features being superimposed on larger, uninteresting topography is even more egregious than the example of

Figure 4.10 because almost any sample surface will be arbitrarily tilted relative to the X–Y scanning plane. Given that AFM height features of interest are usually much shorter than the lateral size of the image, even a $\sim 1°$ tilt results in a substantial change of Z across the image. Thus, the full gradient of color or range of brightness may be consumed by the rendering of the change of Z across the image. This is typical in large images (tens of micrometers across) but can be substantial even in fairly small images: a 5×5 μm region tilted by 1° corner-to-corner, for example, would produce a $\approx$120-nm change of Z. To perceive the features of interest, one wishes to remove this tilt. Typically, this is done using one of three mathematical routines and, moreover, may be performed "on the fly," as the data are being captured and rendered to the computer screen (dependent on details of the AFM system's software). The routines are: (1) a line-by-line adjustment, whereby each scan line of data is fit with a mathematical line (i.e., $Z = c_1X + c_2$, varying parameters c_1 and c_2), and a different equation for each line of data (each value of Y); (2) a full plane adjustment, whereby the entire image is fit with a mathematical plane (i.e., $Z = c_1X + c_2Y + c_3$, varying parameters c_1, c_2, and c_3); and (3) taking a differential δZ between adjacent points along X, whereby the overall image slope common at both adjacent points is subtracted out, leaving only those variations that depart from this slope. Figure 4.12 compares (a) raw height image to (b) the same data following the subtraction of a mathematical plane that was fit through the entire image; the two images invoke the same 500-nm range of contrast from dark to bright. The rectangular strips are approximately 50-nm-tall gold and aluminum lines grown by lithographic procedures on a silicon substrate. In the presence of the arbitrary surface tilt (a), these lines are not visible near the top and bottom of image, but they are distinctly visible from top to bottom of image (b).

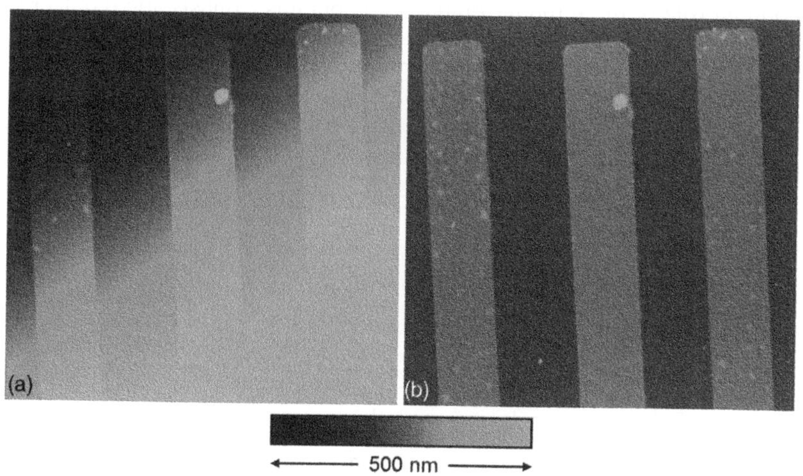

FIGURE 4.12 (a) Raw 35×35-μm height image of gold and aluminum strips lithographically grown on silicon. (b) Same image after a best-fit mathematically plane has been subtracted from (a).

It is worth pointing out that the various vendors' software may use *different terminology* for the preceding mathematical algorithms, even inconsistently within a given software package. For example, in one commonly used version, the term "Flatten" is used for line-by-line adjustments when performed during post-processing ("offline"), whereas the same software refers to this mathematical operation as "realtime planefit" if performed on the fly during data acquisition. (This latter term is simply a *misnomer*: a mathematical plane is not being fit nor subtracted.)

Routines 1 and 2 often benefit from a more nuanced approach. A subset of the points within an image may correspond to a truly planar portion of the surface, whereas other points correspond to protrusions or valleys, or for that matter higher or lower terraces, that clearly do not lie on this plane (such as the rectangular strips in Figure 4.12). In Section 4.2.5.2, we will describe more detailed approaches to mathematical fitting over selected domains of an image, where certain regions are selectively included or excluded before running the mathematical fitting procedure (usually during post-processing). Thus, for example, a plane can be calculated to fit through patches of bare substrate, then subtracted out from the entire image. Even more minimally, a simple three-point selection can be made to determine a mathematical plane, if the user believes these points to lie on a planar portion of the surface. More significantly, we will describe approaches to mathematical fitting that can use higher-order polynomial functions to remove *scanner nonlinearities* as detailed in Section 4.2.5.2.

4.2.3 Choosing Setpoint: Effects on Tip–Sample Interaction and Thereby on Images

Now let us further consider forces during imaging. First, one should recognize the all-important role of the setpoint. Figure 4.13 contains a quasistatic (contact mode)

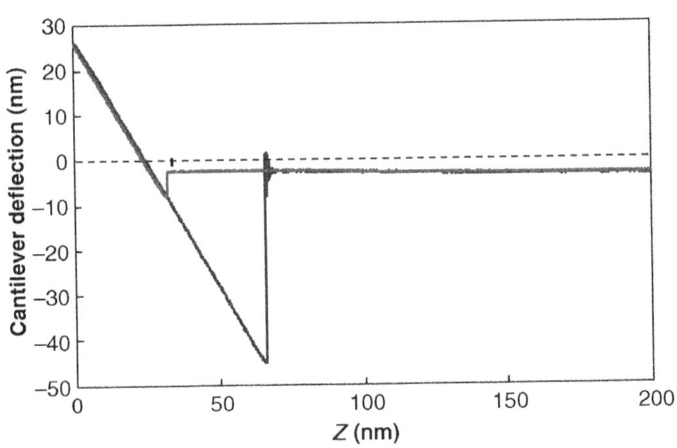

FIGURE 4.13 Quasistatic (contact mode) force curve acquired at a single X–Y location near the center of the image region of Figure 4.8. Dashed line depicts the setpoint and vertical bar the range of deviation from this setpoint (Error Signal), while collecting the image in Figure 4.8.

force curve acquired at a single X–Y location near the center of the image region of Figure 4.8, following the acquisition of the images of Figure 4.8a and b, to illustrate typical operation; the horizontal dashed line depicts the setpoint used for the preceding image and a small vertical bar depicts the range of deviation from this setpoint (Error Signal) as quantified in Figure 4.13. Among other things, we note that the downward deviations of cantilever deflection are small enough that deflection remains positive relative to the undeflected state (e.g., the horizontal data trend preceding jump-to-contact), the *zero* of force, meaning the cantilever spring force is always *pushing* the tip to the surface. This ensures stable operation. In many cases, however, one can operate such that not only deviations of cantilever deflection but even the setpoint itself (the mean deflection) remain *below* this zero. This means that during imaging the cantilever is continually *pulling* the tip in a direction away from the surface! In such a case, tip–sample contact is preserved by the *stronger adhesive forces pulling tip to sample* (i.e., opposing the cantilever spring force). This usually works if scan size is small (micrometers or submicrometer, not tens of micrometers), surface is not very rough (say a few nanometers of peak-to-valley height variations), and the magnitude of tip–sample adhesion is not highly variable across the imaged region (i.e., the minimum in Figure 4.13 remains below the setpoint at all locations). But why would one wish to operate this way? We will see in Section 6.2 that one adhesive contact mechanics model predicts a tip–sample contact diameter that reduces by approximately a factor of two if imaging at a setpoint near the minimum of the force curve (break of contact) such as that in Figure 4.13. Thus, pulling on the tip can measurably *improve resolution*, as well as reduce both the contact force and contact pressure (possibly preventing sample damage). These same contact mechanics models also predict variations in contact diameter, and thus adhesion and contact pressure, due to variations in local surface curvature (i.e., atop a hill versus in a valley). It is important to note that these models, however, do not describe sliding contacts; that is, the size and shape of the tip–sample contact area, and thus adhesion, may be additionally affected by the *dynamics* of the tangential scanning (sliding) process and thus the scan velocity. Thus, the assessment of tip–sample adhesion obtained during a single approach–retract cycle, such as that in Figure 4.13, may not exactly apply to the case of the sliding tip–sample interaction. In Section 7.4.2, we will discuss setpoint ramping in the context of contact mode, wherein the break of contact under a pulling force (strength of adhesion) will be measureable *during the sliding process*.

Adjusting the setpoint to maintain a light positive touch, for the purpose of stable, nondestructive imaging, or reducing setpoint to produce a pulling force and thereby optimal imaging resolution, are not the only choices useful to a broad range of applications. It is not long before the curious operator discovers the usefulness of an *elevated* setpoint. Whether one is studying a thin-film sample by design or investigating the presence of surface contaminants, it is often possible to selectively scratch or even remove a film leaving an exposed substrate. As an example, we return to the polycrystalline Al surface as imaged in Figures 4.8 and 4.9. The same surface imaged at a high setpoint (upward deflection of 170 nm, times a nominal spring constant of $k = 3\,\text{N/m}$ to give an applied load of $\approx 500\,\text{nN}$), initially

FIGURE 4.14 (a) 1 × 1-μm height image collected during high-setpoint contact-mode imaging. (b) Subsequent low-setpoint height image of 2 × 2-μm region with same image center. (c) Differential height rendering of the same data as in (b).

produced a stable 1 × 1- μm height image as shown in Figure 4.14a. Upon lowering the setpoint to a positive deflection of 7 nm and zooming out to collect 2 × 2- μm height and deflection images (Figure 4.14b and c), a square trench is revealed, apparently due to abrasion and clearing of the 1 × 1- μm region during the preceding high-setpoint scanning. That the texture of this surface is modified from the original—containing generally smaller and more uniform grains—is clear in the differential height image of Figure 4.14c. In this case, the smallness of the cleared grains is comparable to the finer grains imaged on the original surface (Figure 4.8c), indicating that the tip shape and sharpness did not measurably degrade during high-setpoint scanning. A reasonable interpretation is that the abraded film is organic in nature, easily torn by the tip at higher forces. But the fact that pristine, reproducible images were obtained at lower force, and a lack of any "box" upon zooming out, indicates that the film is not highly sensitive to abrasion. Such is a hallmark of AFM: an ability to identify a load threshold above which a particular sample component is torn while other components are unaffected. We will return to this system in Chapter 5 to examine with other imaging modes such as friction force and phase, to obtain materials-contrasting signals that can aid to identify film and substrate, complementing the topographic phenomenology of Figure 4.14.

The preceding abrasion, which was controllably explored by the user, may *unavoidably* occur in many cases, unfortunately. For example, on (i) soft synthetic materials, (ii) most biological samples, (iii) objects weakly adhered to substrates, and even (iv) hard surfaces contaminated by organics, one may find that after scanning a small region in contact mode and zooming out, irreversible changes have occurred: a "box" is clearly visible. Often this is unavoidable in contact mode even if operating at strong pulling forces where the net contact force is minimal. Thus, intermittent-contact and noncontact methods have found great utility and, nowadays, may be the most commonly used modes of operation. The main reason is the ability to *remove shear forces*. It is empirically well known that shear forces are a bigger problem than compressive or tensile forces in AFM of soft or weaker adhered samples. Often a material's shear *strength* (resistance to yield) is lower than its tensile or compressive strength. The shear strength may be exceeded in contact mode, causing plastic deformation, although the normal stress remains elastic (i.e., unproblematic). Intermittent-contact methods, such as dynamic mode (treated more fully in Chapter 5) and force-curve mapping, cycle the distance between tip and sample such that the tip is not in contact with the surface for the large majority of each cycle. The amount of X–Y movement during each brief contact is very small—nanometer-scale—such that sliding may be prevented. (In Chapter 7, we will probe static friction via the onset of sliding as a function of increasing shear-modulation amplitude.) Moreover, the contact time may be very short compared to the intrinsic relaxation timescale of many soft samples, such that the material or biological surface may benefit from *dynamic stiffening*. These concepts are well known in the mechanical properties of viscoelastic materials and biological tissues [8–10].

Figure 4.15 contains an example of a block copolymer coating surface that was modified by contact-mode scanning but not by intermittent-contact scanning, whether in dynamic mode (repulsive regime, i.e., true "tapping") or fast force-curve mapping, the latter performed such that the largest load was equal to that under which tearing resulted in contact mode. Initial topographic images of a 1×1 μm

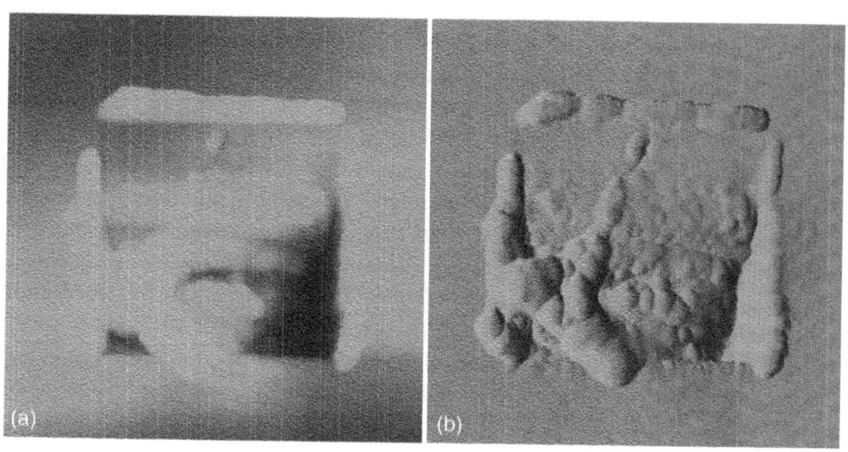

FIGURE 4.15 2×2-μm height (a) and differential height (b) images collected in intermittent-contact following abrasive/destructive scanning of a 1×1-μm region in contact mode.

region were acquired in dynamic, force-curve mapping and contact modes, in that experimental order. In contact mode, repeated scanning produced dramatically changing surface features, a phenomenon not observed during repeated scanning with the two intermittent-contact modes. Larger $2 \times 2\,\mu m$ images were subsequently collected in dynamic and force-curve mapping modes, without further inducing wear, thereby imaging the effects of scanning in contact mode. An obvious box of modified surface topography is observed corresponding to the region scanned in contact mode.

Naturally, most AFM applications, whether in contact, dynamic, or other modes, are performed with parameter values (setpoint and others) selected such that nonabrasive, nonplastic interactions result between tip and sample. But even in this case, one need not explore parameter space for long before discovering the utility of variable setpoint. In Chapter 7, we will vary the setpoint in a continuous fashion in contact mode so as to vary load and thus friction force, providing friction versus load plots wherein the *coefficient of friction* specific to material can be measured. This is a pretty obvious role for a variable setpoint. Less obvious is the role of setpoint in dynamic mode, which is examined in detail in Chapter 5. We will see that the qualitative nature of tip–sample interaction can be exceedingly sensitive to the amplitude setpoint in dynamic mode. In this chapter, our focus, however, is on basic topographic imaging. But even in this ostensibly simple case, the mechanical role of setpoint can be easy to identify as an apparent change of height. This type of phenomenon was seen in Figure 1.19, comparing topographic images obtained on a polymer blend at lower and higher temperatures at which one component softened, such that apparent relative heights changed. Similarly, if one examines a material that is heterogeneous in mechanical compliance at room temperature, one will observe differences in *apparent height* that are in truth differences in *indentation depth*.

Figure 4.16 demonstrates this phenomenon for the case of dynamic ("tapping") mode. Parts (a) and (b) compare images of the same 5×5- μm surface region of a n-alkyl methacrylate polymer blend (75% being $n = 4$, butyl, and 25% being $n = 12$, lauryl) imaged at two very different choices for setpoint amplitude ratio A/A_0, (a) 0.86 and (b) 0.52, where A is the amplitude maintained during imaging (3.82 V and 2.32 V, respectively, in raw read signal) and A_0 is the free-oscillation amplitude (4.45 V) at the resonance frequency (69.53 kHz for a silicon rectangular cantilever with $k = 3$ N/m nominal). See Section 2.4. (The drive frequency was set at 69.42 kHz to help stabilize a dominant repulsive tip–sample interaction, a topic further discussed in Chapter 5.) These settings provide (a) one image that is nearly true topography and (b) a second image that contains stark differences in mechanical compliance. The circular regions are clearly much softer than the surrounding regions: the Z scanner must displace the tip toward the sample surface by an additional $\approx$12 nm to achieve the same reduction of amplitude of oscillating tip, quantified in the horizontal cross section of data overlaid in (c), where the thicker data trace was extracted from (b). Thus, dome-like topographic features a few nanometers tall in (a) reversibly "sink" by $\approx$12 nm in (b). This empirical result is further elucidated in Figure 4.17 in the context of amplitude–Z curves acquired during approach to two (X, Y) locations, one on each of the two polymers. We see that the

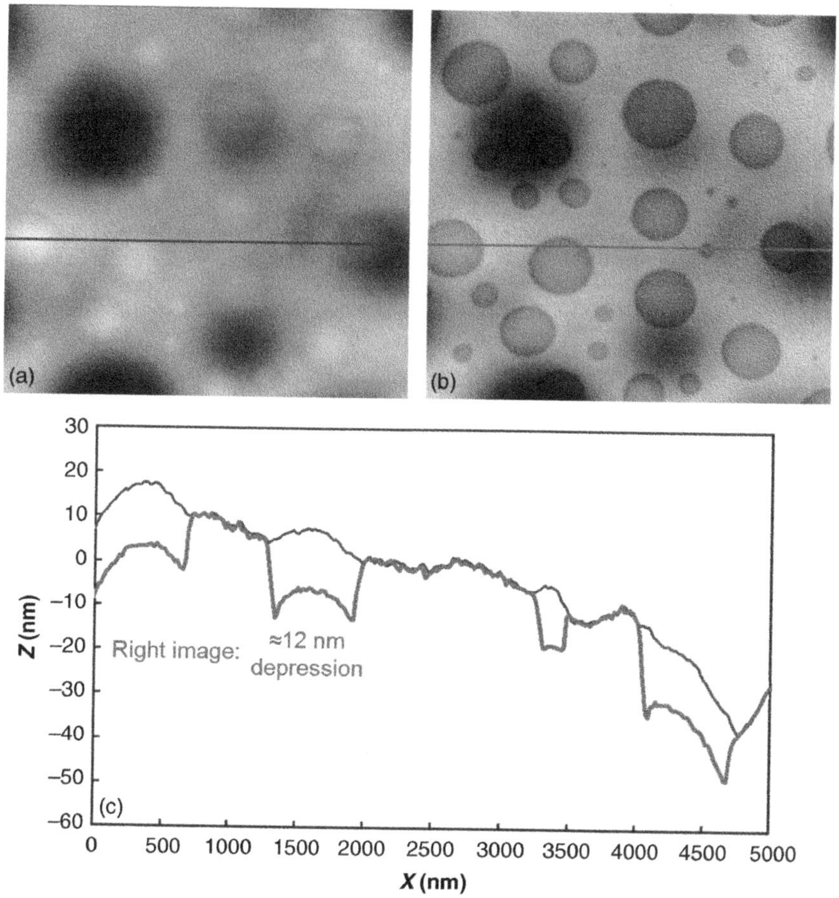

FIGURE 4.16 (a) and (b) Height images of a 5 × 5-μm surface region of a *n*-alkyl methacrylate polymer blend, acquired at dynamic setpoint amplitude ratios A/A_0 of (a) 0.86 and (b) 0.52. (c) Graphical comparison of the height traces designated in (a) and (b).

Z displacement needed to reduce the amplitude from 3.82 V to 2.32 V differs on the two polymers by about 12 nm. This additional *Z* displacement is quantitatively consistent with the 12-nm depressions seen in the "height" trace of Figure 4.16c for the two corresponding amplitude setpoints.

In Chapter 5, we consider phase images acquired simultaneously with the "height" images in Figure 4.16, to examine differences in tip–sample energy dissipation (via Eq. 2.24). We will also examine the same regions imaged using fast force-curve mapping, to compare the impact of each type of interaction on the apparent height: intermittent but *quasistatic* force loading versus *dynamic* tip–sample interaction. Third, although we defer to Chapter 5 a deeper discussion, it is worth previewing another, *exceedingly important height artifact* that often arises in dynamic mode. Although the above discussion of Figures 4.16 and 4.17 addresses

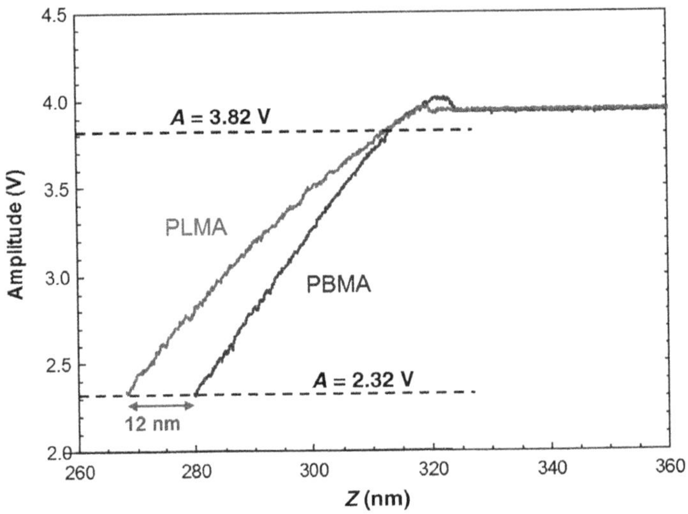

FIGURE 4.17 Amplitude–Z curves acquired during approach to two locations in the region of Figure 4.16, on each of the two polymers.

differences in indentation that masquerade as height differences, dynamic mode also can introduce height artifacts even while operating under extremely delicate tip–sample interaction, where indentation is minimal. This is because *switching* can occur between net repulsive and net attractive interaction regimes, as mentioned in Section 4.1 in the context of Figure 4.2. Upon scanning from one material type to another, or even from valley to hill, a new dynamic state can arise such that the oscillator switches from intermittent contact to an oscillating near-contact state. What has enabled this switch is the feedback-driven Z-scanner displacement, where again is imaged as a (false) change of height. Similar to Figures 4.16 and 4.17, where the oscillator stabilizes in one of two states differing greatly in the magnitude of positive tip indention into sample at nearest approach, the oscillator also can stabilize in one of two states differing in the "sign of indentation" at nearest approach: positive (into a solid sample) and negative (not reaching solid sample). The Z displacement that causes this switch, and thus a height artifact, and can range from $\sim$1 nm to tens of nanometers depending on the nature of the attractive forces at play, from relatively short-range van der Waals to potentially very long-range electrostatic forces.

4.2.4 Finite Response of Feedback Control System

Considering the block diagrams of Figures 4.4 and 4.5, apart from the sign of reactive Z displacement, dependent on mode of operation and change of surface elevation, what can we say of its *quickness*? In the real world, no feedback-actuated control system is instantaneous. Automobile drivers may note that when the car is climbing a steep hill on a freeway, there is a deviation of the cruise control from the setpoint speed, whereas for very gradual variations of road height this deviation is smaller. In other words, the *error* in the response of the cruise control scales with the *slope* of the incline. In AFM, a steeper topography will likewise cause a larger deviation (error) from the setpoint, as seen in Figure 4.8. Because of the nature of the feedback system, the strength of the reaction of Z at any instant scales with the magnitude of this deviation or tracking error as compiled during some preceding time window. The magnitude of deviation obviously depends on the slope of the surface, but it also depends on the *slope of the data relationship between deflection or amplitude and* Z as depicted by the slopes of the diagonal lines in Figure 4.3 and by the data trends in Figures 4.17 and 3.23a. Generally, there are control system *gain settings* (discussed below) that the operator adjusts to allow for variability in this s_Z–Z or A–Z slope, in turn dependent on material type. On a rigid material, the slope is greatest; a change of Z would produce an approximately equal change in deflection or amplitude. On softer materials, the slope will be lower (see Figure 3.6, specifically part b) and thus the operator would need to set the gain higher to achieve the same "quickness" in the feedback.

Most AFM systems use multiple gain settings. These weight different aspects of the error signal to determine the strength of reaction in Z. A common scheme of feedback in many technologies (e.g., temperature control) is the PID controller, where PID stands for proportional-integral-derivative. A PID controller attempts to correct the error between the process variable (e.g., a temperature read or the laser–photodiode-based measurement of deflection or amplitude in the case of AFM) and the setpoint specified by the operator. It does this by employing an algorithm to calculate, and hardware to generate, a corrective action. This corrective action is the weighted sum of three terms. The proportional gain setting multiplies the last measurement of the error signal (dependent on sampling rate), whereas the integral gain setting multiplies a running sum of error signals. The derivative gain, if employed, weights the reaction based on the rate of change of the error signal. These three actions together adjust a control element—such as the foot pedal in an automobile's cruise control system or a current to a resistive heater in a temperature control system—in opposition to the sign of the error signal. Many applications use only one or two control types, say I or PI instead of PID control. The derivative action is exceedingly sensitive to measurement noise, and thus is not typically used in AFM Z-control systems. The significant noise in AFM systems, for example, the thermal noise in the cantilever deflection signal, means that the overall control system is typically engineered to weight the integral signal the heaviest. By integrating over many measurements, the effects of thermal noise in the cantilever are suppressed. Some commercial systems employ a fourth control action, so-called "look ahead"; here, the error at the same pixel number along the fast-scan axis (call this "i"), but in the previous scan line or lines (numbers "j") is weighted by a look-ahead gain before being added to the aggregate reaction. This way, responses that are adjacent in space by not adjacent in time—a necessary aspect of raster scanning—are weighted in the control-system response.

Figures 4.18 and 4.19 are representative of the relationship between gain settings and nanoscale surface tracking for PI and PI/look-ahead control systems.

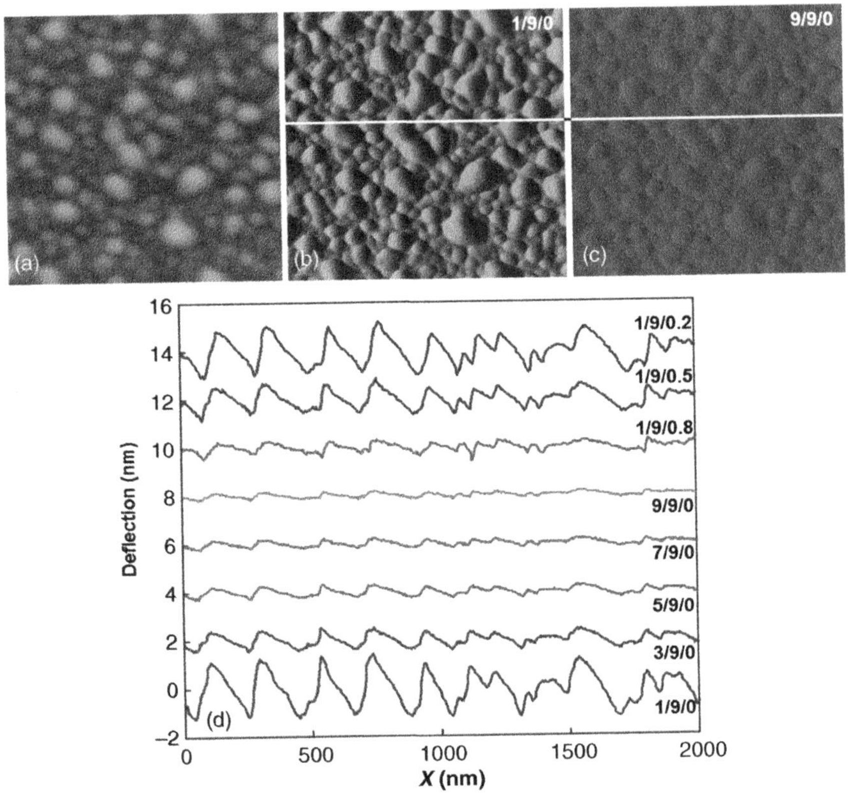

FIGURE 4.18 (a) Height image of (contaminated) polycrystalline aluminum acquired in contact mode at proportional/integral/look-ahead gain settings of 1/9/0. (b) Error signal image acquired simultaneously with (a). (c) Error signal image acquired as in (b) but with integral gain setting at 9. (d) Error signal data acquired at different integral/proportional/look-ahead gain settings as indicated. Data acquired along scan line marked in (b) and (c).

Contact-mode imaging was employed atop a 50-nm-tall polycrystalline aluminum film (with surface contamination) grown on a silicon wafer by lithographic procedures (as in Figure 4.8). Image size is 2×2 μm and scan frequency is three lines per second. Increasing gain settings results in better surface tracking: the reaction of the Z scanner to departures from the setpoint deflection is quicker. Figure 4.18b–d demonstrates the dependence of error signal (deflection relative to setpoint) on integral gain under PI control, varied from settings of 1–9 (higher values induced oscillations in the control circuit). Also, under PI/look-ahead control with the latter setting varied from 0.2 to 0.8 at integral gain of 1. The departure of deflection from a constant value ranges from ± 0.2 nm in the best case to ± 1.3 nm in the worst, for heights ranging over ≈ 18 nm. Combining high integral and look-ahead gain settings further reduces this error as shown in Figure 4.19, a 2×2-μm image scanning over the 50-nm drop from aluminum strip to silicon substrate. Setting to the instrument-

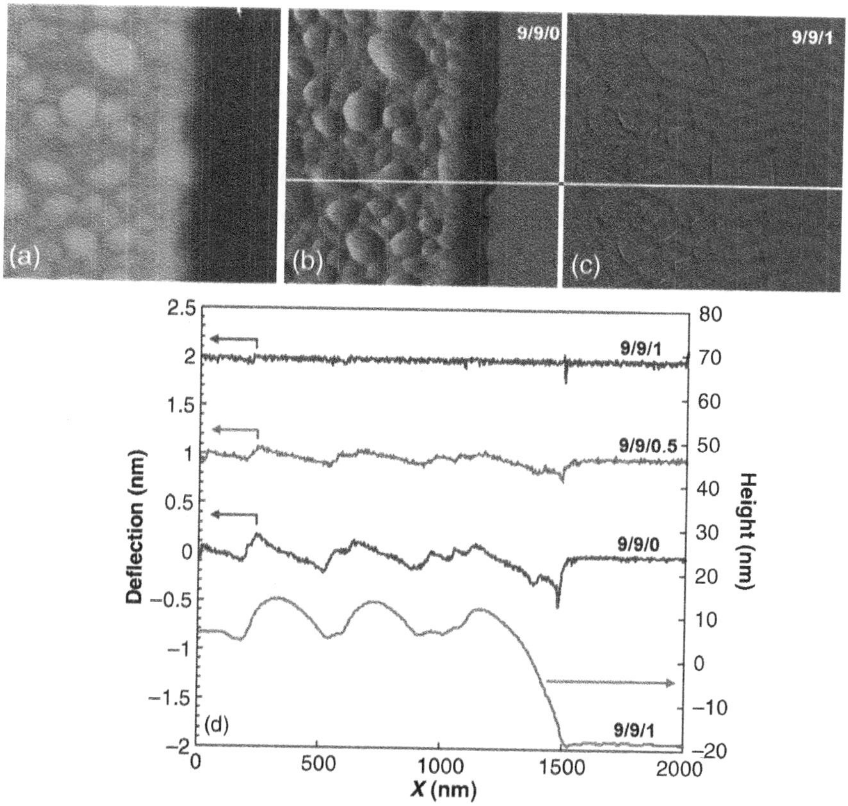

FIGURE 4.19 (a) Height image of (contaminated) polycrystalline aluminum acquired in contact mode at proportional/integral/look-ahead gain settings of 9/9/0. (b) Error signal image acquired simultaneously with (a), contrast scale enhanced relative to Figure 4.18c. (c) Error signal image acquired as in (b) but with look-ahead gain setting at 1. (d) Three error signal traces and one height trace acquired with integral/proportional gains set to 9 and look-ahead gain settings as indicated. Data acquired along scan line marked in (b) and (c).

maximum look-ahead gain of 1, together with an integral gain as high as possible (9, oscillations appearing at higher values), reduces the error signal to nearly below the noise with the exception of a 0.2-nm downward blip corresponding to the steepest part of the film edge and a few other ~0.1-nm deviations at steep locations.

One subtlety is worth noting in Figure 4.19c; steep edges that are oriented vertically in the image tend to be better tracked (have less visible error signal) than edges that are oriented at more like a 45° angle or horizontally. (Look closely at individual grains.) This is because the error in a preceding scan line is a better predictor of how the control system should react in a given scan line. An edge that is oriented vertically in the image is better "anticipated" by the control system than an edge that is oriented significantly off-vertical, because a steep location at a given pixel i in scan line j is more likely to be present at the same pixel i in scan line $j+1$

(or $j - 1$ if rastering in the opposite vertical sense). Thus, error signal may be reduced by judicious orientation of sample relative to raster-scan orientation or vice versa. If utilizing the lateral force to obtain frictional data and contrast (Chapter 7), then the operator's choice of raster-scan orientation is restricted (so that lateral forces during a given fast-scan produce a twist of the cantilever, such that the laser spot moves horizontally at the quad photodiode). In this case, the only remaining setup variable that can affect the quality of tracking over steep edges (such as the edge of the metal line in Figure 4.19) is the user's orienting of the sample.

A conjugate issue to "feedback quickness" is that of scan speed. Obviously, slowing down gives the control circuit more time to react via feedback, and thus results in better tracking. For example, one often observes a remarkable improvement (reduction of error signal) when reducing from 3 Hz to 1 Hz scan line frequency. The most obvious drawback is that a typical 512-line image requires over 8 min to collect instead of less than 3 min. Also note that scanning *size* necessarily determines scanning speed (i.e., in nm/s or μm/s) by the simple relationship

$$\text{Scan speed} = (\text{Scan line frequency}) \times (\text{Scan length}) \times 2.$$

One should recognize that the rate at which the error signal grows as a tip scans up a steep slope depends on this scan speed, not intrinsically on scan line frequency. Thus, zooming out (increasing scan length) by a factor of two is equivalent to increasing the scan line frequency by a factor of two. Indeed, the tracking of laterally small domains can significantly degrade upon zooming out, in addition to the problem that one has fewer pixels available to resolve small features. Scan speed is also conjugate to another issue that is treated more broadly in Section 4.2.7, that is, drift. If the sample is "settling" (e.g., due to adhesive double-stick tape relaxing), then there is appreciable sample movement, at least on the nanometer scale, during the time required to capture an image. Thus, features in the image will be somewhat distorted because an object is moving during the time it takes for the multiline raster scan to traverse the object. In such cases, one may decide to scan faster to reduce the effects of drift (though a better solution is to improve the sample mount).

In many cases, the "option" of scanning faster to minimize the effects of drift (or because one's time is limited) is even less of an option. One reason is the above issue, that the control circuit's tracking of height is degraded by scanning at higher speeds. But another issue is independent of the instrument device performance and applies mainly to dynamic mode: the quality factor or Q of the cantilever, as discussed in Section 2.4.2. Tracking of surface is contingent upon achieving *steady-state dynamic equilibrium* at each pixel location. In the language of mathematical physics, the oscillatory motion of the tip, $s_Z(t)$, should be dominated by the particular solution to the differential equation for Newton's second law, not the complementary solution. The latter is transient and decays exponentially in time, but the rate at which it decays scales inversely with Q. Thus, for high Q, one should scan slower. Or if one insists on scanning faster, this physics leads to the seeming paradox that a high-quality factor can result in low-quality images! Note that viscous fluid immersion, most obviously water immersion, strongly decreases Q because of damping (see Eq. 2.19) and effectively removes the impact of Q on surface tracking, as the instrument device performance becomes the primary issue. Another possibility is to artificially reduce Q with external electronics. Commercially available options, whether part of the purchased instrument or contained in third-party add-ons, are generically termed "Q control."

A final issue is the *role of setpoint in topographic tracking in dynamic mode*. As previously stated, the integration of error signal over some time window plays a strong role in this tracking, per the weighting of the integral gain setting. In contact mode, one usually finds that the growth of the integrated error is *symmetric* with respect to the sign of deviation from setpoint, as follows. If the cantilever deflection exceeds setpoint (while tip travels up a steep incline), the integral errors grows in the positive sense. Similarly, any deviation below setpoint (while traveling down an incline) causes integral error (negative) to grow in magnitude *just as strongly*, even if the applied force becomes negative (pulling on the tip) because of adhesion (see Figure 4.3a). Generally, one does not choose to operate at the most negative applied force possible because variations in adhesion can result in a complete loss of tip–sample engagement, where the tip springs from the surface (requiring a large elevation of setpoint to re-engage). Dynamic mode, however, is very different. While traveling up a steep incline, the amplitude drops below setpoint and this negative error grows. But if traveling down a steep incline *with setpoint amplitude chosen to be nearly as large as the free-oscillation amplitude* (in order to maintain a delicate tip–sample interaction), the amplitude can only increase to the value of free oscillation, no more. Thus, the growth of this positive error signal may be weak, and as integrated will not drive the feedback circuit strongly enough to maintain proper engagement. In cross section, one will note that the amplitude signal becomes flat (at the free value) and the topography signal *straight lines* from high to low with no real topography data contained therein. This problem is sometimes called *parachuting* or *hang-gliding*. One may be tempted to turn the integral gain way up, but this is no solution because there is little error signal to be integrated, and, moreover, the high gain will cause feedback oscillations elsewhere in the image. Slowing the scan speed can help, but, in practice, *one usually must decrease the setpoint amplitude*; this way a deviation above setpoint raises the operative point on the diagonal line in Figure 4.3b (or Figure 4.3d), growing the error signal appropriately and thereby driving the feedback system to properly track surface topography. The tradeoff is, of course, a less delicate engagement.

4.2.5 Realities of Piezoscanners: Use of Closed-Loop Scanning

This section covers a number of piezoscanner nonidealities that routinely impact the quality of results, including both subtle and glaring effects. We also discuss closed- versus open-loop scanners. In recent years, this has transitioned from a more fringe topic to a practical issue in the initial purchase and configuration of AFM systems or in upgrade options.

Of various scanner artifacts, two are encountered almost immediately in topography images because of nonideality in piezoscanners: nonlinearity in Z and creep in

X–Y. Z nonlinearity manifests as an upward or downward *curvature* in apparent topography and X–Y creep as *distortion of shapes* in 2D images. We will detail examples of each case later in this section. In Section 3.5, we discussed nonlinearity in the Z scanner as well as hysteresis (directional and scan-history dependence) in this nonlinearity, in the context of force-curve measurements (Figures 3.18 and 3.19).

4.2.5.1 Creep Before addressing creep in X–Y, we first consider the less severe case of creep in Z. This phenomenon was mentioned in passing in Section 3.5, as a source of leftward or rightward drift of force-versus-Z curves, meaning a gradual offset of Z with elapsed time, diminishing in time. Thus, sudden, significant changes in the Z bias result not only in the immediate response of Z displacement (as desired) but also in a subsequently uncontrolled, slow displacement (undesired) that dies away in time. It was mentioned in Section 3.5 that one can minimize creep in Z by first performing several Z cycles over a particular Z ramp size of interest and at a particular Z offset, before collecting force curves of interest. In topographic imaging, however, sudden displacements in Z occur when scanning over steep locations; here, one cannot somehow "wait" for Z creep to subside before continuing to scan laterally.

We begin by noting that in Figure 4.19d, the bottom (height) trace contains a small amount of creep manifest as a slight upward "recovery" after the tip reached the bottom of the steep vertical drop (while scanning left to right). This phenomenon results from the fact that a rapidly changing Z bias, in order to extend the Z piezoscanner as the tip traveled down the edge, was followed by *further uncontrolled extension due to creep* once the tip reached the silicon substrate surface. The s_Z-driven control system, reacting to an upwardly deflecting cantilever as creep progresses, alters the Z scanner bias so as to reduce the extension; this is equivalent to a surface that is higher as the tip continues to scan rightward, and thus is imaged as such. Z creep effects are more apparent in Figure 4.20a because four times larger steps of 200 nm were scanned. It also becomes more obvious if one carefully compares data acquired while scanning in opposite directions, also shown in Figure 4.20a (vertically offset for clarity). One looks for creep effects immediately following in time a sudden change of height (i.e., different for the two directions, as indicated with vertical arrows). It appears as if there is a slight "overshoot" followed by a relaxation back toward a less extreme (of high or low) value. Figure 4.20b contains similar traces across the same sample using the same tip/cantilever, but plotting the output of a special Z-displacement sensor (inductive) as the height value, instead of using the Z bias as proxy for height. Here, creep effects are absent, as expected, because we are not measuring the feedback-driven voltage applied to the Z scanner; instead, we are measuring actual Z displacement. Close inspection reveals, however, a significantly greater amount of noise in Figure 4.20b, produced by the special Z sensor. (This reality is a common theme regarding optional AFM attachments: although expanding capabilities and/or alleviating certain problems, the usage of special attachments often introduces other complications and nonidealities. It is best to have both standard *and* special capabilities in one's arsenal, usage being dependent on application.)

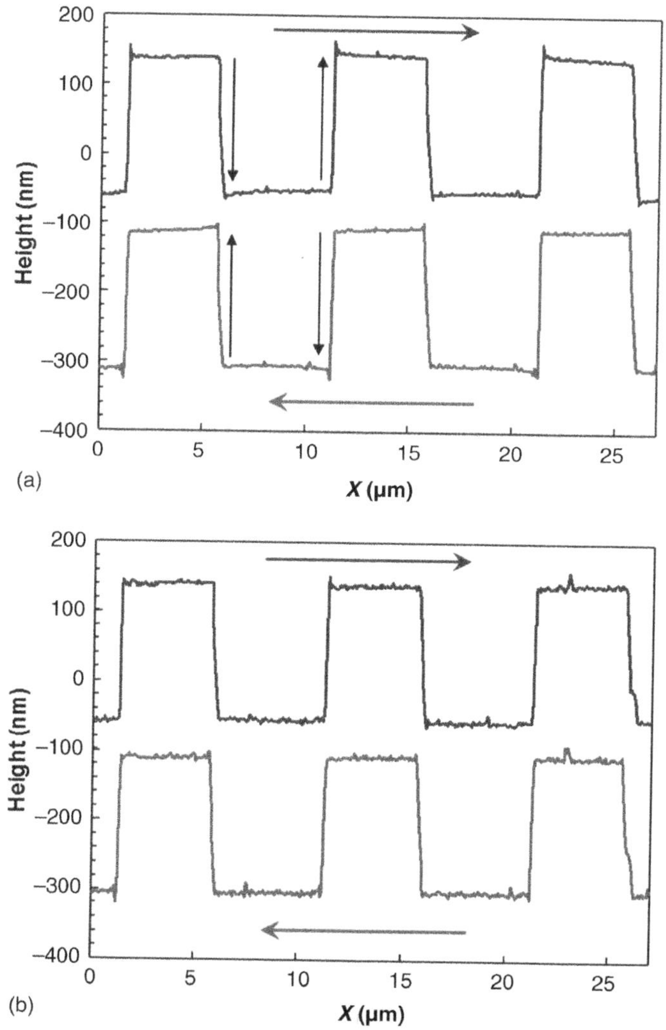

FIGURE 4.20 Comparison of height traces obtained during trace and retrace on a silicon calibration grating containing 200-nm deep pits, via two different methods for measuring *Z* displacement: (a) converting the applied *Z* bias to a "height" using a linear calibration coefficient and (b) using the output of an inductive *Z*-displacement sensor.

Z creep in AFM height images is not noticeable in most cases; that is, most surfaces do not contain tall, abrupt vertical changes followed by flat terraces as in Figure 4.20. In most images, abrupt changes of height are very small fractions of the full *Z* range of the scanner (commonly several micrometers), thus the change of applied bias is usually measured in fractions of a volt, and rarely more than a few volts, both small compared to the full bias range of ~400 V. Creep in *X*–*Y*, by comparison, is perhaps the most commonplace and irksome artifact in AFM, occurring

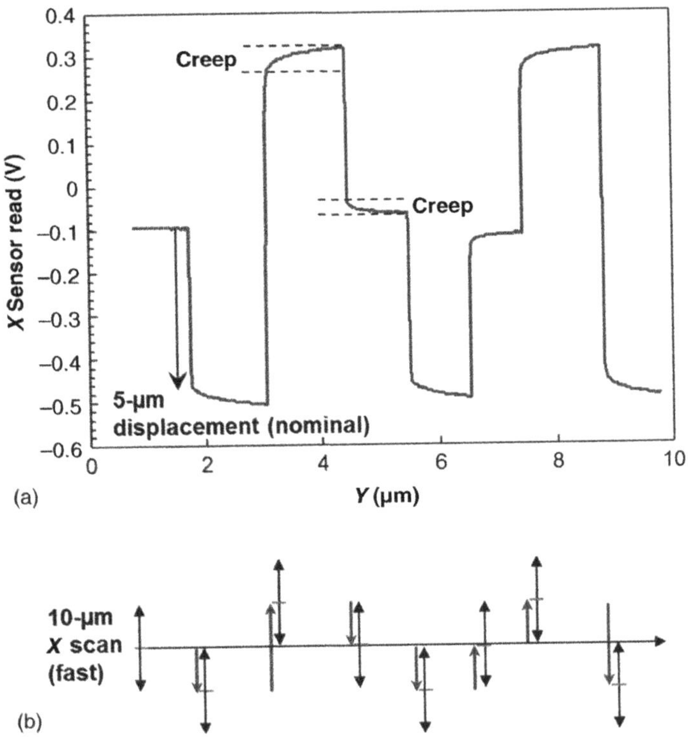

FIGURE 4.21 (a) Measurement of *X* creep upon sudden 5- or 10-μm displacements along *X* as depicted in (b). Sensed movement is the center of a 10-μm scan.

when one suddenly moves the location of an image or significantly changes its size (or both). As with *Z* creep in force curves, one can reduce the impact of lateral creep by allowing elapsed time at the new *X*–*Y* location and/or scan size before acquiring the data of interest. The wait can be long, however, particularly if offsetting a large distance in the *X*–*Y* plane while zooming to a small image size. This is because the change of applied biases for *X*–*Y* often is tens of volts, a significant fraction of the full range (of again ~400 V), and even more importantly, comparable to the range of voltages being ramped in a subsequent image.

This effect is demonstrated in Figure 4.21. During a 10×10 μm raster scan with a ~100-μm-range scanner, the image center was displaced by nominally 5–10 μm in the positive or negative *X* (fast-scan) direction, corresponding to an applied bias change of approximately 20–40 V, respectively. In Figure 4.21a, an inductive *X*-position sensor monitors the true midpoint of the *X* scan (sensor read signal in voltage units) as the raster scan proceeded in the increasing *Y* direction. An initially centered *X*-scan was followed by *X*-displacements of sequentially −5, +10, −5, −5, +5, +5, and −10 μm. Contrary to the idealized depiction of the *X* scan in Figure 4.21b, a substantial creep response is observed in (a) following each *X* offset. The 10-μm offsets produced nearly twice as much creep as the 5-μm offsets, as demarcated by dashed lines for two cases. The direction of creep was always in the direction of the offset that produced the creep.

As already stated, with conventional *open-loop* piezoscanners (i.e., programmatically ramped *X* and *Y* biases), the only recourse is to wait for creep to subside. If frequently zooming in and out over a wide range of image sizes, and/or if frequently offsetting *X* and *Y*, the wait time can be substantial fraction of overall operation time. The only bona fide solution to this problem is a *closed-loop X–Y* scanner (akin to the *Z* case of Section 3.5). Such a device utilizes inductive, capacitive, or light-interferometry sensors to measure scanner movement; two *feedback-control circuits*, dedicated to each of *X* and *Y*, continually adjust the applied scanner biases to minimize the error signal between sensor reads and voltage sets, so as to achieve each desired *X–Y* coordinate within the matrix of points comprising the raster scan.

Figure 4.22 demonstrates the 2D manifestation of *X–Y* creep in AFM deflection (error signal) images of uniformly spaced metal (alternating gold and aluminum) strips lithographically grown on silicon. Closed-loop (left) and conventional open-loop (right) images were acquired while changing back and forth between larger (100 μm) and smaller (30 μm) raster scans. Both images were acquired with fast-scan horizontal and raster-scanning upward. Upon zooming in or expanding back out, rampant *X–Y* creep is observed for open-loop scanning. Creep results in stretching or compression of features, manifest as false curving or incorrect orientations of linear features, because of time-dependent scanner movement (i.e., not controlled by a scanning algorithm). None of this is manifest under closed-loop scanning because the *X* and *Y* positioning sensors are continually measuring the location coordinates of the scanner and modifying the applied *X* and *Y* biases to ensure correct positioning. It should be clear that the benefit of closed-loop scanning is greatest in applications where one is repeatedly zooming in and out and/or offsetting the imaging center in the *X–Y* plane. An example is shown in Figure 4.23 for the same sample as in Figure 4.22; a large, 100 × 100-μm image was first fully acquired, then

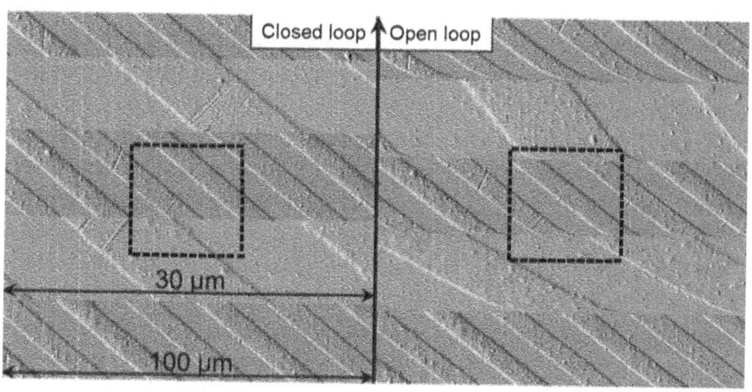

FIGURE 4.22 Comparison of scanner performance while operating in closed versus open loop. Rastering was upward and raster scan size was alternated between 100 μm and 30 μm. Deflection images of a 1D array of metal lines on silicon.

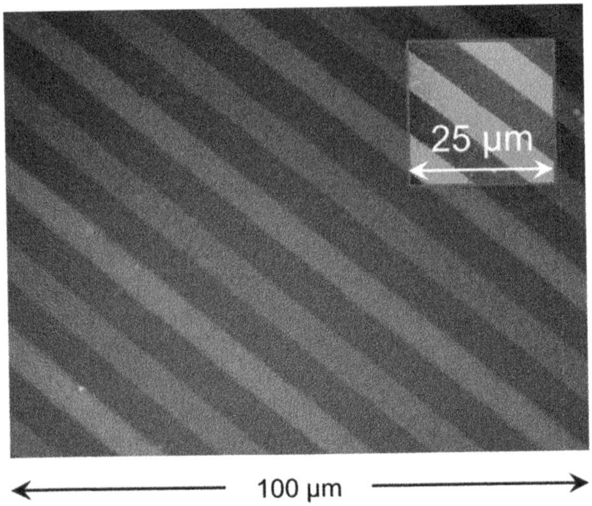

FIGURE 4.23 Demonstration of zooming accuracy under closed-loop scanning. Larger image plus drawn square were captured in a screen shot, whereas zoomed image in upper right was captured with the AFM acquisition software.

a small 25-μm square designated for zooming. A screen shot was used to capture the large image including the position of the zoom square; then the immediately acquired 25-μm zoom image was superimposed (using a different tint scale) inside of this square to check for consistency. The agreement is seen in the positional and orientational alignment of the edges of the metal strips. Many micrometers of creep in *X* and *Y* no doubt take place following such a zoom, but this undesired movement is in fact *measured* by the *X* and *Y* sensors; these measurements drive the control circuits to change the applied *X* and *Y* biases *so as to produce the correct X–Y coordinates*, thereby compensating for the creep. Obviously, if the measurement of angles is critical to an application (say between atomic or molecular steps, i.e., crystallographic information), then a closed-loop scanner can be an important instrument component.

Even if one is simply engaging at a point and zooming out to a fixed image size, closed-loop scanning removes the wait time (often minutes) for creep effects to become minimal. But as with the *Z* sensor, there is usually a measureable noise "penalty" in a closed-loop *X–Y* scanner that is particularly visible in small-scale images (although minimized by reducing the gain settings for the *X* and *Y* feedback). This is exemplified in Figure 4.24, 500 × 500-nm height images of a poly(ethylene glycol)–poly(butylene terephthalate) diblock copolymer acquired in dynamic mode. The image acquired under closed loop *X–Y* (left) contains a greater streakiness compared to the image acquired with conventional open-loop scanning (right). One can think of the extra noise in the closed-loop image as a positional "fuzziness"; for each *X–Y* positioning coordinate, the feedback-driven scanning system introduces an additional lateral "shaking" (as it adjusts the applied voltage to

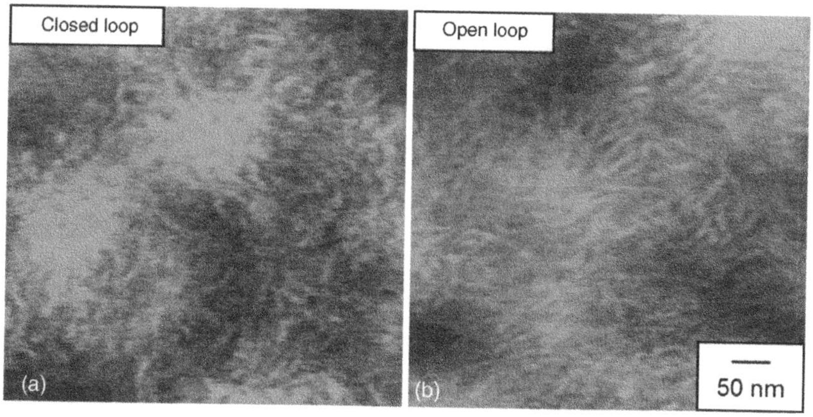

FIGURE 4.24 500 × 500-nm height images of a poly(ethylene glycol–butylene terephthalate) diblock copolymer acquired in dynamic mode, under (a) closed-loop and (b) open-loop *X–Y* positioning, to examine differences in noise.

keep the positional read, with noise, equal to the set), making each point position less well controlled.

4.2.5.2 Nonlinearity, Hysteresis, Cross Coupling Let us now examine the issue of *Z* nonlinearity in greater detail. Although the plane fit correction of Figure 4.12 seemed to do a fine job of leveling the image, closer examination reveals a curved character in this adjusted surface, an artifact that becomes more noticeable for larger images. This is demonstrated in Figure 4.25, 100-μm images of the same sample as in Figure 4.12 (alternating gold and aluminum strips on silicon). Images (a) and (b) are the raw height images acquired during trace and retrace, respectively, as depicted with arrows (rastering up in both cases). Images (c) and (d) are the same data following subtraction of planes determined from the choice of three points on silicon in each image (crosses). There are clearly remaining curvatures due to nonlinearity and clear differences in the character of this curvature due to hysteresis: concave up (brighter in center) and down (brighter near edges), respectively, in images (c) and (d), along *X*. The concavity along *Y* is down for both cases, but up if rastering downward (not shown), exactly analogous to the direction-dependent concavity seen along *X*.

It is worth digressing to note that many AFM post-processing programs allow one to selectively fit mathematical curves or surfaces through *portions* of an image. One example is shown in Figure 4.26 (top). Here, the position of raised terraces (metal lines) resulted in a somewhat incorrect initial plane fit of the entire height image, such that the rightmost end of the image, containing the substrate, was higher than the same substrate at the left end of the image (a). A new plane fit was performed using only those subregions of the substrate contained inside of four rectangles, denoted in (a), to determine the plane. The result is the image in (b). But

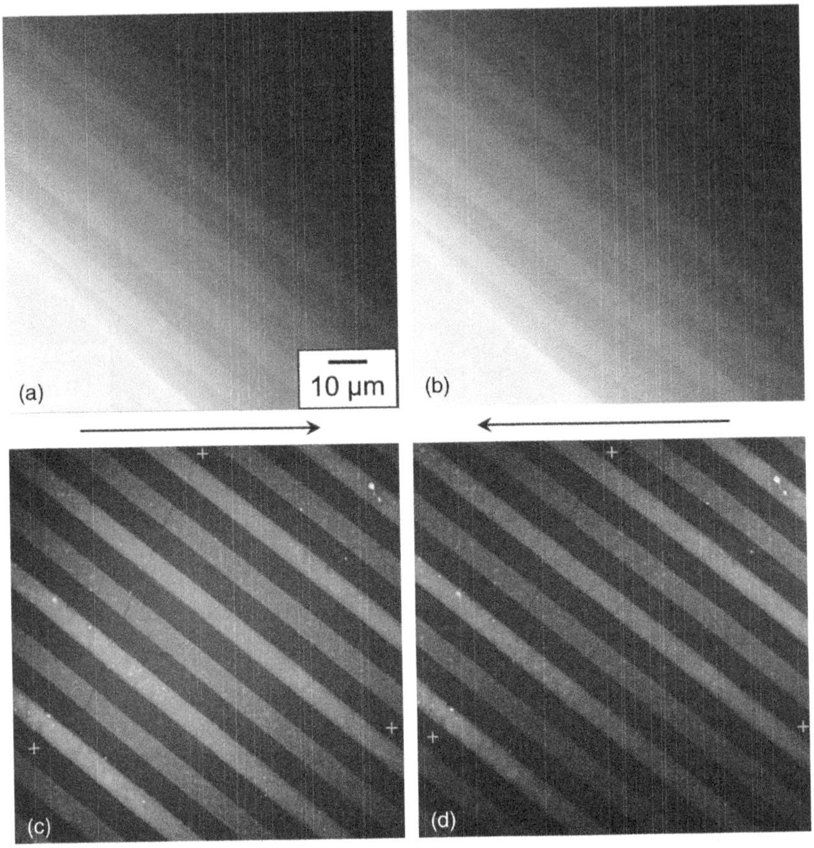

FIGURE 4.25 Hysteretic Z nonlinearity revealed in large images of a 1D array of alternating gold and aluminum lines on flat silicon wafer. Bottom images (c) and (d) were produced by subtracting out the planar tilt and Z offset contained in raw height images: (a) trace and (b) retrace.

now one perceives slight high points at each of the left and right edges, due to nonlinearity in Z (as in Figure 4.25). This is displayed quantitatively in Figure 4.26 (bottom) comparing the same scan line of data. The selected rectangular regions were then fit by a quadratic mathematical surface, and this surface was subtracted from the entire image. The result is seen in Figure 4.26c, and a single scan line of data is shown in the plot of Figure 4.26 (bottom).

To understand how these nonlinearities in Z versus X arise, the first clue is the data of Figure 3.18. Here, we examined a nonlinearity and symmetric hysteresis in the actual displacement of Z as a function of Z bias. Figure 4.27 illustrates how this relationship necessarily produces a nonlinearity in Z versus X (and given Figure 3.18, it is easy to grasp how this nonlinearity is different in sign depending on scan direction, i.e., is hysteretic). Figure 4.27a depicts a flat surface with a finite tilt (exaggerated in Z) typical in AFM as noted earlier, because some finite tilt is always

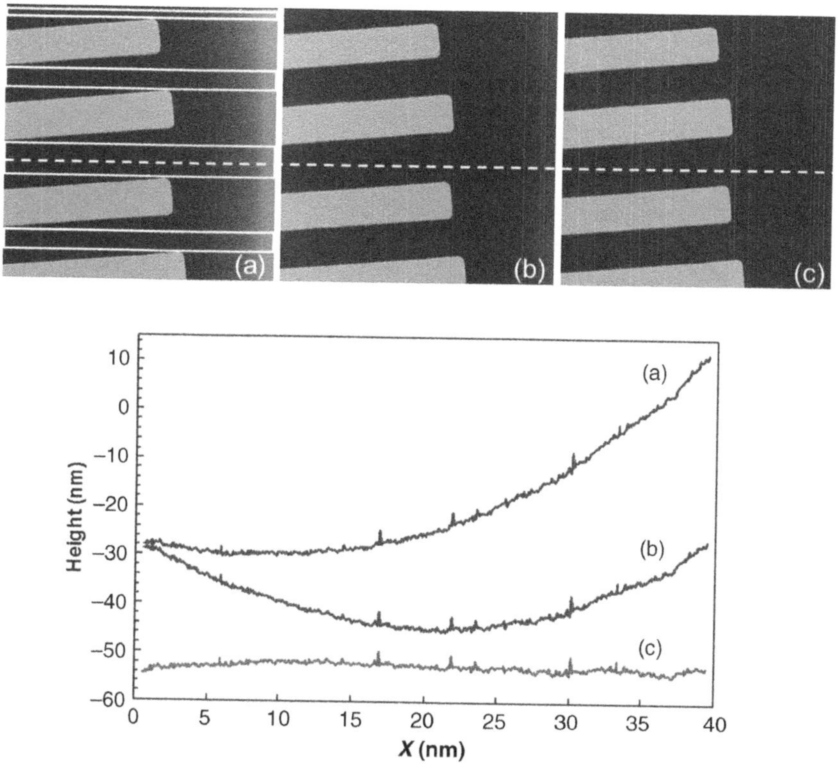

FIGURE 4.26 (Top) 40 × 40-μm height images of metal strips on a silicon substrate at different stages of post-processing: (a) after subtraction of whole-image plane fit; (b) after subtraction of plane fit determined by data within rectangular regions marked in (a); (c) after subtraction of quadratic surface fit determined by data same regions marked in (a). (Bottom) Single trace of data from the scan line marked by dashed line in top images.

present between sample plane and *X–Y* plane of scanner. Evenly spaced *X* displacements (pixel spacing) produce evenly displaced *Z* steps under the usual feedback control. But an intrinsically nonlinear relationship between *Z* displacement and *Z* bias (Figure 4.27b), such as experimentally observed in Figure 3.18 (dependent on ramping direction), means that unevenly spaced *Z*-bias steps must be caused by the feedback control, in order to execute even *Z* displacements. But, in conventional scanners, it is these increments of *Z* bias that are taken as proxies for the surface height; this means the topographic image must contain a nonlinear trend, depicted in Figure 4.27c. Following the fitting and subtraction of a mathematically planar surface, the tilt of this image is removed but not the curvature. The upward or downward sense of this curvature is a strict function of the *X*-scan direction (trace or retrace), because the *Z*-bias ramp is in opposite directions for scanning up or down the sloped surface.

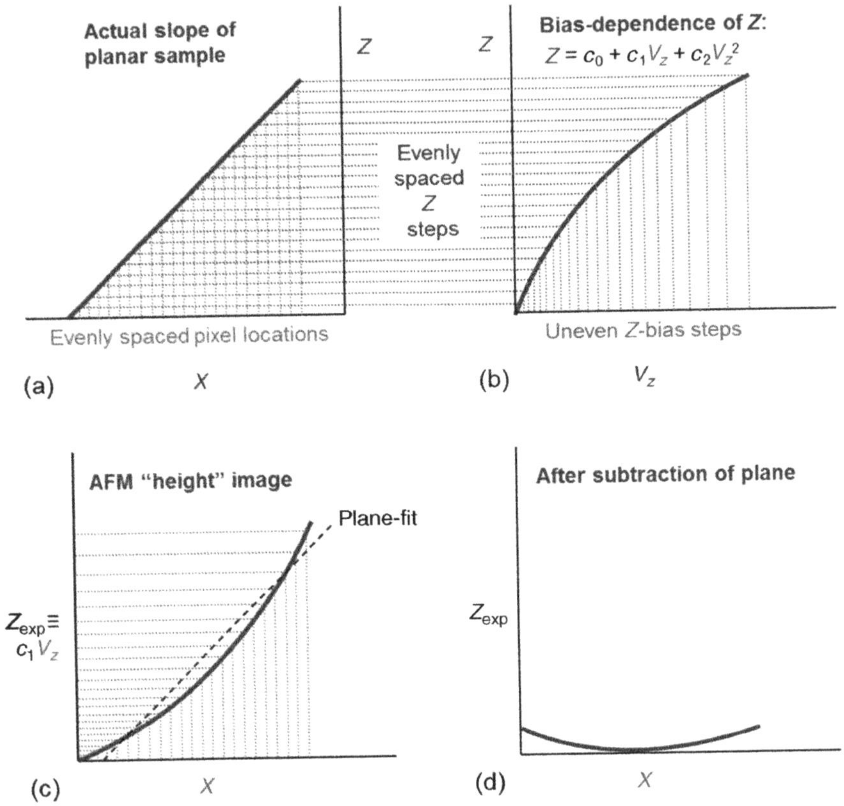

FIGURE 4.27 Explanation of the origin of curved topography due to Z scanner non-linearity. (a) Planar tilted surface tracked in constant increments of X and thus evenly spaced Z displacements. (b) Intrinsically nonlinear relationship between Z displacement and Z bias, thus unevenly spaced Z increments. (c) Resulting nonlinear trace of topography. (d) Final curved topography after subtraction of plane.

Given the symmetric character of the relationship between Z bias and Z displacement, one expects the curvature along X to be equal but opposite in Figure 4.25c and d. Figure 4.28 examines the data along one particular scan line of each image, both before (a) and after (b) the subtraction of a mathematical line. Although the concavity of the trace and retrace data in Figure 4.28b indeed changes sign, the magnitudes are not equal: the downward concavity in the trace scan clearly is of greater magnitude. Moreover, the extra concavity does not look parabolic in the trace scan, suggesting higher-order nonlinearities; it turns out that if one very carefully levels the sample (e.g., using a three-stepper adjustable tripod sample stage[1]) to less than 0.1° tilt of sample relative to X–Y scanner plane, a substantial, higher-order nonlinearity remains in both trace and retrace data. (Most software can handle fitting/subtracting such higher-order

[1] This is a commercial feature on some AFM's, such as the Agilent 5500.

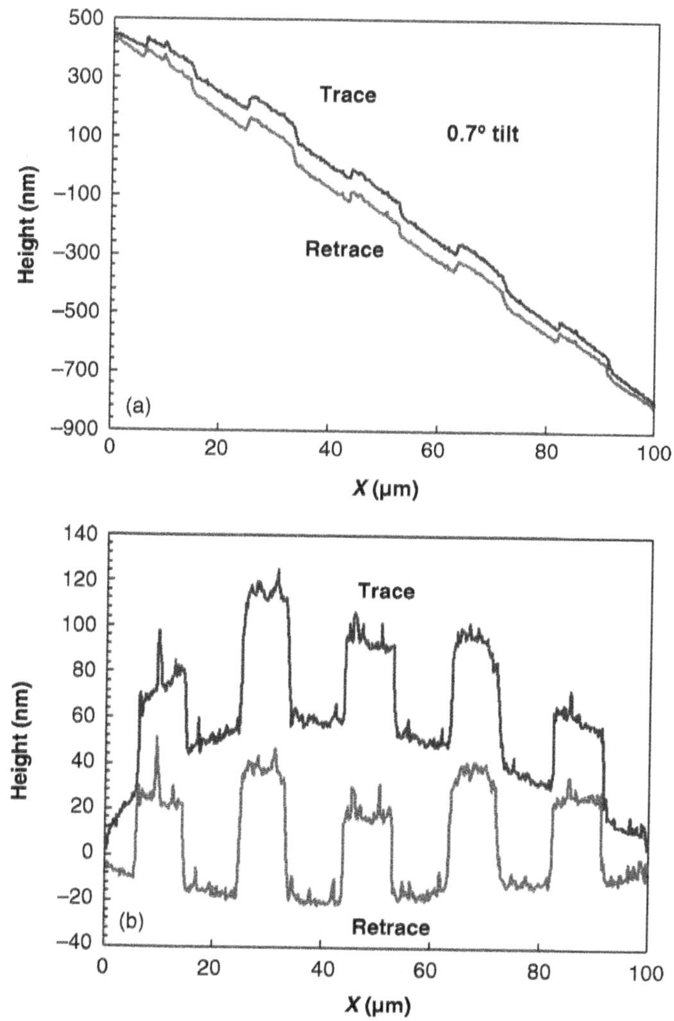

FIGURE 4.28 (a) Height data along one particular scan line of each image in Figure 4.25a and b. (b) Height data along the same scan line after the subtraction of a mathematical plane to yield the images of Figure 4.25c and d.

nonlinearities, at least up to cubic polynomial; the freeware Gwyddion will do as high as 12th order.) Having largely removed the Z displacement versus Z bias nonlinearity of Figure 3.18 from the picture, what is the source of this "extra" nonlinearity?

The source of additional nonlinearity in Z during X–Y scanning is known as *cross coupling* between piezoscanner axes. The physical stacking of X–Y and Z electrodes applied to the piezoelectric tube scanner (the type most commonly used to move tip or sample) is schematically illustrated in Figure 4.29. The bottom part of the piezo-scanner, which produces X and Y scanning, cannot do so without slightly changing

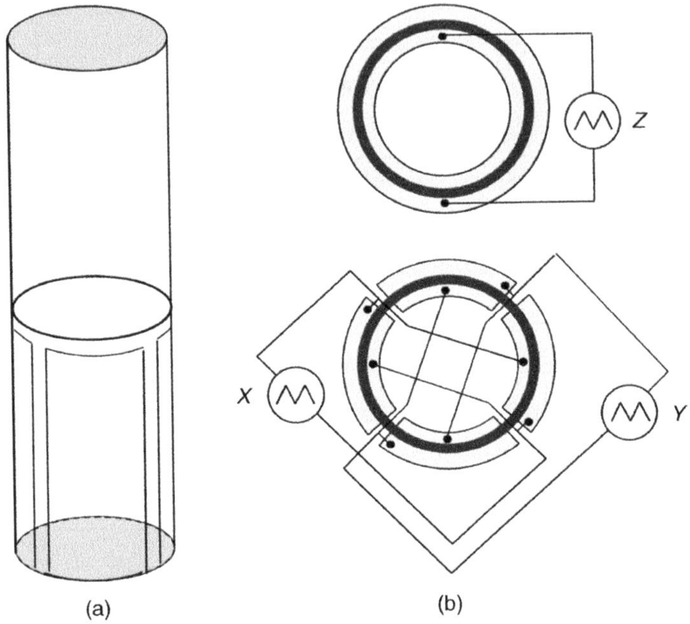

FIGURE 4.29 (a) Illustration of an X–Y/Z piezoelectric tube scanner. The scanner contains top and bottom portions, the top here producing the principal Z movement and the bottom the X–Y movement. (b) For Z motion, two electrodes apply biases to inner and outer faces of the upper piezotube (dark in cross section). For X–Y motion, eight electrodes apply biases to inner and outer quadrants of the lower piezotube (again dark). A given inner segment is biased the same as the outer segment on the opposite side of the cylinder. This produces displacement in X or Y. There can be DC offsets in X and Y as well as AC ramping biases on top of these offsets, so that raster scanning is produced relative to some nonzero image center (X–Y origin), and Z variations can take place atop a Z displacement relative to its midpoint.

Z. This reality derives from the fact that any piezoelectric crystal is of course a 3D object; the spacing of atoms along two crystal axes cannot change without changing atomic spacing along the third axis. Thus, the upper portion of the scanner, which is intended to produce the Z displacement, must also displace to offset (i.e., correct for) whatever Z displacement is produced by the lower portion. This "extra" Z displacement is contained, as always, in the "height" image. Generally, cross coupling in a real piezoelectric scanner can be quite complicated, thus a higher-order nonlinearity is contained in height images.

Earlier we saw that Z creep is actually a less-noticeable problem than X–Y creep. Now that we have examined, in some detail, nonlinearities in Z, what about X–Y nonlinearities? As X and Y biases are ramped to produce a raster scan, what ensures that the actual steps in X and Y are evenly spaced? Figure 4.30a graphs X-positional displacement (via an inductive sensor) versus the X bias. The approximate sensor output is 15 V per 100 μm of movement; thus, the voltage ramp produces a scanning distance of 67 μm. A precisely quadratic relationship is observed, very similar to

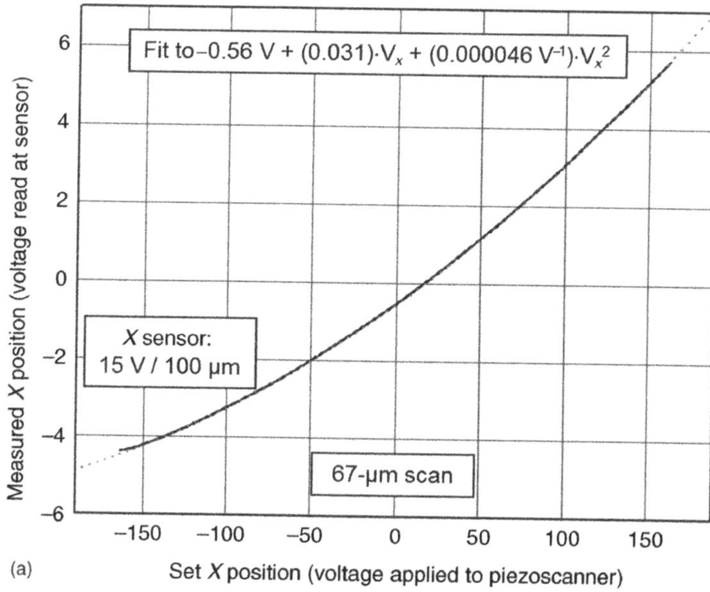

(b)

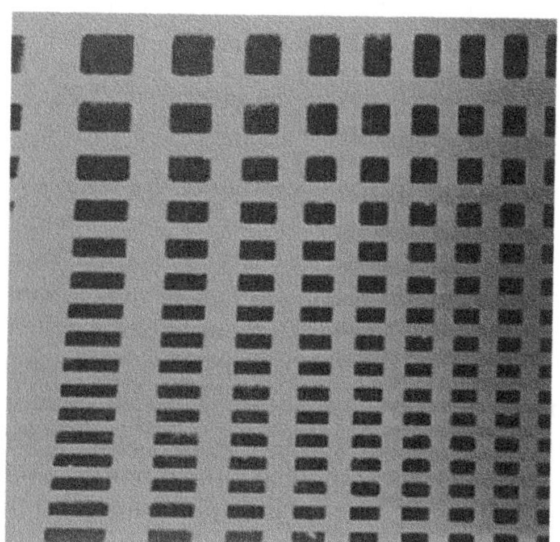

FIGURE 4.30 (a) *X* ramping (left-to-right) for a ramping range ±170 V. The measured *X* position is plotted versus the applied *X* voltage using black symbols (continuous curve). A polynomial fit is shown as a dotted curve. (b) Nonlinear distortion of a calibration grating consisting of evenly spaced square pits, obtained under linear ramping of *X* and *Y* scanner voltages.

that of *Z* displacement (Figure 3.18). If the scanning algorithm were to ramp *X* and *Y* in even voltage steps, one should get unevenly spaced positions for each point location in the image. Thus, unevenly spaced measurement points would be mapped into evenly spaced pixels, distorting image features. Such a result is exemplified in

Figure 4.30b, a height image of a calibration grating consisting of evenly spaced and uniformly sized square pits, obtained under linear programmatic ramping of X–Y.

Closed-loop scanners measure X and Y position and vary the X and Y biases to ensure evenly spaced measurement points, unlike Figure 4.30b. Thus, nonlinearity is removed (along with creep, considered earlier). But what about open-loop scanners? When properly calibrated, one obtains quite accurate reproductions of calibration gratings, meaning evenly spaced features. This feat is accomplished by using a *nonlinear scanning algorithm;* the X and Y biases are ramped in a nonlinear fashion (meaning the increment of voltage from one pixel location to the next is not a fixed value) to produce evenly spaced measurement points. This means there are calibration coefficients in front of both linear and quadratic voltage terms in the scanning algorithm. The proper values of these coefficients are estimated by scanning a grating of known, evenly spaced features (in X and Y), and choosing the coefficients such that evenly spaced image features result. Typically, there is a software routine provided by the AFM vendor to automate this calibration process. Finer considerations in this process include differing calibration parameters for ramp up versus ramp down, compensation for coupling between X and Y axes, and proper orthogonality between X and Y. Most users are unaware of these complexities, and benefit from quite linear X–Y scans thanks to precalibrated scanners. Exceptions are those users/managers charged with the task of recalibrating scanners. Procedures for doing so are typically multi-hour and tedious.

4.2.6 Shape of Tip and Surface

Now let us return to details of force sensing at the sample surface. In reality, our force sensor is not a point; rather a 3D object with a finite breadth and a somewhat ill-defined shape at the bottommost, pointed end (though often modeled as a hemispherical cap or paraboloid). Thus, the sensed forces result from interactions summed over typically several nanometers of lateral extent, even if the forces fall off very steeply with distance. This interaction region may be centered at an inclined part of the surface, such that the aggregate force, being a vector, is not vertical. This is intuitive if one thinks of a contact point at the location of the point sensor of Figure 4.1a. Clearly, a repulsive force normal to the contact site would point in some significantly off-vertical direction. Attractive forces on the tip at this location also would sum to a vector that does not point straight downward. Clearly, if our feedback system is only monitoring the (near-) vertical component of force (or the near-vertical oscillation amplitude), then at sloped surface locations, there is some variability in tracking. Even for an idealized, instantaneous feedback system, the force of tip–sample interaction will not be uniform across the surface. In quasistatic contact mode, this means that the total contact force atop a sloped surface, normal to the surface, will exceed its value on a horizontal surface by an amount that increases with slope (i.e., such that the vertical component of force remains constant). A complete treatment must further include the vertical component of shear (frictional) forces, which necessarily add to or subtract from the topography-derived vertical force component. We return to this topic in Section 7.1.

A more obvious concern, related to the shape of surface and tip, is whether the tip is geometrically excluded from narrow valleys, and at the other extreme, whether tall and narrow surface protrusions are distorted by the shape of the tip. In other words, what is the nature of the intersurface topology mapped out by the moving interface between two geometric objects, tip and surface? Figure 4.31 illustrates the concept in cross section. Clearly, the assembled points of contact between a parabolic tip and a rugged (on the tip's scale) surface will not reproduce the true surface topography; some valleys are too narrow to allow full penetration by the tip, and many protrusions are so narrow that considerable broadening or *dilation* will result in the image. In the latter case, you could say the protrusion images the tip just as much as the tip images the protrusion. Indeed, one can purchase commercial arrays of vertical needles for the purpose of imaging the tip shape, which can be important to quantitative tribological measurements [11]. It can also be important to identify any irregularity in the tip such as a shoulder due to some adhered contaminant, or a bent or blunted tip (usually along with the observation of increased tip–sample adhesion). This is a simple problem in geometric superposition. It is particularly important to recognize the effect when characterizing nanoscale objects such as nanoparticles, nanotubes, nanowires, and others. The inset of Figure 4.31

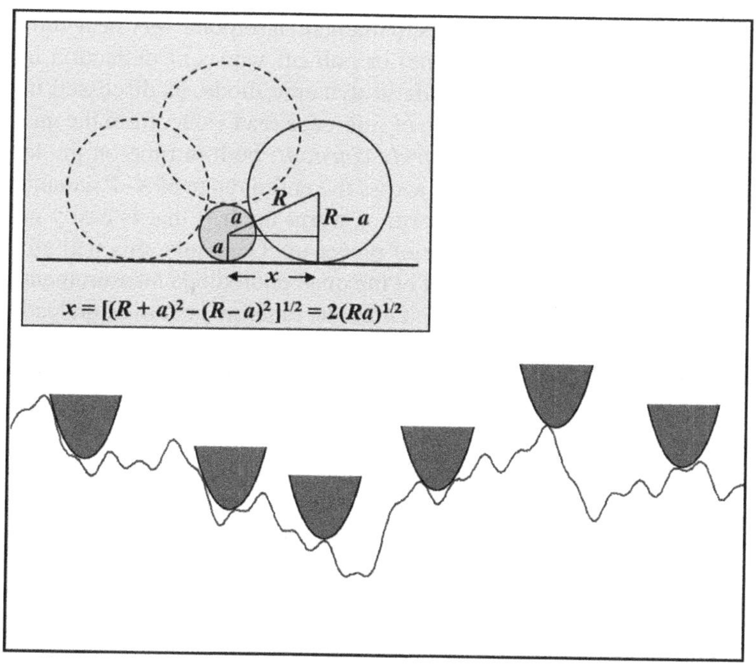

FIGURE 4.31 Depiction of a parabolic AFM tip in contact at various points on a nanoscale-rugged surface. *Inset*: Lateral dilation of a circular cross-sectional object on a flat substrate. An object of radius a is imaged as having a larger radius $2(Ra)^{1/2}$, where R is the radius of curvature of the tip.

quantitatively illustrates how the lateral size of a circular cross-sectional object, sitting on a flat substrate, becomes dilated in an AFM image as a result of geometric superposition. The object's radius a dilates into a radius $2(Ra)^{1/2}$, where R is the AFM tip's radius of curvature.

Similarly, steep slopes will be imaged at reduced steepness due to the shape of the tip. If such locations are deep enough, the angled sides of the tip (usually a three- or four-sided pyramid) will make contact with the surface rather than the rounded end of the tip. In such cases, one may end up measuring "surface" slopes that are actually those of the tip. One should bear in mind that these angles are not necessarily the same for all sides of the tip; most significantly, because of cantilever tilt, the side of the tip opposite the cantilever will be less steep, and its opposing side more steep, than either of the other two sides. Again, one can use gratings with tall vertical edges (e.g., that of Figure 4.20) to image cross-sectional tip shape including rounded end, angled sides and modified angles (due to cantilever tilt). Conversely, ultrahigh aspect-ratio tips are commercially available, such that vertical sidewalls in, say, microelectronic devices can be imaged as (nearly) vertical.

4.2.7 Other Realities and Operational Difficulties—Variable Background, Drift, Experimental Geometry

There is another concern if one wishes to maintain a setpoint very near certain absolute values, such as the zero (baseline) or pull-off values of deflection in contact mode or the free-oscillation amplitude in dynamic mode, as discussed in Section 4.2.3. The concern is whether the zero or reference read value, from the quad photodiode and associated electronics, is *truly constant*: both in time on the laboratory scale (minutes to hours) and in space across the entire range of X–Y scanning.

Many instruments exhibit some form of warm-up drift that is easily noticeable on the scale of an hour from the time of power-up. Generally, this will affect hardware components such that the output of the quad photodiode measurement changes slowly in time, usually monotonically. Often the baseline or zero of deflection shifts downward during warm-up, such that any chosen absolute setpoint in contact mode corresponds to a *gradually increasing loading force* of tip to sample (because the setpoint deflection, in its raw voltage read, is increasing relative to the drifting voltage that corresponds to the zero of deflection). As per Section 4.2.3, this can mean that the tip is indenting farther into the sample, thus the "height" data will contain an increasing amount of sample compliance with passing time. It also can mean that an initially benign loading force condition becomes a destructive condition, where sample is tearing and/or tip is dulling during imaging. This problem is removed if the instrument continually corrects for this drift of deflection baseline, as is the case of force-curve mapping under relative triggering (i.e., methods that measure the maximum deflection *relative to baseline* within each approach–retract cycle, detailed in Chapter 6). In dynamic mode, typically the resonance frequency and phase offset (which ensures 90° phase lag at resonance, the correct mechanical behavior of a damped, driven oscillator) will drift during warm-up. Relative to any selected drive frequency, the frequency dependence of amplitude and phase

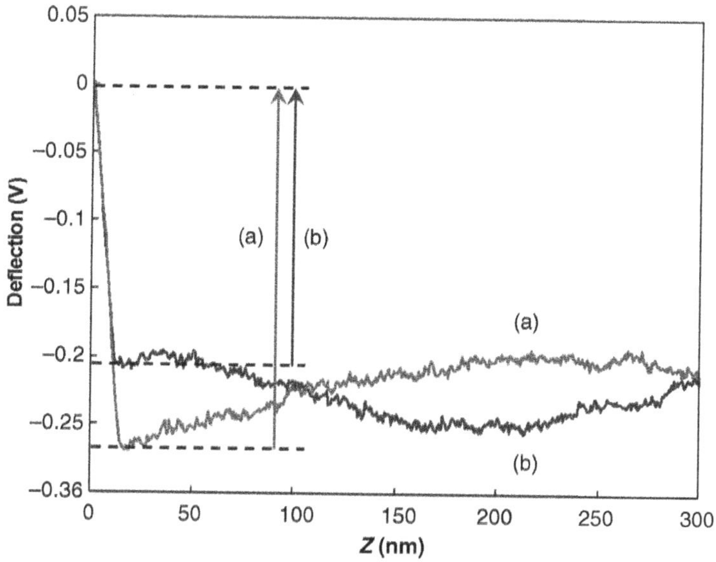

FIGURE 4.32 Overlay of two approach force curves acquired at locations (a) and (b) on poly(butyl) methacrylate, to illustrate differences in the horizontal position of the wavy optical background. This translates into a 3-nm height artifact during imaging for a setpoint at deflection = 0 V: case (a) produces a surface image lowered by 3 nm relative to case (b).

(approximate Lorentzian and inverse tangent functions, as in Figure 2.7) will shift. This case will be further examined in Chapter 5 in the context of phase imaging.

Variations of the zero of deflection from one *X–Y* location to another are exemplified in Figure 4.32 with the aid of force curves. Two approach force curves are overlaid, acquired at surface locations on a glassy polymer (poly butyl methacrylate) spaced by a couple of micrometers. In each case, a wavy optical background (see Figure 3.15 and related discussion) is apparent over the ~280 nm of precontact *Z* displacement. (A rigid cantilever was employed such that no jump-to-contact instability results.) In case (a), the wavy baseline is at a relative minimum at the point at which tip contacts surface, whereas in case (b), the baseline is horizontally displaced such that it reaches a relative maximum at the point of tip–surface contact. Thus, for a fixed absolute deflection signal of 0 V, the physical amount of cantilever deflection differs for the two cases, approximately 0.27 and 0.21 V, corresponding to a difference of about 3 nm in this case (inverse sensitivity $S^{-1} \approx 50$ nm/V). This produces differences in indentation that are well below 1 nm (i.e., a small effect on topography if imaging at this setpoint), but more significantly it forces the *Z* scanner to approach by an additional 3 nm in order to reach a deflection setpoint of 0 V, or 3 nm of height artifact. In other words, location (a) would be imaged 3 nm "falsely low." Even more advanced systems (e.g., fast force-curve mapping) that continually account for vertical drift of the deflection baseline cannot fully account for the leftward or rightward displacement of the wavy background such as illustrated in Figure 4.32.

Apart from the internal sources of drift and variable background considered above, are there extrinsic sources of such problems? Section 4.2.5 covered quite a variety of potential image distortions due to scanner creep and nonlinearity. But there are other causes of time-dependent tip or sample movement during an image, analogous to scanner creep, and reasons why the spacing of point locations within an image may systematically deviate from a constant, analogous to scanner nonlinearity. One must remember that the tip and sample are not isolated objects. They are coupled to the environment by a number of mechanisms, some of which may vary in time, or in space as the sample or tip perform a raster scan.

One should not presume that a coupling interface or material, such as adhesive tape in the case of the sample, is absolutely stable. Indeed creep is an everyday fact in many polymeric materials including adhesive tape. As with piezoscanner creep, simply waiting long enough for creep to subside is one strategy. But the relaxation time for creep in a macroscopic piece of tape can be very long. Because AFM images typically require minutes, and the distances explored may be resolved to nanometers, the impact of creep in tape can be huge. Even if the majority of this creep is vertical, it is highly unlikely that the nominally vertical direction is exactly parallel to the axis of the Z scanner. Thus, features of a surface may move on the nanoscale in X–Y during the time it takes for the AFM scan system to raster scan across them. This movement, however, is independent of the AFM scan direction, and thus provides clues that the source of drift is external (i.e., tape) and not intrinsic creep of the piezoscanner. Features would appear "stretched" during the upward raster scan but "compressed" during the downward raster scan, for example, with such a source of external creep.

Another external source of drift that *may* be affected by the scan direction is unintended touching of scanning and stationary hardware. Depending on instrument design, the stationary objects (e.g., tip/cantilever, chip, and the clip that secures it, for a scanned-sample system) may be spatially very close to the scanning components (here sample, tape periphery). It does not require excessively sloppy preparation for the stationary and scanning components to end up touching each other. In this case, the directionality of the fast scan will have an impact on the nature of the "smear" seen in images.

The design of liquid cells may also introduce coupling between stationary and scanned components. Perhaps the most notorious example is a closed "fluid cell" comprised of a top piece that is stationary and a bottom surface that is in fact the scanned sample, plus a soft rubber O-ring forming the sidewalls. Obviously, for large lateral scans, the stationary O-ring may induce a substantial shear stress on the moving sample, such that the latter may move a distance that is only a fraction of the scanner displacement. Thus, in some cases, egregious distortions and length-scale discrepancies can result.

In other cases, the crowded nature of experimental geometry does not introduce drift to otherwise normal operation; it *precludes operation* entirely. For the case of samples that are devices or macroscale objects (e.g., cut pieces of devices), while attempting to approach the tip to a lower surface, the cantilever itself, or the chip to which it is attached, or the clip that is securing the chip, may come in contact with

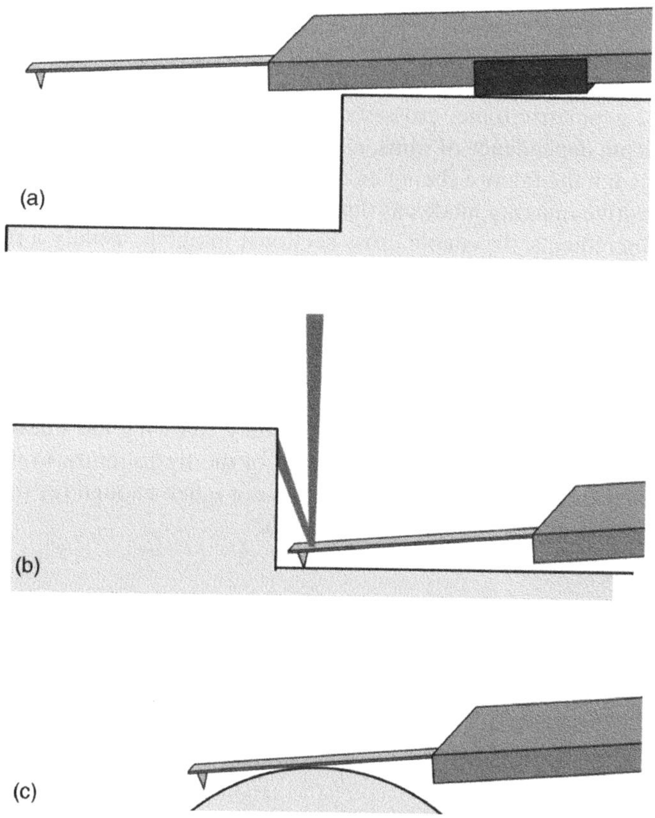

FIGURE 4.33 Examples of how crowded experimental geometry can prevent AFM operation. (a) Sample touching the chip to which the cantilever is attached, or the clip that secures this chip. (b) Sample blocking cantilever-reflected laser beam. (c) Sample touching cantilever.

some portion of the sample. Given the fact that these pieces extend off to one side of the tip, a judicious reorientation of the sample may enable a clean approach of tip to surface. An additional complication, however, is the need to have an unobstructed laser path from the top surface of the cantilever to the photodetector. These considerations are illustrated in Figure 4.33a and b. In orienting the sample such that the higher surface is positioned left of the tip, away from the cantilever/chip/clip, one may now find that the laser path is obstructed. Such situations may require trimming the sample to remove the obstruction to the laser path.

Another fairly common example of geometric "exclusion" occurs when imaging over large portions of curved surfaces (Figure 4.33c). If scanning down the incline of a curved surface, in such a direction that the high point of the surface is beneath the cantilever, one may eventually "run aground": the cantilever may come into contact with the surface, precluding imaging. This can only happen if the Z range explored is greater than the height of the tip (which may require using the Z motor to expand the Z range, per Section 4.2.1) or because the initial engagement was located to that particular side of the high point. Obviously, one will not suffer this problem if engaging and imaging in the opposite direction, such that the high point of the surface is not underneath the cantilever.

As a final geometric topic, cross-sectional AFM can be a powerful method to look at the depth dependence of films, although in these cases, topographic information is usually not the interest (being an artificial surface created by cutting). Rather, property-sensitive imaging modes as discussed in Chapters 5–7 may reveal the segregation of ingredients. To enable cross-sectional imaging, usually a jig or custom clamp has been purchased or machined, two jaws that can screw together to hold the sample upright, on edge. Often one wishes to image all the way out to the edge of the cross-sectioned sample, as this is the original surface location. Scanning over the edge is not necessarily a problem, unless one scans far beyond the edge such that the end of the tip descends far lower than the cross-sectional surface. Upon scanning back, the feedback may work adequately, allowing the sidewall of the tip to scan up the sample edge, and thus the end of the tip to return to the top of the cross-sectioned surface. But if the feedback is not quick enough (or the operator is scanning too fast), damage to the tip may result.

REFERENCES

[1] Liptak, B., *Instrument Engineers' Handbook: Process Control*. 1995, Radnor, PA: Chilton Book Company, 20–29.

[2] Tan, K.K., Q.-G. Wang, and C.C. Hang, *Advances in PID Control*. 1999, London: Springer-Verlag.

[3] Israelachvili, J.N., *Intermolecular and Surface Forces*. 2nd ed., 1991, San Diego: Academic Press, 450.

[4] Butt, H.-J., B. Cappella, and M. Kappl, Force measurements with the atomic force microscope: Technique, interpretation and applications. *Surf. Sci. Rep.*, 2005, **59**: 1–152.

[5] Garcia, R. and R. Perez, Dynamic atomic force microscopy methods. *Surf. Sci. Rep.*, 2002, **47**: 197–301.

[6] Bocek, D., About "Bocek" degrees used in the extender package on digital instruments nanoscope scanning probe microscopes. *Microscopy Today*, 2001 (June): 24–25.

[7] Haugstad, G., et al. Atomic force microscopy of AgBr crystals and adsorbed gelatin films. *Langmuir*, 1993, **9** (6): 1594–1600.

[8] McCrum, N.G., B.E. Read, and G. Williams, *Anelastic and Dielectric Effects in Polymeric Solids*. 1967, London: Wiley.

[9] Ward, I.M. and D.W. Hadley, *An Introduction to the Mechanical Properties of Solid Polymers*. 1993, Chichester, UK: Wiley. 334.

[10] Fung, Y.C., *Biomechanics: Mechanical Properties of Living Tissues*. 2nd ed., 1993, New York: Springer-Verlag.

[11] Hammerschmidt, J.A., G. Haugstad, and W.L. Gladfelter, *Probing polymer viscoelastic relaxations with temperature controlled friction force microscopy*. Macromolecules, 1999, **32**: 3360.

5

PROBING MATERIAL PROPERTIES I: PHASE IMAGING

This and the following two chapters describe commonly utilized AFM modes for probing material properties [1,2]. These modes proceed in parallel with topographic imaging and are offered in many commercial base packages, or can be added to base systems for an extra few percent to tens of percent in price. Properties that can be revealed include rigidity (elastic modulus), viscosity and viscoelasticity, frictional response, surface energy and potential, polarization and charge state, and magnetization. Given appropriate hardware, the modes can be operated under environmental control such as relative humidity or liquid immersion including controlled pH and ionic concentration. Variable sample *temperature* is another simple add-on that in conjunction with variable measurement *rate* opens fundamental studies of kinetic and thermodynamic behavior in condensed matter or, from a more practical viewpoint, unveils heterogeneities that may not be visible at room temperature. Indeed, all of the methods described in Chapters 5–7 have strong motivations (and expenditure justifications) both in microscopy—producing revealing images—and in quantitative measurements.

An optimal pedagogy for introducing topics within property-sensitive modes is not obvious. One approach is to start with quasistatic and then "advance" to dynamic AFM methods. We choose not to follow this pedagogy. One reason for instead beginning with dynamic AFM in Chapter 5 is that the auxiliary phase imaging is probably the most commonly utilized materials-contrasting method. This is because dynamic (i.e., AC or "tapping" mode) AFM is the most commonly used

Atomic Force Microscopy: Understanding Basic Modes and Advanced Applications, First Edition.
Greg Haugstad.

topographic imaging scheme, in turn, due to its broad success: the widest range of materials and biological entities can withstand the tip–sample interaction of dynamic mode without tearing or other deleterious responses. A second reason, however, for beginning with dynamic AFM is that phase measurements are *needed for proper topographic imaging* in dynamic mode, unlike the relationship between friction imaging in contact mode or adhesion imaging in force–curve mapping modes. This is because one must use phase to *diagnose* the regime of tip–sample interaction, dominant attractive or repulsive, and thereby ensure that so-called height images are reasonable approximations of true topography. In other words, our treatment of topographic imaging in dynamic mode in Chapter 4 is *incomplete* without examining phase numbers in greater depth than we have in the preceding chapters. We have pushed the topic into this dedicated chapter on phase imaging because, first, it is not enough to measure phase at a single point on the surface. One can be deceived into thinking that the tip–sample interaction regime is dominant attractive or repulsive everywhere, only to find by imaging phase or spatially sampling phase–Z curves that this is not the case, as demonstrated in Section 5.1.

Second, phase turns out to be the most complicated imaging mode in many respects, being both scientifically rich and methodologically meticulous (although not always performed as such!). Covering only the diagnostic role of phase vis-à-vis topographic imaging (in Chapter 4) was not really an option. Indeed, while imaging phase one often encounters fascinating variations related to sample heterogeneities, begging a treatment of physical meaning. Thus, there is not a natural way to separate the use of phase in the aid of topographic imaging from the use of phase in property mapping, unlike, say, topographic and frictional imaging within contact mode. We will find, however, that the *materials contrast mechanisms* of phase, sliding-friction and adhesion hysteresis, are in fact similar; each relates to energy dissipation during motional tip–sample interaction, often involving the irreversible (hysteretic) imposition and release of stress. For phase, this was discussed briefly in Section 2.4. We will see in Chapter 7 that lateral forces sensed as the tip slides along in contact mode can be dissected to examine dissipative contributions (i.e., independent of topographic roughness), which may probe molecular responses similar to those revealed by phase measurements and thus may similarly contrast materials. We will also see that adhesion hysteresis maps often provide similar contrast as phase imaging.

Chapter 5 can be considered mandatory to fully understand and broadly use phase imaging in air. Issues are presented roughly in the order that they arise when setting up, operating the instrument, and processing/interpreting data. First, a user wishes to ensure accurate height data. In Section 5.1, we discuss the role of phase measurements and imaging in diagnosing and controlling the dynamic interaction regime, net attractive or repulsive, such that height artifacts can be avoided, or at least understood and quantified. In the process, one must learn of the principal parameters affecting the interaction regime: drive frequency, drive amplitude, and setpoint amplitude. Upon examining the phase data more closely in this context, the user notices certain complications and nonidealities in the nature of the phase data that derive from instrumentation details as well as possible environmental effects. In Section 5.2, we discuss several of these complications and caveats, and recommendations for dealing with them. Next, the user usually wishes to more deeply understand the physical information contained in phase images. This is the subject of Sections 5.3–5.5.

In Section 6.4.3 and Chapter 9, we will discuss dynamic methods that are somewhat "higher order," in that only select users may need to apply these methods, depending on the specifics of the application. This includes approach–retract curve mapping of amplitude and phase with special data post-processing; electrostatic responses probed in interleave (a second pass per scan line); and multifrequency methods that utilize higher vibrational modes of the cantilever.

5.1 PHASE MEASUREMENT AS A DIAGNOSTIC OF INTERACTION REGIME AND BISTABILITY

As already stated, phase plays a crucial role in proper topographic imaging in dynamic (AC/"tapping") mode. It is well established, and exceedingly important, that an oscillating AFM tip/cantilever interacting with a sample surface can be bistable [3–5]. This means that for a given (mean) Z position of oscillating tip relative to sample surface ("fly height" in Figure 4.), as controlled by the feedback circuit (i.e., via the setpoint amplitude), two possible dynamic stable states can arise. We refer to these as attractive and repulsive regimes, as stated in Section 2.4.3 in discussion of Figures 2.12, 2.14–2.16. In some writings, the states are called high- and low-amplitude for a given Z-positional value [5]. This means that during AFM imaging at a fixed setpoint amplitude under feedback control, there must be *two different possibilities for height*: one for which at nearest approach the tip–sample interaction is dominant repulsive and the other for which this interaction is dominant attractive (Figure 4.2). Bistability also means that at any location within an image, the oscillating tip can settle into one *or* the other dynamic steady state, dominated by attractive or repulsive interactions. Sufficient oscillations occur at each pixel location (provided one scans properly slow, $\sim Q$ oscillations per pixel) for a new stable dynamic state to be reached from one pixel to the next. Thus, it has been demonstrated that the oscillator may stabilize in one or the other regime *within specific subregions of an image* in a strictly reproducible way and in precise correspondence to sample heterogeneities (which produce qualitatively different stable dynamic states) [6].

5.1.1 Phase (and Height, Amplitude) Imaging as Diagnostics

The data in Figures 5.1 and 5.2 exemplify this remarkable but not uncommon situation, on an ultrathin film of (semicrystalline) polyvinyl alcohol (PVA) cast on cleaved mica from aqueous solution. Such films exhibit three characteristic phases (another use of the term) that differ markedly in their interaction with the AFM tip in several imaging modes [6,7]. A first layer completely covers the substrate and is strongly adhered as evidenced by AFM scratch resistance [7]. Discontinuous

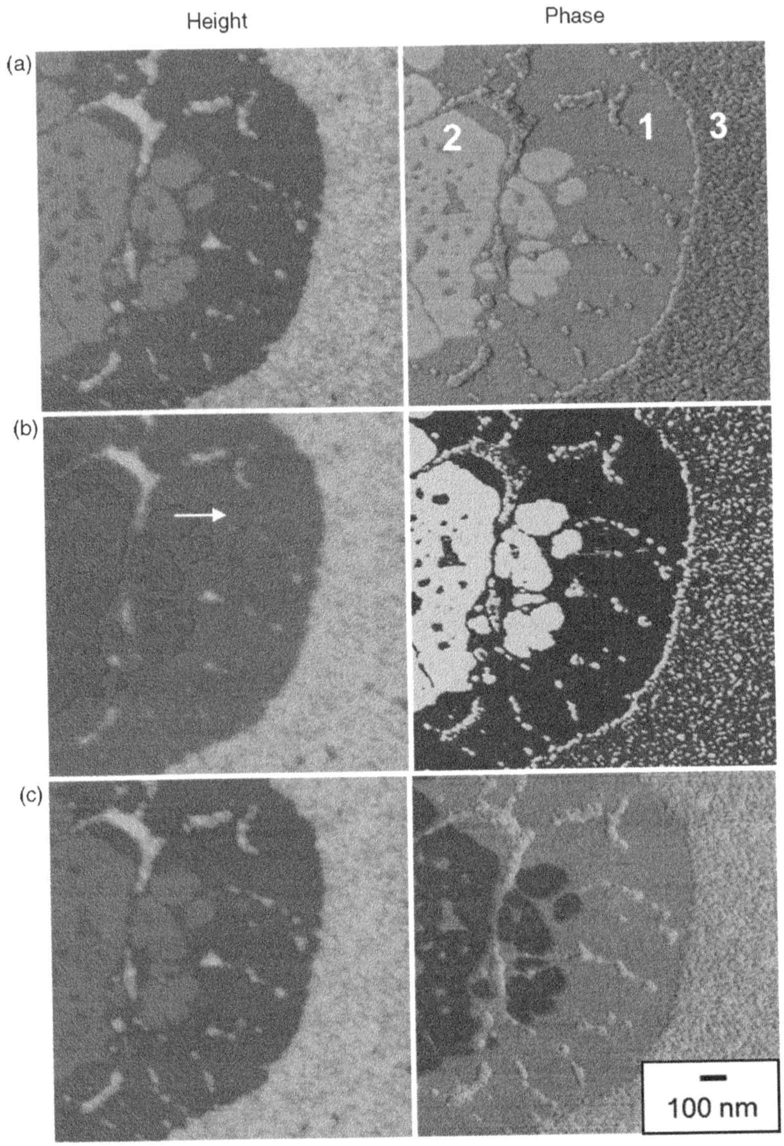

FIGURE 5.1 Height and phase images of a polyvinyl alcohol thin film containing three characteristic components. Three drive amplitudes produce free oscillation resonance amplitudes (A_0) of (a) 14, (b) 10, and (c) 4 nm corresponding to uniform repulsive, coexisting, and uniform attractive regimes, respectively.

second-layer domains (which can be AFM raster-abraded in contact mode to reveal the first layer) differ in crystallinity; a minority phase of uniformly 1-nm-thick domains apparently is highly crystalline, whereas a more variably thick (~3-nm) majority phase is itself highly heterogeneous. In Figure 5.1a, we label these layers

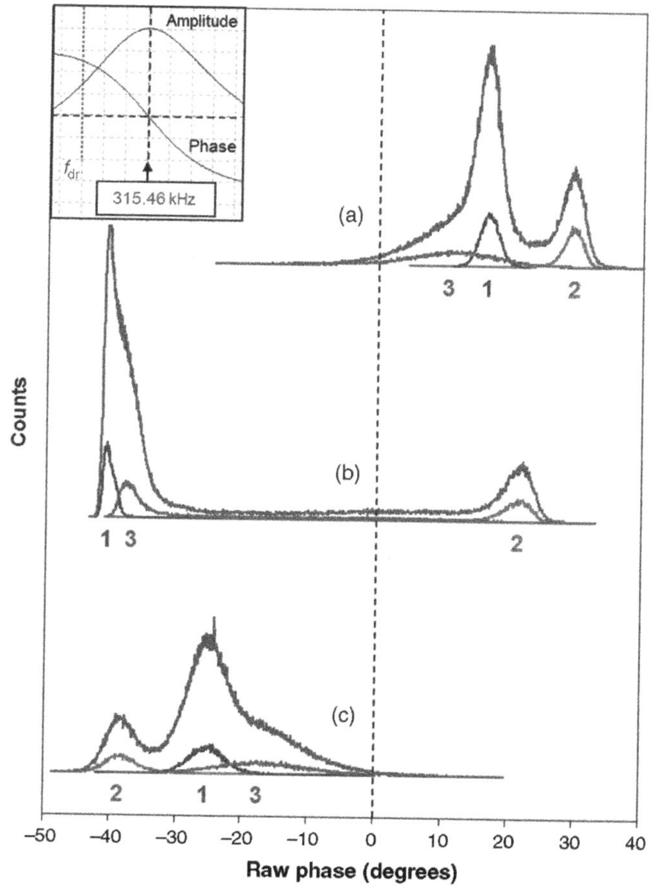

FIGURE 5.2 Histogram representations of the raw Bruker Extender phase data from the images in Figure 5.1 for film components #1–3. Multipeak histograms are from entire phase image; smaller single peaks derive from individual film components. Inset shows the operational drive frequency f_{dr} with respect to resonance-centered amplitude and phase versus drive frequency curves.

1, 2, and 3, respectively. Height and phase images are shown for three different magnitudes of cantilever excitation (drive amplitudes A_{dr}) producing the following amplitudes of free oscillation at resonance, A_0: 14, 10, and 4 nm. (These values were converted from the raw voltage reads to nanometer units via the slope of amplitude (V) versus Z (nm) measurements on silicon, assuming negligible indentation; see Section 3.2 and Figure 4.3.) In each case, the setpoint amplitude was maintained at 70% of A_0 while driving at a frequency $f_{dr} = 315.16$ kHz, 0.30 kHz below the resonance frequency of 315.46 kHz, such that the setpoint amplitude was 90% of the free-oscillation amplitude at this reduced drive frequency. (This a common choice of operational parameters with the intent to stabilize the repulsive regime as further discussed in Section 5.1.) In each phase image, the color or tint was scaled and offset to provide visual contrast (i.e., removing dark or bright offsets, further discussed in Section 5.2.1), as is the standard practice (Figure 5.1).

The phase histograms in Figure 5.2 quantify the phase values in Figure 5.1, in raw degrees as output by commercial electronics (Section 5.2.5). These data are presented without any further offset (in software), apart from the initial "zeroing" of phase at free resonance designated in the inset where the dashed lines cross. (This step is discussed in detail in Section 5.2.1.) Note that this *conventional* zero (i.e., in instrument software) is physically the state where the tip motion lags the driving signal by 90°, at resonance, as per Figure 2.7. As with the height image data considered in Figures 4.9–4.10, the histograms in Figure 5.2 simply count the number of pixels that denote phase measurements lying within narrow increments or bins. Selecting small subregions within each of the three surface domain types (1, 2, and 3) enabled the quantification of peak position (mean dissipation) and breadth (dissipative heterogeneity) for each domain type. The mean values of the three peak positions in Figure 5.2 are raw phase values that can be calibrated as per Section 5.2.5 and in principle converted to approximate energy dissipation per tip–sample interaction using Equation 2.23. We will discuss the physical ramifications of this analysis in Section 5.3 (Figure 5.2).

Figure 5.1 moreover demonstrates what is arguably the most troublesome aspect of dynamic (AC, "tapping")-mode AFM: significant changes in apparent relative heights, at least on the nanoscale, related to the interaction regime. Not only quantitatively erroneous but even qualitatively misleading information may be obtained. Figures 5.1b and 5.2b represent just such a situation; here a wide range of phase values indicates the coexistence of both dominant attractive and repulsive dynamic states within the image. By comparison, the histograms of Figure 5.1a and c are much more restricted in the range of phase within the attractive or repulsive regimes. The bright phase domains concentrated at middle left of Figure 5.1b, region #2, correspond to height domains that are sunken by about 0.3 nm compared to the adjacent regions. In Figure 5.1a and c, these same domains are instead elevated by about 1 nm compared to the adjacent regions. This latter measurement of region #2 island height is consistent with observations in contact and pulsed force modes (which do not tear this particular sample at low loads). So something is wrong with the height image in Figure 5.1b. The Z scanner is not displacing so as to bring tip and sample together upon moving from higher region #2 to lower region #1, as it would if properly tracking topography from left to right over this 1 nm step. Instead, it is displacing apart, as if region #1 is higher than region #2; these displacements are of course rendered as "height" values, because Z-scanner displacement is our means of measuring height. To reiterate, this is because AFM "height" or "topography" data are *not* height or topography *per se*; they are displacements of the Z scanner (or more precisely, as covered in Section 4.2.5, the *applied voltages producing these displacements*) to maintain a read signal, in this case cantilever oscillation amplitude, at some (nearly) constant value. We hope the Z displacement is due to a real height change, but we cannot assume this to be the case if the phase is correspondingly changing by large amounts.

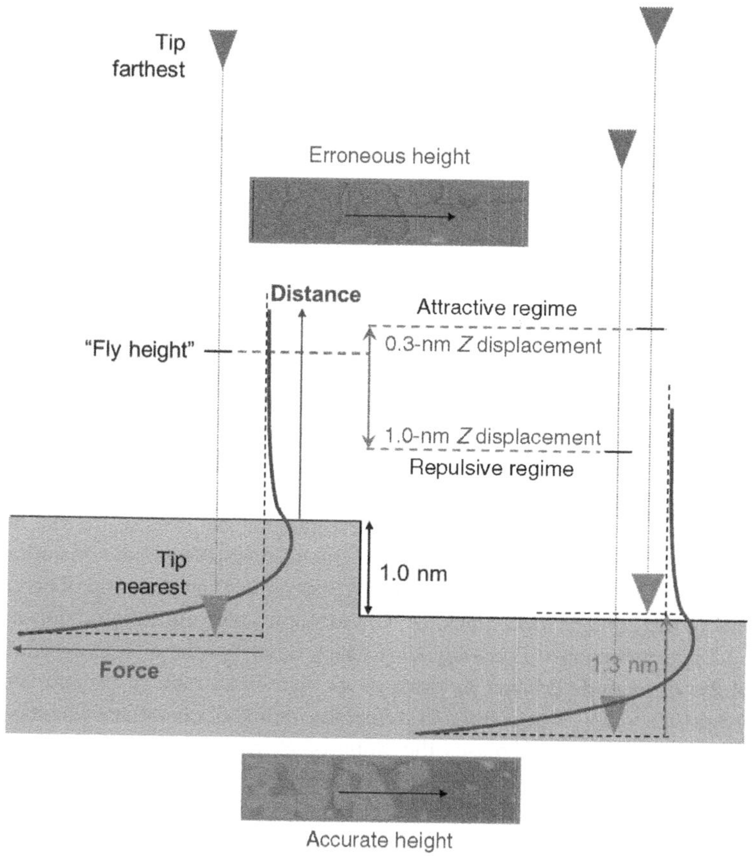

FIGURE 5.3 Illustration of how an oscillating tip can stabilize in the repulsive regime such that the "fly height" accurately tracks a 1-nm step (bottom image), or erroneously switches from repulsive to attractive regime upon traversing the step. The latter generates a 0.3-nm increase in "height" (top image).

Figure 5.3 illustrates the situation in Figure 5.1. A cartoon surface cross section is overlaid by a force–distance relationship that schematically depicts a transition from attraction to repulsion, with the onset of repulsion at the inflection point of the curve. At left is depicted a vertically oscillating tip in the repulsive interaction regime, where at nearest approach the tip is pushing appreciably into the material. (This distance is exaggerated.) Upon scanning to the right, a preservation of the repulsive regime enables a tracking of the topography; the midpoint or "fly height" of the vertical oscillation steps 1.0 nm downward, reproducing the actual 1.0 nm step in the surface topography. This is the case in Figure 5.1a. If, however, the oscillator transitions and stabilizes in the attractive regime (after hundreds of oscillation cycles), then the new fly height has displaced away from the surface, in this case by 0.3 nm. Thus, a 1.0-nm-tall island is imaged as a 0.3-nm-deep trench. This is the case of Figure 5.1b. What is missing in this cartoon is an accurate depiction of how the force–distance curve changes from the left to the right region of the sample; that is, possibly, the *physical origin* of the change in oscillator behavior that may drive the transition from the repulsive to attractive regime. Even if we were to probe this force–distance relationship by acquiring quasistatic force curves, it may not explain the transition between regimes, because the *dynamic* steady-state equilibria within the attractive or repulsive regime may reflect velocity-dependent differences between the two surface regions, such as viscous character. In Section 6.4.3, we examine the change in the Z-dependence of amplitude and phase on region types 1–3, to further address physical differences. Our emphasis in this section is diagnostic.

Region-specific bistabilities—spatial domains of dominant attractive or repulsive interaction regimes—do not necessarily result from heterogeneities in material properties (e.g., soft vs. rigid) or molecular structure (e.g., amorphous vs. crystalline). How can this be? Answer: topography. If there are valleys with concavity similar to the tip, then the relative strength of attractive forces can be much greater within the valleys than atop the hills. Thus, parameters chosen to stabilize the repulsive regime may work just fine except in narrow valleys, where the oscillator switches to the net attractive regime and concomitant higher flying height. This is most problematic on surfaces containing valleys with radii of curvature *nearly as small* as that of the tip. This quantitatively affects the contact mechanics, as, for example, seen in the Hertzian model of Section 2.3 and particularly Equation 2.11, which describes the effective radius of curvature of the tip–sample contact in terms of the individual radii of tip and surface. In this expression, a valley is quantified as a negative radius of curvature, whereas a hill is positive. A negative term increases the value of R from what it would be on a flat surface, effectively making the tip blunter. This increases the area over which the tip is close enough to the surface to sense attractive forces, as well as increasing the contact area and thus reducing contact pressure. Together these effects grow the impact of attraction relative to repulsion. Thus, an oscillator stabilized in the repulsive regime on a flat surface may transition into the attractive regime in a valley. Conversely, an oscillator stabilized in the attractive regime on a flat surface may transition into the repulsive regime at a narrow hilltop; here the contact and near-contact area are reduced, thereby increasing contact pressure and decreasing the area over which the tip is close enough to the surface to sense attractive forces.

Figure 5.4 exemplifies the effect of topography on attractive/repulsive interaction regimes, on a polycrystalline gold surface. Parameters were initially chosen to stabilize the repulsive regime (by driving below resonance, 250.92 kHz, as depicted in (a), with a relatively free-oscillation amplitude between 50 nm and 60 nm using a nominal $k = 40$ N/m cantilever), but in the resulting height–amplitude–phase images (b), certain narrow valley locations designated with circles are where the oscillator slipped into the attractive regime. At the edges of these locations, the oscillator suddenly moves away from the surface; thus, the amplitude (error signal) image, which is in essence differential height ($\delta Z/\delta X$), helps to highlight these locations. Such regions of net attraction can be greatly expanded by driving the cantilever above resonance, the bottom set of images. Boundaries demarcating regions of net attractive or repulsive interaction are also highly visible in the amplitude image.

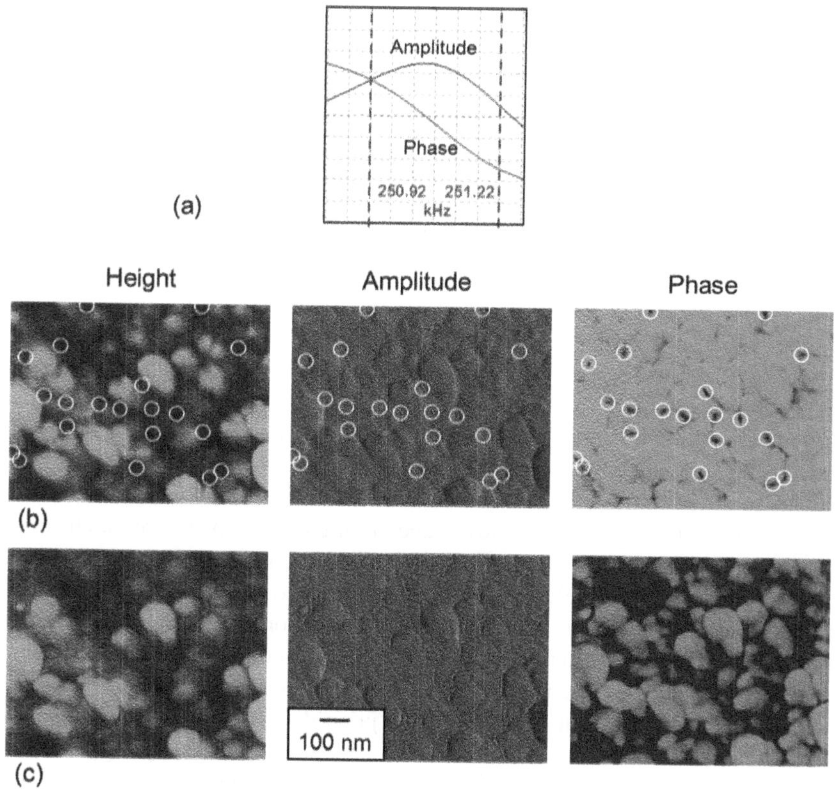

FIGURE 5.4 Coexisting attractive and repulsive imaging regimes on a polycrystalline gold surface, driving at frequencies below and above resonance depicted in (a). Height, amplitude (error signal), and phase images reveal features characteristic of regime switching in both (b) below-resonance and (c) above-resonance cases.

One notes that the dark regions in phase (net attractive) appear to be raised in the height image, the result of the higher "fly height" as per Figure 5.3.

At this juncture, it is important to note that *long-range electrostatic forces* also can have a significant effect on the overall interplay of attractive and repulsive forces in dynamic AFM. In Chapter 9, we discuss methods in which a voltage bias is applied between tip and sample to intentionally produce such long-range electrostatic forces, related to the capacitance of the tip–sample system. In most cases, one is explicitly investigating related properties such as dielectric constant. In general, however, the freedom to bias tip or sample also can be important to ordinary dynamic AFM, *even if interested only in surface topography*—especially on highly conductive materials such as the polycrystalline gold surface of Figure 5.4. This is due to an unavoidable fact: the material being studied usually has a different surface potential (i.e., work function for metals) than the material comprising the tip. Thus, as tip and sample approach each other, an electric field is generated. If electrons are free to move within the tip and/or sample, the electric field can cause charges to accumulate on tip and/or sample surfaces resulting in a significant charge–charge or charge–dipole forces. Necessarily, this means a significant force gradient and thus a shift of resonance frequency. If one is attempting to image within the repulsive regime, this phenomenon can lessen the regime's stability. Indeed, the presence of net attraction in the valleys in Figure 5.4a is in part due to electrostatic forces derived from different tip and sample potentials. Thus, the application of an external bias to tip or sample, chosen to *precisely offset the intrinsic surface potential difference*, can be an important optimization strategy (e.g., removing the toggling to net attraction in the valleys of a surface such as that in Figure 5.4a). On the other hand, if one is attempting to image within the attractive regime, the difference in surface potentials can strengthen attractive forces and thus stabilize this regime. The longer range of these forces, however, can result in a significantly increased "fly height" and thus a smearing of the topographic image. For attractive regime imaging, an optimization strategy may be to *include* electrostatic attraction to improve stability against the repulsive regime, or it may be to *remove* electrostatic attraction to reduce fly height and thus improve resolution. The user needs to weigh advantages and disadvantages in the context of the research problem and the realities of the surface being imaged (e.g., challenging surface topography).

The most quantitatively elucidating measurements to identify changes in interaction regime are amplitude and phase versus Z curves. Some will refer to amplitude–Z as a force curve because this measurement is the exact analog to the deflection–Z force curves collected in (quasistatic) contact mode. But the phase–Z measurement provides important additional insight, often a jump discontinuity at some value of Z demarcating repulsive and attractive regimes. This is exemplified in Figure 5.5, five overlaid amplitude and phase versus Z curves obtained during approach at five different free-oscillation amplitudes A_0 ranging from 7.7 nm to 23.6 nm (by using five corresponding drive amplitudes of the same relative magnitudes, 100–300 mV on a Bruker Nanoscope III Multimode), on an untreated silicon wafer using silicon tip and rigid rectangular cantilever (k~40 N/m, Q≈400). Here, the oscillator was driven at its resonance frequency $f_0 = 315.46$ kHz, as designated in the inset, such that the raw phase reading for Z beyond tip–sample interaction is zero. In each case, during approach, the amplitude decreases upon tip–sample interaction (earlier for larger amplitudes of course); later the amplitude suddenly jumps up by ~1 nm, then decreases again with slope somewhat steeper than before. The phase data shows a dramatic jump from above to below 90° (note inverted scale) at the same value of Z at which the amplitude jumps. This tells us that the transition to ~1-nm higher amplitude yields a transition from attractive to repulsive regime. (Thus, the alternative names of low- and high-amplitude states [5].) In Section 5.3, we will further discuss changes in energy dissipation for cases of both continuous and jump discontinuities of phase with decreasing Z. In this section, our emphasis is on diagnostics and control of interaction regime via parameter settings.

Among the five cases in Figure 5.5, the jump between regimes occurs at very different values of A/A_0: large in the case of large A_0 and small in the case of small A_0. Thus, the repulsive regime is dominant for a large range of amplitudes when the cantilever is strongly driven, and a small range of amplitudes when lightly driven.

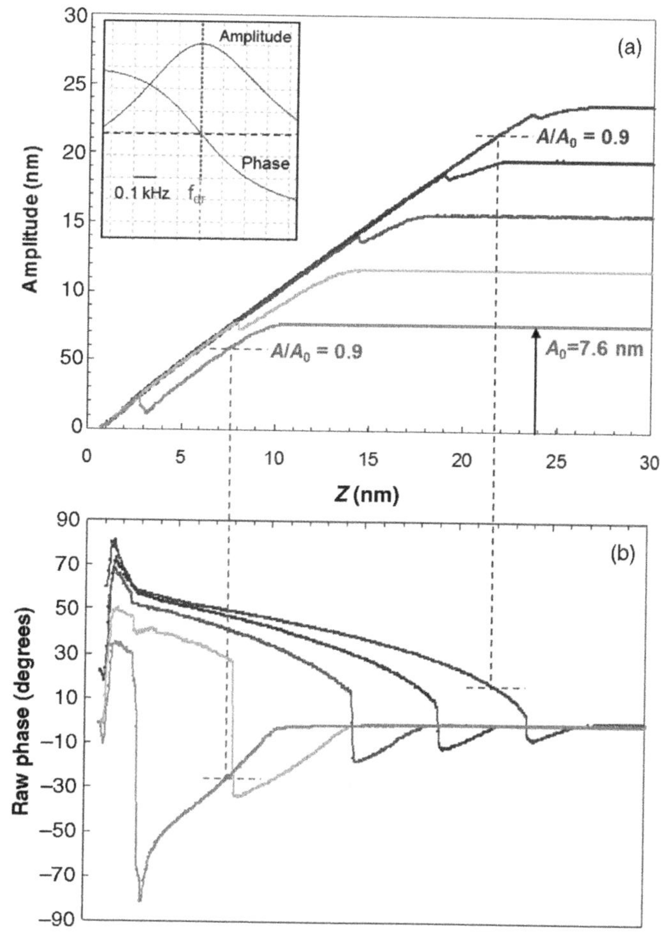

FIGURE 5.5 (a) Amplitude–Z and (b) phase–Z curves on a silicon wafer driving at resonance frequency and at five different drive amplitudes, producing free-oscillation amplitudes A_0 ranging from 7.6 nm to 23.8 nm. Dashed horizontal lines denote typical setpoint amplitudes for repulsive-regime (top curve) and attractive-regime (bottom curve) imaging states at a setpoint amplitude ratio $A/A_0 = 0.9$.

For each of the five cases, one could choose a setpoint amplitude A that results in the net attractive or repulsive regime. (Stabilizing the repulsive regime would be difficult for the lowest-excitation case, and stabilizing the attractive regime would be difficult for the highest-excitation case, because the applicable range of the setpoint A is very small for each.) Also note that for a particular fixed value of $A/A_0 = 0.9$, marked with horizontal dashed lines in Figure 5.5 for the lowest and highest A_0 cases, the attractive *or* repulsive regime results, as indicated by the corresponding phase values to either side of 90° (at which the vertical dashed lines cross the phase–Z curves). This is a common approach to stabilizing regimes while keeping A/A_0 close to unity (i.e., to minimize the mechanical compliance artifact, e.g., as examined in Figure 4.15): large A_0 for repulsive, small A_0 for attractive.

It should also be apparent that the presence and location of transitions between attractive and repulsive regimes, as seen in amplitude and phase versus Z, must be affected by the application of a voltage bias between tip and sample. A net potential difference, in general due to both applied bias and intrinsic surface potential differences, will produce long-range electrostatic forces that act to suppress a transition from attractive to repulsive regime as seen under purely van der Waals interactions—shifting the jumps in Figure 5.5 to lower Z and thus setpoint. However, the application of a bias chosen to precisely offset the intrinsic surface potential difference will shift these jumps to higher Z or setpoint, expanding the repulsive regime. Thus, for a given free-oscillation amplitude, one may be able to stabilize high setpoint operation in the attractive regime, or lower setpoint imaging in the repulsive regime, by applying an appropriate tip or sample bias.

One must recognize, however, that on a heterogeneous surface, the amplitude/-phase–Z curves may depend on the (X,Y) location; thus, the *value of amplitude at which the oscillator transitions from attractive to repulsive may vary with (X,Y) position*. One usually needs to choose a setpoint amplitude that is not close to the transition, anticipating that the transition amplitude will change a bit, perhaps with local surface curvature (e.g., Figure 5.4) depending on (X,Y) location. If the transition is (inadvertently) reached, say repulsive to attractive, then the oscillator may remain "locked in" that new state until a very different kind of location is reached (less lossy or stiffer material, steep hill). This "locking in" is due to hysteresis as exemplified in Figure 5.6 for two amplitude–Z curves ($A_0 = 7.7$ and 15.6 nm, the

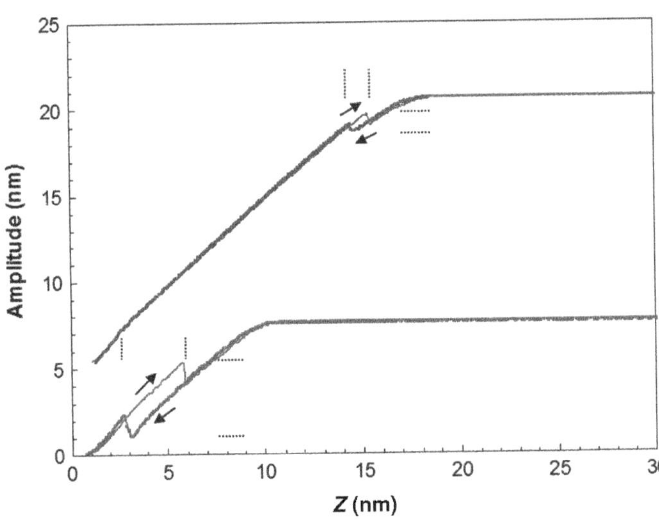

FIGURE 5.6 Full approach–retract amplitude–Z curves on a silicon wafer, demonstrating hysteresis in the transition between repulsive and attractive regimes for two different free-oscillation amplitudes. Upper curve displaced for clarity.

latter displaced upward for visibility) but over full approach–retract *cycles*. (Note slight additional Z-scanner hysteresis such that the curves do not exactly overlap outside of the physical hysteresis loop; see Section 4.2.5.) The hysteresis loop is much larger for the smaller A_0 curve, both horizontally and vertically. This means that the locking of the oscillator into one state or the other (attractive or repulsive) may be stronger for the case of lower A_0.

The images of Figures 5.1 and 5.4b were acquired while driving the oscillator somewhat below the resonance frequency. This is a common setting in the attempt to stabilize the repulsive regime. Here again, amplitude/phase–Z curves can be used to examine the situation more quantitatively. This can be seen by comparing the data in Figure 5.5 to that of Figure 5.7, the latter acquired in the same session with the same

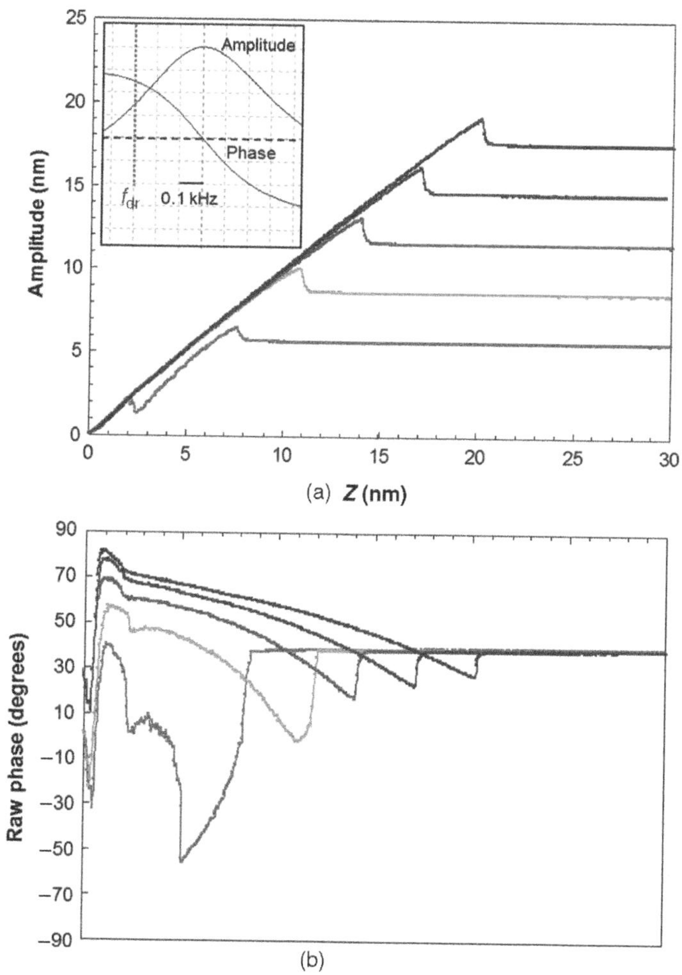

FIGURE 5.7 (a) Amplitude–Z and (b) phase–Z curves during approach to a silicon wafer driving below resonance (inset) and at the same five drive amplitudes as in Figure 5.5.

tip but driving 300 Hz below the resonance frequency of 315.46 kHz, as marked in the inset, instead of at resonance. The same set of five drive amplitudes were explored, and data were acquired during approach. Except for the bottommost curve, corresponding to the smallest value of A_0 (7.7 nm), all raw phase data at reasonable setpoint amplitudes are positive, and thus the oscillator is always in the repulsive regime. Thus, it is the *combination* of driving the oscillator below resonance and avoiding a small A_0 that will most reliably stabilize the repulsive regime. Also note that, in principle, we may operate at setpoint amplitude exceeding free-oscillation amplitude at this off-resonance drive frequency, A_{free} (but less than A_{free} at resonance, A_0). This enables a lighter tip–sample interaction than if $A < A_{\text{free}}$, but with the risk of loss of engagement, analogous to operating at negative applied force in contact mode.

It can be revealing to explore a broader and more detailed set of drive frequencies. Figure 5.8 overlays phase–Z curves during approach in (b) for 15 different drive frequencies in the vicinity of resonance as denoted in (a) along the free-oscillation amplitude curve. These data were acquired on a PVA film similar to that in Figure 5.1 (region 1), using a similar tip–cantilever as described in association with that figure. When driving at resonance frequency f_0, as in Figure 5.5, there is a jump discontinuity from attractive to repulsive regime as Z decreases, at $Z \approx 28$ nm. When driving at the lowest frequency f_- denoted in Figure 5.8a, about 0.25 kHz below resonance, a sudden decrease of phase is seen upon tip–sample interaction during approach, but phase then continues to increase as Z decreases akin to Figure 5.7. When driving at the highest frequency f_+, the phase decreases monotonically as Z decreases, indicating a lack of transition to the repulsive regime. The behaviors at the intervening drive frequencies systematically interpolate between the extremes of drive frequency f_+ and f_-. So for this point on this sample with this tip/cantilever, one can well stabilize either interaction regime by choice of drive frequency, to the left or right of the resonance peak. But again the question is whether such a choice alone will adequately stabilize a given regime across the imaged area. Figures 5.1 and 5.2b exemplify how a choice of drive frequency to the left of the peak does not necessarily stabilize one regime at all points across the surface.

5.1.2 Comments on Imaging in the Attractive Regime

One of the most remarkable successes of dynamic AFM is the ability to image *liquid-phase* condensed matter within the attractive regime. One example is the imaging of water droplets condensed at high humidity, described in Section 5.2.4. A related example is the imaging of chemical patterns via *decoration by a nonvolatile liquid phase* (e.g., glycerol) that chemically adheres to one type of subregion within the pattern but not others. More common applications are thin films containing at least one liquid-phase ingredient.

Even certain solids can be problematic to image because of *capillary* interactions (detailed in Section 6.3), but the problem can be alleviated with attractive-regime imaging. The most common example is nanoparticles (quantum dots) that are weakly adhered to substrate, Figure 6.13. In addition to attractive-regime AC mode, low humidity is often necessary. This helps to prevent *capillary transfer*: the

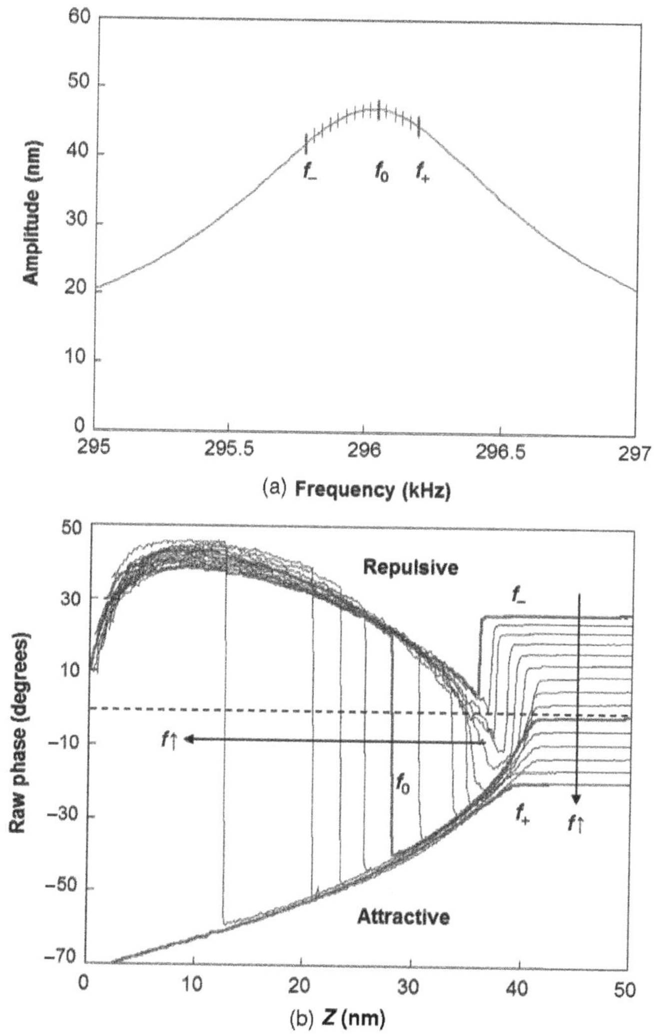

FIGURE 5.8 (a) Designation of 15 drive frequencies relative to a depicted resonant cantilever response. (b) Phase–Z curves obtained at the 15 values of drive frequency in (a). Arrows indicate the effect of increasing drive frequency on both the free-oscillation phase value (right) and the discontinuous transition from attractive to repulsive regime (left).

mobilization and vertical displacement of the nano-object from substrate to tip (discussed in Section 6.3.5). Contact mode is usually an abysmal failure in attempting to image such objects, simply plowing them around, but ambient dynamic mode can fail miserably as well. One can quickly obtain a "popcorn ball" of nanoparticles adhered to the end of the tip, such that each isolated nanoparticle sitting on a flat sample substrate is imaged in topography as a cluster of features (i.e., sample imaging tip as much as tip imaging sample).

5.2 COMPLICATIONS AND CAVEATS REGARDING THE PHASE MEASUREMENT

In Section 2.4, the nature of phase data was introduced in ideal terms, divorced from measurement issues, that is, purely in terms of physical meaning. As discussed in that section, phase is sensitive to shifts of the resonance frequency of the oscillator, in turn related to the tip–sample "interaction stiffness" (positive or negative derived from repulsion or attraction), and more generally sensitive to energy dissipation during tip–sample interaction. This sensitivity makes phase a revealing mode for contrasting materials and biological entities, as exemplified in Chapter 1 and in this chapter. Quantitative, and in some circumstances even qualitative, interpretations, however, can be troublesome. In Section 4.2.1, the instrumentation for phase measurement was introduced as an element in a block diagram schematic, Figure 4.5 (which principally addressed the feedback control system). In the following subsections, we will look in greater detail at the realities and caveats of this measurement.

We discuss six classes of issues deriving from both instrumental and physical artifacts affecting the collected phase data. These arise from (1) hardware and software offsets in the time lag between the sinusoidal signal that is generated to drive the cantilever and the actual motion measured at its tip-end, (2) drift in the instrumentation, resulting in time-variations of oscillator amplitude, resonance frequency, and phase offset, either intrinsic (machine warm-up) or extrinsic (lab environment), (3) variations in oscillator behavior as a function of distance *beyond* tip–sample engagement (e.g., that take place as the stepper motor performs the coarse approach of tip to sample), (4) coupling between topography and phase, primarily ways that surface slope affects the measured phase, (5) nonidealities in the electronics used to perform the phase measurement (dependent on vendor design), such that the phase lag of a driven harmonic oscillator with damping (Section 2.4) may not be exactly what is measured, with implications toward the energy dissipation interpretation of phase, and (6) nonidealities in the resonance lineshape, amplitude, and phase as a function of drive frequency. Topics #1–4 and 6 are often immediately apparent to the observant operator upon first use of dynamic (AC/"tapping") mode with phase imaging. Topic #5 will be systematically important to the interpretation of phase data in the context of energy dissipation. The following six subsections treat topics #1–6 in order.

5.2.1 The Phase Offset

Measuring the time lag between a sinusoidal vibration imparted to the base of the cantilever and the motion of its tip-end is simpler in concept than in practice. An electrical AC driving signal (i.e., a sine wave) must travel from the electronic-waveform source device through electrical contacts to a vibrational element (e.g., a piezoelectric bimorph); the generated vibration must then propagate through bulk materials as well as material interfaces until it reaches the chip and thereby the base of the cantilever. Because of the high frequency of this AC signal, there can be appreciable phase shifts due to capacitance, in principle, as well as other phase-affecting elements along this path. One makes no attempt to quantify each of the contributing phase shifts. Instead, one determines an overall apparatus phase shift (including possible electronic shifts on the backside, between oscillating tip and final measured phase) that must be subtracted out to yield the mechanically correct phase lag between cantilever base- and tip-ends at resonance, 90° (although the instrument software may call this "zero," as discussed below).

It is a simple matter to determine the resonance frequency; one measures the amplitude of tip excitation as a function of drive frequency and locates the peak. For driving at this resonance frequency, the phase is offset by some amount to force the phase lag of the tip-oscillation sine wave to be 90° relative to the driving sine wave, as if all the other instrumental contributions to the overall phase shift have been eliminated. The software typically refers to this as "zeroing the phase," because a 0–180° actual range of phase is mapped into a +10 to −10 V scale for the data acquisition system, meaning 0 V (center scale) at 90° phase lag. This +10 to −10 V scale is typically converted by the software to degree units, +90° to −90° (i.e., 9° per volt); this means that one would compute the actual, physical phase lag of tip relative to cantilever base as 90 minus the software phase.

Figure 5.9a and b displays the measured frequency dependence of amplitude and phase of a freely oscillating (i.e., tip far from sample), driven cantilever using (a) incorrect and (b) correct phase offsets, the latter after invoking the needed shift to "zero the phase" at the resonance frequency. If this step is not properly completed, the reference value for phase can differ from the correct value, such that phase measurements are quantitatively erroneous; if the reference value is far from correct, phase images even can be qualitatively wrong, that is, inverted in contrast: a material that should appear relatively dark in a phase image (from physical considerations within, e.g., the repulsive regime, say because of softness) may appear relatively bright, leading to incorrect assignment. This is exemplified in Figure 5.10a and b, 3 × 3-μm height and phase images (with different color offsets) obtained on a 75:25 blend film of poly(butyl methacrylate) (PBMA) and poly(lauryl methacrylate) (PLMA), exhibiting a phase contrast inversion from (a) incorrectly to (b) correctly zeroed phase. The circular regions are the minority PLMA. The images were acquired in the dominant repulsive regime with $A/A_0 = 0.79$, $A \approx 60$ nm, using an aluminum-coated silicon cantilever (with Si tip), $k \approx 40$ N/m, and driving at a frequency below resonance as denoted by the vertical dashed lines in Figure 5.9. (Under such conditions, one expects the PLMA to be the more energy dissipative and thus exhibit a less positive net phase shift relative to the zeroed phase at resonance out of tip–sample engagement. We will revisit physical aspects of phase for this system in Section 5.3.) Both the qualitative phase (sense of image contrast) and the quantitative magnitude of phase differences (1.3° va 2.3°) markedly differ in Figure 5.10a and b.

To ensure a correctly zeroed phase, one typically performs some software variant of the following steps, in sequence (although possibly automated in some AFM systems).

1. Examine the response amplitude while repeatedly sweeping the drive frequency.

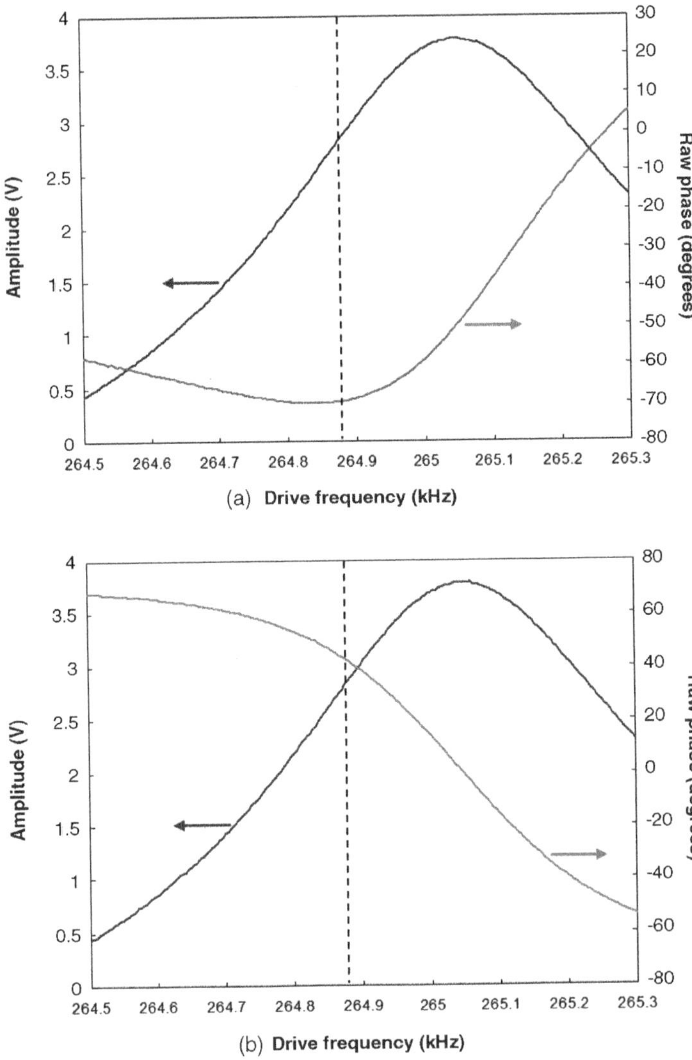

FIGURE 5.9 Measured free-resonance spectrum of an aluminum-coated, rectangular silicon cantilever for cases of (a) improper and (b) proper zeroing of phase.

2. Locate the most prominent peak.
3. Center the frequency sweep window (e.g., in Bruker Nanoscope software) or place a frequency-specifying cursor (e.g., in Agilent Picoview software) on this peak.
4. Zero the phase at this frequency (e.g., a button click in Nanoscope and Picoview).

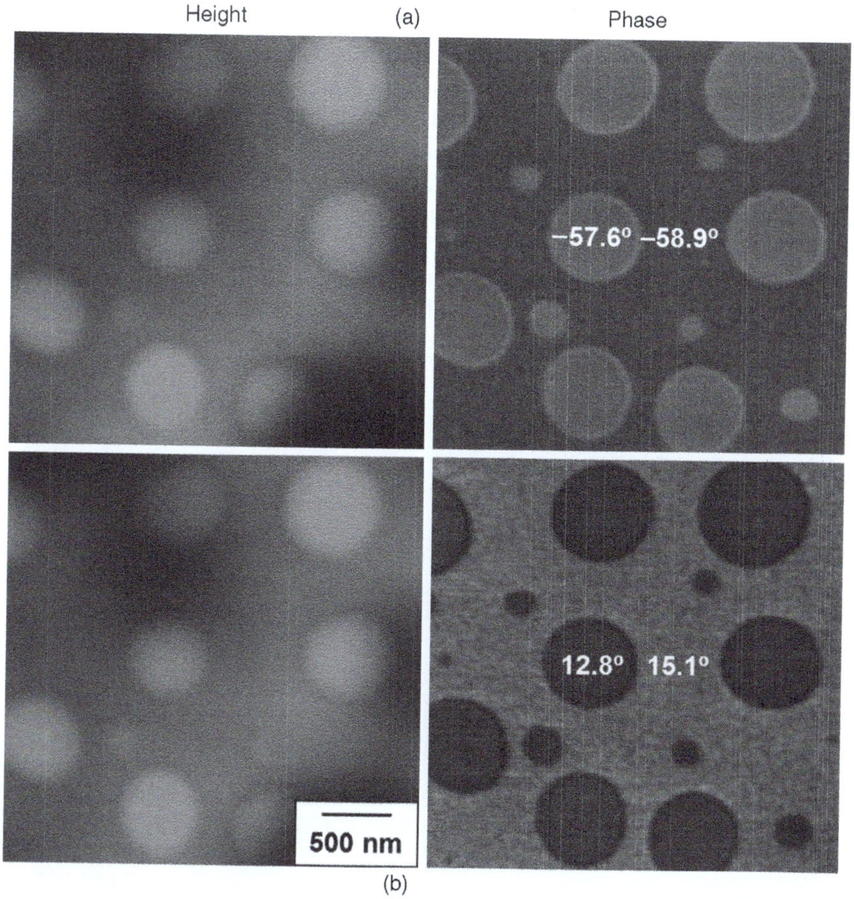

FIGURE 5.10 Height/phase image pairs acquired with the phase offset (a) improperly and (b) properly zeroed as per Figure 5.9.

Potential mistakes are to forget to zero the phase (i.e., likely using a value from a previous user), or to zero phase at some frequency other than resonance (e.g., perform the steps in the wrong order). The result can be an image such as that in Figure 5.10a.

We will see in the coming sections that it is also essential to keep track of *absolute* phase values for proper quantification, for example, to control imaging regime (as discussed in Section 5.1). AFM software is often by default configured to render the real-time image to the control computer's screen only after the data values have been *manipulated for viewing clarity*. The simplest manipulation is a data offset (not to be confused with the phase offset discussed above to yield *physically* correct numbers), which positions the mean value of the data at the center of the range of colors or tints. Given the real-time nature of this manipulation, it is often performed for each scan line of data independent from the previous lines. Typically, the color or brightness scale is further squeezed, that is, the *data range* setting in the software (or some equivalent name), meaning the range of rendered color or brightness, is selected to span the range of data values rather than the full 180° scale. The result is often an image containing both bright and dark colors or tints (i.e., interesting contrast), whereas if rendered over the full 180° scale, the image may appear entirely bright or dark, with only fine shades of differences (e.g., corresponding to a several-degree range of phase values). Moreover, the line-by-line nature of the adjustment may result in false transitions of color from one line to the next that appear as image "bands."

Figure 5.11 uses the phase image data of Figure 5.10 to illustrate the preceding concepts. The phase images of Figure 5.11a and b render exactly the same data as in

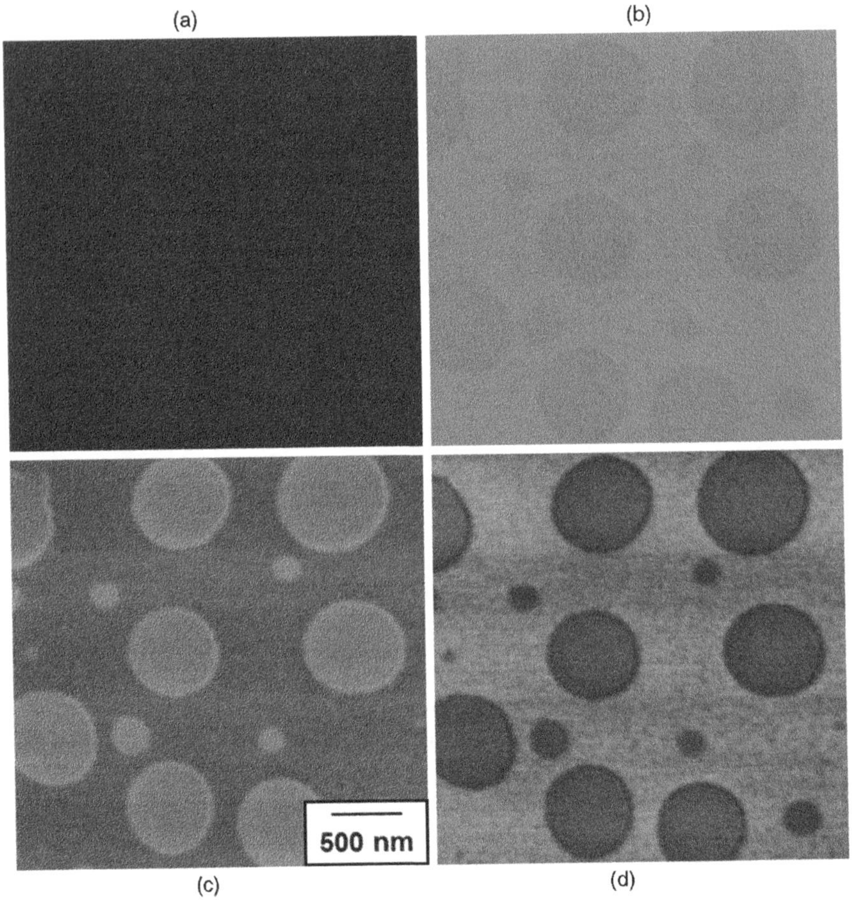

FIGURE 5.11 Phase image data from Figure 5.10 but rendered differently as follows. (a) and (b) Applying the full color range to the raw phase data from Figure 5.10a and b, respectively. (c) and (d) Following a line-by-line adjustment whereby each line of data is shifted numerically higher or lower (brighter or darker) such that the mean value equals the middle of the color scale, resulting in false bands of dark or bright in the images.

the phase images of Figure 5.10a and b, but assign the complete range of color-/brightness to the full 180° scale (i.e., raw), and without offsets to center the data at the midpoint of the color scale as done in Figure 5.10. Given that parameters were selected to stabilize the repulsive regime, meaning a "bright" raw phase image, viewing the images in this way immediately draws suspicion to the dark image of Figure 5.10a (whereupon the operator should go back to the free oscillation frequency sweep measurement to check the phase offset). The phase images in Figure 5.10c and d also contain the same data as Figure 5.10a and b but are shown as commonly viewed during the acquisition stage, with line-by-line offsets of the data such that each line's mean value independently corresponds to the center of the color/brightness range. Here, the overall bright or dark character is not viewed, and thus the interaction regime is not easily assessed in the image. Moreover, bright and dark bands, artifacts of this data manipulation, are evident outside of the circular regions. The bright bands seen in Figure 5.11d are the result of lower data values within circular regions present along these scan lines. These values lower the mean of a given scan line, such that upon subtraction of the mean from each datum along the line, the numbers outside of the circular region are not as low as they would be in the absence of PLMA circles (i.e., compared to lines of data where the mean value is higher). Somewhat fainter dark bands are visible in Figure 5.11c for the analogous reason but opposite in sign. The phase images in Figure 5.10 also were treated with a line-by-line adjustment in off-line post-processing (often known as "flattening"), but importantly the circular regions were *excluded* (or "masked") by the operator within the software routine. Thus, the mean values calculated and subtracted per line of data reflected only those measurements atop the PBMA, not the PLMA. The result is a lack of band artifacts in the Figure 5.10 phase images, unlike Figure 5.11c and d.

5.2.2 Drift in Resonance Frequency, Phase Offset, Quality Factor, and Response Amplitude

Properly zeroing the phase does not guarantee quantitatively correct or consistent results. Thermal warm-up generally may affect the resonant frequency and Q factor of the oscillator, the instrumental phase offset, and the magnitude of response amplitude for a given drive amplitude. After a cold start, and usually through the first hour and perhaps as long as several hours of operation, one may need to continually adjust the drive frequency and phase offset to account for this drift. Failure to do so may produce at least several degrees and perhaps tens of degrees of drift in the phase measurement. Figure 5.12 illustrates thermal warm-up effects (on a Bruker Nanoscope III system). The total change of phase offset, as measured at the initial resonance frequency, is $-16°$ over the course of $2\frac{3}{4}$ h of total warm-up time since a cold start. This was comprised of a -24 Hz shift of the resonance frequency, producing a $-7°$ shift of phase offset due to the phase–frequency relationship (i.e., generally a larger effect for higher Q factors), and another $-9°$ intrinsic shift of phase offset. This latter value is relative to the zero of phase if rezeroed at the -24 Hz shifted resonance frequency as depicted with arrows. Figure 5.13a plots the time dependence of both

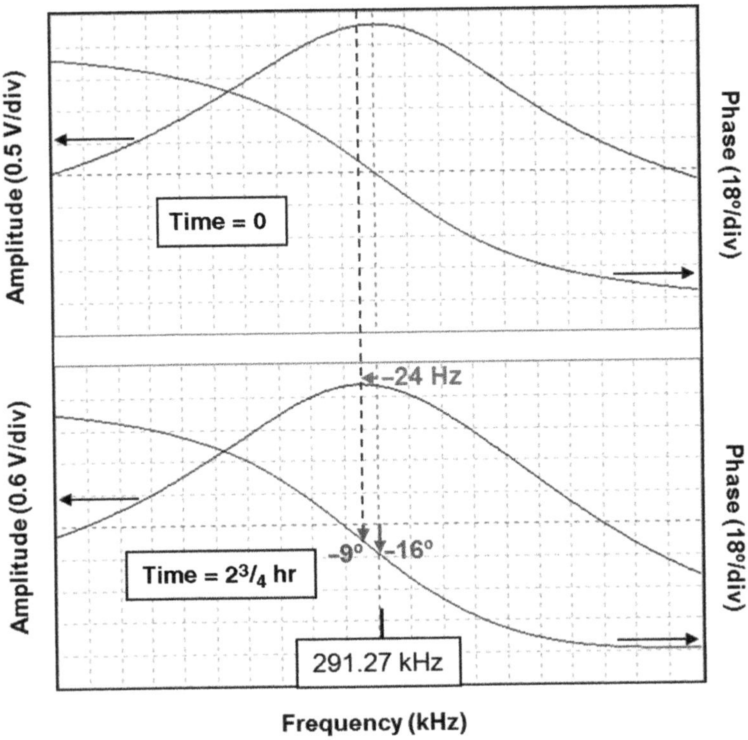

FIGURE 5.12 Amplitude and phase resonant response initially (within a few minutes of power-up) and $2\frac{3}{4}$ h later. Warm-up-induced shifts in resonance frequency and phase offset are depicted with arrows.

the resonance frequency and this intrinsic phase shift (i.e., rezeroing at each new resonance frequency). Interestingly, one sees a monotonic lowering of resonance frequency that has not reached steady state even after $2\frac{3}{4}$ h, and a non-monotonic change of phase offset. One would need to continually monitor and compensate for each of these shifts to maintain quantitative control over phase measurements.

The drift of the resonance frequency necessarily changes the free-oscillation amplitude at any selected drive frequency, given the resonant relationship between free-oscillation amplitude and drive frequency (e.g., as in Figure 5.12). Thus, for example, if one is imaging at a drive frequency somewhat below resonance as in Figure 5.9b (a strategy to stabilize the repulsive regime as discussed in Section 5.1), then a leftward shift of the resonant relationship must produce an increase in free-oscillation amplitude. For a fixed setpoint amplitude A, the ratio A/A_{free} decreases (although the physically more significant A/A_0 remains constant, see Eq. 2.24). But another, independent source of potential drift in free-oscillation amplitude is an

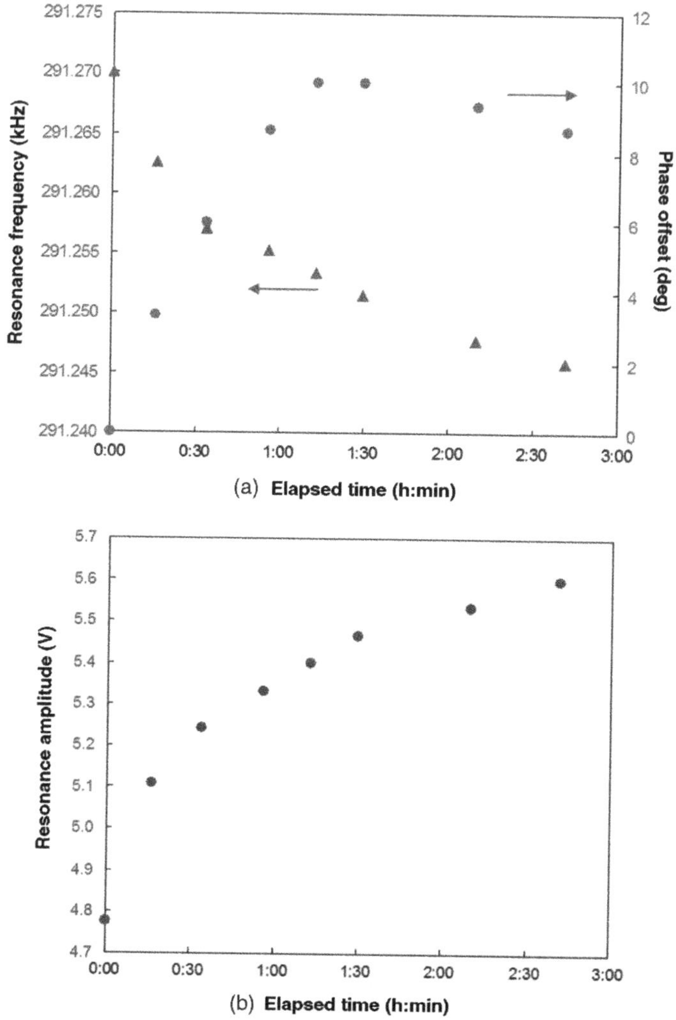

FIGURE 5.13 Time plots of thermal drift in the oscillator's resonance response.

actual change in A_0. For a fixed drive amplitude setting, the entire amplitude–frequency curve may drift upward or downward, that is, the energy of the driven oscillator may increase or decrease, perhaps as the mechanical coupling of the vibrational signal to the cantilever chip changes during warm-up. Indeed, this was observed during the course of collecting the measurements in Figure 5.13a. (Note the change of vertical scale in Figure 5.12.) The time dependence of resonance amplitude is shown in Figure 5.13b. Over the course of $2\frac{3}{4}$ h, a monotonic increase in free-oscillation amplitude A_0 of 17% is observed, corresponding to a 37%

increase in oscillator energy given the relationship $E = 1/2kA_0^2$. On soft materials or biologicals, one expects significant changes in both height and phase images due to this drift. If imaging at fixed setpoint amplitude A, the ratio A/A_0 also would have changed by 17%. The change of energy dissipation, and thus phase, due to the change of A_0 alone (i.e., apart from the drift of resonance frequency and phase offset) would be substantial given the relationship Equation 2.24.

In addition to warm-up induced shifts in the resonance amplitude and frequency as well as the instrumental phase offset, one will generally find greater instability in these parameters if the laboratory environment is unstable, whether room temperature or humidity. This also means that different behaviors (e.g., the separation of interaction regimes within parameter space) might result at different seasons of the year, dependent on the stability of a facility's environmental control. The viscosity of air is the principal determinant of the cantilever quality factor Q. Air viscosity increases with temperature within a few degrees of room temperature—roughly 1% change per 4°C of temperature T [8] (but is essentially constant vs. the range of possible water vapor content near room temperature [9]). Because Q varies inversely with the velocity coefficient b in Equation 2.19 (and thereby dynamic viscosity [10]), its percentage decrease with increasing T approximately equals the percentage increase in air viscosity. Note that the resonance frequency ω_0 is also affected by damping, per Equation 2.18. A mathematical power series expansion of this expression, assuming the damping-derived shift in resonance frequency $\frac{1}{2}(b/m*)^2$ is very small compared to ω_0, gives a negative frequency shift that scales *quadratically* with b and thus $1/Q$. Hence, the resonance frequency decreases (with increasing temperature) even more significantly than the quality factor decreases.

Additionally, the coupling of the driving vibration to the cantilever chip may be affected by temperature and humidity, and thus the relationship between drive and response amplitudes as well as the phase offset. Finally, one should note that metal-coated cantilevers can exhibit an effective change in properties versus T and RH, due to differences in thermal expansion in the two materials as well as variable surface tension due to adsorbed water on the coated versus uncoated surfaces. Taken together, variations in ambient environmental conditions may translate into variations in ω_0, Q, and φ_{offset} by at least a few percent. Given the strong functional dependence of phase on these three parameters, phase variations can reach tens of degrees. These issues are even more egregious if experiments are being conducted as a *function* of humidity and/or temperature as this usually means wider ranges. In this case, the continual measurement of A_0, F_0 $(\omega_0/2\pi)$, Q, and φ_{offset} during the course of environmental changes is strongly recommended.

An interesting display of the effects of variable air temperature can be performed by breathing near the cantilever during frequency-sweep measurements. Substantial resonance curve shifts, translating into tens of degrees of phase, are measurable. Another rather inadvertent example is to monitor the change of resonant parameters through the course of a group demonstration in a small lab space, where temperature may increase by a few degrees and humidity by more than 10% due to crowded bodies. Phase images may vary in contrast as parameters drift.

5.2.3 Change of Phase and Amplitude During Coarse Approach

Even if one properly locates the free resonance frequency and zeroes the phase at this frequency, after coarse approach and engagement one observes that the phase is in fact *not* correctly zeroed (e.g., by examining phase–Z curves just beyond tip–sample interaction). This is due to long-range distance dependence in the resonance behavior (quantified in F_0, A_0, Q): changes that are measurable over hundreds of nanometers and up to tens of micrometers, as demonstrated in Figure 5.14. This is mostly due to a phenomenon termed *squeeze film damping* [11,12]. It turns out that

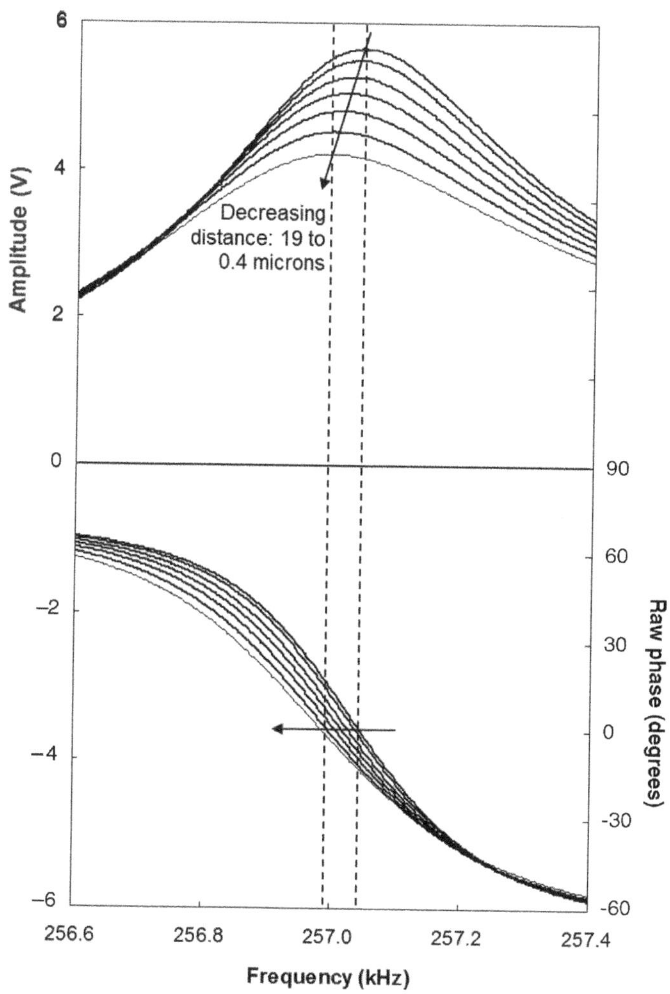

FIGURE 5.14 Frequency dependence of raw amplitude and phase signal at different distances actuated by the Z motor. Highest to lowest resonance amplitudes A_0 ranging from 75 nm to 54 nm were measured at individual distances of 19, 10, 5, 3.2, 2, 1.2, and 0.4 μm.

the air drag on the cantilever increases as the distance between cantilever and sample surface decreases. This is because net work is done (cantilever kinetic energy is extracted) in pushing air out of the gap between the cantilever and the sample surface during each oscillation cycle. The maximum distance to which this effect is appreciable is comparable to the lateral dimensions of the cantilever. Some studies have investigated the phenomenon in detail [11–13], but it is easily observed during routine AFM usage as suggested above in the context of phase–Z curves. Figure 5.15

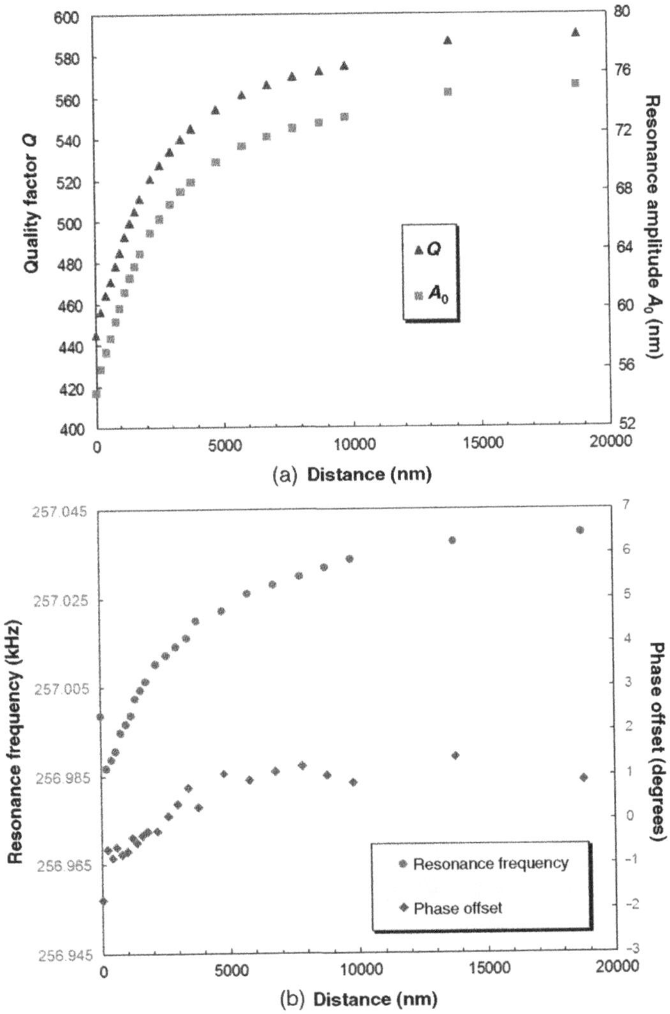

FIGURE 5.15 Distance dependence (via Z motor) of (a) quality factor Q and resonance amplitude A_0, and (b) resonance frequency and (intrinsic) phase offset. The change of trend at the leftmost data point in (b) was used to estimate the zero of the distance scale.

derives from analyzing a larger set of frequency-sweep curves. The strong distance dependence in damping is definitively seen as the reduction of quality factor Q and resonance amplitude A_0 with approach in (a). Because resonance frequency varies with Q (see Eq. 2.20), its distance dependence tracks that of Q (b). (Both resonance frequency and A_0 changed out to 50 μm distance, beyond which changes were within uncertainty.) What is not predictable is the *intrinsic* change of phase offset; the inflection point of the phase–frequency curve does not exactly track the shift of maximum of the amplitude curve, A_0. The plotted phase offset is the additional phase shift needed to align this inflection point with A_0. Although measurable, this effect is significantly smaller ($\approx 2°$) than the change of phase due to the shifting resonance frequency ($\approx 12°$). (It resulted in part from thermal drift during the course of measurements, but was principally a function of distance.)

For the AFM operator, the difficulty is that even a slight change in resonance frequency relative to a fixed drive frequency, as well as a slight change in phase offset, can translate into several degrees of change in phase. This can produce inconsistent results when the task of choosing drive frequency (relative to resonance) and zeroing phase is *not performed at a consistent cantilever–sample distance*, during the session(s) from which data is to be compared. For detailed and quantitative work, one should carefully examine amplitude–Z and phase curves while adjusting drive frequency and phase offset. Figure 5.16 shows such curves

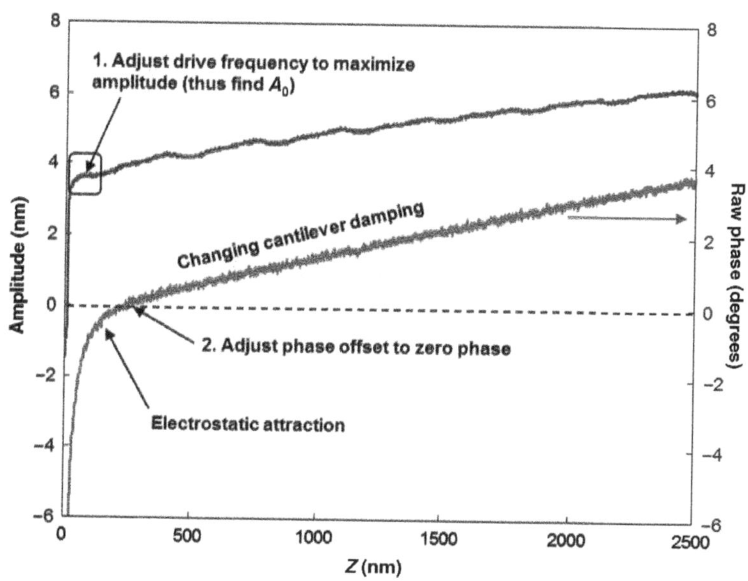

FIGURE 5.16 Amplitude–Z and phase–Z curves during approach from 2.5 μm away from tip–sample engagement, to distinguish increasing air-damping effects on the cantilever oscillation from the onset of tip–sample interaction.

over a large excursion of Z, 2500 nm, to demonstrate the change of amplitude and phase resulting from the increase of damping with decreasing distance. (Waviness in the amplitude data derives from the same laser-light interference artifact as discussed in Section 3.4.) This change is essentially linear for such a "short" distance compared to that of Figure 5.15 (there 20,000 nm). A procedure is to first adjust the drive frequency until the amplitude is maximal, thus finding A_0, then adjust the phase offset to zero the phase. The latter may require distinguishing between the part of the phase–Z curve that changes linearly with distance (due to changing cantilever damping) from a more curved part due to the onset of electrostatic attraction between tip and sample, which one would not want to "zero away," because it is physically relevant to the sought measurements (whereas the change in cantilever air damping with distance is not relevant). This procedure should be repeated during the course of a session to compensate for thermal drift.

5.2.4 Coupling of Topography and Phase

Upon first examining a phase image, the user may notice that topography seems to be "leaking" into the phase image. This may be especially obvious if monitoring the amplitude (error signal) image at the same time. The phase image, in addition to containing material contrast, may in some ways resemble the height and/or amplitude image. Why would this be? There are three principal mechanisms of coupling between phase and topography. The first was already mentioned in the discussion of Figure 5.4: the fact that the strength of attraction increases in valleys having curvature comparable to the tip and decreases atop hills with such tight curvature, opposite in sign. Thus, dark phase (attraction dominant) can correspond to dark height (low spots in narrow valleys) and bright phase can correspond to bright heights (high spots atop narrow hilltops).

The second and third mechanisms of coupling between phase and topography relate to surface slope. The first of these is instrumental rather than physical. The effect derives from the imperfect tracking of topography; the error signal varies as the slope or first derivative dZ/dX of topography, meaning the amplitude is deviating from its setpoint in correspondence with the local slope. We have seen in amplitude/phase–Z measurements (e.g., Figure 5.5) that there is generally variation in phase as amplitude varies. Indeed, this variation is often strong and monotonic. In some cases, a *fraction of a nanometer* deviation from the setpoint amplitude may produce a change in phase of *several degrees* (see, e.g., the bottom phase curve in Figure 5.5). Moreover, this inverts in sign upon moving from a positive to negative surface slope, enhancing the visual effect by another factor of two.

To demonstrate, we compare an amplitude image on the PBMA–PLMA blend system of Figures 5.10 and 5.11 to the phase and height images, all collected during left-to-right scans. One must note that the conventional amplitude image is *inverted* in sign, brighter being a deviation below setpoint and darker a deviation above setpoint, so as to drive the Z-feedback control circuit correctly, in the same sense as contact mode (as discussed in Section 4.1). We also compare single line traces along the fast-scan axis. Here, the usual correspondence between height and amplitude (error signal) is clear (although inverted relative to contact-mode height/deflection): the amplitude signal varies with local slope dZ/dX of topography. The phase trace contains about 1.1° of materials contrast (per dashed lines) obtained in the repulsive regime (very "light tapping" while driving below resonance), with PLMA circular regions higher in phase, plus a slope-derived component similar in magnitude. A deviation of amplitude below setpoint on an upward surface slope results in lower phase, and a deviation of amplitude below setpoint on a downward surface slope results in higher phase. This is because the phase–Z trend is downward as amplitude decreases, as in the bottom curve of Figure 5.8b. Even on the PBMA majority surface (i.e., outside of the circular regions), there is a rolling topography that produces a variable background in the phase image. The slope-derived component can be minimized by optimizing gain settings and scanning slowly. Another

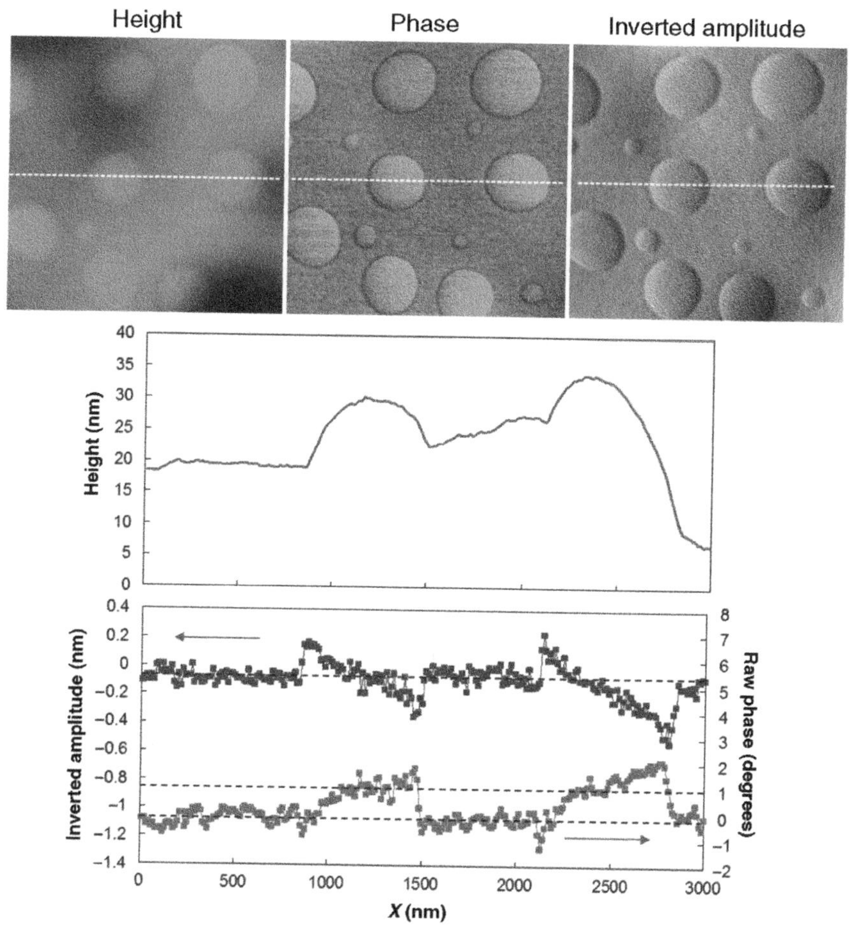

FIGURE 5.17 Relationship between topographic slope along the horizontal (fast-scan) axis and imaged changes in phase under left-to-right scanning.

approach is to collect phase over trace and retrace scans and add the two images, such that the deviations tend to cancel out and the material contrast doubles. (A caveat is that *X* hysteresis can produce a "double image," unless steps are taken to offset one of the two images along *X* prior to adding the images.) It is also worth noting the presence of dark rings around the PLMA domains in phase in Figure 5.17. These likely relate to the local concavity around the base of the PLMA dome-shaped domain, increasing the role of attractive interactions, mechanism #1 for topographic coupling to phase images.

The third mechanism by which topography seemingly "leaks" into phase images again relates to real differences in energy dissipation (as for mechanism #1, the curvature effect). But this is another surface slope effect, and is dependent on cantilever orientation. The images in Figure 5.17 were acquired with the fast-scan axis perpendicular to the long axis of the cantilever. At any point, the local surface slope measured *parallel* to the cantilever long axis, which in Figure 5.17 means dZ/dY, affects the energy dissipation. This is a result of the cantilever tilt that is commonly engineered into AFM instruments to enable the laser optical path. A roughly 12° tilt of the cantilever means that the tip's oscillating motion has a lateral component and thus produces shear dissipation when the tip contacts the sample surface [14]. The increase of dissipation due to shear may exceed the reduction of dissipation due to slightly decreased tip velocity normal to the surface and slightly lesser penetration into the sample (although, in principle, the opposite relative impact could be manifest, dependent on material properties). Thus, a surface slope that is downhill along the direction from tip to base of cantilever often may exhibit higher dissipation, and thereby a change of phase signal, compared to a surface location, where this slope is zero or uphill in the direction from tip to base of cantilever. In the images of Figure 5.17, this effect, if present, is very weak, the result of the slope dZ/dY being less than 0.1. When the effect is strong, there can be additional problems with the Z-feedback circuit ringing out of control. These artifacts are exemplified in the following.

In Figure 5.18, we explicitly compare height and phase traces along the slow-scan, long-cantilever axis direction (vertical) in attractive-regime phase images of water droplets hundreds of nanometers tall, containing slopes dZ/dY of nearly 0.4, much greater than on the circular PLMA domains in Figure 5.17. The water droplets condensed on a micrometers-thick coating of a biodegradable commercial polymer Polyactive™—the semicrystalline block copolymer poly(ethylene oxide–butlyene terephthalate) (PEO–PBT)—at the 94% relative humidity imaging conditions in an environmentally controlled Agilent 5500 AFM system. (This interfacial phenomenon is reversible, i.e., the droplets appear and disappear or grow and shrink with humidity cycling. The two blocks differ in that PBT is hydrophobic and highly crystalline while PEO is hydrophilic and amorphous.) The observation of interest to our discussion here is the pronounced slope in the phase cross section. Higher phase near the top edge of the imaged droplet (right side of droplet in cross section) corresponds to higher energy dissipation in the attractive regime, the anticipated correspondence. The effect is superimposed on an overall higher phase and thus dissipation atop the droplet compared to the surrounding Polyactive surface.

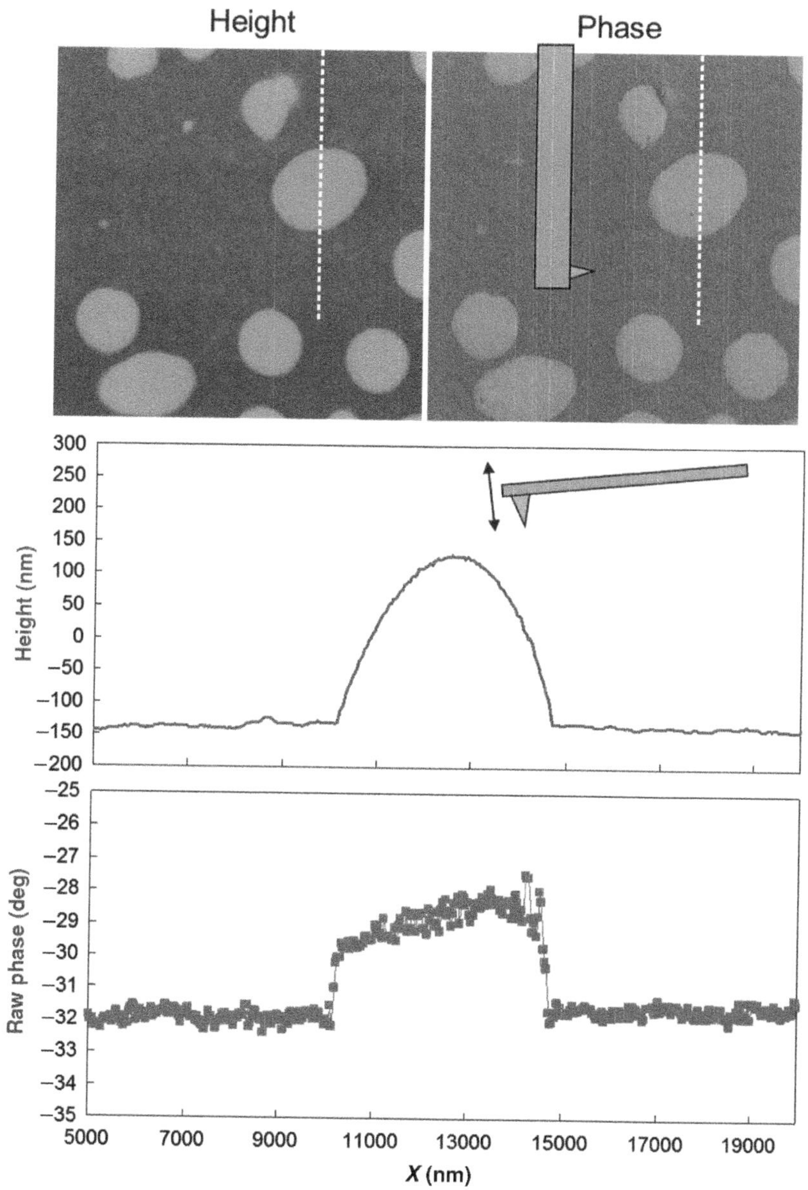

FIGURE 5.18 Relationship between topographic slope along the horizontal component of the cantilever's principal axis and phase measurements, with the fast-scan axis chosen to be perpendicular to cantilever.

Often on highly dissipative materials (e.g., sticky) and/or at locations of steep slope dZ/dY downward toward cantilever base, it is difficult to avoid oscillations in the Z-control feedback circuit. This is exemplified in Figure 5.19 over the same

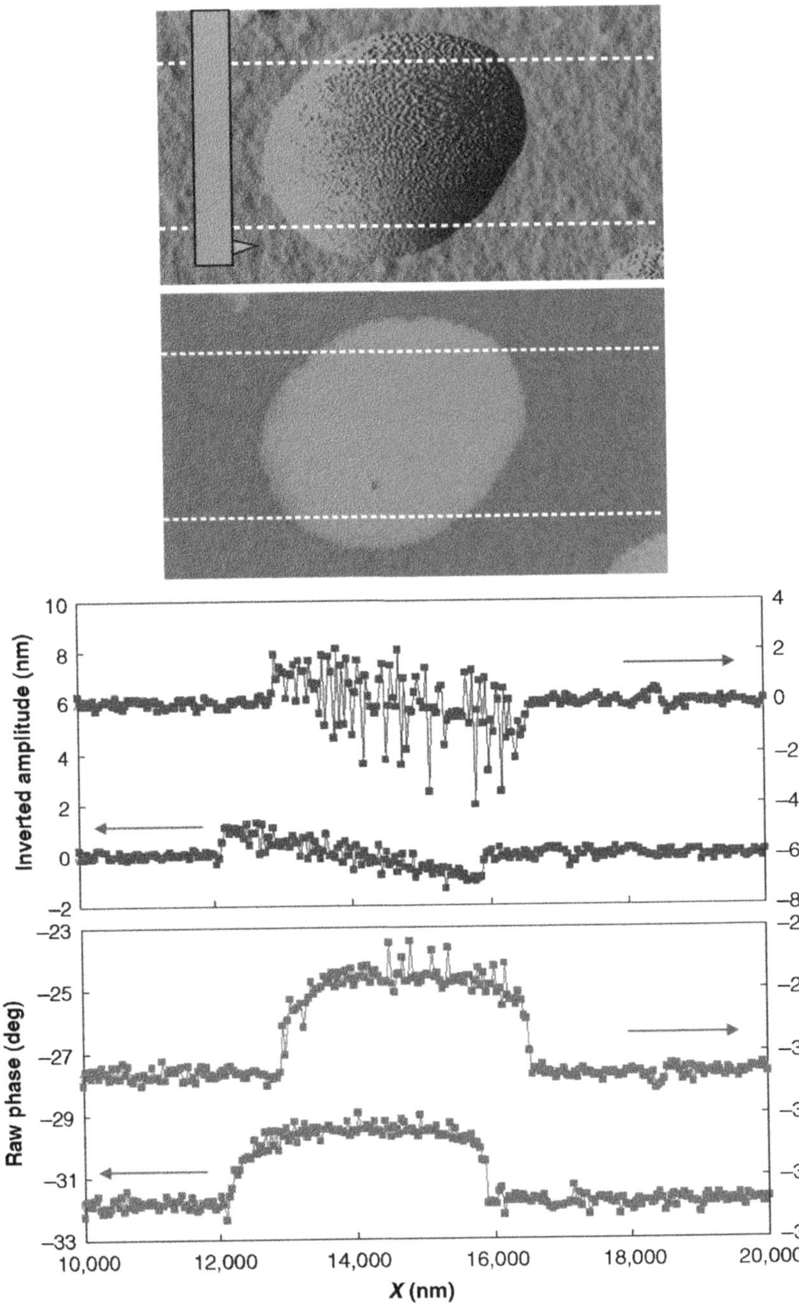

FIGURE 5.19 Relationship between topographic slope (top image and graph) along the horizontal (fast-scan) axis and phase (bottom image and graph) when uncontrollable Z-feedback oscillations are present dependent on the slope along the slow-scan axis.

droplet as examined in Figure 5.18 but along the fast-scan direction (X). Two traces are compared in both amplitude and phase, on roughly equally steep locations but on opposite sides of the droplet, the bottom of the two traces being on the lower-dissipation side of the droplet per the preceding discussion. The amplitude traces exhibit the expected slope because of the variations in Z tracking as discussed in Figure 5.17, but the top trace exhibits much more noise. This noisy tip–sample engagement is manifest in the phase data as well. Close inspection of the amplitude (or phase) image reveals a quasiperiodic signal much as one sees if Z-control feedback gains are set too high, but only on the highly dissipative material and/or localized to the highly dissipative geometry (steep slope). Reducing the gain settings does not remove this quasiperiodic signal.

The above is a not uncommon artifact on sticky or highly sloped surfaces of certain materials: uncontrollable Z-feedback oscillations. The phenomenon can be roughly understood as follows. A highly dissipative surface may dampen the cantilever oscillation more quickly than the reaction time of the Z-feedback control. Once the integrated error signal has grown to a large negative number due to the amplitude falling far below setpoint over many oscillation cycles, the feedback dramatically displaces Z in the positive sense such that tip and sample fully disengage (i.e., such that the cantilever amplitude increases to the full free-oscillation value). Then, once the integrated error signal has grown to a large positive number due to the amplitude reaching values far above setpoint, the feedback strongly displaces Z back the other way, such that tip and sample reengage, only to have the amplitude drop well below setpoint again due to strong damping. After many oscillation cycles, the feedback once again yanks the sample and tip apart, amplitude well above setpoint, then yanks back toward engagement and below setpoint, and so forth. Often the only way to eliminate or at least minimize this artifact is to give the oscillator more energy, meaning choosing a higher drive amplitude to attain a higher free-oscillation amplitude. The situation can be further aided by operating at setpoint amplitude that is nearly as large as the free-oscillation amplitude (i.e., A/A_0 nearly one). Operating within the attractive regime on liquidy samples such that tip and liquid do not make contact—meaning under purely "long-range" van der Waals and possibly electrostatic interaction—is sometimes the only way to make things work. In this case, the fly height may be nanometers off of the surface, thus the measured height of droplet or domain may be greater than reality.

To complete this topic, we must note that localized, material-specific oscillations in the Z-control circuit instead can be related to a *steep amplitude–Z* ($A(Z)$) data trend. This is not uncommon on mobile materials and/or when polymeric contaminants are present on the tip; there can a wicking effect that aggressively dampens the cantilever oscillation such that amplitude drops more steeply with decreasing Z than the tip–sample distance would dictate on a stable solid. To see this effect, one can choose a few point locations on specific material domains and collect $A(Z)$ curves. An even more elegant method, previewed here and further treated in Section 6.4.3, is to collect a large set of $A(Z)$ curves over a grid of locations in the X–Y plane (albeit a data acquisition routine typically requiring longer periods of time than conventional images). Some commercial AFM hardware/software

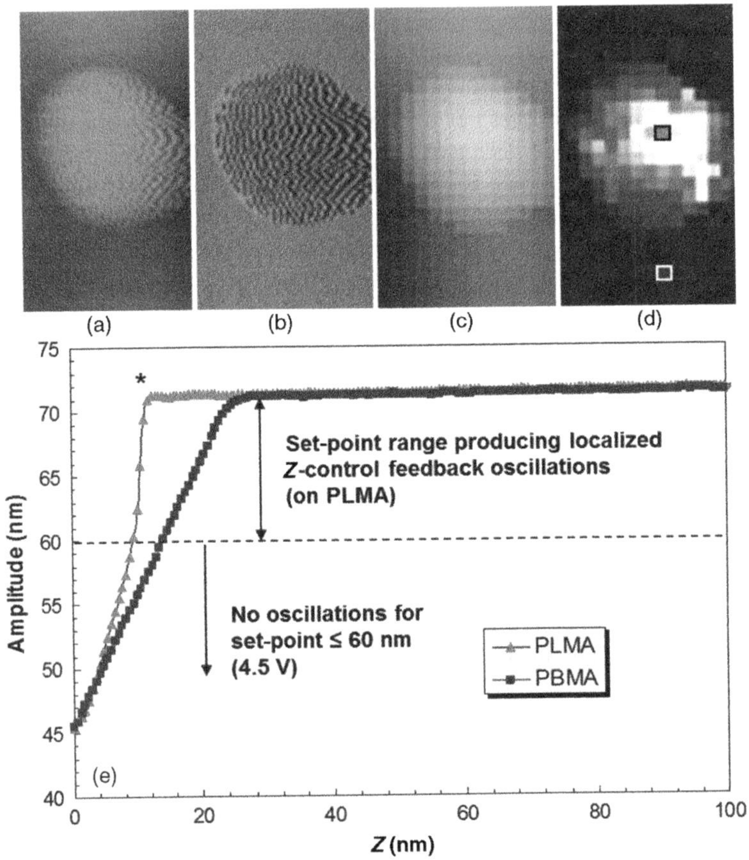

FIGURE 5.20 *Z*-control feedback oscillations seen in conventional (a) height and (b) phase images specific to liquidly sample domains, caused by an overly steep amplitude–*Z* relationship, as probed in *A*–*Z* curves acquired in an array of locations as rendered in (c) height and (d) *A*–*Z* slope.

systems enable such *data-cube* acquisitions within dynamic (AC/"tapping") mode. Our first example was Figure 1.21, although in that case the *time-averaged deflection* was collected, becoming quasistatic contact following reduction of the oscillation amplitude to zero at nearest approach. Now we consider results on the same PBMA/PLMA film discussed with Figure 5.17, obtained on a Bruker Multimode using the so-called *force volume* routine, but measuring $A(Z)$ instead of time-averaged deflection versus Z. The algorithm is triggered to reverse the Z scan during approach when the amplitude has decreased from its value far from the surface (here $A_0 = 5.4\text{ V} = 71\text{ nm}$) by a prescribed amount (here $2\text{ V} = 26\text{ nm}$), a *relative trigger*.[1] In Figure 5.20, we show side-by-side problematic conventional height

[1] Details of triggering and other aspects of approach–retract curve mapping are treated more fully in Chapter 6.

(a) and phase (b) images containing localized Z-control feedback oscillations, as well as coarse-resolution height (c) and *A*(*Z*) slope (d) images, on PLMA and PBMA domains obtained during the data-cube acquisition. Each square pixel in (c) and –(d) is centered at the location of an acquired *A*(*Z*) curve. (Custom software was used to compute the slope *dA*/*dZ* centered at the 18th data point from the reversal of *Z* during approach at each location, and render in grayscale in (d).) Two representative *A*(*Z*) curves are displayed in (e), one from each polymer domain at the locations of two squares shown in (d). The *Z* scale is zeroed at the nearest approach point at which the reversal is triggered (i.e., not according to surface height). A steep *A*(*Z*) trend is observed at the onset of tip–sample interaction (*) on PLMA, whereas a near-unity slope is observed on PBMA. Below about 60 nm, the slope of *A*(*Z*) on PLMA is more moderate. Conventional AC-mode images contained feedback oscillations at setpoints above this 60-nm threshold but were well-behaved below 60 nm. Thus, an *A–Z relationship that is much steeper than unity* is an indicator of problematic behavior commonly encountered on liquidy samples and/or when mobile species have contaminated the tip.

5.2.5 The Phase Electronics and Its Calibration

The presence and nature of the fifth difficulty is highly dependent on the specific electronics used to measure phase. The most common source of this difficulty is the so-called Extender electronics package sold with many (older) Bruker Nanoscope systems. We single out this case for a detailed treatment because of its wide presence in both academic and industrial labs still today, and the wide and ongoing reporting of data acquired with it, whether in technical journals, conference presentations, internal corporate reports, or contract analysis. The same vendor engineered a more expensive add-on that removes the difficulty of older systems and made this improved electronics standard on newer systems. But it is a common observation that the Extender remains heavily utilized for AFM phase measurement. To the author's knowledge, no detailed discussion of calibration procedures has been published. The following is an attempt to fill this gap.

A bit of history helps to understand how the problem arose. The Extender was designed for purposes other than to measure phase *per se* [15]. Instead, a *phase-like* signal (in frequency dependence) was needed to drive a second feedback-control circuit, to enable the measurement of frequency shifts of the oscillating tip/cantilever in the presence of long-range force gradients in electrostatic force microscopy and related methods. (We will discuss these in Chapter 9.) In other words, what was needed was a signal that varied fairly linearly with frequency near resonance, as is the case with phase in Figure 2.7. A measurement of *true phase*, however, requires a lock-in amplifier or similarly sophisticated device that utilizes *phase-locked loop electronics* to track the relative offset in time of two sine waves, and moreover with a frequency range extending up to hundreds of kilohertz. Such devices are expensive and actually overkill for the original purpose: to track frequency shifts near resonance. Thus, historically, a different circuit was engineered [15], which simply *multiplied the driving and response sine waves* and made use of a trigonometric identity,

$$
\begin{aligned}
A \cdot \sin(\omega t + \varphi_A) \cdot B \cdot \sin(\omega t + \varphi_B) &= AB/2 \cdot (\cos[(\omega t + \varphi_A) - (\omega t + \varphi_B)] \\
&\quad - \cos[(\omega t + \varphi_A) + (\omega t + \varphi_B)]) \\
&= AB/2 \cdot (\cos[\varphi_A - \varphi_B] - \cos[2\omega t + (\varphi_A + \varphi_B)]),
\end{aligned} \tag{5.1}
$$

where A and B are the amplitudes of drive and response (tip oscillation) signals, φ_A and φ_B the corresponding phases of these two sine waves, and ω the common frequency. In the resulting expression, the second cosine term is double frequency and filtered out to play no role. The first cosine term is DC and contains the sought phase difference, $\varphi_A - \varphi_B$, but as the argument of a cosine function. Thus, this term is not proportional to phase difference alone. Figure 5.21a overlays a theoretical amplitude signal near a resonance at 300 kHz with both true phase and the cosine of phase, the latter expanded by a factor of 90 so as to span the same 180° scale (and 1—cos φ used to give the same directionality of the data trend). The departure of the cosine-derived signal from true phase reaches nearly 20° at the extremes of frequency plotted in Figure 5.21a, meaning a ~70° excursion from 90° instead of the ~50° excursion expected for true phase. Figure 5.21b shows the same comparison of true phase (thick line) to the cosine-derived signal but as a function of amplitude. We see that the cosine-derived signal lies outside of the true phase double-valued curve. Recall that this thick curve, being for free cantilever oscillation, marks the zero-dissipation envelope; under tip–sample engagement, additional dissipation pushes the phase data below the upper thick curve or above the lower thick curve. (Zero dissipation refers to no tip–sample interaction, whereas dissipation of cantilever motion by immersion medium is always present. See Figure 2.13 and associated discussion.) Thus, the two-branch envelope "corrals" physically possible phase data during AFM operation. Any data values located outside of the thick curve are physically impossible. *Thus, the observation of such data is a clear indicator that the AFM electronics is not providing true phase measurement*, regardless of the nature of the electronic device used.

In the case of the Extender electronics, it seems at first sight that one could convert such a cosine-derived data output to actual phase (i.e., "correct" the data) by using an inverse cosine. But here is where the hidden realities of electronic circuitry are important. The factor B is not strictly a constant. The tip oscillation signal is run through an automatic gain control circuit to make it approximately constant, but nonideal behavior in this circuit is expected for values of $\varphi_A - \varphi_B$ beyond about ±15°; the result is that the "phase" signal provided by the above circuit, in so-called units of Bocek degrees [15], contains both phase *and intrinsic amplitude* contributions.

To examine the intrinsic role of amplitude, Figure 5.22a compares the frequency dependence of amplitude and Extender phase output at five different drive amplitudes yielding several different resonant free-oscillation amplitudes A_0 ranging roughly from 10 nm to 100 nm (read signals of 0.64–6.4 V). The phase curves are

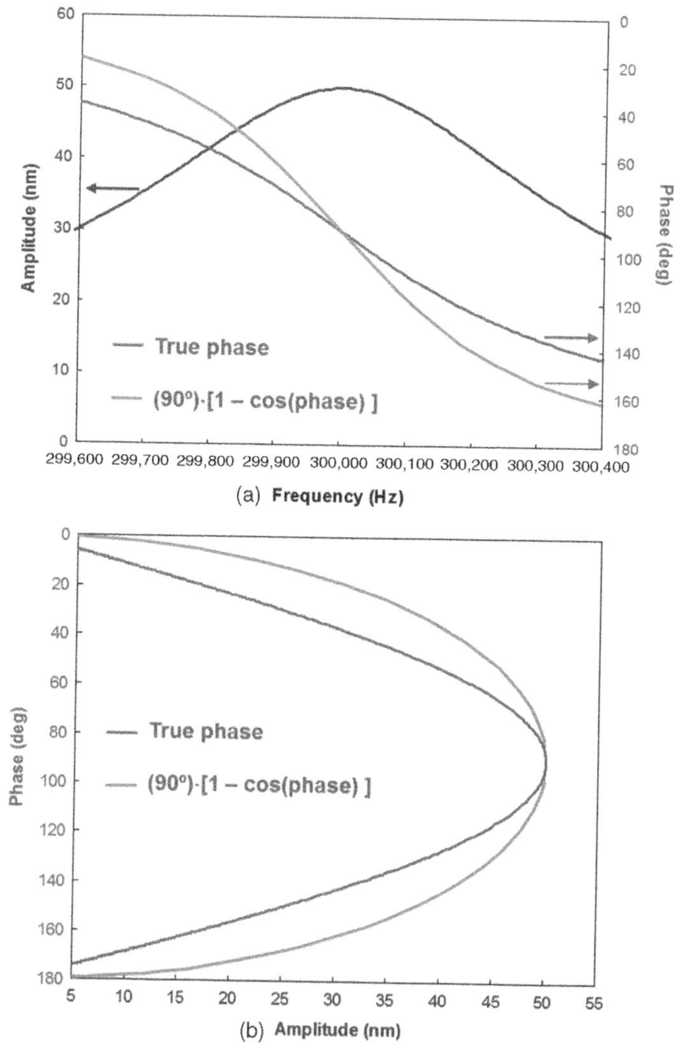

FIGURE 5.21 (a) Theoretical frequency dependence of amplitude and phase (thick curves) and an expression that scales with the cosine of phase. (b) Plots of the two phase curves of (a) versus the corresponding amplitude curve; the cosine-derive curve lies in the physically forbidden zone beyond the true phase curve.

inconsistent beyond a window extending about 30° above or below the zero (dashed lines). This is more evident in Figure 5.22b, the same data but plotting phase versus amplitude with the latter normalized to the free-oscillating amplitude at resonance. Three data sets are nearly identical out to ≈50° above or below the phase zero, corresponding to cases $A_0 = 6.4$, 3.2, and 1.6 V, and the first two of these are essentially identical for all values plotted. The data sets for lower values of A_0 depart

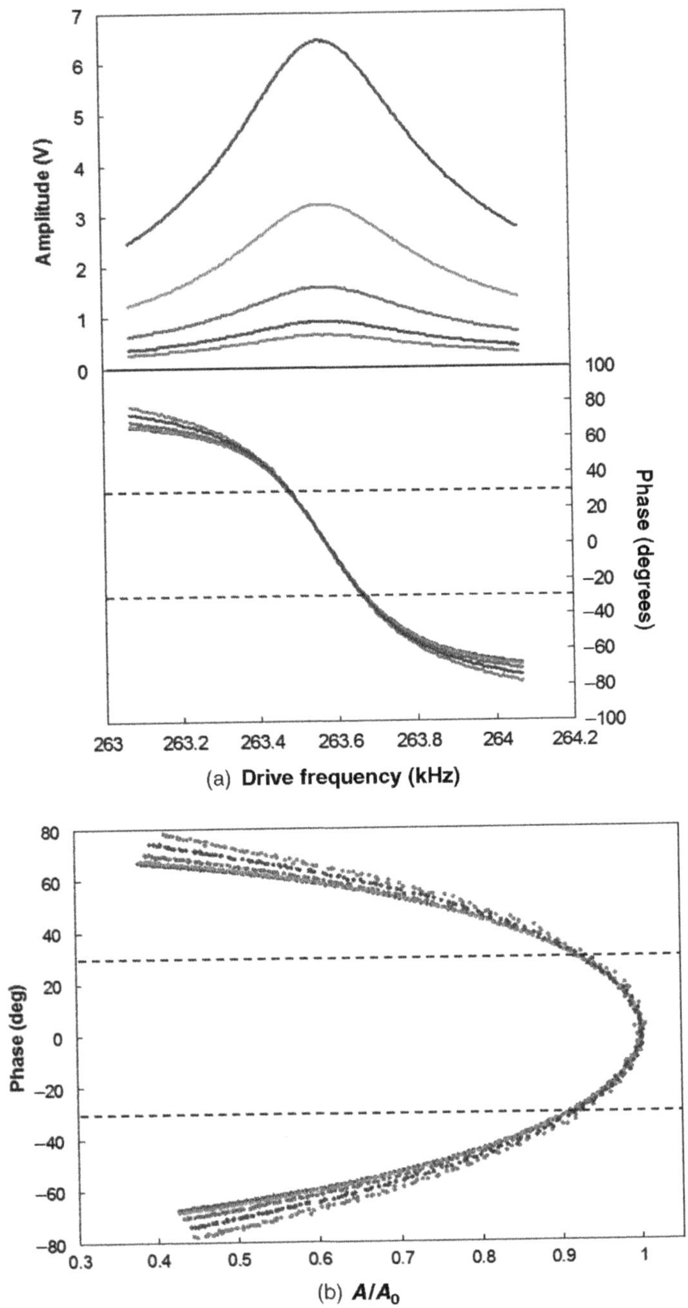

FIGURE 5.22 (a) Frequency dependence of amplitude (A) and Extender phase for five drive amplitudes. (b) Corresponding five plots of phase versus reduced amplitude A/A_0, where A_0 is the amplitude at resonance.

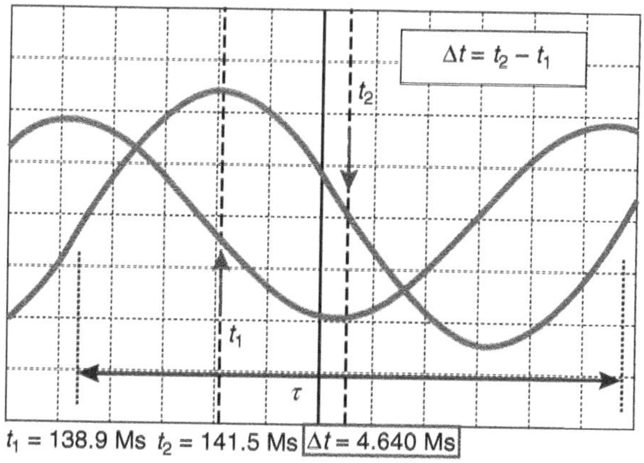

FIGURE 5.23 Oscilloscope screen shot showing the measurement of the time shift Δt between the driving sine wave and the sinusoidal cantilever response, as well as the period of the sine wave τ.

from those for higher A_0 in the *direction* consistent with a cosine-derived signal as shown in Figure 5.22b (i.e., away from the zero), but much less in *magnitude* than would be the case for the cosine-derived signal.

Thus, it is clear that there is no simple inverse-cosine correction. We wish to develop a calibration of phase by measuring actual phase shifts and comparing to the output signal of the AFM phase electronics, in this case, the Extender package. The most readily available, and conceptually direct, method for doing so is to use an oscilloscope to examine the driving and response signals (accessed with an often optional but highly recommended and relatively inexpensive BNC interface available from the AFM vendor).[2] Figure 5.23 shows an oscilloscope screen shot of such a phase shift measurement in time domain, $\Delta t = 4.64\ \mu s$ for a ~60-kHz drive signal, determined by placing vertical dashed cursors at points at which each sine wave crosses the horizontal axis. This value can be converted to the sought phase shift in degrees via

$$\varphi = 360^\circ \cdot \Delta t/\tau, \tag{5.2}$$

where τ is the period of the sine wave. Figure 5.24a and b plots sets of such phase shift values (circles) obtained over a range of frequencies near the resonances of both (a) "soft" ($k \approx 0.6$ N/m) and (b) "rigid" ($k \approx 40$ N/m) cantilevers. (The diameters of the circles approximately represent the measurement error bars.) The plotted values were obtained after subtracting a constant phase offset for each cantilever so

[2] Oscilloscopes ranging up to megahertz frequencies are quite common and considerably less expensive than high-frequency lock-in amplifiers. Most labs will not have even a low-end lock-in amplifier available, much less one with frequency range extending to hundreds of kilohertz.

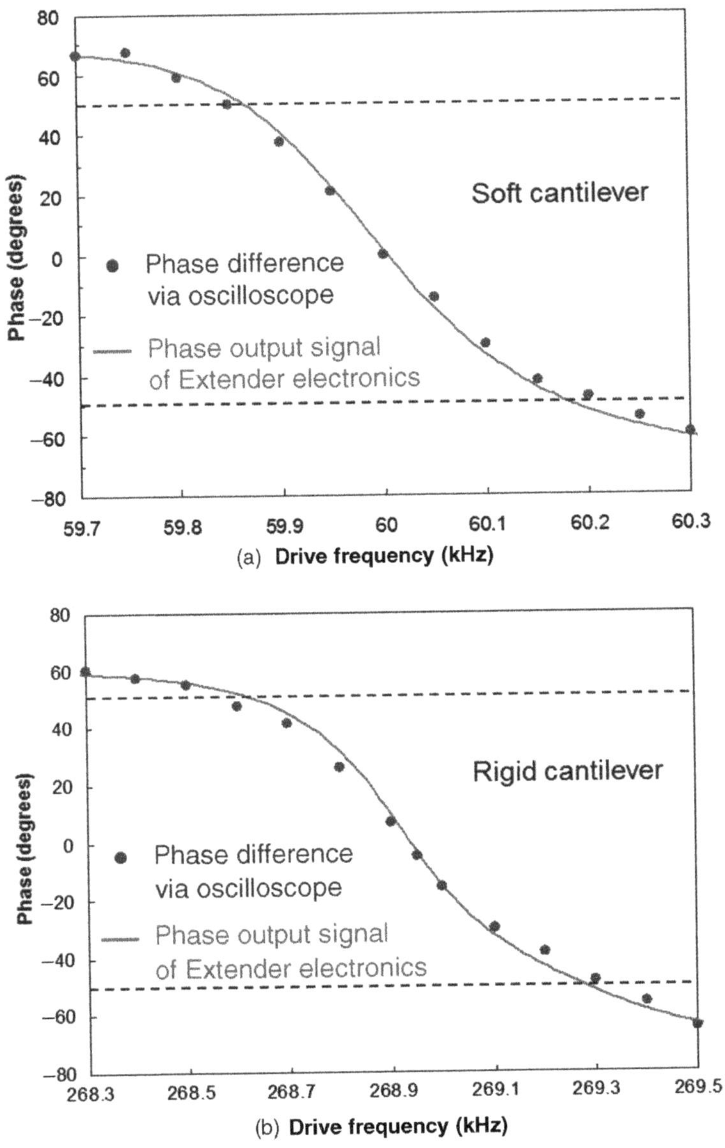

FIGURE 5.24 Frequency-dependent measurements of the phase shift between drive and cantilever-response signals, acquired with an oscilloscope and the Bruker extender electronics using (a) soft and (b) rigid cantilevers.

as to zero the phase at resonance for comparison to the output "phase" signal of the extender electronics (smooth curve). Although the discrepancies between extender electronics and oscilloscope do not appear to be large at first sight, we typically find that for phase differences ranging within $\pm 50^\circ$ of the zeroed signal at resonance (i.e., bordered by dashed lines in Figure 5.24), the extender signal overstates the departure of phase from zero by a few degrees, except for values very near zero. These departures tend to be small, however, compared to what a cosine of phase signal (multiplied by 90 for scaling) would produce (Figure 5.21). The departures are indeed in the *direction* predicted by the cosine formulism within ±50° of zero. Beyond this phase window, the departure of the extender electronics phase signal from that of true phase first becomes smaller, around 60°, then become bigger and usually opposite in sign, with Extender values being smaller in magnitude than true phase values (not shown in Figure 5.24).

One is not really seeking the calibration of phase as a function of frequency, rather, as a function of amplitude (ultimately setpoint). In Figure 5.25a, we plot the same phase data as in Figure 5.24b but versus the free-oscillation amplitude values provided by the AFM control system, normalizing each amplitude A to that at resonance A_0. Much of the "envelope" of the zero-dissipation true phase (oscilloscope values, circles) lies inside of the corresponding phase values output by the Extender electronics, which is physically not allowed. (This data is rather asymmetric with respect to the zero of phase because the resonance lineshape was asymmetric with respect to the resonance frequency, further discussed below.) But this is for a freely oscillating cantilever. It seems worthwhile to compare oscilloscope phase to extender phase *under tip–sample engagement*, and as a function of setpoint amplitude. This is the measurement for which we wish to be calibrated, of course. Figure 5.25b contains such a comparison obtained while imaging a 500 μ 500-nm region on the PBMA subcomponent of Figure 5.10, using a different cantilever of the same nominal type as that used to collect the data in Figure 5.25a. Again we find that the Extender phase output resembles the true phase measured with the oscilloscope (circles), but upon close inspection we observe a discrepancy of a few degrees in much of the data.

In developing a simple calibration for Extender phase, we have two essential observations: (1) Extender phase is not hugely different from true phase within 50° of the zero and (2) calibration of phase more than 50° beyond the zero would be complicated because of the intrinsic amplitude dependence (i.e., Figure 5.14). It turns out that all of the Extender phase values in Figures 5.24 and 5.25 within 50° of the zero can be adjusted to agree notably better with the oscilloscope phase by simply multiplying by approximately 0.9. (But checking with the oscilloscope is recommended in general.) Upon doing so and examining the numbers closely, one may find that invoking a second fitting parameter produces a substantially better (nonlinear) calibration. In particular, we have multiplied the Extender phase P by

$$c_1 \cdot \left(1 + \left[(P/c_2)^2\right]^{1/2}\right), \tag{5.3}$$

using $c_1 \approx 0.7$ and $c_2 \approx 130$ (and higher values of c_2 for low-amplitude cases, e.g., $c_2 \approx 220$ for the outermost data set in Figure 5.22b is equivalent to $c_2 \approx 130$ for the

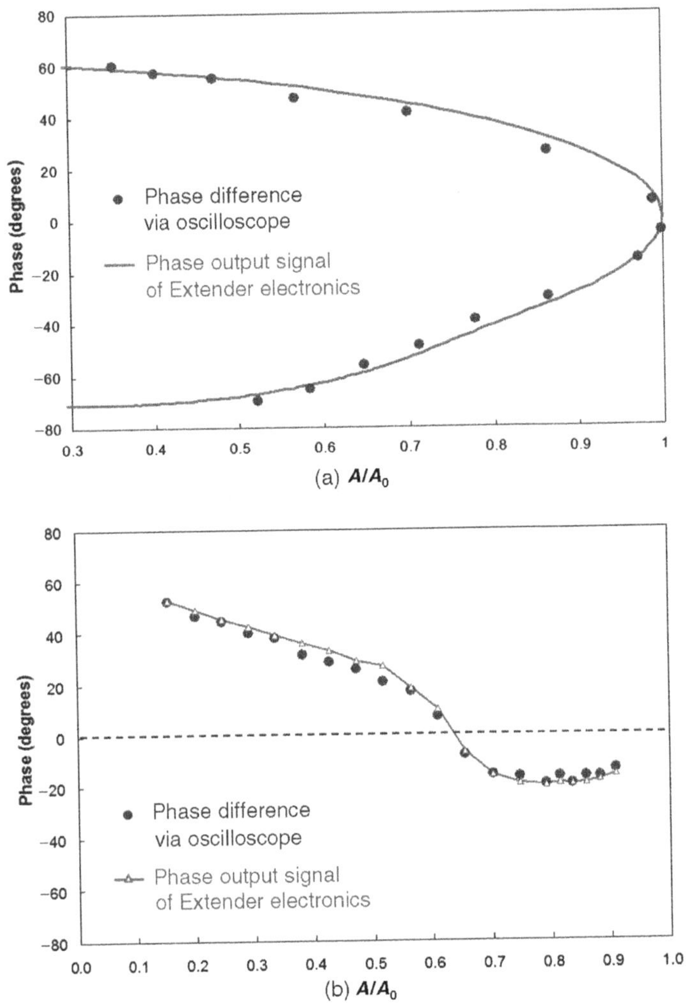

FIGURE 5.25 Phase shift between drive and cantilever-response signals, acquired with an oscilloscope and the Bruker Extender electronics, versus reduced amplitude A/A_0 (a) in free oscillation and (b) during phase imaging on the PBMA/PLMA blend of Figure 5.10.

innermost data set). (The square followed by square root allows treatment of both positive and negative Extender phase.) To demonstrate one result, in Figure 5.26 we plot phase–Z curves taken in the attractive regime during retraction from a polycrystalline gold surface, consecutively acquired using the Extender package and a Stanford Research Systems SR530 lock-in amplifier (via BNC signal access) for phase measurement. (A nominal $k = 3$ N/m uncoated silicon cantilever was used, with resonance frequency of about 64 kHz to allow usage of a lock-in amplifier limited to 100 kHz range.) We also overlay the Extender phase data

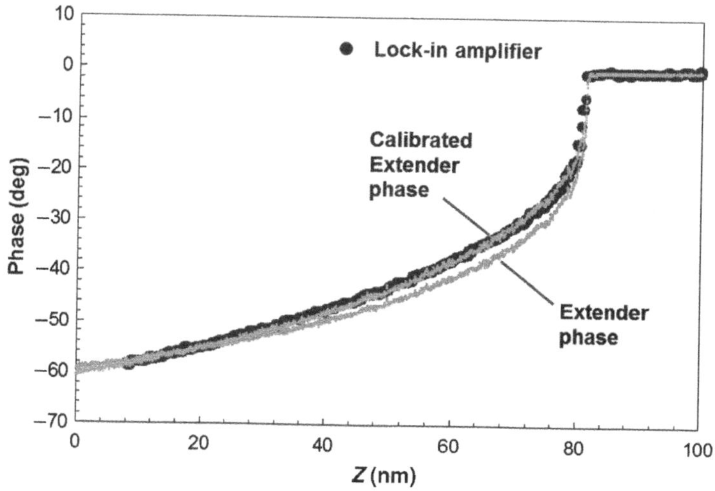

FIGURE 5.26 Phase–Z curve comparing phase measurements using a lock-in amplifier (circles) to raw and nonlinearly calibrated measurements (lines) using the Bruker Extender electronics.

recalibrated using the above two-parameter fitting expression. The agreement is very good within 60° of the zero of phase.

Those who have access to a lock-in amplifier for calibration purposes should not necessarily use it for regular data acquisition (i.e., in lieu of the Extender electronics), whether the application is approach–retract curves or images. The reason: in addition to the likely limitation in frequency range, common lock-in amplifiers are limited in *measurement rate*. The phase–Z curve in Figure 5.26 was acquired slowly, 2 s per approach–retract cycle, and with the lock-in amplifier's time constant minimized; yet a slight approach–retract hysteresis was observed in some of the lock-in acquired data while absent in the Extender-acquired data (on the nonviscoelastic gold surface). More notably, the measurement delay can be apparent in images. The height/phase image pair in Figure 5.27 was acquired on the PBMA/PLMA blend film of Figure 5.10, exhibiting phase contrast between two polymer phases. The line frequency was a slow 0.5 Hz and the slow-scan axis was disabled to illustrate the horizontal offset between Extender-acquired phase and lock-in acquired phase as the input signal was alternated between the two devices. The phase domains are displaced slightly rightward under left-to-right scanning due to the measurement time delay in the case of the lock-in amplifier. (Note the lack of this horizontal offset in the height data.) In summary, although the simplicity of the Extender electronics results in "untrue" phase data requiring recalibration, its quickness is a positive attribute. But if one prefers the correctness of a lock-in amplifier to the calibration procedure suggested in this section, one should be sure to purchase a fast (more expensive) lock-in amplifier.

We also reiterate that for values of amplitude less than 1 V raw (roughly 15 nm), the above Extender calibrations are not so good for phase values more than 30°

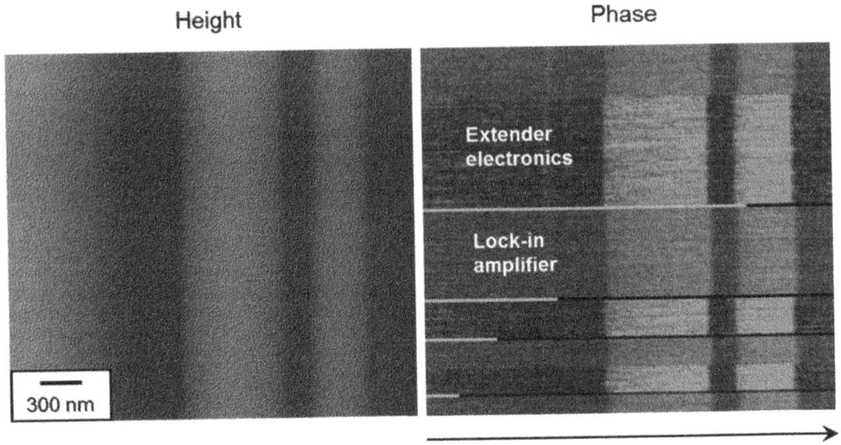

FIGURE 5.27 Height and phase images acquired on PBMA/PLMA in disabled vertical scan (repeated 1D horizontal scan) during which the phase measurement was alternated between a lock-in amplifier and the Bruker Extender electronics.

above or below the zero (Figure 5.22). But in most cases, this is not an important caveat; such small amplitudes are primarily invoked to stabilize the attractive regime when using stiff cantilevers (to keep the kinetic energy $\frac{1}{2}kA^2$ small), often in conjunction with a high setpoint amplitude ratio (e.g., $A/A_0 > 0.9$) such that the intrinsic amplitude dependence is minimal. In this case, the above phase calibration procedures are fine.

5.2.6 Nonideality in the Resonance Spectrum

A key assumption in the energy dissipation interpretation, the focus of the following Section 5.3, is harmonic behavior. The derivation integrates motion over a single oscillation cycle assuming *single-frequency sinusoidal excitation and response*, both in the absence and presence of tip–sample interaction. Much recent activity in the literature regards the excitation of higher oscillatory modes of the cantilever by using multiple driving frequencies (Section 9.2) [16–18]. If driving at a single frequency very near the fundamental resonance, as is conventional, it has been shown that nearly all of the energy resides in the fundamental mode (for typically used cantilevers) [5,19]. The energy-dissipation derivation considers only this energy. But one discovers that very near the fundamental resonance peak, the cantilever excitation spectrum can be complicated. Often asymmetry in the form of "shoulders," and sometimes double peaks, are observed. As a result of this nonideality, the phase trend near resonance can depart even further from an inverse tangent function (i.e., in addition to the nonidealities discussed in the preceding sections). This is a spectral nonideality the new user is almost certain to encounter, perhaps wondering whether the quantitative validity of images will be compromised by a "messy" resonance spectrum.

The principal source of these complications is the drive signal itself. As stated in Section 5.2.1, in between the sine-wave generating electronics and the base of the cantilever is the vibrating device itself (e.g., piezoelectric bimorph) and material interfaces as well as bulk solid media. A very complicated frequency dependence in the amplitude of the cantilever-driving vibration may result. It is easy to approximately measure this spectrum; one simply moves the laser spot back from the tip-end of the cantilever onto the chip surface, which lies at nearly the same angle of inclination (some adjustment of the position-sensitive detector may be needed), and increases the drive amplitude. The obtained spectrum may not be exactly the same as the vibrational spectrum at the base of the cantilever, but should resemble it. In particular, the chance presence of features *near the cantilever resonance frequency* may help to explain anomalies (e.g., shoulders) in the observed cantilever resonance spectrum. Figure 5.28 overlays the vibrational spectrum on the chip (scaled up by using a 20-V drive amplitude) and the corresponding cantilever response spectrum (0.35-V drive amplitude). The correlation of features is notable, especially if one recognizes that the cantilever excitation spectrum should consist of the driving spectrum *convolved* with the cantilever's Lorentzian response function. This means that features in the driving spectrum near the resonance frequency will be amplified much more in the cantilever response spectrum than will be features away from resonance. Thus, the prominent shoulder on the right side of the resonance peak, a seemingly minor increase of the drive spectrum from approximately 256 to 259 kHz, apparently results in a right shoulder in the resonance spectrum.

At this point, however, one must think harder about the *relevance* of these spectral features. During imaging as well as amplitude/phase–Z measurements, the drive frequency is *fixed*. Thus, in contemplating Figure 5.28, at first sight one may think, "So what?" Answer: The relevance of the response *spectrum*, as opposed to the response at just one frequency, is that the Lorentzian peak shifts horizontally (in frequency) under the influence of cantilever damping (Section 5.2.3) as well as tip–sample interaction. In the course of operating the instrument, different discrete drive frequencies often are explored, whether to stabilize attractive versus repulsive regime or to compensate for thermal drift. Thus, the amplitude and phase data, for different drive frequencies in different images, effectively depends on the excitation spectrum. Changing drive frequency at a fixed drive amplitude setting can in fact *change the drive amplitude that reaches the base of the cantilever*, as suggested on the vibrated chip (Figure 5.28). But within a given image or amplitude/phase–Z curve, the changes of the data should only reflect the shifting and broadening of the Lorentzian and inverse tangent response functions for amplitude and phase (of the fundamental eigenmode), respectively. Thus, the energy dissipation interpretation should be valid. But comparing energy dissipation between data acquired at *different drive frequencies* may be complicated due to the complexity of the excitational spectrum reaching the cantilever as exemplified in Figure 5.28.

It is worth underscoring that the calibration of Extender phase using a frequency sweep comparison, as in Figure 5.24, is legitimate. Even if the frequency

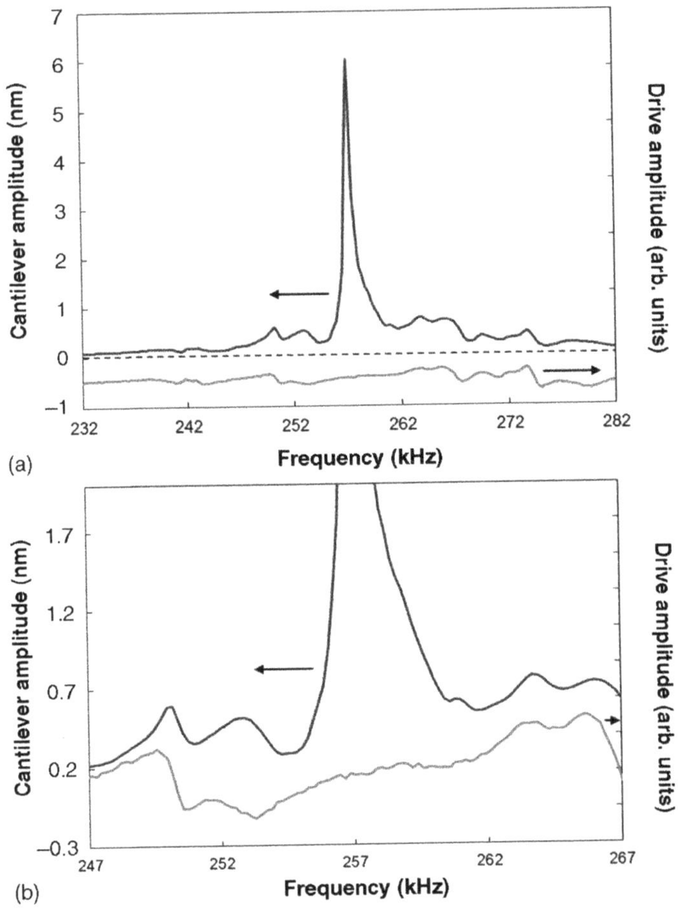

FIGURE 5.28 (a) Wide and (b) zoomed comparisons of the driven cantilever (top spectrum in each graph) and the driven cantilever chip (bottom spectrum, displaced downward for clarity), the latter obtained using a 60-times larger excitation signal.

dependence of the phase (within 50° of the raw phase zero) departs from an exact inverse tangent function, the data are still phase measurements, that is, the Extender phase differs from the oscilloscope phase by various amounts at different raw phase values. That these raw phase values were obtained by changing the drive frequency by a small amount, and thus possibly the drive amplitude per the excitation spectrum reaching the cantilever, should not matter (assuming the response amplitude is not too small as discussed in Section 5.2.4).

In liquid media, the frequency dependence of the driving amplitude is potentially more complicated and dependent on fluid cell design [20,21]. Vibrational excitations can reach the cantilever from all directions via the medium, meaning further possible coupling to solid surfaces. There are also changes in the physics, of course, because of immersion in a much more viscous medium, such as water compared to air, as well as the added inertia due to the mass of the liquid being displaced. The cantilever resonance frequency and quality factor dramatically decrease (see Eqs. 2.20 and 2.21). Theory predicts a reduction in resonance frequency by approximately a factor of 3.8 for a plain rectangular cantilever [22], but a somewhat lesser decrease may be observed [23], likely reflecting the role of the tip (such that the fractional change of inertia is not as great as it would be in the absence of a tip). A large reduction in quality factor produces a large reduction of response amplitude. Thus, achieving the same response amplitude requires increasing the drive amplitude by a correspondingly factor. The "background" spectrum discussed above, reflecting the variable propagation of the driving vibration to the cantilever through solid and fluid media, may be similar at the high frequencies used in air but notably larger at the lower frequencies used in liquid, for the same drive signal amplitude. As a result, the *heights of peaks due to environmental resonances are comparable to the height of the cantilever resonance peak*. The term "forest of peaks" is sometimes used to describe this observation [20,24,25]. In particular, it suggests that non-cantilever resonance operation is possible, that is, by choosing a drive frequency near an environmental ("apparatus") resonance. In this case, much of the richness of the data is lost, because the tip–sample interaction is not modifying the cantilever resonant response function in a measurable way. But in some cases, non-resonant operation may be advantageous, say to remove the nonlinear coupling of tip–sample interaction with the response.

An elegant solution to the problem of a complicated excitational spectrum in liquid is the replacement of the usual vibrational drive scheme with magnetically driven cantilevers [25,26]. Commercial variants use a magnetically active coating on the backside of the cantilever. The source of magnetic excitation is typically a solenoid positioned either below the sample or above the cantilever chip. Thus, a magnetic driving force is imparted *all along the cantilever*, leading to a different mathematical description of the problem. (An alternative is to glue a magnetic particle to the cantilever backside, but still differing in mathematical description from the case of a vibration imparted to the base of the cantilever.) Importantly, in liquids, the frequency dependence of the excitational force does not produce a "forest of peaks"; instead, a smooth and monotonically decreasing background results. Generally, the "purity" of the resonance lineshape is improved under magnetic excitation compared to vibrational excitation, even in air, because the transfer function of the drive signal is not complicated by numerous mechanical interfaces, as it is under acoustic excitation.

Regarding utility, we note first that magnetic excitation enables *off-resonance excitation over a continuum of frequencies*. In comparison, off-resonance vibrational excitation in water via the "forest of peaks" is only possible within certain narrow frequency bands. In air, one could choose to operate at, say, 10 kHz instead of a 100 kHz resonance frequency, possibly providing greater viscoelasticity sensitivity because the tip velocities are lower and contact times may be closer to intrinsic material response times. Second, some experimental results suggest more delicate imaging under magnetic excitation, ostensibly because vibrations are not being coupled throughout the liquid cell [27].

5.3 ENERGY DISSIPATION INTERPRETATION OF PHASE: QUANTITATIVE ANALYSIS

In this section, we consider more quantitative and physical aspects of phase imaging and phase–Z (along with amplitude–Z) curves. That the phase measurement relates to energy dissipation during tip–sample interaction is, needless to say, not obvious to the casual user. As stated in Section 2.4.4, Equation 2.23 can be derived by considering power balance (Appendix 3). Power delivered to the oscillator by the driving vibration is balanced by cantilever dissipation (partially internal but mainly due to fluid medium resistance, unless operating in vacuum) as well as repeated tip–sample dissipation (heat). Thus, one can calculate the tip–sample energy dissipation as the difference of the average input energy from the driver over one oscillation cycle minus the average cantilever damping energy per cycle. The latter is characterized using the free-oscillation behavior to measure Q, which to good approximation is measured as the resonance frequency f_0 divided by the full width of the resonance peak, Δf at which the amplitude is reduced to $A_0/\sqrt{2}$. (One should note two important caveats: in Section 5.2.3, we learned that Q upon tip–sample engagement is reduced to roughly 75% of Q tens of micrometers away from engagement, and in Section 5.2.6, we discussed the fact that the resonance lineshape may be distorted, e.g., a shoulder could result in a calculated value of Q that is falsely reduced by of order 10%.) The result of the power-balance derivation is Equation 2.23; we rewrite here in a convenient form similar to Equation 2.24 but with the energy prefactor being the total oscillator kinetic energy *at the operational setpoint amplitude*, $\frac{1}{2}kA^2$,

$$E_{\text{tip-sample}} = \frac{2\pi}{Q}\left(\frac{1}{2}kA^2\right)\left(\frac{\sin\varphi}{\bar{A}} - \frac{\omega}{\omega_0}\right), \tag{5.4}$$

where $\bar{A} \equiv A/A_0$ is the setpoint amplitude normalized to free-oscillation amplitude at resonance. Considering that the ratio ω/ω_0 is strictly unity if driving at resonance and generally between 0.999 and 1.001 in the follow examples, we treat this ratio as 1. Thus, the approximate *fraction of oscillator energy dissipated per tip–sample encounter* is given by a simple expression:

$$\frac{2\pi}{Q}\left(\frac{\sin\varphi}{\bar{A}} - 1\right). \tag{5.5}$$

In air, the coefficient $2\pi/Q$ is very small, usually of order 0.01. This means that the fractional energy dissipation per encounter is very small unless $\bar{A}$ is tiny, meaning one cannot image at such miniscule setpoint amplitude ratios as can be easily verified by trial and error.

In the following subsections, we consider examples of this quantitative analysis in both attractive and repulsive regimes. In the process, we examine the effects of tip–sample interaction on height images, in the context of energy dissipation. In the first example, we explicitly cover the practical steps to analyze phase data acquired with a Bruker Nanoscope/Extender system, attending to issues of Section 5.2.

5.3.1 Variable A/A_0 Imaging

Changing A/A_0 from nearly 1 to ~0.5 in dynamic AFM imaging is useful to vary depth sensitivity, our first example of variable A/A_0. For phase-segregated systems in the presence of a surface, there is potential for variable segregation as a function of depth. One anticipates a change in both the absolute and relative (between components) dissipations revealed in phase images at different values of A/A_0. To exemplify, we revisit the system of Figure 1.12, a micrometer-scale thick film (spin-coated onto silicon) of the ABA triblock copolymer poly(styrene–isobutylene–styrene) (total length 2A of polystyrene (PS) 27% of the molecule) that was further solvent vapor annealed for a day to aid segregation. At room temperature, PS is glassy and polyisobutylene (PIB) rubbery. The expected bulk phase segregation is cylinders of PS in a matrix of PIB; but given the lower surface energy of PIB, segregation to the surface is thermodynamically favored [28]. Studies of a similar triblock system [29,30] have found a preponderance of PS cylinders oriented either horizontal or vertical with respect to the surface in thin films.

Figure 5.29 compares height and phase images on the SIBS film acquired in the repulsive regime under (a) strong and (b) weak dissipation as follows. Driving approximately 0.2 kHz below resonance ($F_0 \approx 307$ kHz, $k \approx 40$ N/m) and with a relatively large setpoint amplitude $A = 3.0\,\text{V} \approx 40\,\text{nm}$ (conversion using amplitude–Z curve), cases (a) and (b) correspond to drive amplitudes A_{drive} of 150 and 75 mV (of the signal applied to a piezoelectric bimorph oscillator), respectively. The latter was chosen by setting the setpoint amplitude at 3.0 V and reducing A_{drive} until engagement was lost, then incrementally increasing until engagement was regained and minimally stable. At this point, $A/A_{\text{free}} \approx 0.97$ (determining A_{free} from amplitude–Z curves). The ratio of A_{free} to A_0 was determined from the initial frequency sweep measurement with tip and sample roughly 50 μm apart, shown in Figure 5.29c (keeping track of the zero of amplitude). This enabled A_0 to be estimated just beyond engagement (accounting for the leftward 0.05 kHz air-damping shift of the resonance peak) and thus the ratios A/A_0 of (a) 0.42 and (b) 0.84. The former is a heavy dissipation case resulting in substantial "height" contrast (differences in indentation) ranging approximately 2.0 nm, whereas "height" variations in the latter are an order of magnitude smaller.

Figure 5.30 compares line traces of data from the designated locations in the phase images of Figure 5.29. The same two data sets are plotted in four ways: (a) raw phase data (as output by the Extender electronics, zeroed at the resonance frequency), (b) processed phase data using a nonlinear calibration as discussed in Section 5.2.6, then subtracted from 90 and scale visually inverted, (c) converted to energy dissipation but leaving out the nonlinear calibration from (b), and (d) converted to energy dissipation following the calibration in (b). In the absence of Extender calibration, the energy dissipation values are substantially underestimated for the weak case, and dissipation contrast is substantially underestimated for the strong case. We again see that in the repulsive regime, brighter phase corresponds to lesser energy dissipation.

Of physical interest is the minimal height and dissipation contrast under weak-dissipation imaging conditions in much of the imaged region. Together these strongly

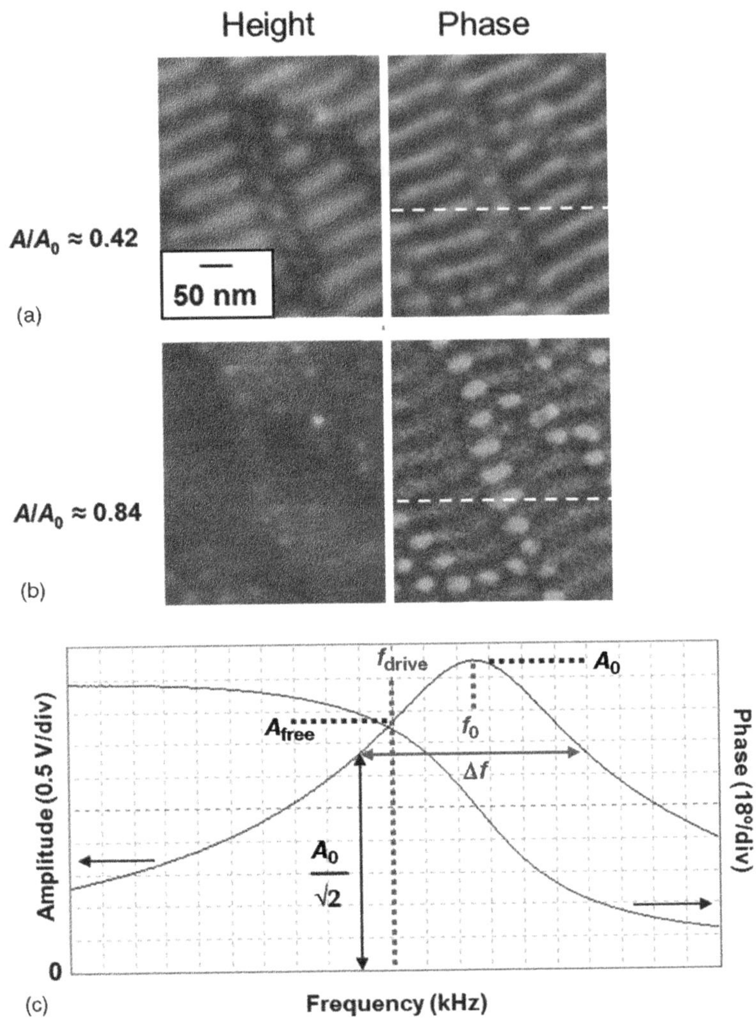

FIGURE 5.29 Height and phase images of a poly(styrene–isobutylene–styrene) film acquired at two setpoint amplitude ratios corresponding to (a) high and (b) low energy dissipation per tip–sample encounter. Drive frequency was below resonance as indicated in (c) along with other parameters relevant to the discussion in the main text.

suggest that PS cylinders oriented parallel to the surface are covered by a thin layer of PIB as follows. These cylinders are barely perceivable in the phase image of Figure 5.29b, corresponding to about 5% difference in energy dissipation compared to adjacent PIB-rich domains. By comparison, the more circular domains, presumably vertically oriented PS cylinders, are easily seen under weak dissipation settings, being significantly less dissipative. Importantly, under strong dissipation conditions,

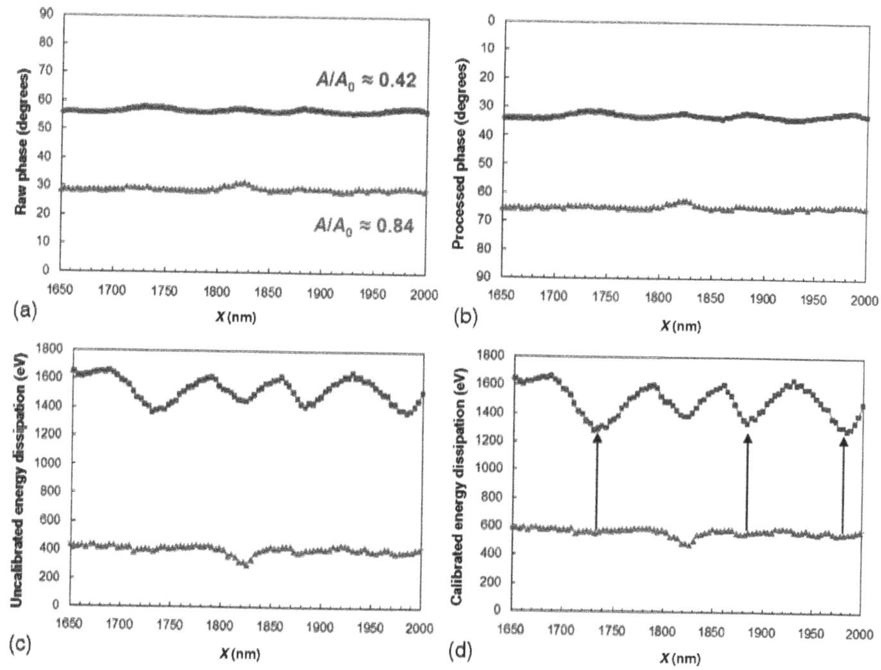

FIGURE 5.30 Line traces of (a) raw and (b) processed (calibrated) phase data extracted from the scans marked by dashed lines in Figure 5.29 for high and low dissipation imaging as related to the displayed setpoint amplitude ratios. Corresponding energy dissipation values derived from the (c) uncalibrated and (d) calibrated phase data. Vertical lines mark transitions of certain domains from minimally to strongly contrasting upon change of setpoint.

the horizontal PS cylinders transition (arrows in Figure 5.30d) to slightly lesser dissipation than the vertical cylinders and 20% lesser dissipation than the adjacent PIB-rich regions. It is likely that the lower-surface energy of a PIB skin produces less adhesion hysteresis with the tip than PS, whereas the internal viscoelastic deformation losses under strong mechanical interaction are similar whether the PS is present at the surface or just below a skin of PIB.

What is hopefully apparent in Figure 5.30 is (1) the more noteworthy and transparent comparison of energy dissipation numbers rather than phase numbers and (2) the quantitative superiority of calibrated phase data. Detailed modeling of differences in tip–sample interaction is best compared to energy dissipation numbers. More routinely, the user may be interested in empirical comparisons, but nevertheless as quantitative as possible. Variations in quality factor from cantilever to cantilever, for example, can make meaningful sample comparisons difficult if examining solely raw phase data. Magnitudes of energy dissipation may also vary with radius of curvature R from tip to tip, but likely within a simple scaling relationship (i.e., mechanics as, e.g., in Eq. 2.8 and also adhesion hysteresis as discussed in

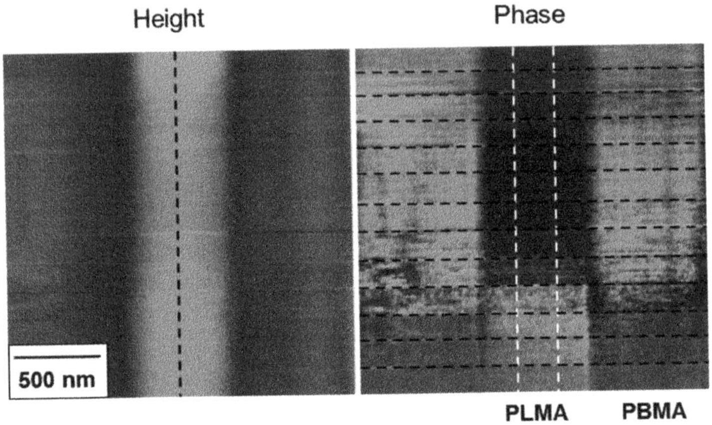

FIGURE 5.31 Height and phase images on PBMA/PLMA acquired in disabled vertical scan (repeated 1D horizontal scan) during which the setpoint amplitude A was decreased 12 times at the locations designated by horizontal dashed lines (bottom to top).

Chapter 6). Imaging tip shape using grids with vertical edges can aid in gauging radius of curvature and thus make the preceding quantitative analysis even more meaningful across a set of tips.

Our second example of imaging at variable A/A_0 involves changing setpoint amplitude A at fixed drive amplitude A_{drive} and thus A_0. We return to the PBMA/PLMA blend film of Figure 5.10. Here, we drive at resonance to examine both attractive and repulsive regimes including the amount of energy dissipation as this boundary is smoothly traversed, unlike the PVA and silicon wafer cases of Section 5.1 (examined in further detail in Section 5.3.2). To explore a large number of A/A_0 values, we *stepped the setpoint within two "images" under a disabled slow scan.* (The 1D scan crossed the centermost PLMA domain of Figure 5.10.) One of these two height/phase image pairs is shown in Figure 5.31, following a "zero-order flattening" whereby each line of data values (height or phase) is shifted numerically up or down such that its average is centered in the color range. The change of setpoint amplitude was from high to low in $-0.2\,\mathrm{V}$ steps bottom to top in the images, with changes of setpoint indicated by horizontal dashed lines in the phase image. The most interesting feature that is immediately apparent is the qualitative change of phase contrast, from PLMA brighter to PLMA darker than PBMA, followed by the lessening of the latter contrast at the lowest amplitudes (top of image). What is much less obvious, in the absence of quantitative examination (next paragraph), is a corresponding change of height contrast from large to small to large.

Figure 5.32 displays the setpoint amplitude ratio ($\bar{A} = A/A_0$) dependence of (a) calibrated phase (following the nonlinear adjustment of Section 5.2.5, valid above $A/A_0 \approx 0.2$) and zero-dissipation envelope, (b) energy dissipation values calculated from the phase data, and (c) the height of the center of the PLMA domain relative to the immediately adjacent PBMA surface height. These plots derive from

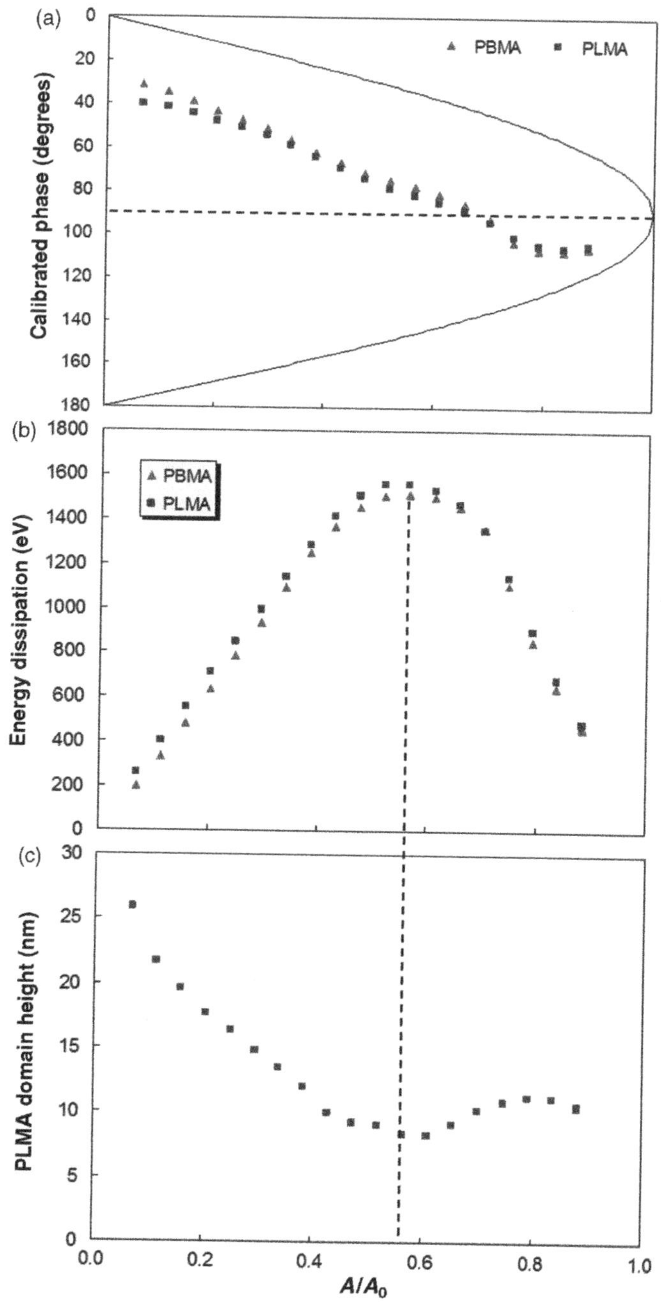

FIGURE 5.32 A/A_0 dependence of (a) calibrated phase histogram peak positions derived from the subregions of the images in Figure 5.31 prior to "flattening," as well as corresponding (b) energy dissipation values and (c) apparent PLMA domain heights.

the phase data prior to the line-by-line adjustments ("flattening") that produced Figure 5.31 (setpoint amplitudes ranging from 3.9 V to 1.5 V) and a second similar image (setpoint amplitudes ranging from 1.3 V to 0.3 V and finishing back at 3.9 V). The PLMA phase data were averaged from within the two vertical dashed lines shown in Figure 5.31. The PBMA phase data were averaged over corresponding regions to the right of the PLMA regions. We see in (a) that the phase data transition *smoothly* from the attractive to repulsive regime with decreasing A/A_0, with the *PLMA data symbols always lying farther from the* 90° *midpoint* of the phase scale, at which this transition occurs, indicating consistently higher energy dissipation on PLMA. This is born out in the full energy dissipation calculation results in (b). Moreover, we find a maximum in energy dissipation centered around $A/A_0 = 0.55$. The smooth, nearly parabolic nature of the change of energy dissipation with A/A_0, and its trend toward zero dissipation at $A/A_0 = 0$, is striking and physically reasonable, and certainly *not apparent in the phase plot*. Corresponding with this maximum is a minimum in the measured height of the PLMA relative to PBMA, meaning maximum indentation of the tip into PLMA when energy dissipation is maximal. This important physical correlation is *hidden to the user if only examining phase data in the absence of energy dissipation interpretations.* What is further revealed by this analysis is an intriguing change of apparent height at low A/A_0, suggesting possibly a strong viscous resistance of PLMA to tip penetration that is absent or far weaker on PBMA.

5.3.2 Fixed A/A_0 Imaging

Most commonly, one performs dynamic AFM imaging at amplitudes around 90% of the free-oscillation amplitude while driving below resonance. This keeps dissipation moderate and usually minimizes inaccuracies in height images due to mechanical compliance (e.g., the "true" height of the PLMA domain in Figure 5.32c is ≈10 nm) and/or bistability (Section 5.1). But one may choose to explore interaction regimes (cognizant that bistability may result), perhaps seeking elucidating contrast, by varying both A and A_0 in tandem *so as to keep $\bar{A}$ constant.* As per Equation 5.1, this analysis also allows one to examine *nonlinearity* in material response—a departure from a linear relationship between energy dissipation and total oscillator energy—which may be important to fundamental studies as well as providing material properties contrast. Nonlinear dissipative response would be analogous to non-Newtonian rheological response [31] (i.e., a velocity-dependent viscosity) or non-Amontons frictional response (i.e., a load-dependent coefficient of friction) [32].

We return to the bistable case of an ultrathin PVA film (introduced in Section 5.1 in the context of height artifacts) to exemplify this procedure as well as examine the relationship between bistability on energy dissipation. First, the cantilever was driven below resonance and phase images collected at proportionally varied A and A_{drive} (such that $r_{\text{sp}} = A/A_0$ was constant). These images then were converted to histograms (as in Figure 5.2) whose centroid peak positions are plotted versus oscillator energy $\frac{1}{2}kA^2$ in Figure 5.33a, following the nonlinear calibration of Extender phase per Section 5.2.5. Figure 5.33b plots the corresponding energy dissipation

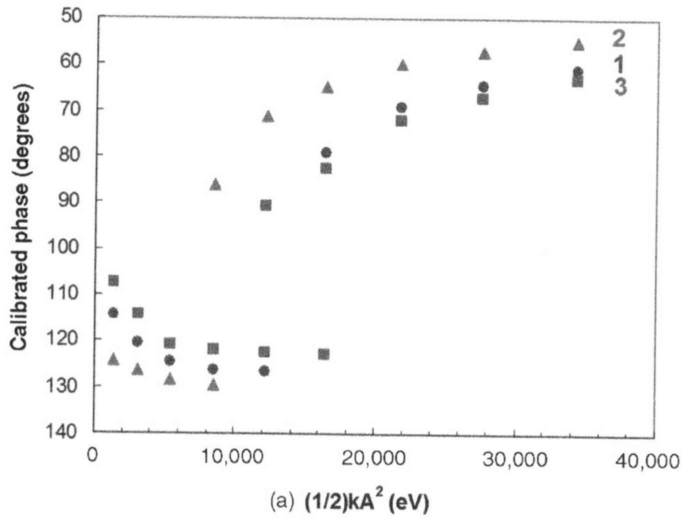

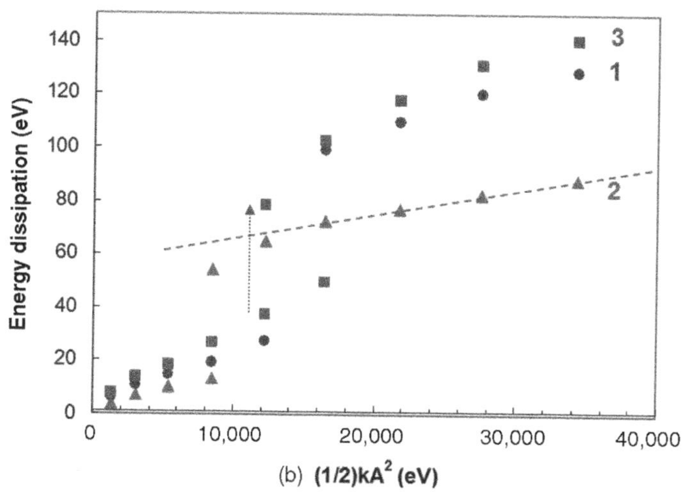

FIGURE 5.33 (a) Calibrated phase versus oscillator energy $\frac{1}{2}kA^2$ for the three polyvinyl alcohol film components of Figure 5.1 over nine drive amplitudes at fixed A/A_0. (b) Corresponding values of energy dissipation exhibiting characteristic slopes per film component; dashed line depicts one case.

per cycle (symbols). The dissipations in the repulsive regime are much lower than in Figure 5.32 in part because the present amplitudes only ranged up to 16 nm as compared to 52 nm in that previous example (similar cantilever spring constant). In the attractive regime, the energy dissipation values are markedly lower than in the repulsive regime. But there are *characteristic slopes of energy dissipation versus oscillator energy* specific to each of the three film components (1–3; see Figure 5.1). Moreover, the relative magnitudes of these slopes, among the three film components, are consistent whether in the attractive or repulsive regime. One might refer to this slope as the *coefficient of dissipation*, analogous to a coefficient of friction (Chapter 7, where we reexamine this PVA thin-film system) or coefficient of viscosity. A similarly central role for this coefficient of dissipation is implied, for the characterization of materials, thin films, surface-modification treatments, etc.

The vertical offset from the trend at low energy in the attractive regime to that well into the repulsive regime is approximately 40 eV. Thus, *upon transitioning from dominant attraction to dominant repulsion, there is a large jump in energy dissipation*, increasing by a factor of two to four depending on film component. The data trends also *depart slightly from linear upon entering the repulsive regime* (e.g., compare case #2 to dashed line), becoming more linear at higher energies. This analytical result makes some sense because the changeover from van der Waals attraction to Hertzian-type repulsion means that the "well" in the force–distance relationship should more strongly affect the oscillator's response (tip spends more time at this distance); see Figure 4.2 case (c). The well has high curvature as the force–distance trend transitions from a weak positive slope to a steep negative slope; such a strong curvature is indeed expected to be a source of nonlinear response [33–35]. All of these physical assessments are, again, invisible to the user if examining phase data in the absence of the energy dissipation interpretation.

Now let us digress a bit on this PVA physical system. A significant and unanticipated finding in this example is the strong similarity of relative dissipation between film components (1–3) in *both attractive and repulsive regimes*. The three components are understood to differ in molecular degrees of freedom [6,7]. Component #1 has somewhat reduced freedom due to strong adsorption to substrate. Component #2 has strongly reduced molecular freedom due to highly crystalline ("tighter") conformational states. Component #3 is bulk-like and thus semicrystalline, containing amorphous and crystalline nanoscale domains strongly differing in conformational freedom. The results in Figure 5.33 for component #3 only reflect the average behavior across this broad distribution (i.e., the peak position of histograms such as those in Figure 5.2). In the attractive regime, the tip may be dipping into a surface fluid layer wherein molecular freedom affects dissipation. On the other hand, there may be differences in average Hamaker constant (see Section 2.2) within components 1–3 due to molecular packing differences, such that noncontact interactions differ (although presumably not by much, unlike, e.g., the difference between a metal and a polymer). These interactions do not disappear in the repulsive regime; the tip spends most of its time out of contact. If the contact dissipation did not differ much among sample components (i.e., if viscoelastic responses are very similar), then nonmechanical differences might still be responsible for overall difference of dissipation during a complete oscillation cycle even in the repulsive regime. Differences of adhesion hysteresis might also loom large: stronger attraction implies stronger adhesion, and irreversible (hysteretic) conformational transitions during the formation-to-rupture of contact might be further activated by higher stresses due to this stronger attraction. But this is speculative. In Chapter 6, we will directly examine tip–sample adhesion differences probed in force-curve mapping modes as well as sliding friction in contact mode, for comparison with the preceding results.

5.3.3 Variable A/A_0 Via Z-Dependent Point Measurements

Our final type of energy dissipation analysis via phase measurements implements variable A/A_0, but in the context of point measurements of amplitude and phase versus Z (also known as dynamic "force curves"). This measurement type is perhaps the most exhaustively analyzed across sample types and in comparison with numerical simulation and analytical theory [5,36–40]. Here, we measure the response of the bulk-like PVA thin-film component of the preceding section, driving at resonance as done for measurements on silicon in Figure 5.5. In Figure 5.34, we show amplitude and phase (Bruker Extender data calibrated per Section 5.2.5) versus Z curves during approach for two nearly equal drive amplitudes that produce free-oscillation amplitudes of 15 and 16 nm. (Z is arbitrarily zeroed.) The 15-nm case remained in the attractive regime for all values of amplitude except various small values. The 16-nm case started out in the attractive regime at high amplitude and suddenly transitioned to the repulsive regime at around 13-nm on the Z scale (arbitrarily zeroed). These data sets are replotted as phase versus reduced amplitude A/A_0 (eliminating Z) in

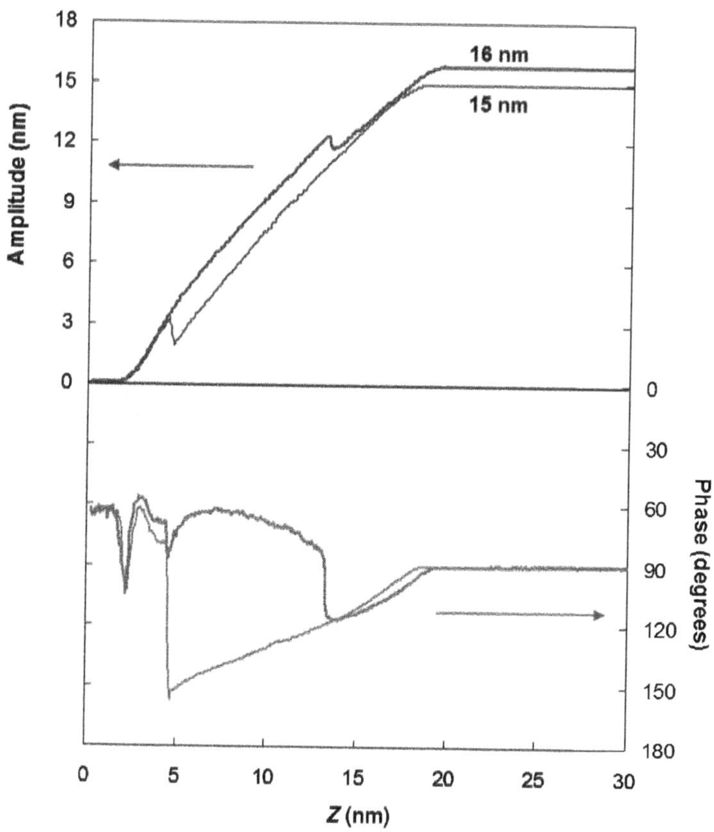

FIGURE 5.34 Amplitude–Z and phase–Z curves on PVA thin-film component #3 of Figure 5.1, using two drive amplitudes at resonance producing free-oscillation amplitudes of 15 nm (thin lines) and 16 nm (thick lines).

Figure 5.35a, and in turn converted to energy dissipation plots in Figure 5.35b. (This conversion is not valid in detail at $A/A_0 < 0.3$ because the calibration procedure breaks down at low amplitude; see Section 5.2.5.) Similar to Figure 5.32b, a peak dissipation is observed roughly centered at $A/A_0 = 0.55$. More notable is the near

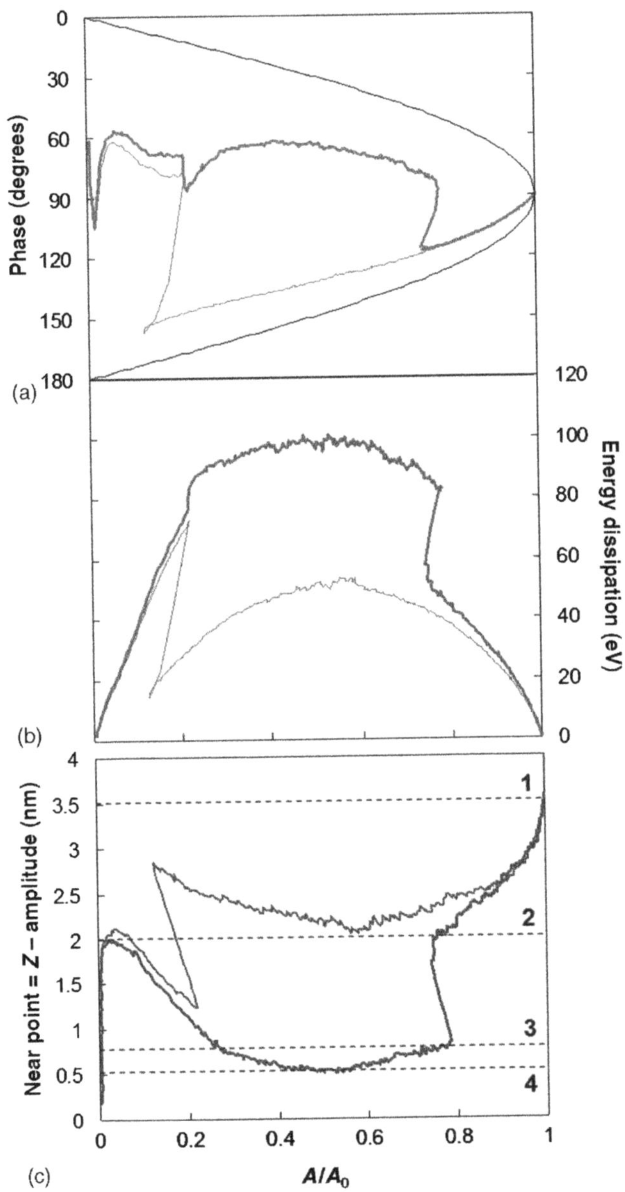

FIGURE 5.35 A/A_0 dependence of (a) phase, (b) energy dissipation, and (c) nearest tip–sample approach point (arbitrarily zeroed) on PVA thin-film component #3 of Figure 5.1, using the two drive amplitudes of Figure 5.34.

doubling of the peak dissipation in the repulsive compared to attractive regime case, even though the oscillator energy $1/2kA_0^2$ was only increased by 12% to force this transition. This is much in contrast with the gradual change see in Figure 5.32 on PBMA and PLMA as the oscillator passed the net-attractive to net-repulsive boundary. Indeed, within the case $A_0 = 16\,\text{nm}$ on PVA, the abrupt change in phase corresponded to an increase of energy dissipation of 50–80 eV per cycle.

It is further elucidating to compare this energy dissipation behavior with the extent of penetration into the tip–sample interaction. One useful empirical analysis simply plots the difference of Z and amplitude A versus A/A_0 as in Figure 5.35c, a measure of the "near point" of tip approach toward sample within each oscillation cycle. (Note that the zero of the vertical scale is arbitrary, given that the zero of Z scale in Figure 5.34 is arbitrary.) That these two curves in (c) are essentially inverted versions of the energy dissipation curves in (b) is remarkable, considering

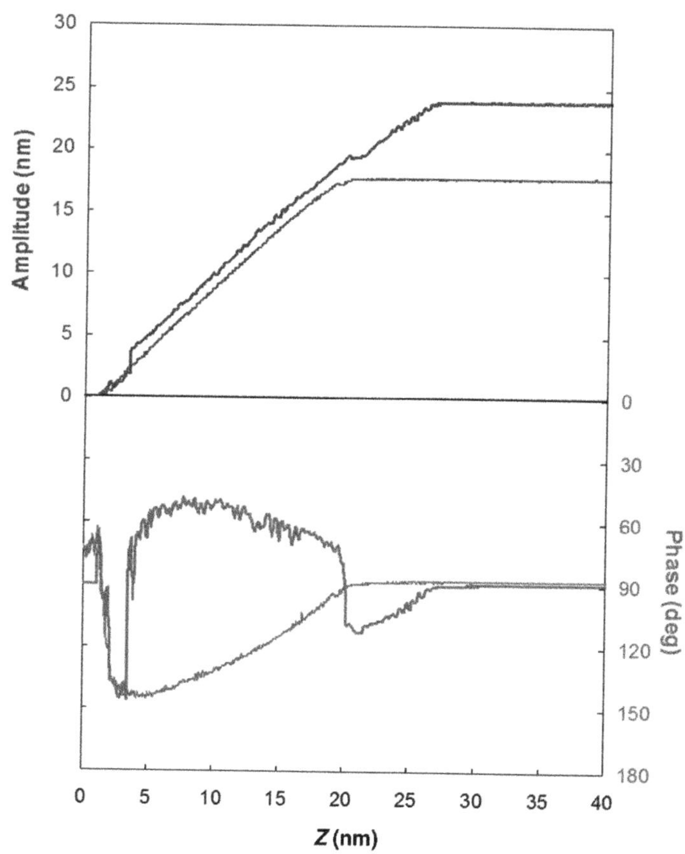

FIGURE 5.36 Amplitude–Z and phase–Z curves on a gold polycrystalline surface, using two drive amplitudes at resonance, producing free-oscillation amplitudes of 17 nm (thin lines) and 24 (thick lines) nm.

that the latter mainly reflect changes in phase data while the former derive entirely from amplitude data. Several interesting values in Figure 5.35c along the vertical scale are marked with dashed lines, numbered from top to bottom as follows: (1) onset of attractive interaction, (2) closest penetration while remaining in the attractive regime, (3) penetration upon transition from repulsive to attractive regime, and (4) deepest indentation into solid polymer. One can treat position #3 as the approximate surface location, although, in truth, the surface is slightly higher because finite indentation takes place upon stabilizing at this dynamic equilibrium position (indeed with markedly higher energy dissipation than at position #2). The change of distance from #2 to #3 is 1.2 nm, a reasonable upper-bound "width of the well" created as the distance dependence of van der Waals attraction is augmented by that of contact repulsion (see Figure 4.2).

Figures 5.36 and 5.37 contain a second example of the data analysis performed in Figures 5.34 and 5.35a and b, for the case of a polycrystalline gold surface, using

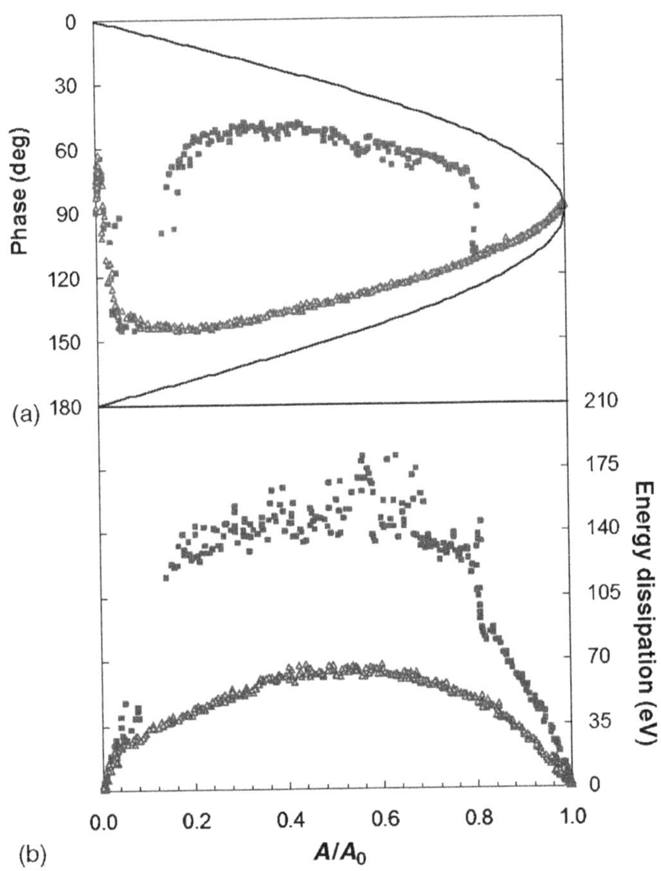

FIGURE 5.37 A/A_0 dependence of (a) phase and (b) energy dissipation on a gold polycrystalline surface, using the two drive amplitudes of Figure 5.36.

a nominally similar tip/cantilever (but true phase measurements with an Agilent AFM). A somewhat larger change in drive amplitude was used to transition from a dominant attractive regime to repulsive regime. We consider this example in part because it highlights a not uncommon observation: rather erratic oscillator behavior (strong increase in noisiness of data) upon nearing and transitioning to the repulsive regime. Second, the dissipation curve in Figure 5.37b has a nearly horizontal trend upon reaching the repulsive regime (apart from spikes). A nearly flat-topped appearance has been simulated for tip–sample dissipation due to adhesion hysteresis [41] (modeled as different interfacial energies during approach and retract), whereas a repulsive but viscoelastically dissipative interaction has been simulated to produce a more parabolic trend in the dissipation-versus-A/A_0 curve as seen in Figure 5.32 for both methacrylate polymers [40]. Purely noncontact attractive force hysteresis modeling produced a parabolic-type trend, as observed here in the attractive regime.

5.4 VIRIAL INTERPRETATION OF PHASE

A final subtopic in the interpretation of phase measurements is much less known than the energy dissipation interpretation of phase, and perhaps more esoteric. It provides a way to relate phase to the *time-averaged* interaction force via the *virial theorem* [42]. The latter is a fundamental relationship between the time-averaged kinetic energy K and the virial, which is the time average of force times displacement in the direction of the force vector, provided that the total force is comprised of (i) conservative forces and (ii) frictional forces proportional to velocity. The virial theorem for dynamic AFM is

$$\langle K \rangle = -\frac{1}{2}\langle F \cdot s_z \rangle,$$

where the brackets denote the time average, in our case average per oscillation cycle. This leads to the relationship [43]

$$\cos\varphi = \frac{2Q}{kAA_0}\left[\frac{\langle F_{\text{tip-sample}}\rangle^2}{k} - \langle F_{\text{tip-sample}} \cdot s_z \rangle + \frac{1}{2}kA^2\left(1 - \frac{\omega^2}{{\omega_0}^2}\right)\right]. \tag{5.6}$$

Given an oscillation amplitude much greater than the mean cantilever deflection (typical), it can be shown that the first term is negligible. Given that ω/ω_0 is very nearly equal to 1, the third term is also negligible. Then the cosine of phase relates directly to the virial,

$$\cos\varphi \approx -\frac{2Q}{kAA_0}\langle F_{\text{tip-sample}} \cdot s_z \rangle.$$

For interactions that comprise a small fraction of each oscillation cycle near the bottom of the tip oscillatory motion, $s_z \approx -A$ (typical), we may write

$$\langle F_{\text{tip-sample}} \cdot s_z \rangle \approx -A\langle F_{\text{tip-sample}} \rangle$$

giving

$$\cos\varphi \approx \frac{2Q}{kA_0}\langle F_{\text{tip-sample}}\rangle = 2\frac{\langle F_{\text{tip-sample}}\rangle}{F_0}, \tag{5.7}$$

which importantly *relates the cosine of phase to the ratio of mean tip–sample interaction force to driving force* F_0 $(=kA_0/Q)$.

Finally, one can derive an expression that eliminates k and Q—and thus makes the results independent of variations in these parameters from cantilever to cantilever—by using Equations 5.7 and A3.4 relating cosine and sine of phase to $\langle F_{\text{tip-sample}}\rangle$ and $E_{\text{tip-sample}}$, respectively,

$$\frac{2}{A_0\cos\varphi}\langle F_{\text{tip-sample}}\rangle \approx \frac{k}{Q} = \frac{E_{\text{tip-sample}}}{\pi A^2\left(\dfrac{\sin\varphi}{A/A_0} - 1\right)}$$

giving[3]

$$\frac{E_{\text{tip-sample}}}{\langle F_{\text{tip-sample}}\rangle} \approx \frac{2\pi A^2}{A_0\cos\varphi}\left(\frac{\sin\varphi}{A/A_0} - 1\right). \tag{5.8}$$

This relationship allows one to compute the ratio of *energy dissipation to average force* per cycle by measuring A, A_0, and φ. In a sense, this gives an intensive rather than extensive measurement of dissipation. Larger average force would imply larger energy dissipation, other things being equal; thus, by dividing this average force out, one is contrasting a more intrinsic dissipation within an image. This is reminiscent of classic viscoelastic measurements on polymers that examine *loss tangent*, the ratio of loss modulus to storage modulus (i.e., dissipative to elastic response) [44–46].

5.5 CAVEATS AND DATA ANALYSIS STRATEGIES WHEN QUANTITATIVELY INTERPRETING PHASE DATA

One must exercise caution when computing physical quantities, such as energy dissipation or virial (Sections 5.3 and 5.4), from phase data. An image often will include many surface locations that are notably *steep* in topography, sites at which the operative amplitude A *deviates appreciably above or below setpoint* (depending on the sign and magnitude of topographic slope relative to the scan direction, as well as gain settings and scan speed, at minimum). A deviation of amplitude A from setpoint necessary produces a phase value φ that cannot be casually employed in the energy dissipation or virial formalism.

[3] Roger Proksch, personal communication.

To obtain good statistics, one typically needs to analyze many values of phase compiled from subregions of images or even entire images, often as histograms (e.g., Figure 5.2), or by simply averaging values within relevant regions. A histogram *peak position* may provide a good representative value of phase (the statistical *mode*) for locations at which the operational amplitude has deviated little from setpoint. But in some cases, a histogram peak may be ill-defined, because the histogram contains multiple overlapping peaks that are not easily identified (even with peak-fitting numerical routines) or because the histogram simply is odd shaped and thus not well described as a small set of peaks derived from starkly different materials. In these cases, one can in principle perform a weighted average over all measured phase values, the weighting given by the number of pixels (counts) at each histogram bin value. (This is mathematically the same as a simple average performed within the image-analysis software.) Here again, however, caution must be exercised; such a simple statistical weighting *may not be physically justified*. Deviations in operative amplitude, above and below setpoint, *may not produce deviations in phase that "average out"* across the phase histogram in the context of energy dissipation. The reason is worth emphasis:

> A positive deviation in phase of magnitude $|\delta\varphi|$ (due to a deviation in amplitude δA at say a location of large positive topographic slope) will be converted into a deviation in energy dissipation, positive or negative in sign (dependent on imaging regime, attractive or repulsive) and magnitude $|\delta E|$, whereas a negative deviation in phase *of the same magnitude* $|\delta\varphi|$ (i.e., at a location of opposite topographic slope) will be converted into a deviation in energy dissipation of the opposite sign *but not equal in magnitude* $|\delta E|$. Thus, these deviations will not average out.

In this respect, variations in energy dissipation contained in phase data are different from those in other property imaging signals such as lateral force or adhesion, when derived from variations in topographic slope. In the cases of lateral force and adhesion, variations in the measured force due to variations in topographic slope (i.e., due to changes of applied loading force relative to setpoint) are expected to be symmetric with respect to the sign of topographic slope provided that the feedback tracking is good. (For lateral force, this is treated in detail in Section 7.1.) In the case of phase, however—because the phase–dissipation relationship is nonlinear (containing a sine function)—variations in apparent energy dissipation due to variations in topographic slope will generally not be symmetric with respect to the sign of topographic slope.

An egregious case may result under very delicate attractive-regime imaging; negative deviations in amplitude with respect to setpoint can produce phase values that lie below the zero-dissipation value at the nominal setpoint amplitude. If the source of these values is not carefully identified, the result would be nonphysical *negative values of apparent energy dissipation* (under assumption of constant amplitude). This is depicted in Figure 5.38 containing real data for attractive-regime phase φ versus reduced amplitude A/A_0. Considering an

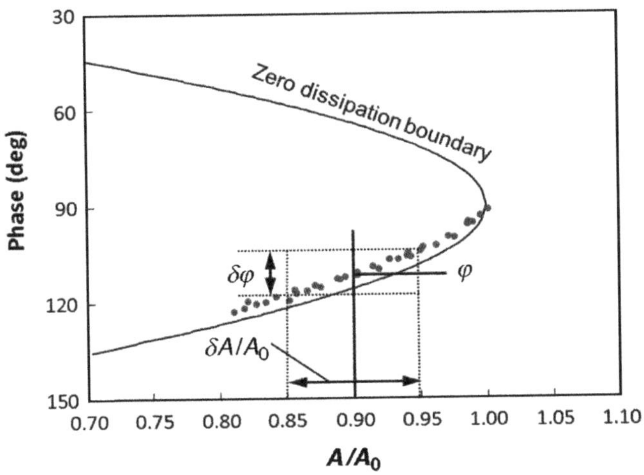

FIGURE 5.38 Depiction of deviations in phase $\delta\varphi$ derived from deviations in amplitude δA with respect to setpoint amplitude ratio $A/A_0=0.90$, due to large topographic slopes. An assumption of constant amplitude results in phase values falling below the zero dissipation boundary in the attractive regime.

imaging setpoint $A/A_0=0.90$ with substantial deviations $|\delta A|/A_0=0.05$ due to challenging topography, the lowest values in the corresponding $\delta\varphi$ window are below the zero-dissipation boundary at $A/A_0=0.90$, which if casually interpreted would imply negative energy dissipation.

A relevant example of phase histogram analysis (which also presents a somewhat novel and illustrative AFM application) is shown in Figure 5.39: time evolving phase values from images of drying ink. At time zero (first image acquired, roughly 10 min after ink deposition) the interaction is dominant attraction (negative raw phase) but quickly develops a repulsive histogram peak that grows strongly with time. Both attractive and repulsive peaks shift further outward during drying, corresponding to less dissipative response, likely due to less viscous and viscoelastic (i.e., more elastic) behavior, respectively. (It is worth noting that if one utilized a phase measurement that is a weighted average of the entire two-peaked histogram, one would make a huge mistake; values of raw phase well below and well above zero, corresponding to relatively *small* energy dissipation, would *average to values near zero*, meaning a *large* dissipation! Clearly, the mathematical complexity introduced by the presence of a sine function in the dissipation formalism is substantial.) After roughly 4 h, the attractive peak position corresponds to negligible energy dissipation (possibly meaning noncontact) such that the *low-phase tailings would correspond to negative dissipation* if taken at face value. But this coating had rugged topography resulting in sizable deviations of amplitude below (and above) setpoint, as discussed above. Given small dissipation in the

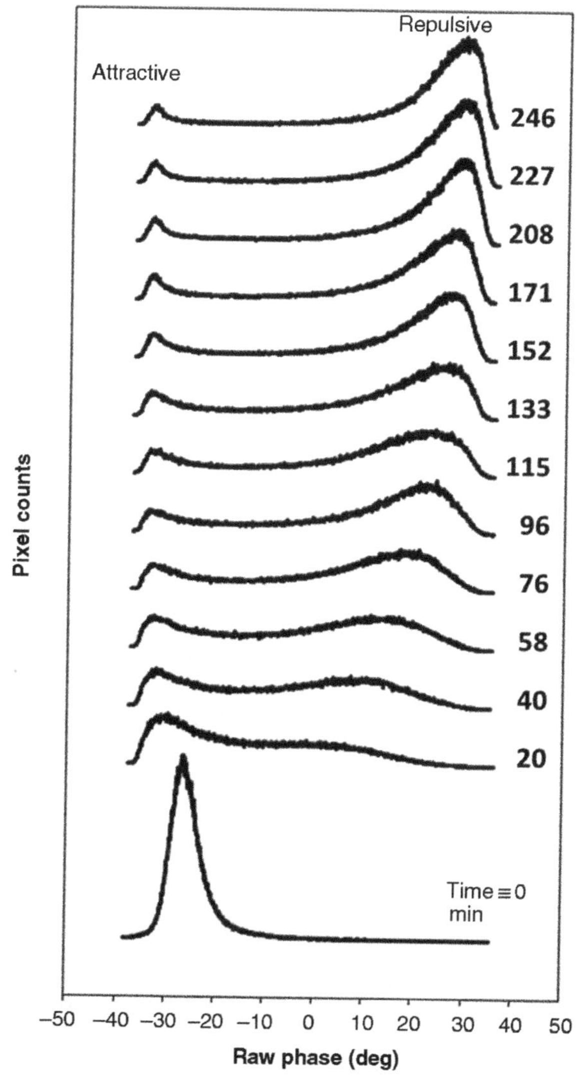

FIGURE 5.39 Summary plot of phase histograms for 13 time points during imaging of drying ink. Dominant attractive regime is replaced by dominant repulsive regime. Both peaks shifting outward indicate reduced energy dissipation.

attractive regime, such deviations guarantee *physically unrealistic values of phase if assigned to the setpoint amplitude*, as discussed in Figure 5.38.

Given the preceding, more careful quantitative work could choose from three strategies: (1) omit data from steep surface locations (often not practical), (2) attempt to account for variable A/A_0 (using the simultaneously acquired amplitude image, i.e., error signal δA) *in the energy dissipation expression per location* at which a phase value is being extracted (also often not practical, especially if analyzing many pixel locations for good statistics), or (3) use the values in the δA image to *correct* for the corresponding phase deviations $\delta\varphi$, assuming a linear relationship between δA and $\delta\varphi$ within the small δA window (i.e., assume $\delta\varphi / \delta A$ equals some constant), as can be confirmed in amplitude and phase versus Z curves. Thus, in (3) one simply invokes an appropriate *scaling factor* c and performs image arithmetic point by point (x,y), replacing the measured φ with

$$\varphi(x,y) - \delta\varphi(x,y) = \varphi(x,y) - c \cdot \delta A(x,y). \tag{5.9}$$

Two examples of this procedure are illustrated with real data in Figures 5.40 and 5.41. In Figure 5.40, a highly crystalline island of polyacrylamide is being contrasted with a more amorphous underlayer via phase data in the attractive regime. The amplitude (error) signal (b) relates simply to the slope of the height signal (a), and is clearly manifest in the phase image (c), a result of weak materials contrast (as is common in the attractive regime). Upon subtracting a scaled error signal δA image, the presence of topographic slope in the modified phase image (d) is greatly reduced and phase contrast between island and underlayer is starker.

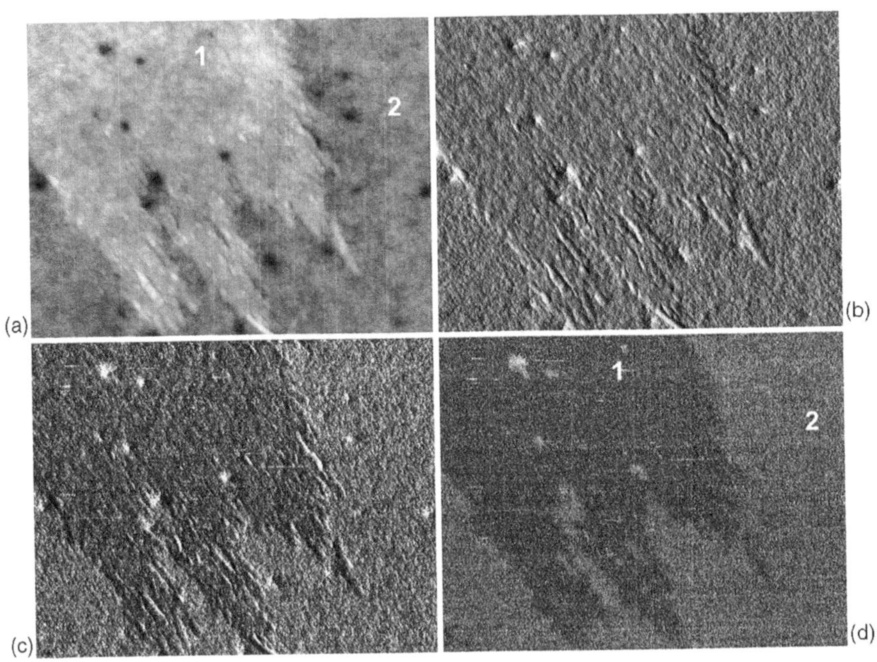

FIGURE 5.40 Height (a), amplitude (b), and phase (c) images of a polyacrylamide thin film containing a (1) highly crystalline island on a (2) more amorphous underlayer. Subtraction of the (scaled) amplitude data from the phase data results in a (d) "corrected" phase image that better contrasts materials differences between 1 and 2 by removing the effects of topographic slope (which causes the operational amplitude to deviate from setpoint).

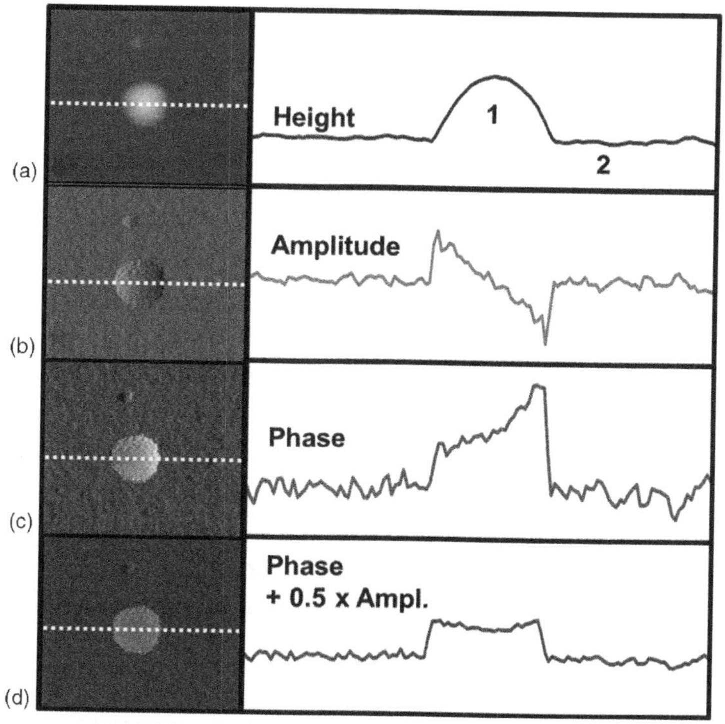

FIGURE 5.41 Height (a), amplitude (b), and phase (c) images of a silanized glass surface at high humidity, at which water droplets form. Deviations of amplitude below and above setpoint result in variations in phase. Upon adding scaled amplitude data to the phase data (d), variable phase is largely removed from both the droplet region and the surrounding film surface, thanks to consistent phase–amplitude relationships in the attractive regime.

Figure 5.41 examines images of a water droplet (that forms at high humidity) on a hydrophobic surface, silanized glass. Cross-sectional data reveal topographic-slope-derived error signal (b) that is substantially removed from the modified phase image (d), in comparison with the original phase image (c).

A potential problem with the preceding type of adjustment, however, arises when the *amplitude–phase relationship* is not even approximately constant across the image on *materially heterogeneous* surfaces. This easily can arise in the repulsive regime on regions of very different in elastic stiffness. Even with feedback tracking good enough to keep δA within some small window, the value of $\delta\varphi/\delta A$ may not be constant across the image. Thus, if one scales a δA image as $c\cdot\delta A$ and subtracts from a phase image φ in the attempt to remove variations in phase $\delta\varphi$ that derive from variations in topographic slope, this subtraction *may provide a good correction at some locations but not others.*

An example of this type of problem is shown in Figure 5.42, the case of a PBMA–PLMA blend (a thick coating as considered in Figures 5.10, 5.11, and 5.17)

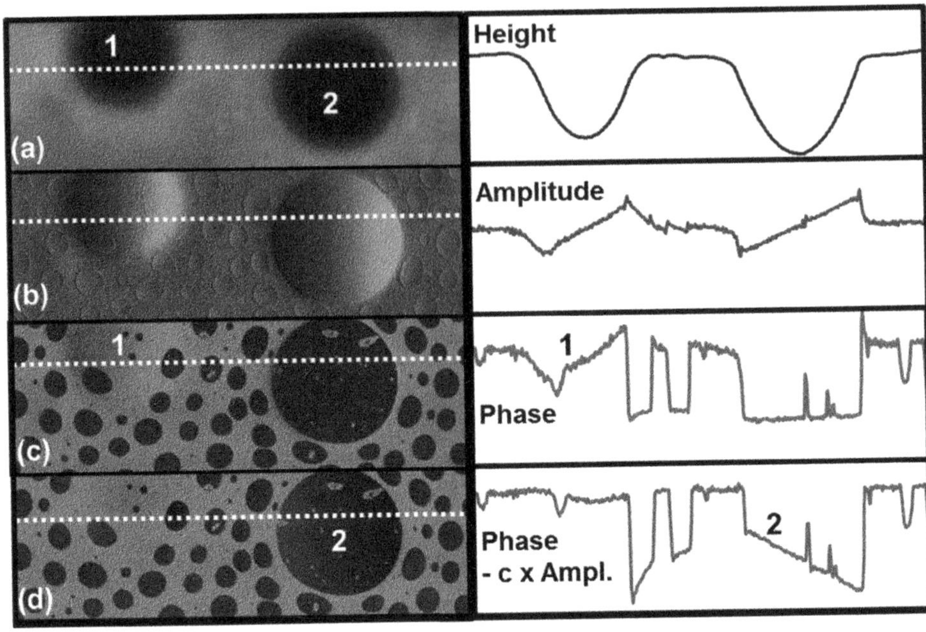

FIGURE 5.42 Height (a), amplitude (b), and phase (c) images of a thick coating of PBMA and PLMA. Strong deviations of amplitude below and above setpoint in the PBMA domains of region 1 result in strong variations in phase, whereas equally strong variations in amplitude on the much softer PLMA region 2 produce little variation in phase, due to a nearly flat phase–amplitude relationship. Upon subtracting scaled amplitude data from the phase data (d), variable phase is largely removed from region 1 but now appears in region 2.

imaged in the repulsive regime. Two deep depressions, in designated regions 1 and 2, are seen in the height image (a): the left (1) corresponding to a mainly PBMA (relatively rigid) surface and the right (2) coinciding with a much softer PLMA surface. The raw phase image (c) shows substantial δA effects (b) within region 1 but not region 2. Following a scaled subtraction of δA, topographic slope effects are largely removed from the adjusted phase image (d) within region 1, but now falsely appear within region 2.

One can imagine using an *algorithm* to choose from *two* values of c depending on the value of φ if starkly different among the materials to be contrasted, in order to produce more accurate phase corrections. This operation could be performed by exporting the 2D data sets into a spreadsheet program that can utilize logic statements to fill each cell (pixel) of a new image with a datum that has been adjusted using an appropriate, material-specific value of c.

A final caution in the quantitative interpretation of phase as energy dissipation, as with the previous issue, is especially important to attractive-regime imaging where energy dissipation values can be very low. Computed values of *dissipation may be strongly dependent on the determined apparatus phase shift*, which can drift with time during instrument warm-up as described in Section 5.2.2. If the utilized apparatus phase offset value is incorrect by just a few degrees, computed energy dissipation values can be miscalculated by many tens of percent, even becoming negative, just as it can in the context of highly sloped topographies. Clearly, both apparatus phase offset and gain/setpoint values (i.e., feedback tracking) can have a considerable impact on quantitative phase results, particularly in the case of a low-dissipation, net attractive interaction.

REFERENCES

[1] Bhushan, B., ed. *Springer Handbook of Nanotechnology*. 2nd ed. 2007, Berlin: Springer.

[2] Bhushan, B. and H. Fuchs, eds. Applied scanning probe methods XII: Characterization, in *Nanoscience and Technology*. 2009, Berlin/Heidelberg: Springer.

[3] Gleyzes, P., P.K. Kuo, and A.C. Boccara, Bistable behavior of a vibrating tip near a solid surface. *Appl. Phys. Lett.*, 1991, **58**(25): 2989–2991.

[4] Anczykowski, B., D. Kruger, and H. Fuchs, Cantilever dynamics in quasinoncontact force microscopy: Spectroscopic aspects. *Phys. Rev. B*, 1996, **53**(23): 15485–15488.

[5] Garcia, R. and R. Perez, Dynamic atomic force microscopy methods. *Surf. Sci. Rep.*, 2002, **47**: 197–301.

[6] Haugstad, G. and R.R. Jones, Mechanisms of dynamic force microscopy on polyvinyl alcohol: Region-specific non-contact and intermittent contact regimes. *Ultramicroscopy*, 1999, **76**: 77–86.

[7] Haugstad, G., Contrasting static-to-kinetic friction transitions on layers of an autophobically dewetted polymer film using Fourier-analyzed shear modulation force microscopy. *Trib. Lett.*, 2005, **19**: 49–58.

[8] LMNO Engineering, R., and Software, Ltd., *Gas Viscosity Calculator*, 2003.

[9] Studnikov, E.L., The viscosity of moist air. *J. Eng. Phys. Therm.*, 1970, **19**(2): 1036–1037.

[10] Ikehara, T., et al., A high quality-factor silicon cantilever for a low detection-limit resonant mass sensor operated in air. *J. Micromech. Microeng.*, 2007, **17**: 2491–2494.

[11] Hosaka, H., K. Itao, and S. Kuroda, Damping characteristics of beam-shaped micro-oscillators. *Sensors Actuators A*, 1995, **49**: 87–95.

[12] Dareing, D.W., D. Yi, and T. Thundat, Vibration response of microcantilevers bounded by a confined fluid. *Ultramicroscopy*, 2007, **107**: 1105–1110.

[13] Pandey, A.K., and R. Pratap, Effect of flexural modes on squeeze film damping in MEMS cantilever resonators. *J. Micromech. Microeng.*, 2007, **17**: 2475–2484.

[14] D'Amato, M.J., et al., Phase imaging and the lever-sample tilt angle in dynamic atomic force microscopy. *App. Phys. Lett.*, 2004, **85**: 4738–4740.

[15] Bocek, D., About "Bocek" degrees used in the extender package on digital instruments nanoscope scanning probe microscopes, in *Microscopy Today*. 2001.

[16] Stark, R.W., T. Drobek, and W.M. Heckl, Tapping-mode atomic force microscopy and phase-imaging in higher eigenmodes. *Appl. Phys. Lett.*, 1999, **74**(22): 3296–3298.

[17] Rodriguez, T.R. and R. Garcia, Compositional mapping of surfaces in atomic force microscopy by excitation of the second normal mode of the microcantilever. *App. Phys. Lett.*, 2004, **84**: 449–451.

[18] Lozano, J.R. and R. Garcia, Theory of multifrequency atomic force microscopy. *Phys. Rev. Lett.*, 2008, **100**: 076102.

[19] Cleveland, J.P., et al., Energy dissipation in tapping-mode atomic force microscopy. *Appl. Phys. Lett.*, 1998, **72**: 2613–2615.

[20] Kokavecz, J. and A. Mechler, Investigation of fluid cell resonances in intermittent contact mode atomic force microscopy. *App. Phys. Lett.*, 2007, **91**(2): 023113.

[21] Motamedi, R. and P.M. Wood-Adams, Influence of fluid cell design on the frequency response of AFM microcantilevers in liquid media. *Sensors*, 2008, **8**: 5927–5941.

[22] Tsukada, M. and T. Watanabe, Theoretical analyses of cantilever oscillation for dynamic atomic force microscopy in liquids. *Jap. J. Appl. Phys.*, 2009, **48**: 035001.

[23] Weigert, S., M. Dreier, and M. Hegner, Frequency shifts of cantilevers vibrating in various media. *Appl. Phys. Lett.*, 1996, **69**: 2834–2836.

[24] Schaffer, T.E., et al., Studies of vibrating atomic force microscope cantilevers in liquid. *J. Appl. Phys.*, 1996, **80**: 3622–3627.

[25] Xu, X. and A. Raman, Comparative dynamics of magnetically, acoustically, and Brownian motion driven microcantilevers in liquids. *J. App. Phys.*, 2007, **102**: 034303.

[26] Han, W., S.M. Lindsay, and T. Jing, A magnetically driven oscillating probe microscope for operation in liquids App. *Phys. Lett.*, 1996, **69**(26): 4111–4113.

[27] Han, W., et al., Strained DNA is kinked by low concentrations of Zn^{2+}. *Proc. Natl. Acad. Sci USA*, 1997, **94**: 10565–10570.

[28] Bates, F.S. and G.H. Fredrickson, Block copolymer thermodynamics: Theory and experiment. *Annu. Rev. Phys. Chem.*, 1990, **41**: 525–557.

[29] Konrad, M., et al., Volume imaging of an ultrathin SBS triblock copolymer film. *Macromolecules*, 2000, **33**(15): 5518–5523.

[30] Knoll, A., R. Magerle, and G. Krausch, Phase behavior in thin films of cylinder-forming ABA block copolymers: Experiments. *J. Chem. Phys.*, 2004, **120**(2): 1105–1116.

[31] Macosko, C.W., Rheology: Principles, in *Measurements and Applications*. 1994, New York: Wiley-VCH.

[32] Carpick, R.W., and M. Salmeron, Scratching the Surface: Fundamental investigations of tribology with atomic force microscopy. *Chem. Rev.*, 1997, **97**: 1163–1194.

[33] Rabe, U., J. Turner, and W. Arnold, Analysis of the high-frequency response of atomic force microscope cantilevers. *Appl. Phys. A*, 1998, **66**: S277–S282.

[34] Lee, S.I., et al., Nonlinear dynamics of microcantilevers in tapping mode atomic force microscopy: A comparison between theory and experiment. *Phys. Rev. B*, 2002, **66**: 115409.

[35] Stark, M., Spectroscopy of higher harmonics in dynamic atomic force microscopy. *Nanotechnology*, 2004, **15**: 347–351.

[36] Chen, X., et al., Optimizing phase imaging via dynamic force curves. *Surf. Sci.*, 2000, **460**: 292–300.

[37] Zitzler, L., S. Herminghaus, and F. Mugele, Capillary forces in tapping mode atomic force microscopy. *Phys. Rev. B*, 2002, **66**: 155436.

[38] Aime, J.P., et al., Influence of noncontact dissipation in the tapping mode: Attempt to extract quantitative information on the surface properties with the local force probe method. *J. Chem. Phys.*, 2001, **114**(11): 4945–4954.

[39] Stark, M., et al., Inverting dynamic force microscopy: From signals to time-resolved interaction forces. *Proc. Nat. Acad. Sci.*, 2002, **99**(13): 8473–8478.

[40] Garcia, R., et al., Identification of nanoscale dissipation processes by dynamic atomic force microscopy. *Phys. Rev. Lett.*, 2006, **97**: 016103.

[41] Israelachvili, J.N., *Intermolecular and Surface Forces*. 2nd ed. 1991, San Diego: Academic Press, 450.

[42] Goldstein, H., *Classical Mechanics*. 2nd ed. 1980, Reading, MA: Addison-Wesley.

[43] San Paulo, A. and R. Garcia, Tip-surface forces, amplitude, and energy dissipation in amplitude-modulation tapping mode force microscopy. *Phys. Rev. B*, 2001, **64**: 193411.

[44] McCrum, N.G., B.E. Read, and G. Williams, *Anelastic and Dielectric Effects in Polymeric Solids*. 1967, London: Wiley.

[45] Ferry, J.D., *Viscoelastic Properties of Polymers*. 1980, New York: Wiley.

[46] Aklonis, J.J. and W.J. MacKnight, *Introduction to Polymer Viscoelasticity*. 2nd ed. 1983, New York: Wiley, 295.

6

PROBING MATERIAL PROPERTIES II: ADHESIVE NANOMECHANICS AND MAPPING DISTANCE-DEPENDENT INTERACTIONS

This chapter explores further intricacies of *distance-dependent probing*, as previewed in Chapters 1–3 and 5, including two elements:

i. greater detail in physical phenomenology and data interpretation;
ii. data acquisition and analysis routines for *mapping* distance-dependent responses.

Most new AFM users (and even many long-term users) are not aware of—or have not utilized—the distance dependence of tip–sample interactions, let alone *X–Y* mappings of this distance dependence. In many AFM applications, these methods, with a little effort, can provide the most penetrating insight into *what is happening at the tip–sample interface*, and, thereby, the best understanding of sample properties while *expanding AFM utility*. Moreover, such "data cube" methods can provide critical information to aid in the *proper interpretation of conventional images* acquired in quasistatic and dynamic modes. A goal of this chapter is to provide an understanding of these enabling methodologies. Prior to discussing the more complex data acquisition methods (Sections 6.4 and 6.5), however, we need a more

Atomic Force Microscopy: Understanding Basic Modes and Advanced Applications, First Edition.
Greg Haugstad.

complete understanding of distance-dependent interactions in the form of *adhesive contact mechanics models* (Section 6.2) and *capillarity* at the tip–sample interface (Section 6.3).

6.1 GENERAL CONCEPTS AND INTERRELATIONSHIPS

In some AFM modes, the making and breaking of tip–sample contact is key to materials sensitivity. Generally, this bonding/debonding cycle (normally non-covalent) is an *energy dissipative* process and thus contributes to phase contrast as discussed in Section 5.3. In this chapter, we revisit the fact that *adhesion*—due to attraction between tip and sample—can be gauged most simply and directly as the force required for pull-off (Figure 1.5). Second, this force can be mapped or imaged to provide materials contrast as previewed in Figure 1.16. We will also discuss a full approach–retract cycle over which one can integrate force times distance—provided that equilibrium conditions are maintained (i.e., no jumps to or from contact)—to obtain the irreversible work or *adhesion hysteresis* as another metric of tip–sample adhesion.

Adhesion further affects other physical phenomena like *sliding friction* (Chapter 7) and *viscoelasticity*, the latter a rate-dependent mechanical response with dissipation. Before getting into details, however, it is worth contemplating both *extensive* and *intensive* ways by which this occurs.

Extensive: Any measurement (e.g., friction force) that scales with the *size of the tip–sample contact area* will scale with the strength of adhesion, given that a greater adhesive force pulls tip and sample together to produce a larger contact area (further discussed in Sections 6.2 and 6.3).

Intensive: Strong adhesion can result from noncovalent *bonding and interlocking at the atomic and molecular level*, as well as molecular-scale rearrangements that produce a more thermodynamically stable interface (lower free energy). Disrupting each interacting element of this interface, whether via normal or shear motion, can require large stresses that further activate molecular motions (or lower activation energy barriers to thermal activation, as discussed in Section 2.6.2). These motions generally produce dissipation: they convert mechanical work by a single degree of freedom (i.e., force times distance) into small bits of energy resident in many degrees of freedom. The latter are moreover *randomized*, such that the energy cannot be somehow recouped into a single degree of freedom. Thus, molecular-scale motions contribute fundamentally to a measured adhesion or friction force, or viscous/viscoelastic response.

We touched on the extensive role of adhesion in the context of the energy-dissipation interpretation of phase imaging in Chapter 5: the idea that two materials with identical modulus, or more generally viscoelastic response, will exhibit phase contrast if tip–sample adhesion differs, because the magnitude of energy dissipation scales with the size of the contact area and deformation volume (i.e., in addition to the intrinsically stronger attraction). Clearly, there is an unavoidable *reinforcement between adhesion and mechanical deformation*: the softer the material, the greater the deformation under attractive force, producing a larger contact area and hence adhesive force, which scales with contact area. The greater the aggregate adhesive force, the stronger the two surfaces are pulled together and thus the greater the mechanical deformation.

Adhesion also can create "reinforcement" between intensive and extensive phenomena. Intrinsically strong tip–sample attraction (i.e., at the molecular scale, energy per molecule) integrates over the contact zone to produce a larger aggregate attractive force, which pulls tip and sample together to grow the size of the contact zone; this, in turn, increases the number of molecules interacting and thus the magnitude of the total force needed to pull tip and sample apart or shear their interface (whether in steady-state sliding or the onset of sliding, i.e., kinetic or static friction). A larger total magnitude of attractive force can produce higher levels of stress within the contact zone, which may *activate characteristic molecule motions that may not otherwise be activated*; even cause a threshold for plastic yield to be exceeded and thus the onset of wear or surface patterning. So what at root are intrinsic, characteristic phenomena at the molecular level may result from the overall magnitude of the adhesion force—here again, a reinforcement between intensive and extensive phenomena. On certain classes of soft materials this is sometimes observed: blunting or contamination of the AFM tip produces a larger contact and thus a larger adhesion force, and, in turn, a larger friction force, resulting in wear during contact-mode scanning that did not occur when the tip was clean and sharp. Another common observation is that the accumulation of molecules on the tip can produce an intrinsically stronger interaction that increases stress and thereby causes wear.

We know from Chapter 2 that attractive forces are not only present at the immediate tip–sample contact interface. van der Waals forces have a range measured in nanometers and thus can extend laterally beyond the periphery of the tip–sample contact, as well as between noncontacting tip and sample atoms below or above the tip–sample interface. Moreover, van der Waals and hydrogen bonding are manifest in air via the formation of a capillary bridge just beyond the periphery of the tip–sample contact zone, whether due to adsorbed/condensed water molecules or mobile low-molecular weight organic species originating from the sample or (contamination of) the tip. We will exemplify the importance of capillary forces to adhesion by varying the relative humidity (RH). To help understand these results, we will also discuss the incorporation of capillary forces in attractive force models.

We also know from Chapter 2 that attractive forces can extend to much greater distances than the nanometer-scale range of van der Waals forces. Some finite attraction beyond van der Waals distances is often present due to electrostatic forces between tip and sample, whether due to surface potential (work function) differences or net charging of sample, and possible spatial variations thereof. Moreover, the application of a tip–sample bias can produce dramatic differences in the electrostatic attraction of the tip to different sample components, reflecting differences in dielectric response, capacitance, or charging. Most commonly these effects are probed in dynamic AFM via interleave or other special data acquisition routines. An interleave scan (along the fast-scan axis) is a second pass with Z displaced a fixed distance from the usual height (Z) values collected in the first past (treated in Section 9.1). During this second pass, a bias may be applied. This bias may be stepped during an image or multiple images may be acquired at different biases; phase or resonance frequency shifts are recorded for each value of bias. Similarly, these measurements can be performed at variable interleave Z.

6.2 ADHESIVE CONTACT MECHANICS MODELS

6.2.1 Overview and Disclaimers

Before considering the details of data acquisition in the modes outlined above, one needs a conceptual framework for understanding what is measured. To this end, this section broadly introduces *adhesive contact mechanics models.* A grasp of these models is not only beneficial to the post-processing and analysis of data; a preunderstanding can lend insight into *appropriate measurement procedures*, such as the ranges and increments of force or rate to explore, even the sharpness of the probe to use. But be forewarned, these models can be complicated and confusing. In journal publications in recent decades, one may track the evolution of increasingly complex contact mechanics modeling and simulations, which treat phenomena such as attractive forces at and beyond the contact interface, a capillary meniscus, viscoelasticity, plasticity, viscous flow/sintering, nano- or atomic-scale roughness at the tip–sample interface, thin-film coated or more complicated layering or gradation of sample properties, stress-modified thermal activation of molecule motions, and more (to be discussed). On top of a blizzard of physical phenomena, the derivations, language, and symbols employed can be inconsistent. And during more than two decades of AFM applications in property analysis, experimentalists generally have not replaced the "crude" models with the more "sophisticated" ones in their analytical toolboxes. At this writing, one may find very recent AFM studies employing simple Hertzian analysis (Section 2.3), ignoring attractive forces altogether; yet one can go back more than a decade and find the usage of the much more complex Maugis–Dugdale model (Section 6.2.4), which incorporates interfacial attraction both within the contact zone and beyond its periphery. (Even detailed finite element analysis of AFM contact mechanics dates back many years.) It is easy to become confused over which model one should use and how to avoid unfounded conclusions, whether by current standards or those of one or two decades past.

Thus, in the following treatment, rather than simply list formulae and formal considerations in applying them, we will also briefly touch on practical relevance: whether modeling the complex details of an interfacial/nanomechanical system is in the end a net advantage to the analyst. The continuing viewpoint of this author is that we must include the *nonidealities and caveats* of a measurement when considering the usefulness of theoretical concepts. Purely formal considerations, such as the order of magnitude of modulus, surface energy, and tip radius of curvature, cannot be our only concern. Then there is the "inverse problem" for the experimentalist: if one wishes to apply a more complex model so as to extract more physical information (e.g., *both* modulus and surface energy of a sample), how should the analyst design and conduct the measurement?

In Section 2.3, the Hertzian contact mechanics model was introduced, providing a relationship between force, contact radius, indentation depth, tip radius of curvature, and elastic constants—modulus and Poissons ratio of sample and tip. Importantly, the Hertzian model does not consider attractive forces between tip and sample, in addition to its restriction to linear elastic response with no intrinsic time or rate dependence (i.e., no viscoelasticity). In Chapters 2–5, we have touched on a number of manifestations of attractive forces. Indeed, the large role, even dominance, of attractive forces is perhaps what most distinguishes atomic force microscopes from nanoindentors (especially given that some of the latter tools now include imaging/mapping capabilities, i.e., an ability to "take pictures" no longer distinguishes these techniques). Thus, a quantitative description of AFM nanomechanics—whether to measure elastic constants such as Young's modulus, or examine the contact-area dependence of friction, or measure the yield stress at the onset of plastic deformation or wear—requires more sophisticated contact mechanics models incorporating attractive forces. But this does not come without complication! Some of the most general and rigorous treatments use Lennard–Jones type interactions to describe both (i) the transition from attraction and repulsion within the contact zone and (ii) attraction within a ring just outside of the contact periphery. This complexity requires *numerical simulation* [1]. First-line simplifications of these models treat the attraction as constant within some circular region, dropping to zero beyond this region. Such a model also describes the effects of a *capillary meniscus*. In either case, this simplification as a circle of constant attraction yields analytical expressions, but these must be solved *parametrically* as will be discussed. This requires numerically varying one number over a range of values until multiple equations simultaneously hold true. Other models account for viscoelasticity, that is, intrinsically rate-dependent responses. The present treatment will not consider these latter models in detail; the reader is directed to articles on the subject [2–4].

Even for the simplest models of the tip–sample interface, there are several parameters involved: the radius of curvature of the AFM tip, the modulus and Poisson's ratio of the material indented, the intrinsic strength of attractive forces between tip and sample (in turn related to their individual surface energies, i.e., polarizabilities), and the range of loading forces considered. We will first introduce these simple models and their dependences on this set of parameters. But it is important to note that even if we exclude the complications of viscoelasticity and capillary forces, these are still grossly simplified models of real contacts, because they strictly apply to (i) contacting *spheres* and (ii) contact areas/deformations *much smaller than the radii of these spheres*.

As discussed with respect to Equation 2.11, a model of contacting spheres allows one to treat the tip–sample interaction atop hills or in valleys, in addition to flat locations (infinite radius of curvature). But it also assumes that the tip can be reasonably approximated as a sphere instead of some other shape, and that both tip and sample are smooth. Just how smooth? Smooth enough that the inevitable roughness—even if only at the atomic scale—effectively vanishes because of a *conformal* contact. This produces a single, continuous contact interface instead of a "multitouch" or *multiasperity* contact. The contact interface hopefully has an overall shape that is close to what it would be if both surfaces were perfectly smooth spherical caps. But such a conformal contact is not likely between rigid materials at low forces. Thus, the attractive forces as well as mechanical deformation may depend on the nanometer-to-Angstrom regime shapes of contacting asperities within the overall contact zone. Note that this is the second time we have pointed out a problem with simple mechanical models for AFM of *rigid materials*. (Recall the discussion in Section 3.2.3 of sliding that may occur even under vertical displacement, because of cantilever tilt and a much greater compliance in the cantilever compared to the tip–sample contact.) If one adds capillary effects—and noting that each asperity may have its own scale of capillary meniscus—it is likely that upon increasing load, the asperities deform such that menisci coalesce into larger menisci. Such details are beyond the scope of the treatment in this book.

6.2.2 JKR and DMT Models

Two simple models named after the authors of the respective publications—the Johnson–Kendall–Roberts (JKR) [5] and Derjaguin–Muller–Toporov (DMT) [6]—account for only one locus of attraction or the other: solid–solid attraction *within* the contact circle or attractive forces in a ring just *outside* of the contact zone, as depicted in Figure 6.1. We consider these models first because they are relatively simple and the most commonly invoked in the analysis of AFM force curves and friction–load plots when attractive forces are significant. The JKR model treats only those attractive forces at the contact interface, whereas the DMT treats only those attractive forces outside of the contact zone. We will see that the JKR model best describes a contact where intersurface forces are short range in comparison to the deformations they produce—soft materials (low tip–sample system modulus K, Eq. 2.12), strong adhesion (high work of adhesion γ, related to surface energy and Hamaker constant, to be discussed) and large system radius of curvature (R, Eq. 2.12). The resulting contact geometry is distinctly different from Hertzian [5]. The DMT model is conversely appropriate to long-range intersurface forces and Hertzian contact geometry; this means stiff materials, weak adhesion, and small radius of curvature. This model also has been used to describe the effects of capillary interactions acting outside of the contact zone, as marked with arrows in Figure 6.1 and discussed in Section 6.3.

The DMT model is relatively simple in that it produces a constant adhesive force $F_{\mathrm{adh}} = 2\pi R\gamma$, where γ is the tip–sample work of adhesion in units of energy per

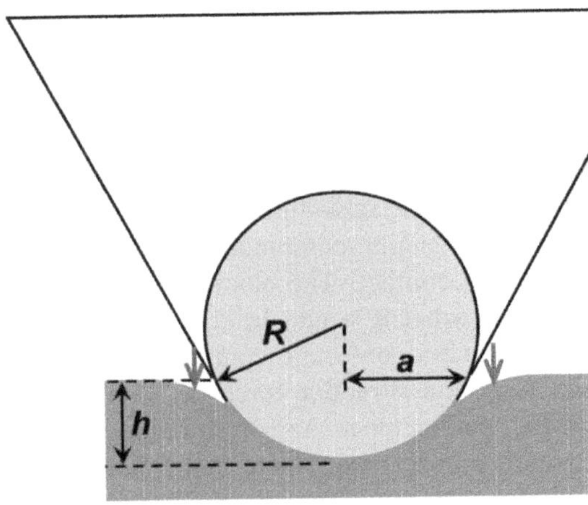

FIGURE 6.1 Locus of attraction between spherical-ended tip and deformed surface. The JKR model considers attraction at the tip–sample contact interface, whereas the DMT considers attraction just outside of the contact zone.

unit area (often stated in mJ/m^2, where the first 'm' stands for milli and the second for meter) or force per unit length (mN/m). In the DMT model, the adhesive force adds to the externally applied force F to produce deformation, but the total force retains the functional form of the Hertzian model ($F = Ka^3/R$). Consequently, the model is sometimes called *Hertz plus offset*. This relationship between contact radius a, applied force F, system modulus K (Eq. 2.12) and reduced radius of curvature R (Eq. 2.11) is

$$F = \frac{Ka^3}{R} - 2\pi R\gamma. \tag{6.1}$$

One can easily solve Equation 6.1 for a to see that the adhesion-derived term $2\pi R\gamma$ adds to the applied force F to *increase* the contact radius a. Also note that for zero applied force F, the equation predicts a finite contact radius

$$a_0 = \left(2\pi R^2\gamma/K\right)^{1/3}, \tag{6.2}$$

whereas in the Hertzian case, because $F = Ka^3/R$, zero applied force gives zero contact radius. This makes intuitive sense; attractive force pulls tip and sample together, producing a nonzero contact area in the absence of an external loading force (or growing the contact area beyond that due to a finite load). Not surprisingly, the expression for indentation depth h is the same as the Hertzian case,

$$h = \frac{a^2}{R}. \tag{6.3}$$

A key element to the DMT model is the lack of a treatment of adhesion along the contact interface itself. The so-called adhesive force is the aggregate van der Waals attraction between a sphere and an infinite half-space (or more generally between two spheres) over the domain extending *beyond the contact circle*. (The relationship $F_{\text{adh}} = -2\pi R\gamma$ is identical to that predicted by the Bradley model from the 1930s [7] that treats *nondeformable* spheres by integrating van der Waals attraction over both spheres.)

The JKR expressions are more complicated:

$$F = \frac{Ka^3}{R} - \sqrt{6\pi\gamma K a^3} \tag{6.4}$$

$$h = \frac{a^2}{R} - \frac{2}{3}\sqrt{\frac{6\pi\gamma a}{K}}. \tag{6.5}$$

Here again we see that an adhesion-derived term $\sqrt{6\pi\gamma K a^3}$ acts like an additional force, but unlike the DMT form, this term contains the contact radius a. Thus, the greater the area of the contact circle (πa^2), the greater the adhesion force. This adhesion force is not linearly proportional to contact area, rather, the $\frac{3}{4}$ power of area. But at least qualitatively the above makes sense for a model that accounts for attraction at the tip–sample interface.[1]

A common equivalent JKR expression solves for contact radius as a function of force (an exercise in using the quadratic equation). For analyst convenience, we state the result:

$$a = \left[\frac{R}{K}\left(F + 3\pi R\gamma + \sqrt{6\pi R\gamma F + (3\pi R\gamma)^2}\right)\right]^{1/3}. \tag{6.6}$$

This equation may be solved for the case of applied force $F = 0$ to obtain

$$a_0 = \left(6\pi R^2\gamma/K\right)^{1/3}, \tag{6.7}$$

which again indicates a nonzero contact radius but larger than that of the DMT model by a factor of $3^{1/3} = 1.44$. As an example, for a tip of radius $R = 10$ nm, work of adhesion $\gamma = 50$ mJ/m^2 and $K = 3$ GPa, we compute $a_0 = 3.2$ nm.

The JKR equations also predict finite contact area at the minimum (most negative) stable load, the force at which pull-off occurs, $F_{\text{adh}} = -3/2\pi R\gamma$. At this critical force, the contact diameter decreases to

$$a_C = \left(3\pi R^2\gamma/2K\right)^{1/3}$$

or $(1/4)^{1/3} = 0.63$ times its value at zero applied force. This indicates that a substantial improvement in resolution can be achieved, in principle, by operating at a

[1] Note that in Equations 6.4 and 6.5, if the work of adhesion γ is set to zero (i.e., no attraction), terms drop out such that one regains the Hertzian expressions (Eqs. 2.9 and 2.10).

pulling force that is almost large enough to snap the tip from the surface. (This is easy to confirm on a very flat glassy polymer, where the smallest resolved "grains" in topography become noticeably smaller.)

Figure 6.2 overlays plots of the above JKR, DMT, and Hertzian expressions for (a) contact radius versus applied force and (b) applied force versus indentation depth, for the case of a rigid tip of 10-nm radius on a sample with a Young's modulus of 3GPa and Poissons ratio of 0.4, and a tip–sample work of adhesion (for JKR and DMT models) of 50 mJ/m^2. The critical force at which contact is lost, discussed in the preceding paragraph, is the point at which the *a–F* trend becomes vertical.

It is also interesting to examine the exerted pressure in the contact zone in the JKR model, as done in Section 2.3 for the Hertzian case. For the JKR model, this is a two-term expression [8],

$$p(\rho) = \frac{2aE^*}{\pi R}\left(1 - \frac{\rho^2}{a^2}\right)^{1/2} - \left(\frac{4\gamma E^*}{\pi a}\right)^{1/2}\left(1 - \frac{\rho^2}{a^2}\right)^{-1/2},$$

where ρ is the radial coordinate and E^* is the reduced modulus as defined in Section 2.3. We note the (nonphysical) divergence of stress at $\rho = a$, a weakness of the JKR model. Of greater interest is the central stress ($\rho = 0$) at zero applied force, where $a = a_0$ is given by Equation 6.7, for a tip of radius $R = 10$ nm indenting a material of reduced elastic modulus $E^* = 2.2$ GPa and with a tip–sample work of adhesion $\gamma = 50$ mJ/m^2: $p_0 = 230$ MPa. This is nearly identical to the Hertzian stress computed in Section 2.3 for the same tip and sample modulus but an applied load of 1 nN. Thus, we see that simply contacting a tip, with zero applied force, to a material with properties like, say, a glassy polymer, can generate a stress of the same order of magnitude as the plastic yield point of the material.

6.2.3 Ranging Between JKR and DMT: The Transition Parameter λ

The JKR and DMT models actually are *limiting cases* within a continuum of parameter space that spans from rigid materials, small R and small γ at the one end (DMT) to soft materials, large R, and large γ at the other (JKR). As one can imagine, there are many *other* cases within this 3D parameter space, such that the DMT and JKR limits may not strictly apply to many real systems. Nevertheless, these models have been frequently employed in the literature, together with the even simpler Hertzian model or Oliver and Pharr method (Section 3.6.1), in the analysis of AFM mechanical and tip–sample adhesion data.[2,3] The value of the radius of

[2] Nonadhesive contact mechanics models can be implemented if the applied forces are much higher than the adhesive forces.

[3] Although the JKR model is usually the more appropriate for soft materials—where attraction presses the materials together to deform and grow the contact area such that this attraction dominates that beyond the contact periphery—one must always bear in mind that soft materials often are better described as viscoelastic. The JKR model does not account for this time-dependent (and thus loading-rate dependent) response.

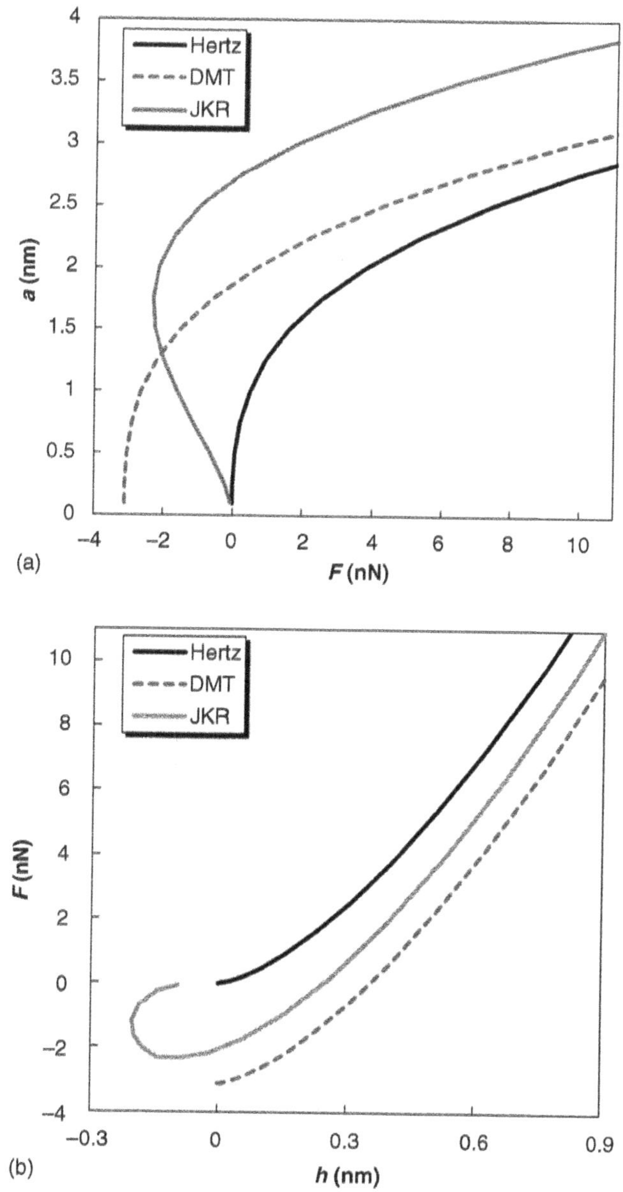

FIGURE 6.2 Overlay of Hertzian, JKR, and DMT relationships between (a) contact radius a and force F, and (b) force F and indentation depth h, for a 10-nm radius spherical tip interacting with a 3-GPa modulus flat material and 50 mJ/m^2 work of adhesion between tip and sample.

curvature R also affects the relative magnitude of attraction at the contact interface and that beyond its periphery. The dimensionless *elasticity parameter* λ expresses these relationships:[4]

$$\lambda = \frac{2.06}{\xi_0}\left(\frac{R\gamma^2}{\pi K^2}\right)^{1/3}. \tag{6.8}$$

The constant ξ_0 is the equilibrium interatomic distance (nucleus to nucleus of contacting atoms) between tip and sample atoms at the contact interface, usually set to 0.16 nm (Section 2.2.1). The DMT model applies if $\lambda \sim 0.1$ or smaller, whereas the JKR is most appropriate if $\lambda > 5$. Within these two rather extreme regimes, each respective model is equivalent to a more general Maugis–Dugdale model (next section), which is also appropriate for $0.1 < \lambda < 5$ [9].

It is important for the AFM analyst to estimate the elasticity parameter λ for the system investigated when intending to perform detailed nanomechanical analysis at low applied loads (i.e., cases where adhesion generates a significant portion of the overall contact force). For AFM, such cases normally involve small to moderate values of K, in order to get measurable indentation (as discussed in Chapter 3). By default, these measurements involve small R compared to other mechanical measurement tools, given much sharper AFM tips, say compared to common Vickers or Berkovich indenters [10]. (An exception is the case of an AFM tip being replaced by a microsphere, typically glued to a tipless cantilever.) Moreover, γ is usually small to moderate for ambient conditions (i.e., where high-surface energy materials, including of course silicon tips, oxidize to become much lower-surface energy materials), corresponding to tens of milli-Joules per square meter (again the rather confusing unit mJ/m^2). Thus, for a nanoscale contact, one does not anticipate a large value of λ unless K is exceedingly small. But we discussed in Section 2.3 the fact that for very soft entities, such as gels or cells, the indentation depth exceeds the hemispherical end-cap of the tip, invalidating geometric aspects of the above models. Thus, rigorous nanomechanical measurements on extremely soft samples should employ a cantilever-attached microsphere, and thereby a much larger numerator applies to Equation 6.8. Together with the small value of K, we anticipate the appropriateness of the JKR model in this case.

Let us calculate order-of-magnitude estimates for λ. A conventional silicon tip of sharpness $R \sim 10$ nm indenting a glassy polymer such that $K \sim 2$ GPa and $\gamma \sim 50\ mJ/m^2$ gives an elasticity parameter $\lambda \sim 1$. Thus, force-displacement data in this case are not appropriately analyzed *in detail* with either of the simplified models, DMT or JKR. The same tip against an oxidized silicon surface such that that $K \sim 100$ GPa and $\gamma \sim 100\ mJ/m^2$ would give $\lambda \sim 0.1$, just barely appropriate to DMT. Going to the other extreme, a glass microsphere of sharpness $R \sim 10\ \mu m$ indenting a gel such that $K \sim 10$ kPa and $\gamma \sim 10\ mJ/m^2$ gives $\lambda \sim 10{,}000$, far into

[4] The similar Tabor parameter [Tabor, D., *J. Colloid Interface Sci.* **58**, 2 (1977)] is sometimes referenced in the literature: $\mu \equiv (1/\xi_0)(16R\gamma^2/9K^2)^{1/3}$.

the JKR regime. The same microsphere indenting a stiff rubber such that $K \sim 200\,\text{MPa}$ and $\gamma \sim 30\,\text{mJ/m}^2$ gives $\lambda \sim 40$, still well within the JKR regime, whereas a conventional silicon tip on the same rubber gives $\lambda \sim 4$, at the boundary of the JKR regime.

Thus, the reader should recognize that nanomechanical data acquired with conventionally sharp AFM tips (as used in imaging) at low load can be properly analyzed with the DMT or JKR contact mechanics models *only in certain cases*: rigid inorganic surfaces in the case of DMT, or rubbers or softer materials (that are at worst weakly viscoelastic), but not so soft that the contact extends beyond the spherical end cap of the tip (or even approaches the boundary of this end cap, strictly), in the case of JKR. Glassy polymers or other stiff organic systems such as crystalline small-molecule materials and thin films (e.g., organic semiconductors, ferroelectrics, self-assembled monolayers), with moduli $\sim$1–10 GPA, tend not to fall into either the DMT or the JKR regime when using conventional AFM tips.

Notwithstanding the preceding statements on rigor, it is nevertheless instructive to use the JKR model to back-calculate and compare the "look" of AFM force curves for glassy and rubbery polymers. Figure 6.3a compares theoretical JKR force–indentation plots for a 20-nm radius tip interacting with 2.2-GPa and 100-MPa polymers, the latter being somewhat lower in surface energy (25 compared to 33 mJ/m^2) as might be the case, say, for a less *conjugated* polymer (i.e., one dominated by single bonds rather than double, triple, or aromatic bonds), not uncommon for many rubbers compared to glassy polymers. Figure 6.3b reconstructs AFM force curves—cantilever deflection versus Z-scanner displacement—from the plots of Figure 6.3a assuming $k = 3\,\text{N/m}$ to get deflection from force and using

$$Z = D - s_Z = -h - s_Z$$

plus an arbitrary constant chose to zero the Z scale at the point of zero deflection in both force curves. This is, of course, the *inverse* of what the experimentalist must do in nanomechanical characterization (i.e., convert from Z to indentation and from deflection to force) as discussed in Section 3.2. Figure 6.3b depicts AFM force curves during retraction assuming a calibrated deflection scale (i.e., volts converted to nanometers using a rigid sample). One sees that on the glassy polymer, the force curve has nearly unity steepness (i.e., most of the compliance is in the cantilever, not the sample), whereas the rubbery polymer force curve is roughly *half* as steep. Second, the pull-off on the rubber occurs *after Z has passed the point of minimum deflection*, whereas on the glassy polymer, pull-off is *at* the point of minimum deflection as normally seen on rigid materials. Although the JKR model is not really appropriate in detail to describe the glassy case, per the above disclaimers, the difference in the basic "look" of the two force curves in Figure 6.3b is valid. Thus, in the process of collecting force curves on, say, a blend of two such polymers, the analyst should be able to identify these two distinct behaviors "on the fly"; detailed data processing is not required.

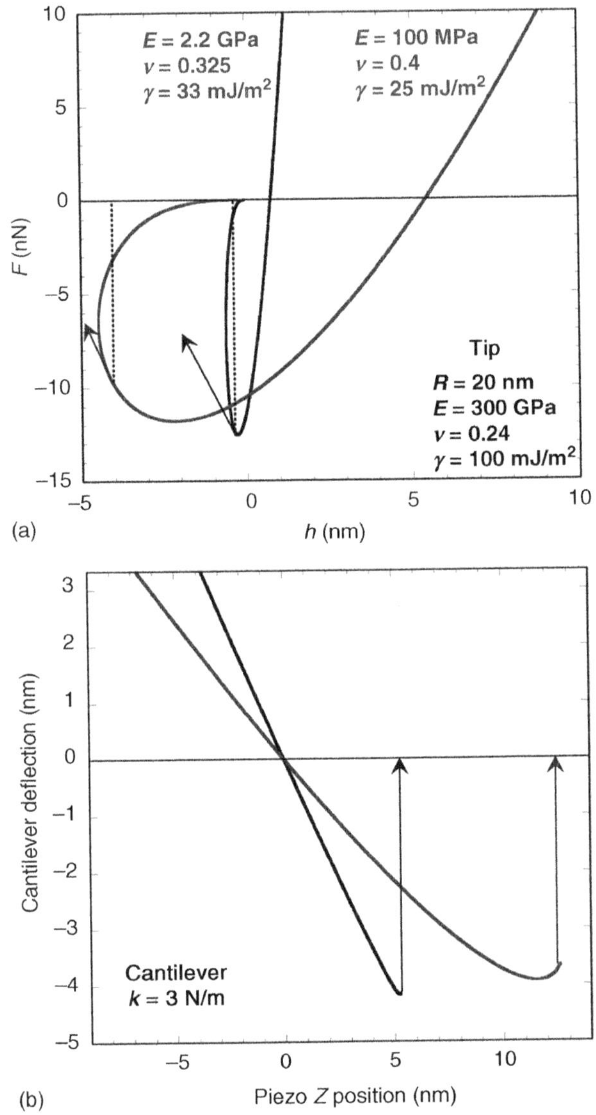

FIGURE 6.3 (a) JKR relationships between force F and indentation h for Si_3N_4 tip interacting with flat surfaces on bulk samples of glassy and rubbery polymer. Arrows denote the locations and directions of jump from contact for a 3-N/m cantilever. (b) Reconstructed force curves corresponding to the data in (a), zeroing Z at zero deflection.

6.2.4 The Maugis–Dugdale Model

Clearly, many systems fall into the intermediate Maugis–Dugdale adhesive contact mechanics regime. This model accounts for both long- and short-range attraction by approximating a Lennard–Jones (Section 2.1) type of interaction with a *square well*: a constant attractive force out to a fixed distance followed by no interaction at greater distances [9]. As a result, attractive forces exist both *within* the tip–sample contact and to some small distance beyond the contact periphery. A Lennard–Jones type of modeling of course would be more physically correct, but has to be done numerically [1]. Unfortunately, even the square-well approximation of the Maugis–Dugdale model does not produce a single, closed-form analytical expression relating the material constants K and γ, system radius of curvature R, and either the contact radius a or indentation depth h, as do the DMT and JKR models. Instead, two *parametric* equations must be solved, together, in order to relate any two of these quantities. There are a total of three parametric equations.

$$F = \frac{Ka^3}{R} - \lambda a^2 \left(\frac{\pi\gamma K^2}{R}\right)^{1/3} \left[\sqrt{m^2-1} + m^2 \arctan\sqrt{m^2-1}\right] \tag{6.9}$$

$$h = \frac{a^2}{R} - \frac{4\lambda a}{3}\left(\frac{\pi\gamma}{RK}\right)^{1/3}\sqrt{m^2-1} \tag{6.10}$$

$$\begin{aligned} 1 &= \frac{\lambda a^2}{2}\left(\frac{K}{\pi R^2\gamma}\right)^{2/3}\left[\sqrt{m^2-1} + (m^2-2)\arctan\sqrt{m^2-1}\right] \\ &\quad + \frac{4\lambda^2 a}{3}\left(\frac{K}{\pi R^2\gamma}\right)^{1/3}\left[1 - m + \sqrt{m^2-1}\arctan\sqrt{m^2-1}\right]. \end{aligned} \tag{6.11}$$

The parameter conventionally called m is dimensionless and unknown in value *a priori*, but it is quantitatively meaningful; it is the ratio of the outermost radius at which attractive forces are felt in the Maugis–Dugdale model (conventionally called c) to the contact radius a. Thus, m relates the lateral extent of the region of total interaction tip–sample interaction—meaning both contact forces (i.e., atom-to-atom electronic orbital interactions) and noncontact (i.e., van der Waals) forces—to that of purely contact forces. That is, the larger the value of the parameter m, the farther the total tip–sample interaction zone is modeled to extend beyond the direct contact zone.[5]

To solve any two of the parametric Equations 6.9–6.11, one must let m vary appropriately between limits that ultimately depend upon λ and *iteratively* solve the equations (i.e., numerically back and forth). This was done to produce the plots in Figure 6.4 (using a common spreadsheet program), overlaying the relationships from the above three models for (i) contact radius versus applied force and (ii) applied force versus indentation depth, for the example case of a rigid tip of 10-nm radius on a sample with a Young's modulus of 3 GPa and Poissons ratio of 0.4, and a tip–sample work of adhesion of 50 mJ/m^2.

[5] Note, however, that for the limiting case $m = 1$, meaning zero attraction beyond the contact circle, Equations 6.9 and 6.10 reduce to the Hertzian equations, not the JKR (and Eq. 6.11 becomes a meaningless $1 = 0$).

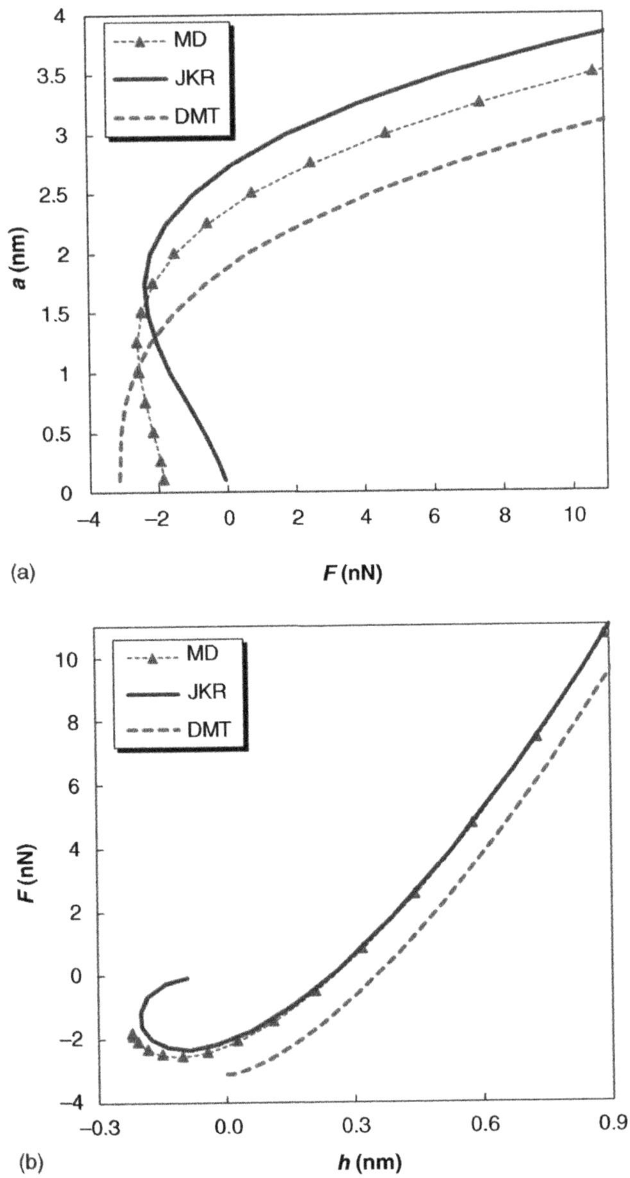

FIGURE 6.4 Overlay of Maugis–Dugdale, JKR, and DMT relationships between (a) contact radius a and force F, and (b) force F and indentation depth h, for a 10-nm radius spherical tip interacting with a 3-GPa modulus flat material and $50\,mJ/m^2$ work of adhesion between tip and sample.

To invoke the quantitative form of the Maugis–Dugdale model without the full complexity of Equations 6.9–6.11 and the need for parametric solution, investigators have utilized a nearly equivalent but simpler formulism developed by Carpick, Ogletree, and Salmeron (COS) relating to contact radius a, and by Piétrement and Troyon (PT) relating to indentation depth h [11,12]. Also, under somewhat different assumptions, other analytical expressions have been derived to produce a–F and F–h relationships similar to those of the MD model [13–15], and of course, finite element analysis has been used to analyze these relationships [16–18]. At this writing, both theoretical and simulation methods applied to adhesive contact mechanics continue to undergo significant development.

6.2.5 Other Formal Relationships Relevant to Adhesive Contact Mechanics

In applying the JKR, DMT, or MD (or COS/PT approximate) formalism to analyze data in the form of force curves (processed as per Section 3.2), or for that matter lateral contact stiffness or friction force versus load plots (subtopics of Chapter 7), the user most commonly will attempt to determine both K and γ based on the shape of the data trend, varying these parameters to fit theory to experiment. But there may be occasions when the user seeks reliable tabulated numbers for K or γ to reduce the number of free fitting parameters. One usually cannot "look up" a reliable number for the work of adhesion γ between two arbitrary materials—any reported numbers will likely depend on the nature of the measurement used. But there are related numbers, in some cases theoretically determined, that one may find tabulated: *surface energies* (or surface tensions) for each material, tip and sample. The theoretical relationship between the surface energies γ_1 and γ_2 of tip and sample and the thermodynamic work of adhesion γ needed to separate them [19] is expressed as

$$\gamma = \gamma_1 + \gamma_2 - \gamma_{12}, \tag{6.14}$$

where γ_{12} is the *interfacial energy*. Thus, in some cases, an appropriate work of adhesion value γ to insert into the desired equations (e.g., Eqs. 6.9–6.11) may be approximated, guided by what is known about surface and interfacial energies for the materials of interest or similar materials. In other cases, published works of adhesion on similar probe-sample systems may exist. In many cases, reasonable approximations are available for the surface energies γ_1 and γ_2; thus, finding a reasonable value of γ may critically depend on the availability of a number for interfacial energy γ_{12}. Most such tabulated numbers are for liquids, because the numbers are defined for interfaces at thermodynamic equilibrium, which is usually not the case for solids in contact [19]. For solids, one may find theoretically determined interfacial energies, though perhaps for specific crystal faces, whereas the tip or sample surface may not be crystallographically well defined. In cases where the interaction may be treated as due to primarily *dispersion* forces (induced dipoles, i.e., not permanent dipoles or hydrogen bonding; see Section 2.2.1), the interfacial energy can be calculated from

$$\gamma_{12} \approx \gamma_1 + \gamma_2 - 2\sqrt{\gamma_1^d \gamma_2^d}, \quad (6.15)$$

where the quantities under the square root are the dispersion force contributions to the surface energies [19]. This expression has been shown to predict the water–alkane interfacial energy to within 5% of experimental values—even though water has the strong propensity for hydrogen bonding—because it does not engage in hydrogen bonding with *saturated hydrocarbons* (unconjugated). Thus, even for a polar tip terminus such as SiO_2, interacting with an apolar sample, the approximation in Equation 6.15 may be appropriate. Then the remaining question is whether data exist to separate the dispersion force contribution from the overall surface energy [20].

Working in dry conditions may reduce the possible complication of a *capillary meniscus* (considered further in Section 6.3) and thus enable the meaningful determination or control of the work of adhesion. Immersing in a liquid medium is often the best-controlled experiment, however, and thus more likely to allow the use of well-defined values of interfacial energies. But one must recognize that under liquid immersion, there are *three* relevant interfaces: tip–sample, tip–liquid, and sample–liquid. A relationship between the work of adhesion for separating materials 1 and 2 in medium 3 is

$$\gamma = \gamma_{13} + \gamma_{23} - \gamma_{12}, \quad (6.16)$$

where γ_{ij} is the interfacial energy between media i and j [19]. For the case of vacuum (and air, to good approximation), γ_{13} and γ_{23} become γ_1 and γ_2, such that Equation 6.16 reduces to Equation 6.14. One may have better success finding or determining values for interfacial energies between various solids and the immersion liquid than the surface energy values in air. For example, water contact angle measurements have been performed on many materials, or can be performed by the analyst in parallel with AFM work. Even commercial *micro*contact angle measurement instruments are available.

6.2.6 Summary Comments and Caveats on Adhesive Contact Mechanics Models

Given that the various surface and interfacial energies γ_i and γ_{ij}, discussed in the previous section, are defined within the strictures of *equilibrium thermodynamics*, the corresponding thermodynamic work of adhesion γ refers to work performed in a *quasistatic, reversible* process. *This is not strictly what AFM measures.* Although we refer to force curves acquired in contact mode as *quasistatic* force curves—meaning at each point in the approach–retract cycle a force balance is achieved (unlike approach–retract curves in dynamic AFM)—truly quasistatic conditions are *not* maintained on an atomic or molecular scale. That is, the AFM user can do nothing to assure the *reversible* formation of interatomic or intermolecular interactions at the tip–sample interface. We have already discussed the presence of hysteresis in various forms in AFM. During the formation and rupture of tip–sample contact—as well as the intervening change of stress as loading force increases then decreases—atomic to molecular-level rearrangements generally occur, driven by the minimization of free energy. Moreover, the formation and breakage of a capillary meniscus may take place in air (Section 6.3). Charge exchange, and even the exchange of atoms or molecules, also may occur (see Figure 3.14 and related discussion). To break the interface (and meniscus if present) usually requires a greater stress—experimentally a greater pulling force—and thus net work is done, energy which has been transferred from the potential energy of the cantilever spring. The two surfaces are not in the same atomic-level configuration before and after the complete touch-push-pull-release cycle. *This is fundamentally what adhesion hysteresis means.* In most cases, the work of adhesion γ employed to utilize equations in this section will be a fitting parameter, that is, a quantity the analyst is seeking to determine via force-curve characterization. Thus, to account for adhesion hysteresis, *two* different values of work of adhesion will need to be determined, specific to approach and retract data sets. The difference is then a measure of adhesion hysteresis. *This is empirically what adhesion hysteresis means* in the context of force-curve analysis.

Note that the addition of attractive forces (i.e., γ) in Equation 6.6, for example, acts to increase the contact radius a and thus degrade resolution. So adhesion is *undesirable if high-resolution imaging is sought*, whereas large variations of adhesion across a surface can be *desirable to achieve materials contrast*. There is no way to optimize the system to suit both goals. Indeed, if the principal goal is quantitative materials or interfacial characterization, not imaging resolution, a common approach is to attach a microsphere (colloid probe) to a tipless cantilever, to greatly improve signal/noise and moreover provide a better defined probe radius of curvature, at least for soft materials where a smooth, conformal contact results (i.e., nanoasperities and their unknown radii of curvature do not play an important role[6]). On the flip side, strategies to reduce γ may be employed to improve resolution. This means modifying the terminal chemistry of the tip without substantially increasing its radius of curvature, for example, by chemisorbing short-chain silanes with apolar terminal groups to yield a low-surface energy [21]. Some tips of this sort are commercially available. More relevant to the DMT model, tip modification can reduce the propensity for capillary condensation and thus minimize a meniscus force beyond the contact zone; fully immersing the tip–sample system in liquid of course removes the capillary force entirely, as previously stated. Further choice of liquid medium can minimize the Hamaker constant (Section 2.2.1) and thus the

[6] This may not be true for friction.

tip–surface attraction at the interface, enabling the smallest contact zone and thereby the highest resolution.

It is worth summarizing five common complications to detailed contact mechanics modeling of AFM data as follows.

1. The tip radius of curvature is only known approximately. Even if samples with step edges or nanospikes are used to inversely image the tip geometry, tips of radius on the order of 10 nm or smaller (for maximal imaging resolution) are often as "sharp" as the object used to characterize tip shape, meaning the extracted tip radius is a sum of actual tip and surface radii. Transmission electron microscopy can be used to characterize AFM tip shape, in principle (given a tilt stage to achieve to very different view angles), but this is usually not reasonable for routine work, and impossible for many AFM users.
2. Each surface, on tip and sample, can contain nanometer or even subnanometer-scale roughness such that at the finest detail the contact is between *multi-asperity* surfaces. The fact that the material under investigation is usually soft may result in a *conformal* sample deformation such that the approximation of a spherical end-cap tip and flat sample surface can be invoked. But if the analyst is examining the tensile regime up to the break of contact, whether in force–distance or friction–load modeling, the picture of a conformal interface may be invalid even on soft materials.
3. Because an AFM's high cantilever compliance usually restricts quantitative force-curve analysis to relatively *soft materials* (organic, polymeric, biological), in many, if not most cases, the response of these materials will be *viscoelastic* (explicitly time dependent). Thus, the material "constants" E and γ are not constants at all; rather, they depend on the time frame of the measurement, or in oscillatory methods, on the frequency of oscillation. Although some contact mechanics models have been generalized to include *linear* viscoelasticity (single relaxation time), and analytically applied with some degree of success, most viscoelastic materials respond via a *spectrum of relaxation times*. One usually wishes to characterize and understand this spectral response, whether the motivation is fundamental or applied.
4. The tilt of the AFM cantilever results in net lateral stresses and possibly sliding during what are nominally vertical force–displacement measurements (Figure 3.7) [22].
5. Most AFM characterization takes place in air, where capillary condensation and/or adventitious contaminants produce a *nanomeniscus* at the tip–sample interface, and thereby an additional source of (usually) attractive force as detailed in the next section. Recent years, at this writing, have witnessed intense activity in analytical modeling of capillary forces, but most analyses of AFM data either ignore capillary effects or effectively incorporate into the attractive forces considered by models like the DMT.

6.3 CAPILLARITY, DETAILS OF MENISCUS FORCE

Capillarity, and, generally, wetting and spreading are broad topics in interfacial science and engineering [23–29] that date back centuries. For example, the performance of liquid lubricants, which often wet surface asperities via capillarity and thereby reduce friction and wear, has been critical to countless technologies through the ages, even at human interfaces as in the case of personal-care products (e.g., conditioner wetting hair cuticles) [29]. Capillarity is not always *engineered* into an interfacial system, of course—under close confinement between solid surfaces, water vapor will spontaneously undergo *capillary condensation* into liquid water at the interface, even though the partial pressure of the water vapor is below its gas–liquid transition value in the absence of confinement. Thus, *water menisci* due to capillary condensation at and near the contacting points of solid surfaces are ubiquitous, and over a wide range of spatial scales: macro, micro, and nano. Nanoscale capillarity and interfacial water continue to be very actively researched, as recently reviewed [30–36].

The importance of interfaces in micromechanical systems and nanotechnology—particularly with regard to adhesion, friction, and stiction—as well as surface-analytical techniques that probe these phenomena on micro-to-nanoscales—such as AFM and the surface forces apparatus [19]—has renewed interest in understanding and manipulating capillarity at small length scales. A number of publications in these areas over the past three decades at this writing reflects its elevated importance. It also attests to the elegance of capillarity as a phenomenon that can be exquisitely measured and theoretically modeled [30–36]. But this literature has demonstrated disparate results in the humidity dependence of adhesion at chemically similar interfaces, hydrophilic probes on mica [37–45] or silicon oxide [43–53], and, as a result, some differences in mechanistic interpretations. Theoretical modeling has identified expected differences between macro-, micro- and nanoscale contacts including shape effects. Computer simulations, in some cases down to the level of molecular dynamics, also have identified sources of variance (e.g., tip shape/roughness) and have enhanced understanding (e.g., of solid-like behavior due to structural forces, Section 2.5.2) [54–62]. Given the ubiquity and even dominant role that capillary menisci can have in AFM force measurements and phenomenology, together with a voluminous yet growing literature, we devote a significant section to capillarity.

Even if not of explicit interest to the AFM user, a capillary water meniscus or so-called *pendular ring* is often present at the probe–sample interface in both standard and colloid-probe AFM in air (i.e., nano- to microscale sharpness). Thus, the AFM user needs to understand how capillarity affects the probe. This includes the strength of the resulting tip–sample interaction; the spatial extent of this force, both vertical and lateral (i.e., effects on the distance dependence of force as well as its lateral scale and thereby spatial resolution); tip stability; molecular transfer processes (between tip and sample), as are utilized in dip-pen nanolithography [63–65]; and property-sensitive imaging modes, most notably the *kinetics* of adhesion as well as friction [30,32,50,59,65–73].

6.3.1 Framing the Issues

Before delving into the mathematics of capillary modeling—given its elegance and thus sometimes over-persuasiveness—it is best to inventory the various issues at play, as a preliminary "reality check." One can categorize these issues into (i) and (ii) *forces* that directly or indirectly result from the meniscus (including timescale issues), and (iii) related *geometric* issues.

First, the meniscus usually will act to *directly* pull tip and sample together, producing stronger adhesion than in the absence of a meniscus (i.e., compared to purely solid–solid attractive van der Waals forces). This is the result of both *surface tension* and *Laplace pressure*. Surface tension acts at the *three-phase contact line*—which in an axisymmetric case is a circle—at the wetting perimeter of the meniscus, each air/water/solid interface denoted with long-dashed lines in Figure 6.5. Horizontal components of this tension cancel out around the ring, while vertical components sum to produce attraction. Laplace pressure derives from the *curvature* of the meniscus (both parallel and perpendicular to sample plane, as detailed in the next section), which yields a reduced pressure inside of a water meniscus, drawing tip and sample together. This phenomenon is also the source of the meniscus in the first place; a small annular crack around the periphery of the tip attracts water condensation from the ambient water vapor until the formed liquid meniscus achieves *equilibrium* with the water vapor in the adjacent gaseous environment, as determined by the final curvature of the air–water interface.

An appropriate sidebar is the question of *timescale of capillary condensation.* Some experimental and simulation studies of capillary condensation at ~10-nm radius asperities suggest equilibration times of ~1 ms (and ~1 s for micrometer-scale asperities as used in colloid probe AFM, i.e., scaling inversely with radius of

FIGURE 6.5 Illustration of an AFM tip in contact with a flat surface including a capillary meniscus with contact lines denoted.

curvature) [50,68,72]. Considering the wide range of interaction timescales in AFM—say ~0.01–10 s during conventional (slow) force curves, ~0.01–1 ms during fast force-curve mapping (Section 6.5), and ~0.01–1 μs in dynamic modes, multimodal applications of AFM cannot be tucked neatly into a particular regime of interaction time. Moreover, it is not just a question of *total* contact time; there is also the question of *equilibration at each measurement point* during pull-off while force is changing, as required for a *quasistatic* treatment to apply. Fast force curves with a 50-μs total contact time may only be truly quasistatic if capillary equilibration occurs within each 1-μs time step during pull-off. On the other hand, one expects only slight changes in meniscus shape from one such time step to the next. At this writing, we note that there seems to be little consensus on the existence or lack of nonequilibrium states during pull-off for typical AFM geometries.

Category (ii) of issues involves *surface forces* that may result from the "immersion" of a small end portion of the tip in a meniscus of water (Chapter 2). Thus, the aggregate force on the tip can be modified by the meniscus in an *indirect* sense. As discussed in Section 2.2.1, usually an intervening liquid medium strongly *reduces the Hamaker constant* characterizing the strength of van der Waals forces. Although weaker, the van der Waals force usually remains attractive, as discussed in Section 2.2.1. An aqueous medium also can *ionize* the solid surfaces; in many cases, the charge on tip and sample will both be negative, yielding repulsion (DLVO forces, Section 2.5.1). On polymeric samples, the water may cause some degree of swelling and thus the production of *steric repulsion* between tip and sample (Section 2.5.3); the liquid molecules themselves may become ordered at the interface with one or both solids (sample and tip) and thereby develop a finite elastic modulus (Section 2.5.2), again generating repulsion. Thus, one notes that repulsion is the principal result of these various surface forces. Another term that is occasionally used in the literature in reference to these often repulsive forces (per unit area) is *disjoining pressure* (i.e., opposing of the "joining" of tip and sample). Historically, the term has also encompassed attractive surface forces (though "conjoining" would be more appropriate in this case). As a final possibility, the water may be thermodynamically driven to *wet* one or both surfaces; thus, the forced *solvent exclusion*, in producing a solid–solid contact, may be another phenomenon acting to disjoin tip and sample (provided that the meniscus is a large enough "reservoir" to drive the system to equilibrium in the thermodynamic sense).

For category (iii) of issues, although system geometry is always important to interpreting AFM data, one should recognize that this is glaringly true for capillary phenomena. Uncontrolled humidity can lead to substantial variations in the *scale* of wetting (at probe and sample) and thus variable capillary forces. The scale of wetting may mean that *nanometer-scale asperities* on the tip are important to the measurement, perhaps just a single nanoasperity if very dry (although the meniscus may be unstable) or several asperities (i.e., nanoroughness) at moderate humidities [74]; yet the entire rounded end of the tip may be wetted in a high-humidity experiment [75]. Even the *shank* of the tip may be relevant at high humidities if the meniscus contact line reaches up this far. (Thus, the magnitude of the capillary force could be affected by the coarser tip geometry, thereby impacting mechanical measurements via this loading force, even though the shank may not contact the sample as in Figure 2.5a.) Indeed, both probe and sample surface roughness will likely span a *hierarchy of length scales* from atomic to at least nanometers and perhaps tens of nanometers; thus, capillary wetting may affect a measurement differently from one experiment to another as different tips are used and different samples investigated, and under variable humidity, dependent on the *spectrum* of roughness (Section 8.3.2). This variation is conceivably even worse for colloid probe AFM compared to regular, sharp-tip AFM. These geometry issues also impinge on the kinetics question stated above: the time for equilibration of the meniscus could depend on the size of the tip portion being wetted.

In summary, we see that a larger number of (potentially opposing) physical effects are important to the tip–sample interface in the presence of a capillary meniscus compared to completely dry or immersed systems. The analyst will benefit by preconsidering which of these effects and issues may be important to the system under study, and what their relative influence might be. Obviously, the humidity range is important. Dry conditions (e.g., RH ~10%) may not fully remove capillary effects, but may result in such a small meniscus that disjoining effects are reduced (e.g., polymer swelling, surface charging); of course, the user is behooved to assess tip shape and stability throughout the measurements. Tracking nanocapillarity studies in the literature is also recommended, given that the field is still young and understandings are rapidly evolving at this writing, particularly with recent effort in molecular dynamics simulations.

6.3.2 Basic Elements of Modeling the Meniscus

Capillarity provides several well-known problems in mathematical physics. A classic example is a liquid bridge spanning a gap between two solid spheres [23]. A special case is letting one sphere radius go to infinity, producing the sphere–flat geometry common to AFM. Assuming adequate mobility of water (over the timescale of observation), a meniscus at the interface of the sphere and flat reaches an equilibrium cross-sectional shape as depicted in Figure 6.6 for the cases of solid–solid contact (a) and slightly out of contact (b) [24]. The exact shape of a capillary meniscus between a sphere and a flat surface results from energy minimization, and thus has been derived using *variational calculus* [25]. The name for this shape, for a small meniscus, where the effects of gravity are negligible,[7] is a *nodoid*; simpler treatments assume the somewhat similar geometry of a *toroid* [24,76]. The predicted force–distance relationships, as the solids separate and the capillary meniscus becomes a water bridge, have been derived either directly from force (and

[7] The dimensionless quantity that determines whether gravity plays a negligible to dominant role is called the bond number *Bo* (or Eötvös number, *Eo*). It is calculated as $Bo = \Delta\rho_m g l^2/\gamma_{LV}$, where $\Delta\rho_m$ is the density difference between fluids (here water and air), g the gravitational acceleration, l the characteristic length scale (here the diameter of the meniscus), and γ_{LV} the surface tension of the liquid–vapor (water–air) interface. A bond number less than one indicates that surface tension dominates. For water, the quantity $(\gamma_{LV}/\Delta\rho_m g)^{1/2} = 2.7$ mm; this is also called the capillary length. Thus, for a micro-to-nanoscale value of l, $Bo = (l/2.7\ \text{mm})^2 << 1$.

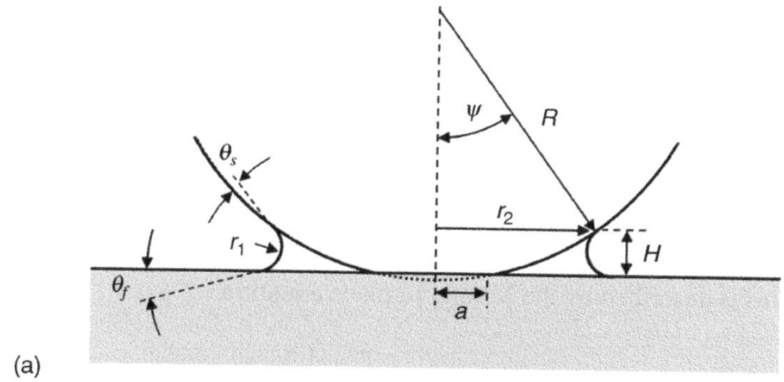

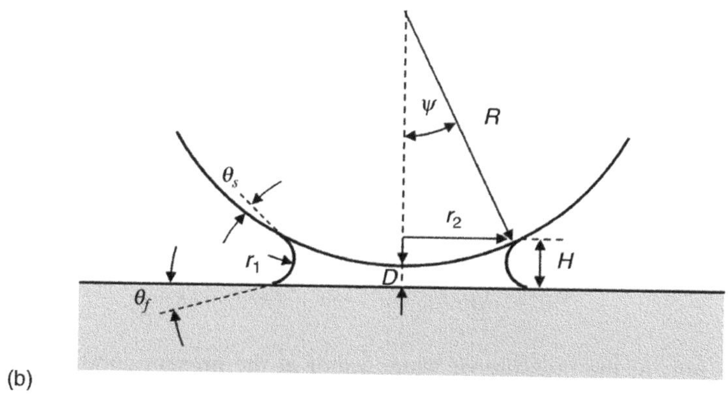

FIGURE 6.6 Cross-sectional shape of meniscus at the interface of sphere and flat in the case of (a) solid–solid contact, where the meniscus comprises a collar, and (b) out of contact, where a capillary neck bridges between solids. See Appendix 4 for finer geometric detail.

pressure) considerations or from an interfacial energy expression that is differentiated with respect to distance to obtain force.[8]

Here we will use a force/pressure description and toroidal geometry. In continuum treatments, this consists of

i. surface tension characterized by two contact angles between a water meniscus and each of two solid surfaces depicted in Figure 6.6, sphere (θ_s) and flat (θ_f), which derive from the respective *surface energies* of sphere and flat and that of water;
ii. Laplace pressure characterized by two radii of curvature of the liquid–vapor interface labeled in Figure 6.6, the azimuthal radius of the water contact circle at the sphere (r_2), and the so-called *meridional* curvature (r_1) viewed by

[8] Note that these do not account for entropy changes (i.e., all components of the Gibbs free energy), which are expected to be relatively small.

cutting a plane through the meniscus perpendicular to the sample surface (i.e., in the plane of the diagram).

In equilibrium, the *total* meniscus curvature (defined below) is a constant reflecting a pressure difference Δp across the liquid–fluid interface, where the liquid meniscus here is water and the surrounding "fluid" is air containing a finite partial pressure of water vapor. (The picture generally applies to the meniscus of any liquid in its condensable vapor, or of a liquid A meniscus phase segregated from a surrounding liquid B.) This is described by the Young–Laplace equation,

$$\Delta p = \gamma_{LV}\left(\frac{1}{r_1} + \frac{1}{r_2}\right), \tag{6.19}$$

where γ_{LV} is the liquid–vapor interfacial energy (essentially equal to the surface energy or "tension" of water, 72 mJ/m^2). The *Kelvin equation* describes a *thermodynamic equilibrium* balancing the vapor pressure on the air side of the air–water interface with the Laplace pressure due to interfacial curvature,

$$\frac{R_g T}{V_m} \ln\left(\frac{p}{p_s}\right) = \gamma_{LV}\left(\frac{1}{r_1} + \frac{1}{r_2}\right) \equiv \frac{\gamma_{LV}}{r_K}. \tag{6.20}$$

The ratio p/p_S of the pressure of water vapor at the present conditions to that at saturation (i.e., partial pressure) is the *RH* expressed as a fraction; V_m is the molar volume of water, R_g is the universal gas constant, and T is the absolute temperature. The sum of the inverse principal radii of curvature of the meniscus, r_1 (negative for concave, in turn the result of hydrophilic surfaces) and r_2 (positive for a circle), defines the inverse of the Kelvin radius r_K, which describes the *total* curvature of the meniscus. Because in most cases r_1 is less in magnitude than r_2, r_K is usually negative, corresponding to a *lower pressure inside of the meniscus* relative to the outside. This pressure differential acts over the horizontal cross section of the meniscus to produce an attractive force.

It should be noted that a circular meridional curvature of radius r_1 (i.e., a constant), at the air–water interface in the vertical cross sections of Figure 6.6, implies a *nonconstant value of* r_K given *variable* horizontal cross section (r_2) from the top to the bottom of the meniscus. Thus, this *circular approximation* [23] is not strictly correct from the standpoint of equilibrium, because in equilibrium there can be only one (hydrostatic) pressure value inside of the meniscus, not one changing from top to bottom of meniscus. This error is, however, typically small [76]. To be more rigorous, one can model using another axisymmetric 3D mathematical shape, the nodoid as mentioned earlier, which provides constant r_K everywhere across the air–water interface. Use of the nodoid, however, introduces greater complexity into the modeling equations, and thus is not utilized in most published work, whereas the circular approximation (toroid) allows for the derivation of simpler expressions.

Before casually applying the above formalism, it is important for the AFM user to understand the quantitative implications of the Kelvin equation. For water at 20°C, one has $\gamma_{LV}V_m/R_gT=0.54$ nm. Then for RH of 10%, 60%, and 90%, one obtains Kelvin radii $r_K\approx-0.23$, -1.0, and -5.1 nm, respectively. Thus, the predicted Kelvin radius at RH = 10% is on the order of the size of *one water molecule*, such that this continuum picture clearly breaks down. The RH = 60% case would correspond approximately to the depicted meniscus in Figure 6.6 for a sphere of radius $R=10$ nm. (Studies with the surface forces apparatus have suggested that for some liquids, this 1-nm scale curvature remains reasonably well described by the continuum picture, although for water this picture may only be appropriate in the several-nanometer regime [41,60].) For this same sphere at RH = 90%, the predicted meniscus would extend slightly beyond the modeled geometry in Figure 6.6, meaning it could reach up to a more conical shank of the tip. This change of effective geometry has implications for the distance and humidity dependences of the meniscus force, discussed below.

If the sphere is lifted slightly above the flat surface (Figure 6.6b), yet the water contact angles with sphere and plate are unchanged (per unchanged surface energies), then the radii of curvature will not remain the same; from geometry one expects r_2 to become smaller and r_1 bigger (compared to contact, Figure 6.6a). Exactly how these radii change further depends on whether *constant volume* of droplet or *constant pressure* in droplet is maintained. Constant volume results in a distance dependence of force that is different from the constant pressure case; the latter means that thermodynamic equilibrium is maintained such that Equation 6.20 is operative (balancing vapor pressure and Laplace pressure). Whether constant volume or constant pressure is a proper assumption for a given probe–sample system relates to the timescale of capillary condensation (and possibly evaporation if Laplace pressure is inconsistent with the Kelvin equation) as discussed in the previous section. Obviously, a choice of model could depend not only on the rate of retraction but also on the volume of the meniscus and thus the RH and surface energies of sphere and flat. Measurements with macro- to microscale surface curvatures suggest that constant volume is the operative condition over a typical timescale of retraction [73,77]. Measurements with sharper AFM tips (40–400 nm) have suggested that even in this case constant volume is the more accurate description during meniscus extension [72]. (A related question is the motion of the contact line. It is well known from contact angle research that the motion of the wetting perimeter during advancing versus receding contact angle measurements can be hysteretic and time dependent [19], and that pinning can occur at surface heterogeneities [78,79].)

6.3.3 Mathematics of Meniscus Geometry and Force

In this subsection, we derive expressions relating force to distance, RH, and other parameters, for a sphere–flat geometry and invoking the circular approximation. We first treat the total capillary force on the sphere, including both Laplace pressure and surface tension terms. In some treatments, the latter has been considered negligible, but for AFM this is not generally true [80], ultimately because the so-called *filling angle* ψ *is not generally small* [32]. The Laplace pressure applies across the horizontal circular cross section of the meniscus and the surface tension applies around the circular perimeter, thus expressed with two terms [32,80],

$$F_{\text{cap}} = \pi(R\sin\psi)^2 \cdot \frac{\gamma_{LV}}{r_K} - 2\pi(R\sin\psi)\cdot\gamma_{LV}\sin(\theta_S+\psi), \tag{6.21}$$

where r_K is negative such that both contributions are negative, to remain consistent with our attractive force sign convention; the rightmost sine function ensures that only the vertical component of the surface tension force is summed around the contact circle (Appendix 4). The filling angle ψ is geometrically related to r_K (Appendix 4) according to

$$\frac{1}{r_K} = \frac{1}{r_1} + \frac{1}{r_2} = -\frac{\cos(\theta_S+\psi)+\cos\theta_f}{D+R(1-\cos\psi)} + \frac{1}{R\sin\psi}. \tag{6.22}$$

Equation 6.22 can be used to eliminate $1/r_K$ from Equation 6.21. Given known liquid surface tension γ_{LV}, and related estimates of the two contact angles θ_s and flat θ_f (assuming nominal sphere and flat surface chemistries), plus an estimate of sphere radius R, we are left with a somewhat complicated expression $F_{\text{cap}}(\psi, D)$. Thus, for a particular sphere–flat distance D (including contact, $D=0$), one need only know ψ to calculate force. What determines the filling angle ψ? In thermodynamic equilibrium, Equation 6.20 is the key because it determines r_K for a particular RH, and thus ψ from Equation 6.22 by numerical solution (simple using an everyday spreadsheet program). Thus, one can plot the humidity dependence of (a) capillary force and (b) size of meniscus: diameter $2R\sin\psi$ and height $D+R(1-\cos\psi)$. Such plots are shown in Figure 6.7 for a sphere of radius $R=10$ nm in undeformed contact ($D=0$) with water meniscus contact angles of 10° at sphere and flat. Note the meniscus diameter ranges from 6 nm to 16 nm for RH = 10–90%, meaning that the filling angle is never very small, indeed 16–56°. One must also note a range of predicted meniscus heights of approximately 0.5–5 nm. Clearly, the use of continuum concepts to describe a meniscus at the low end of this range—a height of about two water molecules—is not warranted. One anticipates some degree of "freezing out" of such a small meniscus due to structural forces under confining conditions.

Given force-curve measurements, it is possible to measure the distance dependence of the meniscus force provided that the cantilever has spring constant large enough to allow navigation through the pull-out from the meniscus (i.e., avoid jump from contact). An open question is the type of constraint determining meniscus shape during such a tensile measurement; as previously hinted, there are two limiting models that have received the most attention.

1. Constant pressure due to thermodynamic equilibrium: assumption of constant meniscus curvature during pull-out, as equilibrated with a fixed vapor pressure far from the meniscus per the Kelvin equation (Eq. 6.20).
2. Constant meniscus volume.

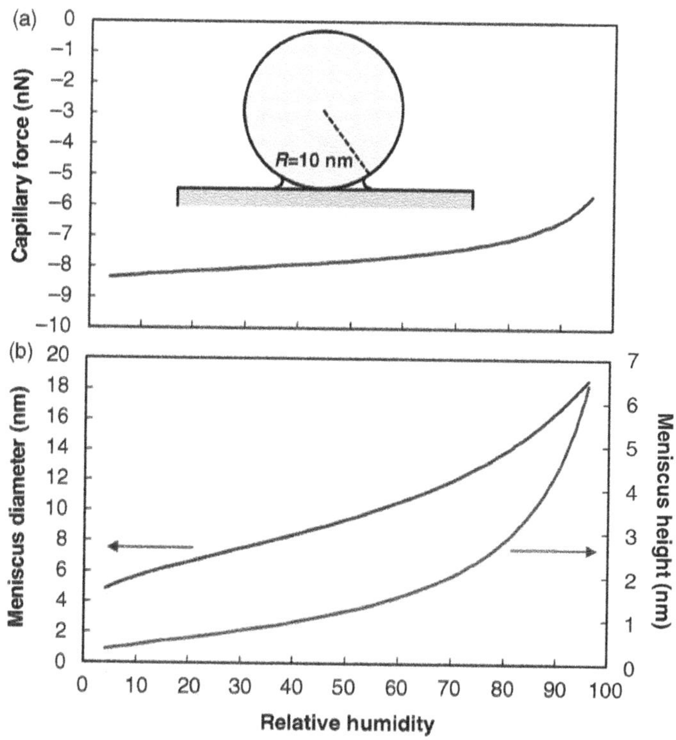

FIGURE 6.7 Humidity dependence of (a) capillary force and (b) meniscus size assuming thermodynamic equilibrium (meniscus curvature determined by partial pressure of condensable vapor) in the sphere–flat geometry. A water meniscus with contact angles of 10° at sphere and flat is assumed.

For Case 2, an approximate meniscus volume can be computed as the *difference* of two volumes (Figure A4.1): that of the full cylinder determined from the meniscus radius and height at the contact circle with the sphere [30],

$$\pi(R\sin\psi)^2[D + R(1-\cos\psi)], \tag{6.23}$$

and that of the cap of the solid sphere that is immersed in the meniscus (i.e., excluding liquid),

$$\pi\frac{[R(1-\cos\psi)]}{6}\left\{3(R\sin\psi)^2 + [R(1-\cos\psi)]^2\right\}. \tag{6.24}$$

Figure 6.8 compares the distance dependence of capillary force in Case 2, constant meniscus volume computed by assuming initial equilibrium at RH = 80% (i.e., retracting so fast that this equilibration no longer holds), to that of Case 1, a

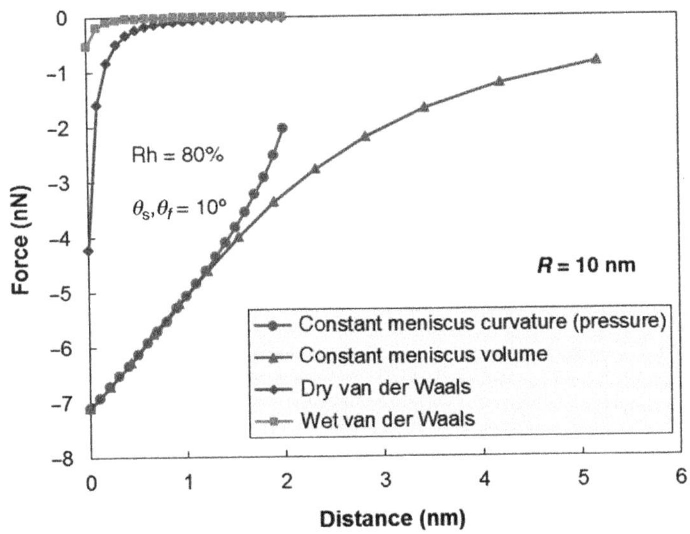

FIGURE 6.8 Distance dependences of water capillary force, assuming constant meniscus curvature or constant meniscus volume, and van der Waals force assuming dry or fully immersed in water with bulk dielectric properties.

constant meniscus curvature (pressure), maintaining thermodynamic equilibrium at RH = 80%; in both cases, 10° wetting angles were maintained and a 10-nm sphere radius was used. The two model dependences become inconsistent beyond a retraction distance at which the capillary force has dropped by roughly a third from its value in contact (−7 nN). In Case 2, constant volume, the force remains significant up to *larger distances* and exhibits the *opposite concavity* to that of Case 1. Both curves are plotted up to the point of an instability suggested by the mathematics; in Case 1, a solution no longer exists for RH = 80% (beyond 2-nm distance) and in Case 2, the predicted overall meniscus curvature changes from negative to positive, which implies collapse (beyond about 5-nm distance).

It is worth reiterating that in the scientific literature one can find derivations of simpler analytical expressions under constraining assumptions; one is small ψ, another is $r_2 \gg r_1$. These may apply to smooth microspheres but not most AFM tips. Given the possible complication of roughness on microspheres, we choose not to present such analytical approximations in this book.

As discussed in Section 6.3.1, we also reiterate that the above-modeled capillary forces are not the only ones present. At minimum, *attractive van der Waals forces* are also present (i.e., Eq. 2.5). The appropriate Hamaker constant is not known, that is, we do not know if we can treat the interacting, lowest portion of the sphere as being effectively immersed in water *with its bulk dielectric properties.* In Figure 6.8, we also plot the distance dependence of both dry and "wet" van der Waals forces for the same 10-nm radius sphere-on-flat geometry (assuming a contact distance of $D_0 \sim 0.16$ nm as discussed in Section 2.2.1). The former invokes a Hamaker constant of silica–silica interaction, $A_H = 6.5 \times 10^{-20}$ J, and the latter a completely water-immersed Hamaker constant, $A_H = 0.83 \times 10^{-20}$ J.

Concerning the shape of the distance dependence of total force, van der Waals plus capillary: if we consider the sum of either van der Waals force trend and either capillary force trend from Figure 6.8, we note that in the constant volume case, the shape of the total force trend would be concave down at all distances, whereas in the constant curvature case, the shape of this trend would be concave down at the smallest distances and concave up at the largest distances (nearing 2 nm in Figure 6.8). This qualitative change of curvature with distance might conceivably provide a check of whether constant meniscus volume or constant meniscus curvature conditions apply. But we must also reiterate that *other surfaces forces* may be present, which may act in a disjoining sense (i.e., reducing the attractive strength of the force), for example, due to quasi-solidification of water near the two solid interfaces (structural forces, Section 2.5.2), manifest at small distances of likely <1 nm, and/or ionization of one or both solid surfaces (DLVO forces, Section 2.5.1), a longer-range effect that presumably would be present at all distances in Figure 6.8. Thus, the shape of the overall force–distance trend could be further modified by these various surface forces. (The first example in the following subsection is perhaps a case in point.) On more mobile molecular solids (e.g., polymers) there might be longer-range steric forces and possibly attractive bridging forces due to individual chain molecules adsorbing to tip and thus spanning between tip to sample (Section 2.5.5), and moreover, these stretching chains may reside within the meniscus or outside of it. (One foresees the potential complexity of in-air force curves!)

6.3.4 Experimental Examples of Capillarity

It is worth examining one of the most commonly studied cases of humidity-dependent adhesion on a hydrophilic surface: atomically smooth mica. It often has been observed that adhesion between a hydrophilic tip (e.g., silicon oxide or nitride) and mica exhibits a nonmonotonic humidity dependence, suggesting that the capillary force modeling in Case 1 above, leading to the monotonic force–humidity trend in Figure 6.7b, is missing information or invoking questionable assumptions. Models applying nonspherical geometries such as flat-ended cylinder, cone, and truncated cone have produced qualitatively different humidity dependences [30,44]. Modeled nanoscale roughness also has strongly affected the humidity dependence of capillary force [30,74,75].

We first examine individual force curves. Figure 6.9a compares force–distance data (i.e., following calibrations of deflection and spring constant, and conversion from Z scanner to distance abscissa) acquired on a freshly cleaved mica surface over complete approach–retract cycles at two relatively high humidities, RH = 80% and 94%, using a silicon tip with native oxide (cleaned with multiple solvents) of nominal radius 10 nm, attached to a rectangular silicon cantilever of calibrated spring constant $k = 4.0$ N/m (Section 3.7). Jumps to and from interaction can be variably present when using such a cantilever, and are included in both force-curve

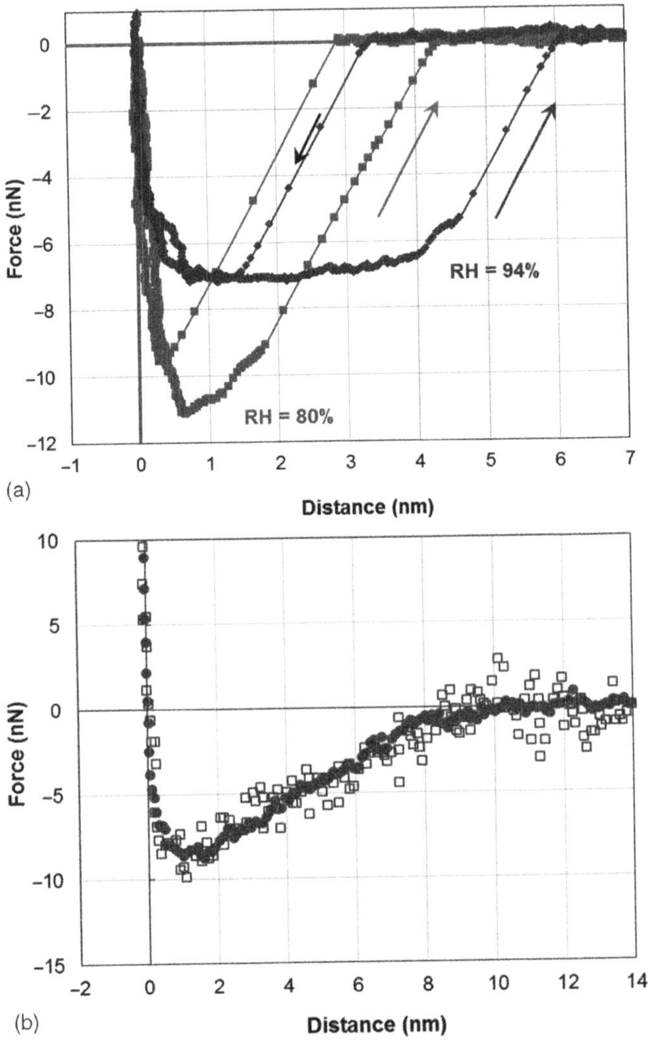

FIGURE 6.9 Experimental distance dependences of force between SiO_2 tip and mica, in the cases of (a) a 4-N/m cantilever at two high humidities and (b) a ~42-N/m cantilever at ambient humidity (~35%). Case (b) further compares a single, noisy force curve to an average of 20 such force curves.

cycles in Figure 6.9a; these are present as characteristic-slope (due to conversion from Z to distance scale, Section 3.2) portions that seem to be "missing" data points. The strength and significant range of adhesion in both cases suggests a predominantly capillary character (see Figure 6.8). There are two noticeable differences in the force curves: a larger magnitude of adhesion force at *lower* humidity (−11 nN compared to −7 nN), and a longer-*range* adhesion force at higher humidity (roughly 5 nm compared to 2 nm at the jump out), indicating a longer capillary bridge. The latter indicates greater water presence in the interfacial system, likely a preexisting layer that may also produce disjoining effects (structural forces) [81] to lessen the magnitude of adhesive force. Detailed modeling should consider the possibility of "complete wetting" (not treated in Section 6.3.3). The essentially constant force from 0.5-nm to 4-nm distance at RH = 94% may relate to a deviation from spherical-cap tip geometry, such as the wetting of a more conical tip shank. Figure 6.7b suggests that the water meniscus could reach beyond a spherical end portion of the tip. (The presence of a preexisting wetting layer would only act to further increase the size of this meniscus.)

The larger magnitude of adhesion force at *lower* humidity in Figure 6.9a is not necessarily surprising given the trend in Figure 6.7a.The *magnitude* of the difference between RH = 80% and 94% is perhaps not expected from a purely capillary picture; but there may be additional disjoining effects due to an abundance of water at 94%, such as ionization. There is also the possible presence of other disjoining effects such as structural forces (Section 2.5.2). We note a departure from smooth force–distance trends, both in the steep negative-slope portion below 1-nm distance (prior to break of "rigid contact") and in the positive-slope portion beyond the force minimum in each curve. Repeatedly acquired force curves (>100) exhibited similar structures. In particular, the force trend in rigid contact (0–0.5-nm distance) contains nearly vertical (high stiffness) segments separated by 0.2–0.3 nm, of the order of a single water molecule, suggestive of "layered" structural forces (Section 2.5.2). At RH = 94% and distances greater than 2 nm, one also discerns structures that appears to be stick-slip in nature, possibly suggesting a degree of solid-like character in the water meniscus, as has been reported [82].

Before moving on to the broader humidity dependence of capillary force, it is worth noting in passing that one can use an order-of-magnitude stiffer cantilever to avoid a jump from contact and "missing" data, the case in Figure 6.9a, but at the expense of signal/noise ratio and thus the ability to discern weak force changes as noted in Figure 6.9a (which may stochastically vary in location). An example of data acquired with a nominal $k = 40$ N/m silicon cantilever (with a silicon–native oxide tip) is in Figure 6.9b on a polycrystalline aluminum film (with native oxide) at ambient humidity of ~35%. One raw force–distance curve is overlaid on the average of 20 such curves acquired at random locations within a 500 × 500 nm region on the polycrystalline film. The signal/noise improvement from averaging is dramatic and provides a better assessment of the shape of the distance dependence of force, most likely capillary in nature in this case. Fine structure such as that in Figure 6.9a is, however, not discernible following this averaging.

We now examine the broader humidity dependence of attractive force. Each datum in Figure 6.10a derives from a complete force curve of a silicon (oxide) tip on a freshly cleaved mica surface. The ordinate value is the magnitude of the most negative force sensed just prior to pull-off (usually called the "adhesive force"). All force curves were taken at conventional "slow" speeds, a total cycling time of 2 s and contact time of ~200 ms. (Two example force curves are those plotted in Figure 6.9a.) The humidity was first set low, around 20% (solid circles), where

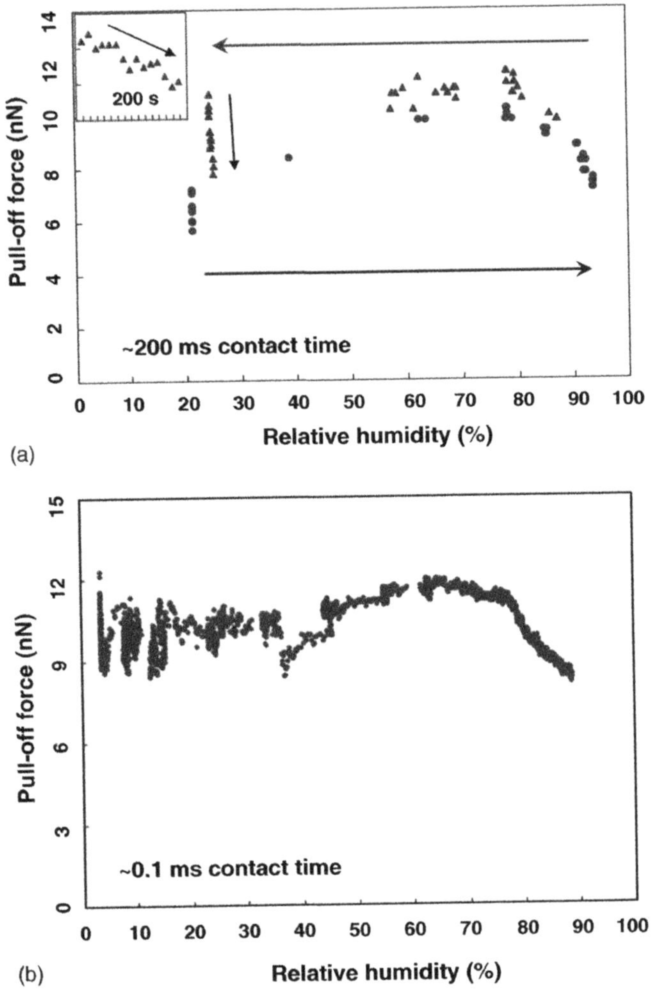

FIGURE 6.10 Humidity dependence of the magnitude of pull-off force measured in (a) slow and (b) fast force curves on mica with a silicon (oxide) tip. Each datum in (a) derives from a single force curve, whereas each datum in (b) averages pull-off from 512 force curves. Variations of humidity were reasonably slow for all cases except for the ~60 to ~25%-RH transition between triangles in (a). The inset in (a) shows the time dependence of the 25%-RH data following this transition.

significant scatter was observed, then raised and equilibrated around 40%, 65%, and 80%, and at several higher values terminating at 94%. Then the humidity was decreased slowly (spanning tens of minutes) to around 60%, followed by a rapid drop (seconds) to about 25%. Here, a strong decrease of adhesion was logged during repeated measurements (depicted by downward arrow) over the course of 200 s as explicitly shown in the inset. At the end of this time, the adhesion values remained at the low end of this trend, consistent with values early in the experiment, but with significant scatter. Thus, an overall increase in adhesion was observed as the RH increased from 20% to around 80%, followed by a significant decrease in adhesion as humidity increased from 80% to 94%. This nonmonotonic trend was largely reproduced upon equilibration as humidity was decreased, with the above-noted time dependence following the rapid humidity decrease from 60% to 25%.

A much larger set of similar measurements was acquired with the same tip in rapid force-curve measurements (contact time ~0.1 ms) conducted in a mapping mode (described more fully in Section 6.5) over a 5 × 5 μm region. Humidity was slowly ramped from 4% to 90% during nine consecutive 512 × 512 point force-curve mappings. Each plotted datum in Figure 6.10b is an average of 512 force-curve measurements, and there are a total of 9 × 512 = 4608 plotted data points, thus deriving from a total of 4608 × 512 = 2.36 million force curves. Adhesion values were observed to undergo significant, time-dependent excursions and scatter at RH < 50%, then largely stabilize in the range RH = 50–90%. This was again observed during a downward change of humidity (not shown). Most likely the instability at low humidity reflects a heightened sensitivity to tip geometry and roughness on the nanometer to subnanometer scale as capillary water decreases in presence and perhaps wets individual small asperities on the tip.

The examples of Figure 6.10 show qualitatively similar humidity dependences over the RH = 20–90% range, even though contact times differed by three orders of magnitude. Slow force curves were also conducted with a much blunter Si_3N_4 tip, producing the adhesion force measurements of Figure 6.11. Again a

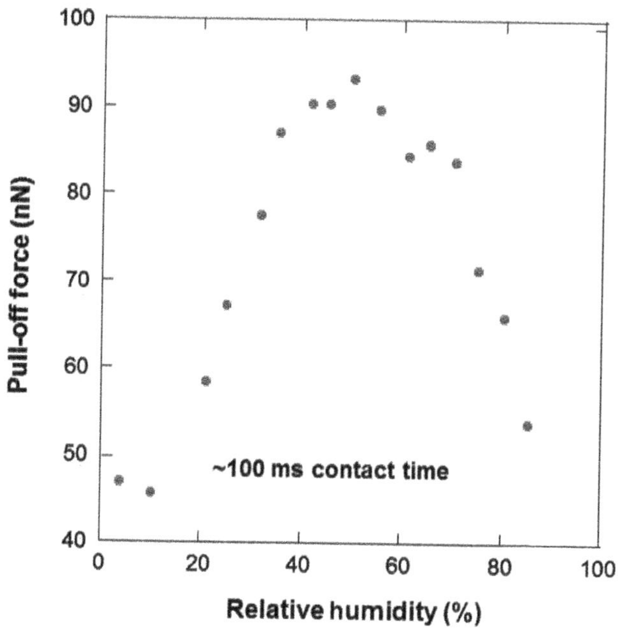

FIGURE 6.11 Humidity dependence of pull-off force for a Si_3N_4 tip interacting with a cleaved mica surface.

nonmonotonic trend is observed, but peaking at lower humidity (RH ~ 50%) and with peak adhesion roughly twice as large as the adhesion at low and high humidities. This stronger nonmonotonic trend may be due to broader scale of roughness on the blunter tip.

Our final example of capillary adhesion involves polymer-melt forces rather than condensed water. It is included to exemplify (1) much larger range and force of capillary adhesion at polymeric liquid films, compared to adventitious capillary condensation, and (2) further strong dependence on film thickness, presumably related to confinement effects. Figure 6.12 compares representative force curves (here in raw deflection-*Z* form) obtained with a Si_3N_4 tip ($R = 20$–40 nm, attached to V-shaped $k = 0.6$ N/m cantilever) on spin-coated polydimethyl siloxane (PDMS, molecular weight 95,000 kDa) films of thickness ranging from 5.5 nm to 205 nm (measured by spectroscopic ellipsometry). Long-range capillary adhesion, a

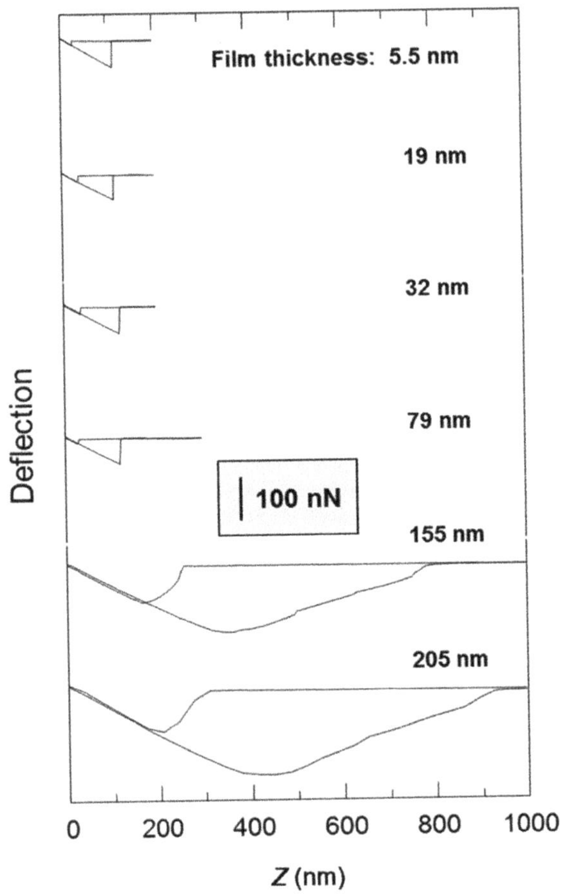

FIGURE 6.12 Force curves obtained with a Si_3N_4 tip on spin-coated PDMS films of variable thickness.

continuous pull-out during retraction, is observed for film thicknesses of 155 and 205 nm (over distances also measured in hundreds of nanometers), whereas abrupt pull-off force curves are observed at thicknesses ranging from 5.5 nm to 79 nm. Interestingly, the 79-nm film is roughly 10 times thicker than the bulk radius of gyration R_g (Section 2.5.3), yet obvious capillary behavior is not seen until 155-nm thickness. By comparison, some SFA studies have indicated that the shear viscosity of confined polymer melts remains bulk-like down to about five times the unperturbed radius of gyration, below which it increases dramatically [83].

6.3.5 Capillary Transfer Phenomena: Difficulties and Opportunities

Our final topic in capillarity involves the transfer of material to or from the AFM tip via a water-capillary meniscus. This phenomenon can be an irritant to users attempting to image small objects weakly adsorbed to a substrate. Physisorbed nanoparticles are easily plowed around by the AFM tip in contact mode; but even in dynamic mode, with minimal shear force, particles can be perturbed and indeed transferred to the tip. In the most delicate cases familiar to the author, one must operate not only in the attractive regime but also *at very low humidity* ($<5\%$) *to avoid this transfer*. The images in Figure 6.13 are of 6-nm diameter PbSe nanoparticles capped with organic ligands. Even with this capping, which must reduce the Hamaker constant, nanoparticle pick-up easily occurs if parameters are not carefully controlled to maintain attractive regime imaging in very dry conditions. Images in Figure 6.13a and b show the result of improper control; a mass of nanoparticles attached to the tip tends to be reverse-imaged by nanoparticles on the substrate (note similar assemblies of dots within each image). More careful control of dynamic-mode parameters and humidity enables a pristine image (c) of close-packed nanoparticles [84].

Perhaps better known is the *advantageous* transfer of material from tip to sample. *Dip-pen nanolithography*—whereby the AFM tip is first dipped in ink to acquire via capillary transfer then scanned in contact mode on a surface to transfer ink to sample (e.g., write lines)—is one well-developed example [63,64]. The role of the water capillary in these transfer processes has been carefully studied. Nanolithography via electric–field-induced oxidation is another example [85]. The capillary meniscus mediates a nanometrically confined oxidation reaction at a semiconductor surface (e.g., silicon) under a voltage-biased AFM tip [86].

Notwithstanding the nanoscale resolution of AFM via inexpensive commercial tips, there have been efforts to develop even sharper appendages—nanotubes, nanowhiskers, etc., grown on or attached to standard tips—to enable a smaller touching zone. This also may provide a spike-like shape to track vertical sidewalls, for example, in microelectronic or microelectromechanical devices. The motivation for sharper tips is pretty obvious, but the expense of these tip constructs can be prohibitive. One route to sharper tips that the average user can attempt "on the fly"—that is, while using the AFM, even on the sample of interest—involves coaxing a hygroscopic polymer (particularly one that tends to self-assemble into rod-like structures) to attach to the AFM tip via water-capillary transfer. Of course, the sought

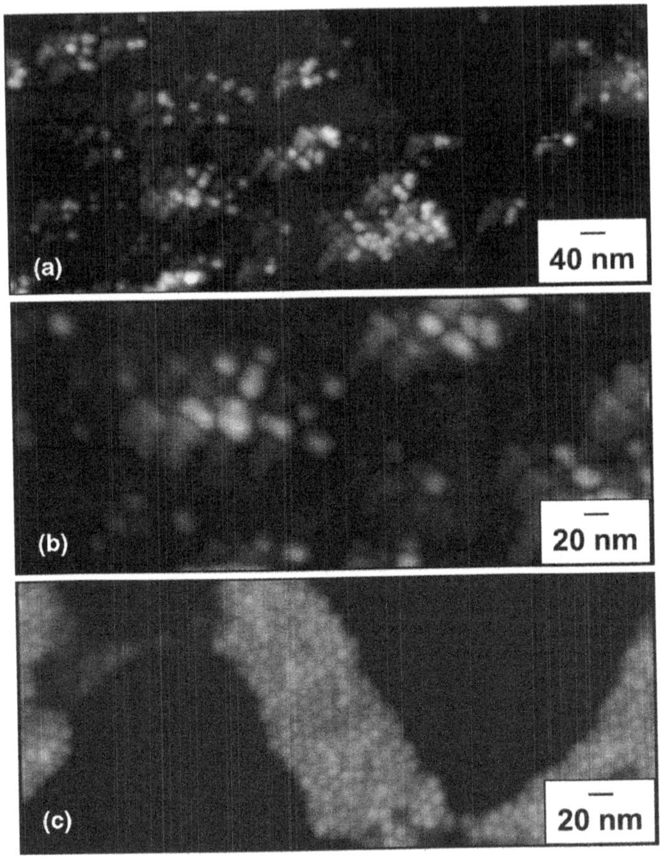

FIGURE 6.13 Dynamic AFM height images of PbSe nanoparticles under (a) and (b) improper imaging conditions, wherein capillary pick-up has occurred, and (c) proper imaging conditions of very low humidity and attractive-regime imaging. All images acquired using a 3-N/m cantilever with silicon (oxide) tip.

"nanofibril" of polymer is not visible, and the user may not have ready access to a high-resolution electron microscope to confirm such a fibril attached to tip. But AFM force curves can be very sensitive to the presence of a polymer bridge between tip and sample, as already discussed.

Figure 6.14 compares two plots of force versus *Z* displacement during retraction, characteristic of a commercial Si_3N_4 tip (thin line, partly off the force scale) and the same tip with a polymer fibril attached (circles) via a scanning procedure that ramps to high loads to scrape the tip into a thin gelatin film (the nominally dry solid, not a gel) and against a mica substrate [87]. During retraction, the tip and sample stick together such that the AFM cantilever bends downward, exerting a pulling force on the tip–sample adhesive contact that grows during retraction until the adhesive contact fails. For the case of the polymer fibril, the greatly reduced adhesive force

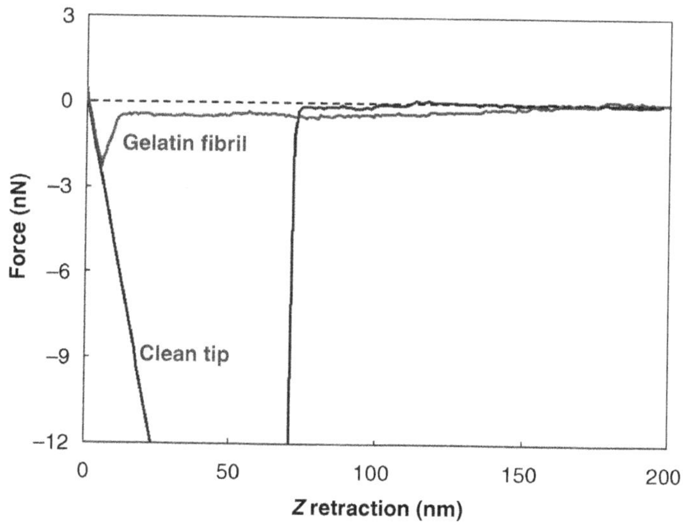

FIGURE 6.14 Force versus Z retraction distance on (dry) gelatin using a clean Si_3N_4 tip and a gelatin fibril attached to tip.

reflects a greatly reduced tip–sample contact area, whereas the long range of attractive force (~160 nm) indicates additional bridging of molecules between solid sample (dry gelatin) and Si_3N_4 tip. The bridge can be detected at all sample locations and even upon switching to a clean inorganic sample surface, attesting to the presence of a fibril attached to the tip. This transfer phenomenon does not take place at low humidity, suggesting that the capillary meniscus mediates the process. Water molecules also are understood to provide hydrogen bonding to stabilize the collagen-fold triple helix conformation [88].

The modified tip is not uselessly transient. It can be stable over the course of many subsequent *images* in contact mode, indeed even for several samples, before ultimately collapsing into a blunt mass that exhibits degraded resolution. Comparing images in Figure 6.15 of a gelatin film before and after (left/right) this tip modification, one finds a transformation from "granular" images (grain size tens of nanometers) to a fibrous image (fiber diameter <10 nm). The grains are about 5–10 times larger in diameter than the fibrils, consistent with much larger tip–sample contact area and thus adhesion. A fibrous morphology is what one expects if nanometer-scale resolution is achieved: gelatin derives from the connective protein collagen that naturally self assembles into a hierarchy of fibrous structures, the smallest being tropocollagen, 1.5-nm in diameter [89]. Consistent with this interpretation of exceptionally high resolution are images of ultrathin porous films less than 2-nm thick that exhibit flat-bottom, 5-nm diameter pores contrasted in frictional response as discussed in Chapter 7 (i.e., tip reaching substrate), indicating an imaging asperity of ~2 nm radius, roughly 10 times better resolution than with the initial bare tip [87].

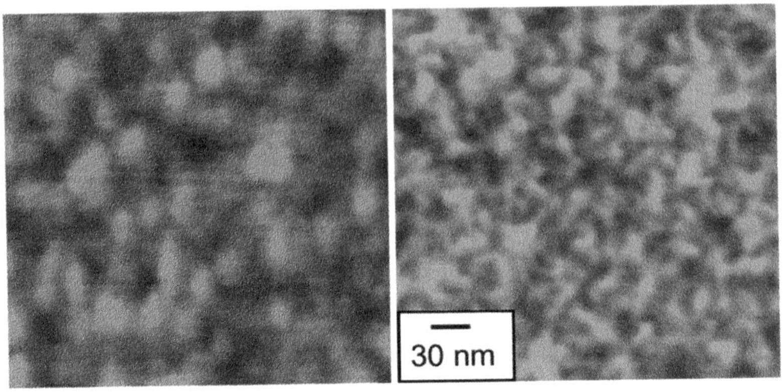

FIGURE 6.15 Topographic images of a (dry) gelatin film using the two tips corresponding to the force curves in Figure 6.14: left image acquired with clean tip, right with gelatin-fibril tip.

6.4 APPROACH–RETRACT CURVE MAPPING

6.4.1 Motivation and Background

Chapters 2 and 3 introduced force curves, which enable the detailed measurement of the distance dependence of force at a point on a surface. In Chapters 4 and 5, both quasistatic and dynamic feedback schemes dependent on this distance dependence were introduced to track surface topography, essentially as iso-force or iso-amplitude contours. (This was with respect to the *Z*-scanner displacement, not the actual tip–sample distance, the latter additionally accounting for vertical cantilever deflection or vertical amplitude, Section 3.2.) In dynamic mode, we further saw that a distance-dependent tip–sample interaction causes a shift in the phase lag of the vertical tip oscillation relative to the drive signal, thereby providing materials contrast (ultimately via energy dissipation). This phase number effectively derives, however, from *integration* over each complete vertical cycle of the tip: from a large distance at which there is no tip–sample interaction, up to the onset of usually attractive interactions (i.e., van der Waals, electrostatic) and often closer so as to make contact and indentation, then the reverse path out wherein physically hysteretic phenomena may occur. Thus, multiple components of tip–sample interaction may be convolved into one number, phase.

It is a natural next step, therefore, to consider data-acquisition schemes that collect the *entire Z dependence* of force, amplitude, phase, or some other quantity *at each location in an image*. (We will often refer to the individual *Z*-dependent subsets as *approach–retract curves* rather than force curves when referring to the *general* data acquisition routine, because the measured quantity may not be quasistatic cantilever deflection, i.e., force.) Thus, in these schemes a *data cube* is generated: a

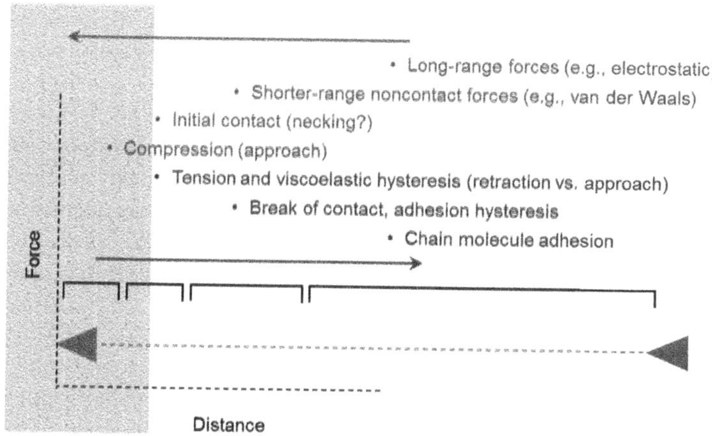

FIGURE 6.16 Illustration of different kinds of physical information contained at different distance locations (positive and negative) during approach and retraction of tip to/from sample.

measurement is made (cantilever deflection, amplitude, phase, or other) over a 3D domain, *X*, *Y*, and *Z*. With this data cube in hand one can in principle render images or at least coarse mappings over *X*–*Y*, contrasting finer or more selective differences in tip–sample interaction within specific *Z* intervals. Figure 6.16 depicts distance intervals dominated by different kinds of interactions and thus information content. One can envision contrasting long-range forces such as those due to electrostatic charging or chain-molecule extension, shorter-range van der Waals or steric interactions, mechanical stiffness or indentation depth, viscoelastic hysteresis (e.g., creep), tip–sample adhesion, and more.

Approach–retract curve mapping methods have been available in some base commercial AFM packages for many years (e.g., the Force Volume acquisition routine in what are now Bruker systems) and as such have demonstrated great utility in materials and biophysical research [90–94]. It is a common observation, however, that in multiuser core labs at universities as well as corporate central analytical labs, most AFM applications do not make use of these acquisition routines; indeed, many "veteran" users seem oblivious to them. Even knowledgeable AFM lab managers may encounter barriers to convincing users to exploit the data cube routines, especially given a broad user base. One barrier is effort: understanding the instrumentation algorithms and the fine points of the acquisition software. Second, the acquisition time to collect a data cube can be much longer than that of common imaging routines. Third, in some cases there is an elevated risk of a tip crash, because the *trigger* for reversing from approach to retract can be corrupted, leading to a fully extended *Z* scanner (hence a huge loading force, and thus a dulled or contaminated tip), as will be discussed. Given the extra time needed to master acquisition concepts and collect large data sets, such an accident can be extremely frustrating. Fourth, and perhaps most importantly, even if the routine runs properly and generates a rich data cube, there may be irksome limitations in data post-processing algorithms (typical with AFM vendor software) and thus little payback for the extra effort.

This section introduces approach–retract curve mapping in general terms. Three commercial implementations (available to the author) are described, followed by examples of data processing as well as several examples of the physical information obtained. Much of the data processing examples utilize software algorithms developed by the author within a particular programming environment;[9] users can do similar things with commercial third-party software packages such as Matlab and Igor Pro.

6.4.2 Traditional Force-Curve Mapping

Two long-available data acquisition routines, Bruker's Force Volume and Agilent's SPS or FlexGrid Imaging, allow one to map thousands of approach–retract curves in typically tens of minutes, creating data files usually spanning ~1–100 MB in size. The most commonly reported numbers of *X–Y* locations have been $32^2 = 1024$ or $64^2 = 4096$. If using an approach–retract cycling period of, say, 0.5 s, this translates into roughly 8.5 and 34 min of acquisition time. But additionally, the routines require time to algorithmically scan back after each scan-line excursion, as well as time to acquire simultaneous height images, such that total acquisition times are considerably longer. (One may choose to shorten this time by operating at shorter *Z*-cycling periods, although not without possible physical implications, e.g., in viscoelastic systems, and usually not without increased *Z*-scanner hysteresis, i.e., instrumental artifacts.)

In Force Volume, the height image is obtained at the same relatively low density of point locations as the approach–retract curve mappings. In SPS and FlexGrid imaging, the height image is obtained at the usual high density of point locations as in conventional imaging, but spanning between each approach–retract curve measurement site. (In SPS, these sites are uniformly spaced, but in FlexGrid, the site locations are user-selected, i.e., the spacing between sites is variable.) In Force Volume, the number of data points acquired during each *Z* scan is fixed (by the operator, e.g., $256 \cdot 2 = 512$); variations in surface elevation are accommodated such that the beginning and ending point, with tip far from surface, is variable from location to location so as to account for surface topography (approximately, i.e., with the usual caveats derived from the finite speed of the *Z*-feedback circuit). In SPS and FlexGrid Imaging, the size of each *Z* scan is variable, and the number of data points correspondingly variable, such that the spacing of *Z* values, and thus the *Z*-scanning velocity, is fixed; the beginning/ending point of each *Z* cycle is a fixed *Z*-scanner position, such that variations in surface elevation are apparent when comparing one approach–retract curve to another (e.g., in the *Z* value at which contact is made), including the overall arbitrary tilt of sample surface relative to *X–Y* scanner plane. The total number of data points in each approach–retract curve (e.g., 2000) is fixed, however, by using zero-value filler data points added to the end of each approach

[9] This software is available in a collaborative interaction.

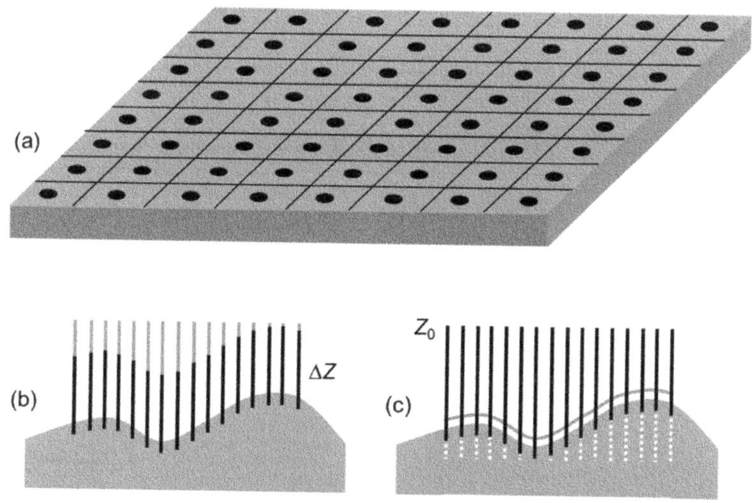

FIGURE 6.17 Illustrations of two commercial approach–retract curve mapping schemes at low point density of sites: (a) an 8×8 grid of sites in X–Y, measurements taking place at dot locations. (b) and (c) Black vertical lines indicating the Z-scan intervals over which data are collected in force volume (b) and SPS/FlexGrid imaging (c) at 16 sites along a single scan line. Gray solid vertical lines in (b) represent the relative Z displacements contained in a simultaneously acquired height image at the same low point density of sites. White dotted vertical lines in (c) depict zero-value filler data inserted into each approach–retract data set such that the total number of data points per Z cycle is fixed (e.g., 2000). Gray solid curve in (c) represents the relative Z displacements contained in a simultaneously acquired height image, at high point density locations between the sites at which Z-scans are performed.

and beginning of each retraction. This dummy data is inserted after performing each approach–retract cycle (i.e., there is no actual, variable wait time at each turnaround point). Figure 6.17 is intended to schematically clarify these rather different data acquisition schemes utilized by Bruker's Force Volume and Agilent's SPS/FlexGrid Imaging. (Some of the difference in scheme may derive from patent restrictions.[10])

6.4.2.1 Aqueous Immersion, Polymer/Inorganic Surface Our first example is also perhaps the richest, with the intention to demonstrate a wide variety of physical behaviors that can be probed in approach–retract curve mapping with post-processing algorithms. This system was previewed in Section 1.5. An ultrathin, hydrophilic polyacrylamide film was adsorbed to a silanized glass substrate via a proprietary technology utilizing ultraviolet light to produce chemical bonds [95]. This surface was immersed in distilled and deionized water, and high-force abrasive scanning was performed in a 500×500 nm region in an attempt to bare the substrate (akin to

[10] The newest version of SPS imaging under the software Picoview is somewhat different from that under Picoscan.

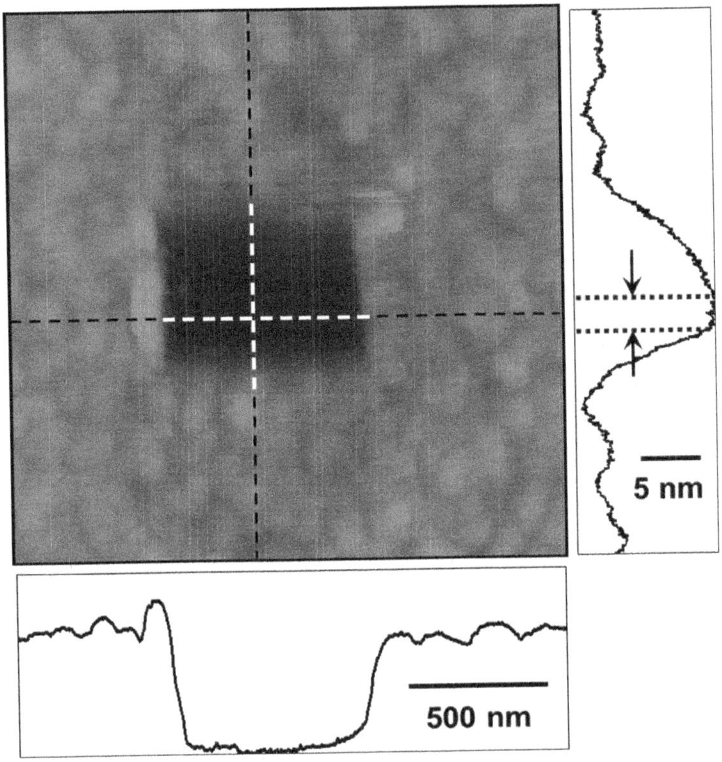

FIGURE 6.18 Contact-mode height image of ultrathin but water-swollen (immersed) polyacrylamide film containing a subregion previously scanned to remove much of the film. Suggested substrate surface is demarcated with dotted lines in the vertical profile at right.

that discussed in Section 8.3.2 in the context of power spectral density analysis). Figure 6.18 contains a conventional contact-mode "height" image of this abrasively scanned region, along with selected horizontal and vertical profiles. The vertical profile suggests that portion of the abrasively scanned region, only 150 nm vertically, was cleared down to the glass substrate approximately 10 nm below the polymer film surface. Force Volume was utilized to examine the properties and behavior of $32 \times 32 = 1024$ sites spread over the entire image region.

In a preview in Section 1.5, a qualitative stiffness map was generated from the slope of all 1024 force curves (Figure 1.16), indicating increased stiffness across much of the 500×500-nm abrasively scanned subregion, along with reduced stiffness along its left edge; this latter behavior was observed at locations of higher apparent surface elevation, indicating a lip of soft, presumably water swollen polymer evacuated from abraded regions. Figure 6.19a compares full force-curve cycles acquired at three representative locations, marked with small squares in the "height" image (Figure 6.19b) that was simultaneously acquired (at the same low-point

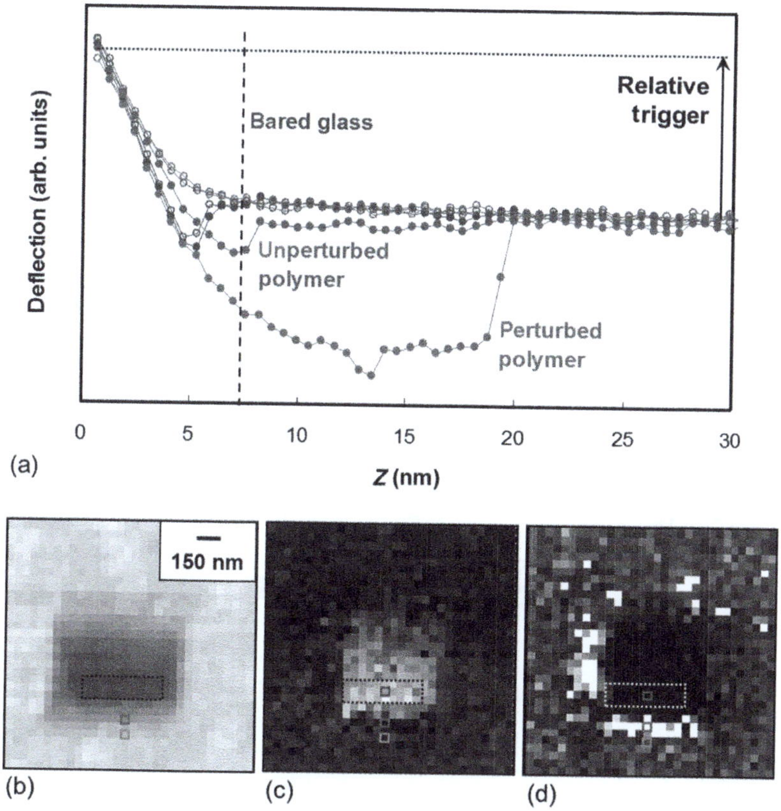

FIGURE 6.19 Polyacrylamide film on (silanized) glass in water, containing a square subregion previously scanned to remove the film (marked rectangular region) or simply perturb the polymer. (a) Representative raw force curves from three locations marked with small squares in (b)–(d), which are 32 × 32 site mappings of (b) height, (c) force-curve slope at the Z-slice denoted with vertical dashed line, and (d) force-curve hysteresis (path integral) over the 30-nm Z stroke shown in (a) (complete Z stroke was 150 nm). The individual force curves in (a) show that on glass an attractive force is present prior to contact, followed by a rigid mechanical stiffness upon contact. On unperturbed and perturbed polymer locations, a steric repulsion prior to contact is followed by relatively soft and rigid contacts, respectively. In retraction, the glass shows almost zero hysteresis; the unperturbed polymer shows significant hysteresis and the perturbed polymer huge hysteresis.

density) as well as two other images *generated* via post-processing: the slope of force curves during approach, just prior to rigid contact (Figure 6.19c), at the *Z* value indicated with a dashed vertical line in (a), and the closed path integral of deflection-*Z* over the force-curve cycle out to the 30-nm value of *Z* (Figure 6.19d), an approximate measure of adhesion hysteresis. A rectangular region is denoted in Figure 6.19b–d that is 500 nm horizontally (the abrasive scan size) and 150 nm vertically, the anticipated size of a clear substrate surface based on conventional imaging (Figure 6.18 plus friction). Thus, the images in Figure 6.19c and d are remarkable, because they reveal a much larger region of predominantly attractive force (bright in (c)) and correspondingly low adhesion hysteresis (dark in (d)). These findings suggest both a lack of steric repulsion (leaving dominant van der Waals attraction) as well as reduced molecular degrees of freedom (suppressing adhesion hysteresis) due to flattened and compacted conformations of polymer. (Note again the higher surface elevations in Figures 6.18 and 6.19b at locations above the rectangular region.)

Figure 6.20 also compares behaviors related to molecular conformation. Nine force curves during retraction are overlaid in (a), acquired at locations marked with small squares in (d); (b) and (c) reproduce the same data as in Figure 6.19, for

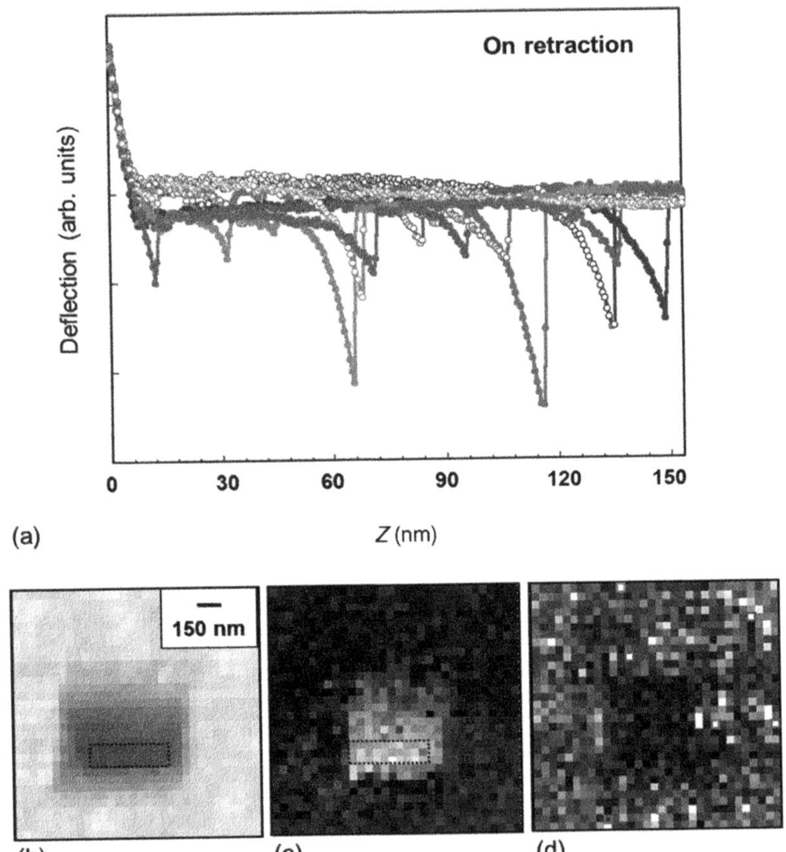

FIGURE 6.20 Same polyacrylamide film on (silanized) glass in water system, and results from same data cube, as in Figure 6.19. (a) Representative raw force curves during retraction only, from nine unperturbed polymer locations marked with small squares in (d). (b) and (c) Reproduced from Figure 6.19b and c for visual reference. (d) Map of approach–retract hysteresis over complete 150-nm Z stroke.

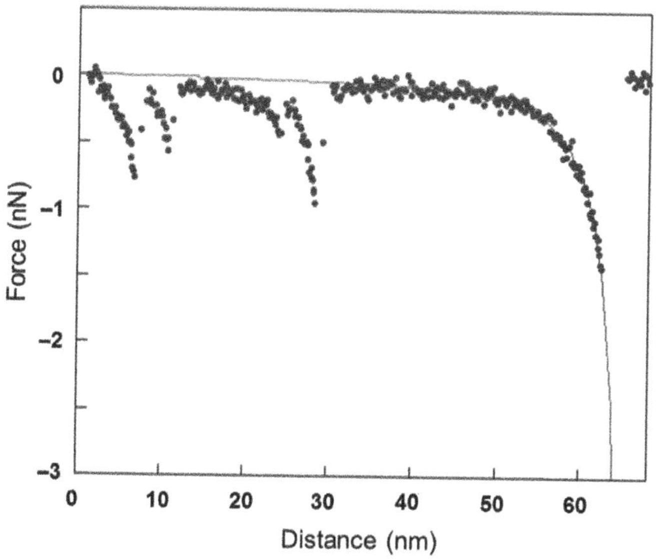

FIGURE 6.21 Analysis of one retraction force curve from Figure 6.20 after conversion to force versus tip–sample distance. The solid curve is a fit of the extended wormlike chain model (Eq. 2.37) to the data (circles) within the distance = 35–65 nm interval.

comparison. The image in (d) is again adhesion hysteresis, but computed over a ≈150-nm *Z* cycle, the entire range of the actual raw data. Here, a much more variable adhesion hysteresis image results, compared to that of Figure 6.19d. This is because both relatively long and short chains extend between tip and sample, depending on the overall length of available chains and the randomness of the molecular locations that physisorb to the tip from one adsorption event to another. The characteristic, presumably entropic sawtooth force–distance trend is seen in the force curves of (a), as discussed in Section 2.5.5. One of these force curves was further processed as discussed in Section 3.2: (1) deflection scale converted from volts to nanometers so as to force unity slope on the bare-glass regions, (2) distance scale computed from the sum of the *Z* and deflection data, and (3) force computed as the product of deflection and an estimated cantilever spring constant. The result is shown in Figure 6.21. Additionally, the finally stretch-snap portion of the force curve during retraction was mathematically fit with a force–distance curve predicted within the wormlike chain model of single-molecule extension (Section 2.5.5). The extracted values of persistence and contour length are 10.4 nm and 68.2, respectively.

6.4.2.2 Nanoindentation of Polymer Blend In the preceding example of Section 6.4.2.1, the spatial presence of very different types of forces—van der Waals versus single-molecule bridging—at different locations across the surface resulted

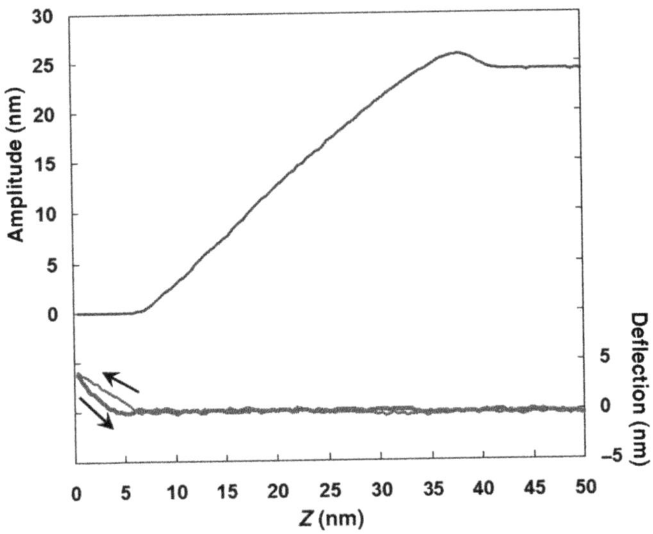

FIGURE 6.22 Two-channel approach–retract curves, measuring both amplitude and time-averaged deflection responses versus Z displacement. While in resonant dynamic oscillation the time-averaged deflection is zero, whereas upon quasistatic contact at about $Z = 5$ nm, the oscillation is extinguished and the time-averaged deflection becomes positive.

from abrasively scanning a polymer film with the AFM tip. In the two examples of Section 1.3, abrasive scanning was at the core of the analytical method for quantifying biofilm cohesion, and it enabled the identification of multilayer polyvinyl alcohol (PVA). But abrasive scanning is just one of many "nanomanipulations" that can be performed with an AFM tip, to obtain depth-derived or other kinds of information. Another example involves *nanoindentation* to assess regions differing in mechanical stability. On soft materials, it is useful to perform this exercise while running in dynamic AFM, approaching to the point that the vertical oscillation amplitude is reduced to zero and a quasistatic contact is made. Thus, the cantilever deflects up to achieve a finite, static loading force, as exemplified in Figure 6.22. We trigger the turnaround from approach to retraction by monitoring the *time-averaged deflection*, as in the bottom plot in Figure 6.22, even though we are "operating" (i.e., in software) within nominally the dynamic (vibrational, also known as "tapping") mode, such that the imaging of resulting indents is done with minimal shear force. In Figure 6.22, upon examining both approach and retract data, we see that a hysteresis loop is observed at positive load. (A nominal 40 N/m spring constant cantilever exhibits little if any adhesion hysteresis, i.e., no jump from contact.) This hysteresis loop derives from plastic deformation of a glassy polymer, driving the tip into the material beyond its mechanical yield point to leave an indent or rupture.

This section's example applies the above method to a polybutyl methacrylate (PBMA) polylauryl methacrylate (PLMA) coating (~1 μm thick) on a silicon

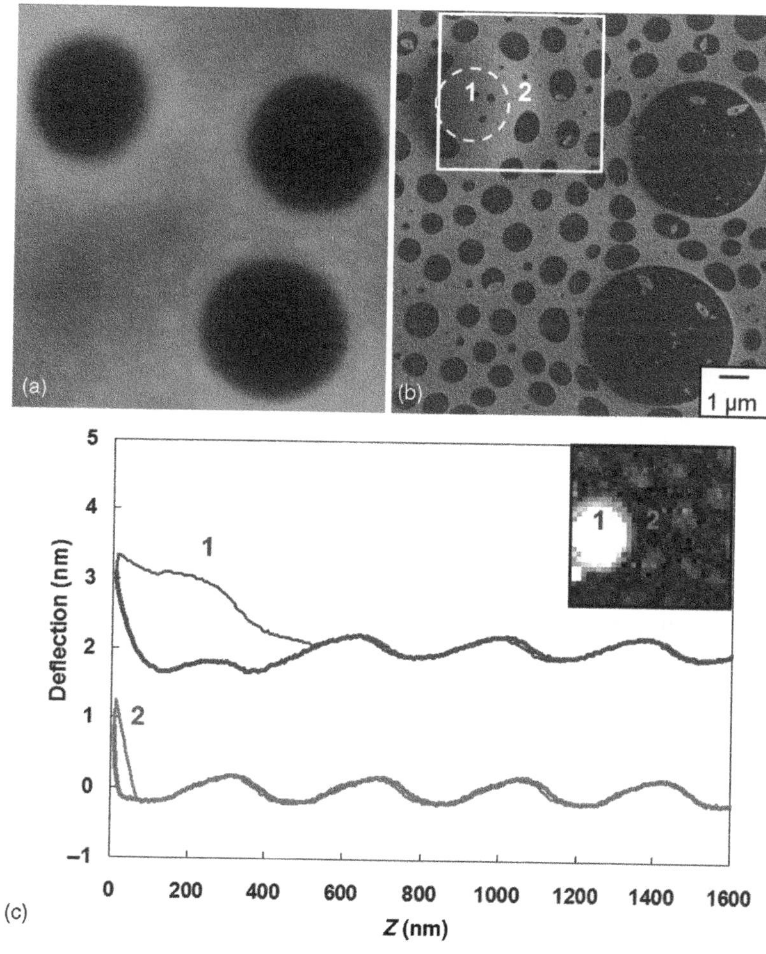

FIGURE 6.23 Dynamic AFM (a) height and (b) phase images on blend film of PBMA and PLMA. (c) Two quasistatic force curves acquired at designated surface locations 1–2 (vertically offset for clarity). Thin line is approach data, thick retraction. *Inset*: contact hysteresis (closed path integral of deflection versus *Z*) over force-curve mapping domain corresponding to square in (b).

substrate. The initial region of interest, displayed in height and phase (net repulsive regime) images of Figure 6.23a and b, contains three deep circular valleys in topography (~250 nm depressions), two of which contain a circular domain of PLMA at the surface as seen in a phase image (dark circles). Force-curve mapping was performed using a trigger for turnaround of 1-nm cantilever deflection and a nominal 40 N/m cantilever (i.e., a ~40 nN trigger). Two representative force curves are shown in Figure 6.23c at sites 1–2 as indicated in Figure 6.23b, locations of PLMA and PBMA, respectively; thin line is approach data, thick retraction. As exemplified

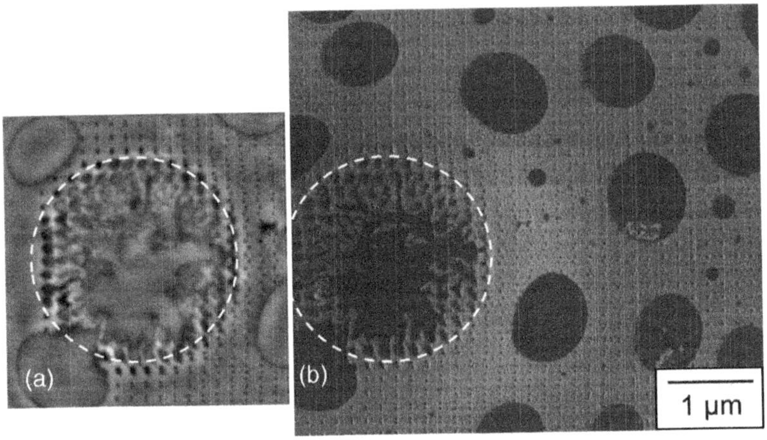

FIGURE 6.24 (a) Differential height (high-order flattened) and (b) phase images following collection of two arrays of force curves that perforated and ruptured a skin of PBMA within the circular region.

here, such small deflection measurements (i.e., due to using a stiff cantilever) can result in relatively large visible variations in the force-curve background signal, thanks to the optical interferometry artifact of Section 3.4. Nevertheless, by comparing the two force curves on each of the two polymers, it is clear that greater adhesion and much greater contact hysteresis is observed on PLMA (inset image), the latter implying much greater plastic deformation—*several hundred nanometers*—into PLMA at the same loading force. Interestingly, the only residual indents seen in subsequent images are in the PBMA regions. Following force-curve mapping over a $5 \times 5\,\mu m$ region (denoted by a square in Figure 6.23b) where a static, high-force contact was made ($\sim$40 nN), one finds a grid of small perforation spots wherever the tip penetrated into the PBMA leaving indentations Figure 6.24a and b. But a much more dramatic tearing of the PBMA occurred within the deep circular valley, seen in the dashed white circular region in both images in Figure 6.24, differential height (a) following curved surface removal via high-order nonlinear flattening (Section 8.1) and phase (b) in the net repulsive regime. This is to be compared with the initial pristine region indicated with a circle in Figure 6.23b. Apparently, a buildup of stress in the PBMA is relieved by rupturing as induced by the indentations. There even appears to be a flow of PLMA (glass to rubber transition temperature $-38°C$) upon rupturing of the PBMA "skin." Thus, this analysis allows one to probe subsurface phase segregation, and to some extent the mobility of the rubbery phase.

6.4.3 Approach–Retract Curve Mapping in Dynamic AFM

Some of the most elucidating applications of *Z*-dependent mapping are fully acquired in dynamic (also known as "tapping" or "AC") AFM. We shall see that on

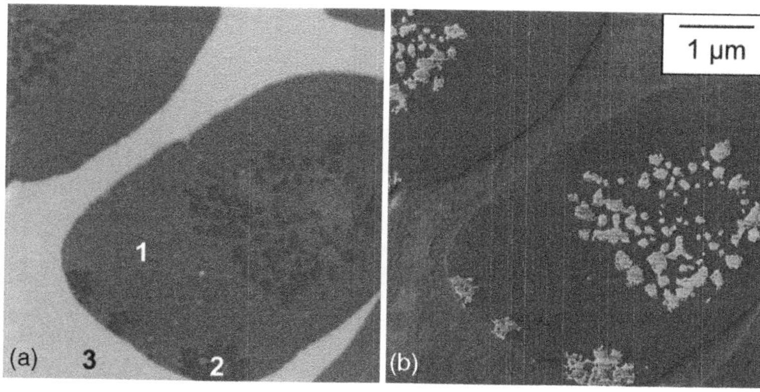

FIGURE 6.25 (a) Height and (b) phase images of a three-component PVA surface.

surfaces with nanoscale variations in height, spatially resolved amplitude–Z curves can be used to *correct* relative height values obtained in conventional images in dynamic mode, and spatially resolved phase–Z curves can provide more revealing understandings of dissipative behavior than single phase values per location (see Section 5.3). We shall also see that both amplitude and phase versus Z measurements can be used to essentially *depth locate* suspended, relatively rigid objects in a soft matrix.

6.4.3.1 Multicomponent Surfaces Our first example system for approach–retract curve mapping is an ultrathin film of PVA (cast from aqueous solution on cleaved mica). Figure 6.25 shows conventional height (a) and phase (b) images acquired with drive parameters chosen to attain strong phase contrast. Three characteristic subregions have been identified [94]: a first layer only ~1-nm thick (#1 in Figure 6.25) and two types of second-layer domains exhibiting relatively low and high dissipation (#2 and #3, respectively, in Figure 6.25). (Scan abrasion attested that these are indeed layers atop a continuous layer #1, and that layer #1 is much more strongly bound to the substrate than the second layer types #2 and 3 are to the underlying first layer.) One might say that the phase contrast in Figure 6.25 is *too* strong: some 80° separating brightest from darkest, with the brightest apparently corresponding to a net repulsive shift of oscillator while the darker regions correspond to a net attractive shift. Thus, the regime-switching phenomenon discussed in Section 5.1 is apparently present in the data of Figure 6.25. Figure 6.26a contains three pairs of raw amplitude–Z and phase–Z curves that are representative of the behaviors of regions #1–3. The trigger for turnaround was a 1.25 V decrease of raw amplitude signal from the free oscillation value of 1.87 V. (Having a measurement of the latter at each location can be very useful if drift, as in Figures 5.12 and 5.13, is taking place and thus the cantilever excitation amplitude is not

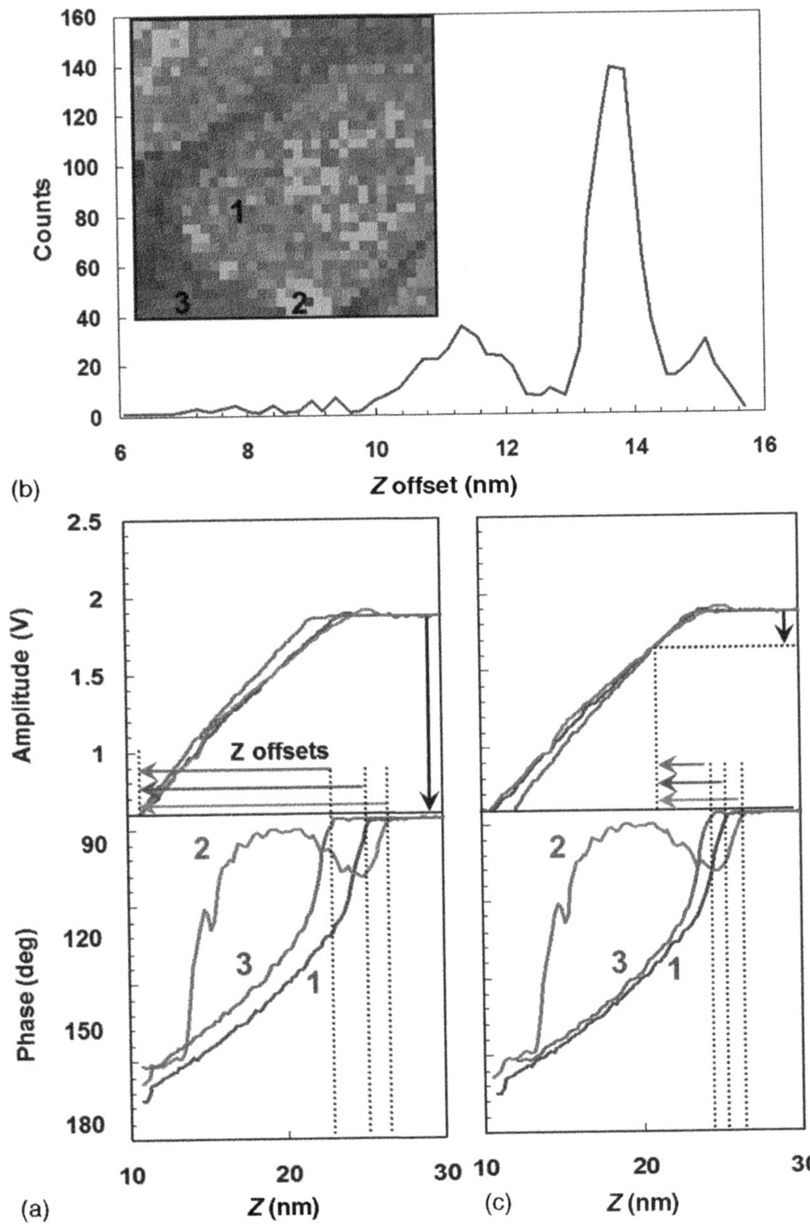

FIGURE 6.26 (a) Raw amplitude–Z and phase–Z curves acquired at the three locations denoted in Figure 6.25. Z offsets indicate *different* Z displacements from the point of interaction onset to the trigger turnaround point. (b) *Inset*: mapping of this variable Z offset. *Graph*: histogram of Z offsets contained in the mapping. (c) Curves from (a) Z-shifted to achieve the same Z value at the amplitude value of 1.65 V, illustrating the origin of height error in Figure 6.25.

constant.[11]) Examining the phase–Z curves more closely, one sees that the Z displacement from the onset of interaction (sudden downward drop in phase with decreasing Z values between 23 nm and 27 nm) to the ultimate approach point is in fact *different* among the three measurement locations: greatest at location #2 and least at location #3, spanning about 4 nm. This variable Z offset is mapped over its $32^2 = 1024$ locations in the inset of Figure 6.26b; the histogram of these measurements further shows three peaks corresponding to the three dominant surface regions (dark, intermediate, bright), centered at Z offset values of 11.4, 13.7, and 15.1 nm, but also Z offset values reaching down to 6 nm and up to nearly 16 nm. This demonstrates that although there is a continuum of Z offset values contained in the height image of Figure 6.25, the values are concentrated in narrower ranges corresponding to each of the three surface region types. Indeed, one can Z offset a simultaneously acquired raw height map (not shown), point by point in the grid, to produce a *corrected* height map. One can also use the data of Figure 6.26c to better understand the conventional height image in Figure 6.25, which was acquired at a setpoint of 1.65 V. In Figure 6.26c, the amplitude–Z and phase–Z curves are shifted horizontally such that the amplitude curves coincide at 1.65 V, corresponding to the feedback condition of Figure 6.25. The relative shifts in the downward drops in the corresponding phase curves in Figure 6.26c quantify the height errors contained at corresponding locations in Figure 6.25, of roughly 1 nm high (#3) or low (#2) relative to region #1, the first layer. Our most notable finding, what appears to be a recessed region #2 (by a fraction of a nanometer), is in reality approximately 1 nm *higher* than the surrounding regions #1, as in the discussion in Section 5.1. One also notes that region #2 was imaged at a net oscillator shift of approximately zero relative to the free oscillation value (phase near 90°), whereas regions #1 and #3 were imaged "deep" in the attractive regime, again qualitatively consistent with Section 5.1. But the data in Figure 6.26a–c provide a more penetrating insight into what is going on during conventional imaging, compared to collecting even a set of conventional images at different setpoints.

Figure 6.27 explores further physical differences contained in amplitude and time-averaged (quasistatic) deflection data cubes obtained on the same PVA film regions. Three amplitude–Z and time-averaged deflection–Z curves are shown in

[11] Importantly, the method of triggering relative to the free oscillation amplitude, as done here, is susceptible to tip crash as follows. If during retraction a long fibril of sample remains adhered to the tip such that the amplitude does not fully regain the free oscillation value, then for the subsequent Z cycle the amplitude reference value (upon approach) is too low, meaning the approach proceeds even farther, to a lower amplitude than the previous cycle; upon retracting from this point of approach, the fibril likely remains attached, and the reference amplitude at farthest retraction is further reduced relative to the free oscillation value. Thus, the following cycle proceeds to even closer approach, and this continues with each cycle until the amplitude is reduced to zero. Of course there is no ability for the amplitude to decrease further during approach, thus the Z scanner continues to extend until reaching the hardware limit, meaning a huge static force, easily micronewtons in scale if using a rigid cantilever. The result is rampant sample tearing and/or tip damage. To prevent such an accident one may instead choose to trigger relative to (decreased from) an absolute amplitude setpoint (but, again, more susceptible to drift).

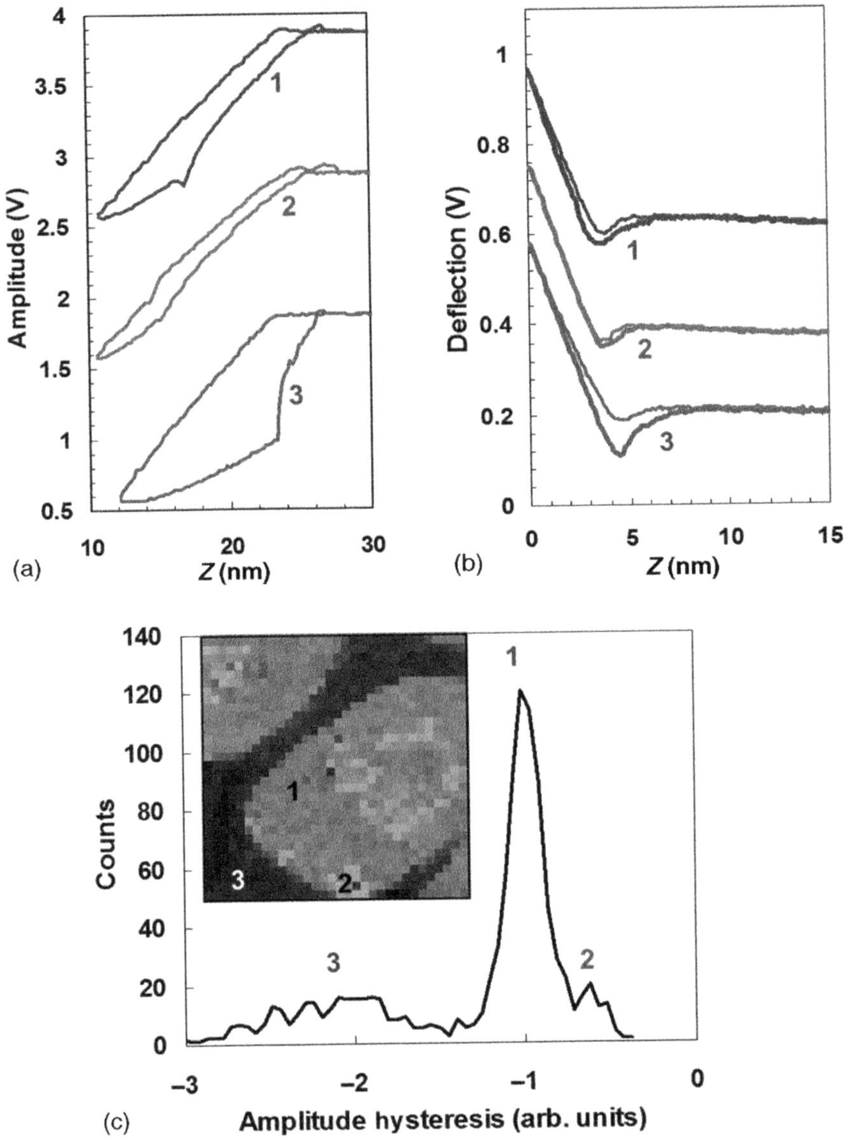

FIGURE 6.27 Representative (a) amplitude–Z and (b) time-averaged deflection–Z curves (vertically offset for clarity) for regions #1–3 of Figures 6.25 and 6.26, exhibiting region-specific hysteresis loops. (c) Mapping (inset) and histogram (graph) of amplitude–Z hysteresis loops.

Figure 6.27a and b, representative of regions #1–3. Most notable is the size of the hysteresis loops specific to regions: largest on region #3 and smallest on region #2. The size of the amplitude hysteresis loops was computed (via path integration) specific to each location and mapped in the inset of Figure 6.27c. The associated histogram data quantifies the wide distribution in hysteresis loop size with three dominant peaks corresponding to the same three regions as in Figures 6.25 and 6.26. In summary, we see that *Z*-dependent mapping followed by post-processing algorithms can be used to clearly identify differences in tip–sample interactive behavior. The examples here suggest that the thickest, bulk-like PVA region #3 is the most hysteretic and thus dissipative, whereas the island regions #2 are the least dissipative, further suggesting the greatest crystallinity in this semicrystalline system [94].

6.4.3.2 Suspended Objects: Iceberg Morphology Although AFM is categorized as a surface microscopy, we saw in Figure 5.29 that the technique has the ability to gauge below-surface segregation of sample components within a couple of nanometers of the surface (the case of a surface-segregated triblock copolymer with less polar polymer thermodynamically driven toward the surface). This is more dramatically the case for rigid objects suspended in a very soft matrix. Given the sensitivity of phase–*Z* measurements to material response, one might expect that phase-*Z* could be exceedingly sensitive to the *depth location* of (relatively) rigid entities. This suggests numerous applications in characterization ranging from biological cells to composite technologies containing soft polymer matrices or gels.

Here we consider the near-surface region of a release (nonstick) face of a commercial peal-off adhesive label system. This medium consists of a micrometers-thick, soft PDMS (silicone) matrix of very low surface energy (i.e., not sticky), plus somewhat stiffer inclusions to affect mechanical integrity: ~200 nm spheres of a diblock-copolymer based styrene-butadiene rubber (SBR). One research question regards the 3D distribution of SBR, preferably not at the surface (minimize stickiness) yet present to help prevent cohesive failure of the matrix (silicone residue transfer). Conventional height and phase images in Figure 6.28a and b suggest the weak presence of SBR spheres at the surface. The analysis of *Z*-dependent mapping of amplitude and phase data indicates, however, that these stiffer domains have a much larger lateral extent when the sensitivity is extended *farther below the surface*, as follows. Thirteen amplitude–*Z* and phase–*Z* curves are overlaid in Figure 6.28c, acquired from the narrow rectangular region designated in the height and phase images (a) and (b). The slopes of the amplitude–*Z* trends are highly variable, indicating relatively stiff and soft behaviors (high versus low slope). Correspondingly, one observes that the *Z* displacement from the onset of interaction during approach to the point of turnaround (relative amplitude trigger of −1.5 V, from 2.2 V to 0.7 V) ranges from 135 nm to 20 nm. We use the steep downward drop in the phase–*Z* data during approach to quantify the *Z* position of interactive onset and then compute a new image (inset) that maps this position. One finds much more of a continuum of values, brightest corresponding to the locations where the SBR spheres seem to be at the surface, based on Figure 6.28a and b. A histogram of these *Z* positions is shown under the amplitude–*Z* curves, peaked around 100 nm and spanning a 130-nm range of values of *Z*. This suggests "iceberg"-type morphologies: SBR peeking out of the surface within only small

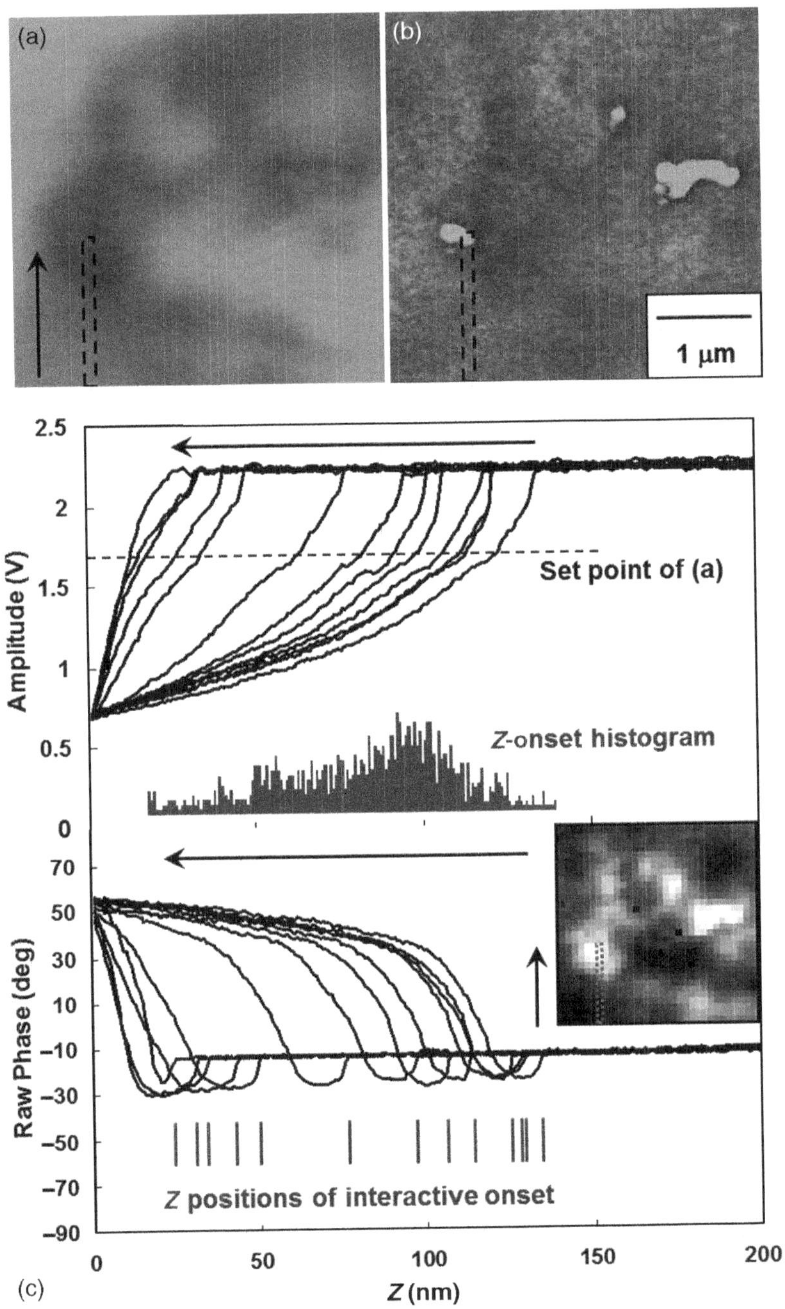

FIGURE 6.28 (a) Height and (b) phase images of silicone-based release (nonstick) surface of adhesive label system. (c) Overlaid amplitude–Z (top) and phase–Z (bottom) curves acquired in the rectangular column of locations denoted in the inset as well as (a) and (b). Inset maps the Z position of interactive onset, brightest being consistent with the leftmost onset marked below the graph, to effectively "see" the depth of rigid additive spheres.

areas, seen in the phase image, but present over a much larger lateral domain if probing below the surface. Clearly, the Z dependence of amplitude and phase can provide rich insights that may be lacking in conventional height and phase images.

6.4.4 Approach–Retract Curve Mapping of Liquidy Domains in Complex Thin Films

Section 3.3 introduced force curves that included the common presence of jump discontinuities during approach and retraction, where the gradient of force exceeds the cantilever spring constant; second, the assignment of the jump-to-contact to *capillary necking*, where mobile adventitious contaminants (generally on sample and tip, such as adsorbed water) are attracted to the nearly contacting tip–sample interface. In force-curve mapping, one occasionally finds that the size of the jump-to-contact is highly variable, even though the surface energies at different locations are not. This suggests a further presence of liquidy matter at some locations but not others. These observations are of great interest in understanding phase segregation of thin films containing one or more liquid ingredients.

One example is model hair conditioner cast as a film from the lamellar gel phase [96], containing an aqueous mixture of PDMS (i.e., silicone) melt, a short aliphatic chain (16 carbons) fatty alcohol, cetyl alcohol, and a similarly short-chain cationic surfactant, cetyl trimethyl ammonium chloride (CTAC). Casting on cleaved mica followed by deionized water rinse leaves a ~1–10-nm-thick continuous film containing characteristically different 2D domains as shown in the height image of Figure 6.29a, acquired in the attractive regime of dynamic AFM [97]. This includes islands several nanometers taller than the surrounding regions. If probed in contact mode, however, islands as seen in Figure 6.29a are not imaged in height, as exemplified in Figure 6.29b; instead, recessed regions ~2-nm deep are imaged in place of the islands. But quasistatic force-curve mapping data reveal the presence of islands indirectly as follows. Two representative force curves are shown in Figure 6.29c, with heavy arrows denoting the Z displacement from the turnaround trigger point to the preceding jump-to-contact. The top curve is representative of behaviors in the recessed regions in Figure 6.29b (i.e., those imaged as higher islands in attractive-regime dynamic mode, Figure 6.29a), a much larger jump-to-contact, indicating a longer-distance capillary necking phenomenon. A 2D mapping and a histogram of these Z displacement values over the same sample regions as Figure 6.29b is shown in Figure 6.29d and e; the 7-nm separation of histogram peak positions is consistent with island heights measured on the same sample in attractive-regime dynamic AFM. Thus, in both conventional contact-mode imaging and in force-curve mapping, the tip penetrates all the way to substrate on regions covered by the liquidy film. We conclude that force-curve mapping is capable of revealing liquidy morphologies that are invisible to contact mode, as well as provide an explanation for their absence (full penetration under quasistatic load).

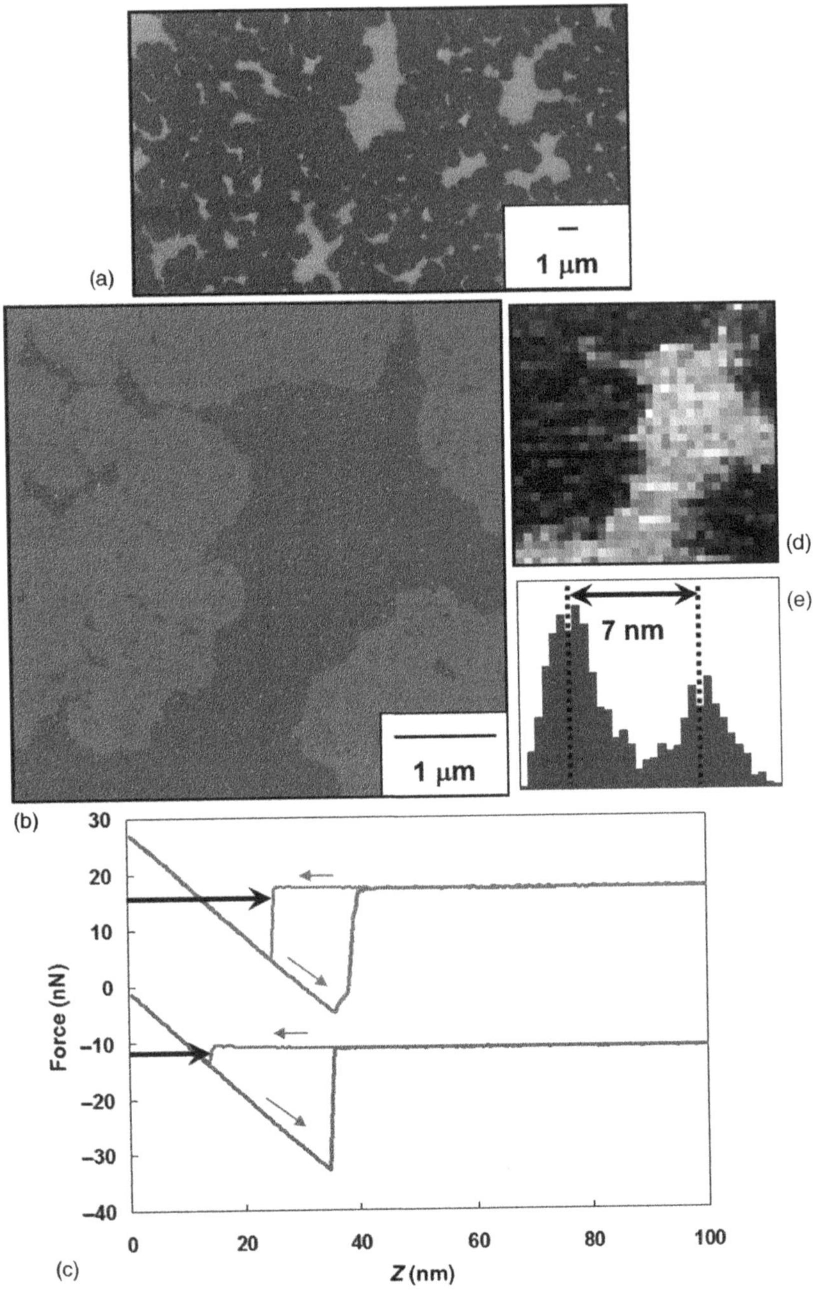

FIGURE 6.29 (a) Attractive-regime height image of complex film of silicone oil and surfactants. (b) Zoomed contact-mode height image of the same. (c) Representative quasistatic force curves from the two regions of (b). (d) Mapping of the displacements from $Z = 0$ to the jump-to-contact (depicted with heavy arrows in (c). (e) Histogram of the 1024 values of Z displacement in (d).

6.5 HIGH-SPEED/FULL SITE DENSITY FORCE-CURVE MAPPING AND IMAGING

The commercial versions of the modes of this section are known, at this writing, as *Pulsed Force Mode* (by Witec, more recently the *Digital* variant) and *PeakForce Tapping* (by Bruker) [98–101]. In order to execute at least one approach–retract cycle for each location in a conventional pixel-density "image" (e.g., 512^2), a much higher Z-cycling frequency (of order 1 kHz) is utilized compared to conventional force curves. Real force signal (via cantilever deflection) acquired using a soft, V-shaped Si_3N_4 cantilever/tip is shown as a solid curve as a function of time in Figure 6.30a. This is superimposed on a large-amplitude dotted sine wave representing the cycling of the Z scanner to approach and retract tip-to-sample or sample-to-tip (dependent on instrument design), with the approach displacement directed upward. The force signal data corresponding to tip–sample contact coincide with the Z sinusoid, for example, within the time interval designated as 1, which is usually a small fraction of the complete period of Z cycling, designated as time interval 2. Outside of contact, the average force signal is zero; but immediately following each break of tip–sample contact, there may be a large ring-down of the cantilever (the "rattling of the diving board" after tip–sample adhesion is broken). Observing and quantifying this ring-down is enabled by a high-speed data acquisition card (e.g., 5 megasamples/s); traditional AFM force curves are not acquired at high enough data sampling rate to time-resolve the cantilever ring-down.

Figure 6.30b examines more closely one contact cycle, together with a flat-line, noncontact force signal immediately before contact and the high-frequency, exponentially damped ring-down just after contact. Although there are capabilities to collect and save complete approach–retract cycles, usually only select points within the cycle are saved to memory or used for tracking topography. The setpoint is the *maximum deflection signal within a subset of the contact time interval.* This setpoint can be an *absolute* read signal (split photodiode differential voltage) or it can be this cantilever deflection signal *relative* to that measured at the zero-force baseline just before contact, denoted by a horizontal dashed line during contact in Figure 6.30b. Deviations from this setpoint cause the feedback circuit to displace the usual Z scanner, which is in series with the sinusoidal Z displacement. (This may be a separate device, or superimposed signals may be utilized within a single Z-scanning device.)

Apart from the force signal utilized by the feedback tracking, other data points may be collected and saved from within each Z cycle, with which images may be constructed. The most obvious datum to use for materials contrast is the lowest point per cycle, which is usually the pull-off force measurement, or in any case (including JKR mechanics, where the lowest point is not pull-off) provides a measure of *tip–sample adhesion.* Another type of measurement is a *differential stiffness,* simply a change of force signal over some specified time differential. It is important to reiterate that by itself this measurement is not contact stiffness, which has units of N/m. Apart from the trivial need to convert deflection signal to

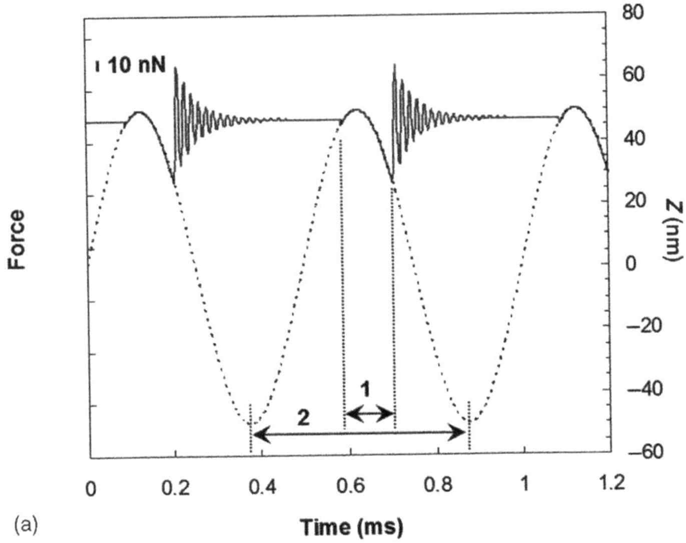

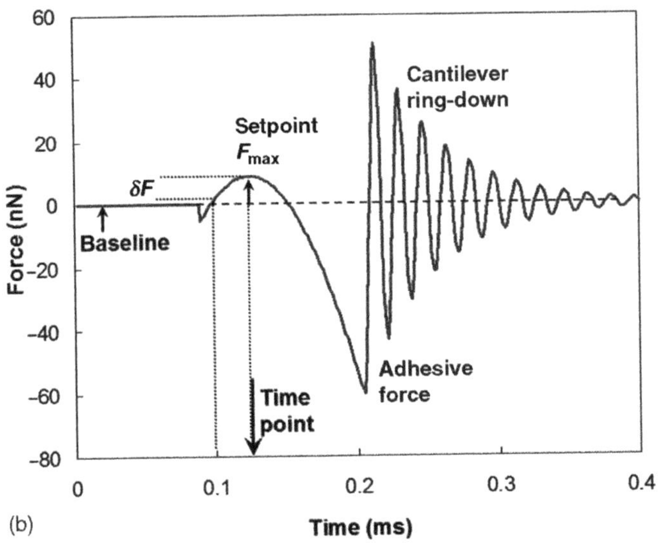

FIGURE 6.30 (a) Wide and (b) narrow time-domain plots of quasistatic force (cantilever deflection, solid curves) during rapid force-curve mapping using a V-shaped cantilever. Sinusoidal *Z* scanner approach and retract cycles are overlaid as dotted curve in (a), with total contact time (1) and total cycle time (2) identified. Relevant operational measurements labeled in (b).

force units (Section 3.2) for the numerator, one must determine the actual *change of tip–sample distance* (i.e., indentation *h*) by computing the change in *Z* over this interval plus the change in deflection in calibrated length units (Section 3.2). Another metric is the *time point at which the maximum force signal is detected.*

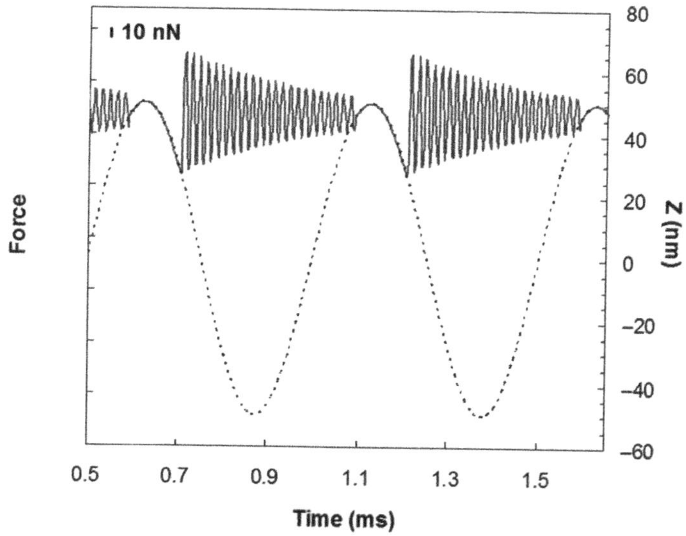

FIGURE 6.31 Same as Figure 6.30a but using a rectangular and higher-Q cantilever of higher spring constant.

This time point, within the contact cycle, may change from one tip–sample contact to another, and this movement may derive from differences in viscoelasticity (i.e., time-dependent stress for an imparted strain).

The character and appearance of the force data per cycle can depend strongly on the stiffness k and quality factor Q of the cantilever used. As discussed in Section 3.3, the magnitude of cantilever stiffness can determine whether jumps to and from contact are observed (low k), as opposed to a continuous force–Z relationship (high k). Second, the magnitude of Q can determine the time duration of the ring-down. In particular, high-Q cantilevers ring-down slowly enough that the cantilever may be still ringing upon the next contact. An example is shown in Figure 6.31 for a rectangular silicon cantilever with no metal coating. Here, the time constant of exponential ring-down is long enough that the decay of amplitude looks almost linear in time, over the (short) time lapse between the break of one contact and the initiation of the subsequent contact. Indeed, on some samples, the cantilever oscillation will continue to be slightly visible even after the formation of the next contact, usually as steeply decaying sinusoid superimposed on the left side of the contact interval (1 in Figure 6.30a). Analysis of the frequency and damping of this signal could conceivably provide a *contact resonance spectroscopy*, the topic of Section 9.2.3.

6.5.1 Liquidy Domains in Complex Thin Films

This example revisits the sample of Section 6.4.4, a film of model hair conditioner cast on cleaved mica from an aqueous mixture of PDMS melt, cetyl alcohol, and

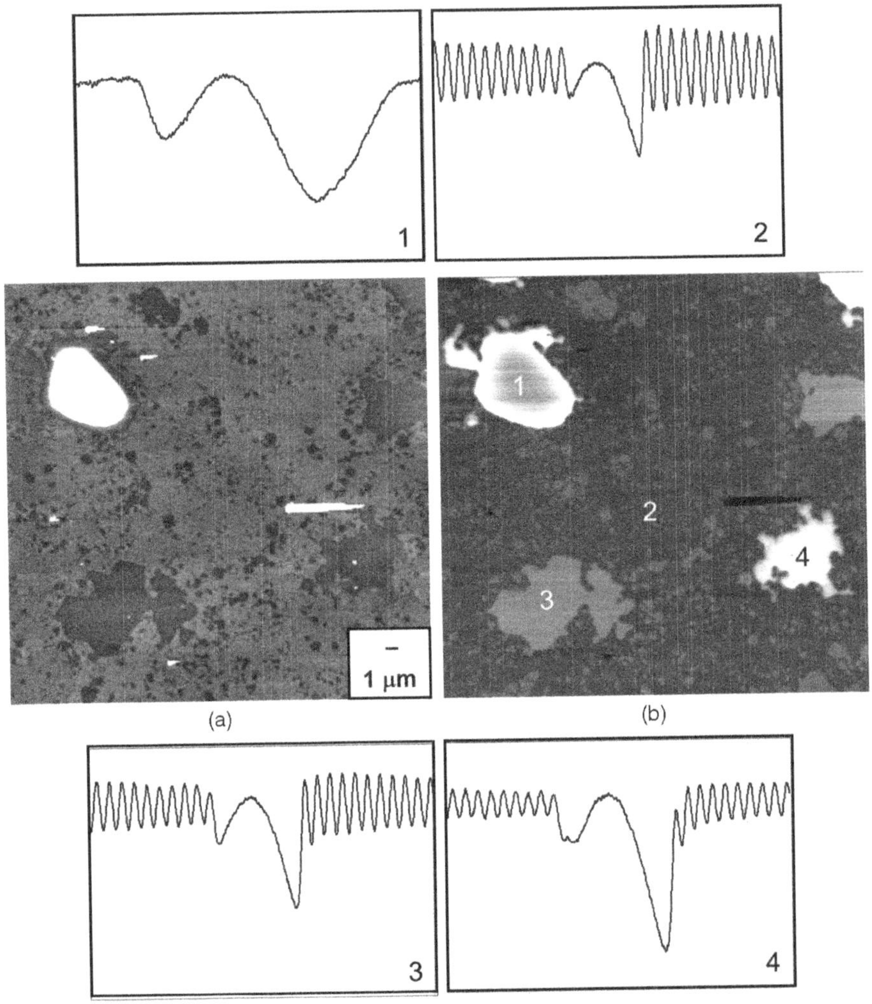

FIGURE 6.32 (a) Height and (b) adhesion (pull-off force) images of a complex film of silicone oil and surfactants similar to that in Figure 6.29. (1–4) Time-domain force curves extracted from a 512 × 512 grid at the denoted locations in (b).

CTAC. Figure 6.32 contains (larger field-of-view than Figure 6.29) height and adhesion images and four selected time cycles of tip–sample interaction, which are characteristic of four different behaviors observed in different surface regions as indicated. Not only is there rich adhesion contrast between the four regions (i.e., variation in lowest force point during retraction), the ring-down behavior is starkly different. Location 1 shows no ring-down at all; instead, a continuous reduction in attractive force during retraction. Location 2 shows the smallest adhesion yet the largest amplitude ring-down, meaning the tip is released from the lowest point. Locations 3 and 4 exhibit increasing adhesion compared to 2 but (counter-intuitively) progressively lesser ring-down amplitude. For this thin-film system, a reasonable interpretation (including a much larger body of work with controls, e.g., formulations with no PDMS) is *increasingly viscous locations* for the progression 2-3–4-1. The ratio of initial ring-down amplitude squared to force minimum (adhesion) squared is a measure of energy retained by the cantilever spring in pulling out of contact. The smaller this ratio (zero at location 1), the more the cantilever's potential energy, contained in the spring force, has been imparted to the sample. If one collects and saves all time cycles (i.e., at every image point), one can in principle construct an image of viscous energy absorption in breaking tip–sample contact.

6.5.2 PBMA/PLMA Blend at Variable Ultimate Load

Figure 6.33 contains images of what could be described as a "canonical" AFM example: materials contrast derived from mechanical and chemical differences. Here, we return to the PBMA/PLMA blend system at room temperature, the former polymer being glassy (E $\sim$ 2 GPa) and the latter rubbery (MPa regime). We compare "height", adhesion, and raw stiffness images acquired at 6 nN (a)–(c) and 30 nN (d)–(f) maximum force per cycle along with overlaid histograms of adhesion (g) within subregions indicated in (b) and (e). In "height" we see large crater-shape features that tend to correspond to high adhesion and low stiffness, presumably indicating phase-segregated domains of PLMA. Also in "height" we see smaller circular regions that are relatively high in adhesion, low in stiffness, and indeed that "sink" in apparent height for the higher maximum force case (d). This depth of "sinking" is about 20 nm, consistent with a modulus of $\sim$80 MPa assuming a tip of 10 nm radius, given the maximum force of 30 nN (see Section 2.3 narrative accompanying Figure 2.5). Correlating with this mechanical compliance is a strong increase in adhesion (pull-off force). For a 30 nN maximum pushing force, an adhesion upon retraction of around 40 nN is observed on the PLMA, whereas only 25 nN is observed on the PBMA. For a maximum pushing force of 6 nN, however, there is very little difference in tip–sample adhesion on PLMA compared to PBMA (about 22 nN). This is therefore a dramatic example of how adhesion hysteresis can be much more than what is (thermodynamically) computed by the relative *surface energies* of PLMA and PBMA (e.g., the JKR model), which would predict differences in pull-off force of less than 10% [102]. Instead, adhesion can be strongly affected by *viscoelastic memory* of the maximum pushing force during approach. The idea is that irreversible, dissipative processes scale with the maximum sizes of the contact area and deformation volume, leading to a dependence on load history, something absent from the elastic, adhesive contact mechanics models discussed in Section 6.2. At this writing, some first attempts have been made to theoretically describe and interpret viscoelastic contact mechanics [2–4].

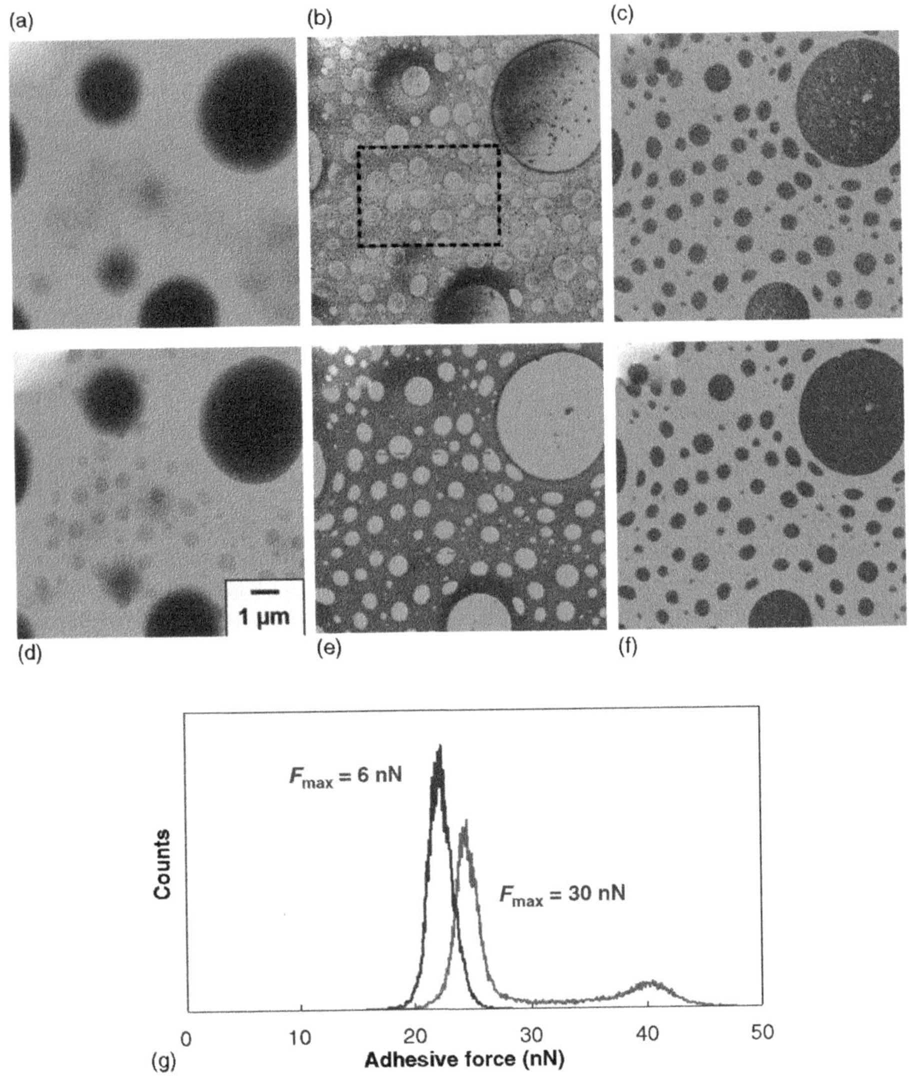

FIGURE 6.33 Height, adhesion, and qualitative stiffness images, respectively, at two F_{max} setpoints: (a)–(c) 6 nN and (d)–(f) 30 nN. (g) Histograms of adhesion compiled over the rectangular regions designated in (b) and (e).

6.5.3 PBMA/Dexamethasone Mixture at Variable Temperature

Figure 6.34 contains height and adhesion images of a 50:50 polymer–drug mixture: PBMA, which is glassy as a homopolymer at room temperature, and dexamethasone, which tends to crystallize in its pure state. This system is considered in Section 8.3.4 (Figure 8.14) in the context of grain-size analysis. As cast, this

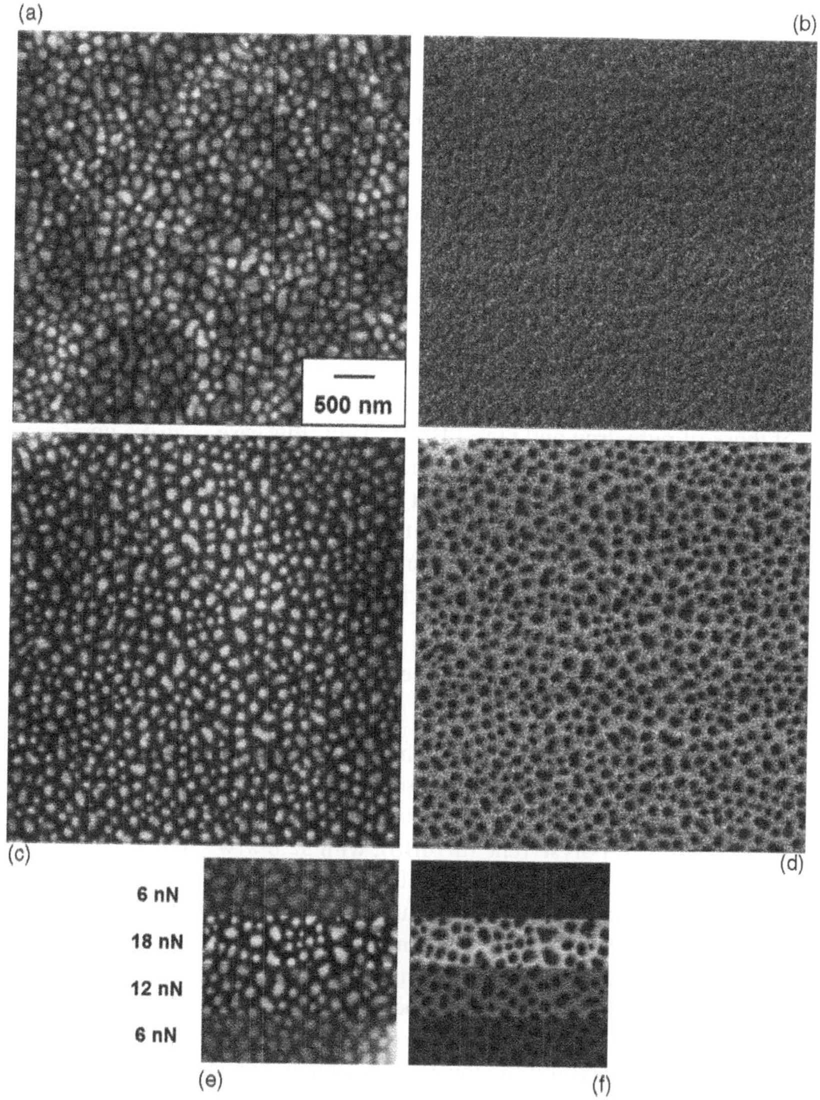

FIGURE 6.34 (a), (c), (e) Height and (b), (d), (f) adhesion images of a thick coating of a 50:50 mixture of PBMA and dexamethasone. Sample temperatures were room (a) and (b) and 95°C (c)–(f). Maximum-force setpoint was varied in (e) and (f) as indicated.

~1 μm-thick coating (spun onto a silicon wafer) suggests a phase segregated, "islanded" morphology in the height image but minimal adhesion contrast (or phase contrast in dynamic mode or sliding friction in contact mode) at room temperature (a) and (b). Raising the temperature to 95°C results in strong contrast in the adhesion image (d) and a slight change in the apparent lateral size of the "islands" relative to the intervening "sea" in the height image (c). Increasing the load setpoint (maximum force) at this elevated temperature further reveals a "sinking" of the sea as seen in (e), indicating that it is relatively soft compared to the islands, and produces a strong increase of adhesion on the sea with much lesser increases atop the islands (f). Given the expectation that the polymer should soften and tackify at this elevated temperature, whereas pure drug domains should remain fairly rigid with much lesser changes in tack, we assign the sea to the polymer and the islands to the drug. Thus, this is a good example of the usefulness of temperature as a "contrast knob."

Upon returning to room temperature, the apparent lateral size of the islands (not shown) reverts to a morphology similar to (a), suggesting that the apparent reduced island sizes at elevated temperature are likely due to softening of the contact felt by the AFM tip near the island edges: material properties that are probably a superposition of those of pure drug and pure PBMA. Thus, this case is also a good example of mechanical differences producing somewhat false morphology, the relative size of different material domains. An attempt to titrate the relative presence of the two materials at the surface would better utilize the room temperature data where mechanical and adhesive property differences are small, whereas the elevated temperature data serve the purpose of identifying the two materials.

6.5.4 Arborescent Styrene–Isobutylene–Styrene Block Copolymer Plus Drug Rapamycin

Our final example illustrates the excellent spatial resolution that can be obtained with fast force-curve mapping, even on soft materials. This case is similar to that of the preceding section, a micrometers-thick coating consisting of arborescent polyisobutylene-block-polystyrene (*arb*IBS), with 32.8 w/w% polystyrene (PS) segments (synthesized by living carbocationic polymerization at University of Akron [103]) and the drug Rapamycin (Sirolimus), together "nanospray" deposited by a proprietary method [99]. Here we encounter a polymer matrix that is itself nanophase segregated, and a drug which in turns segregates into nanoscale domains, but which are somewhat larger than the block phase domains. Figure 6.35 contains height (a) and adhesion (b) images that contain protrusions of two types: "wormy" and circular. The latter exhibit the lowest adhesion and become holes upon aqueous immersion, and thus presumably contain drug, whereas the wormy protrusions contain an intermediate adhesion, with the valleys exhibiting the highest adhesion (and lowest rigidity, not shown). Thus, the wormy protrusions are identified as glassy PS domains and the valleys the rubbery polyisobutylene (PIB) domains that "sink" in part due to mechanical deformation. Within the domains of intermediate adhesion, however, there is an "ultrafine" substructure (i.e., finer than the usual block copolymer nanophase segregation length scale, ca. 50 nm) that has not been observed in similar coatings containing a linear PIB–PS–PIB block copolymer (e.g., Figure 5.29). Presumably, this ultrafine structure is due to the arborescent character of the IBS, that is, a finer and more complex phase segregation than the PS-cylinders-in-PIB-matrix segregation known from studies of the more common

(a)

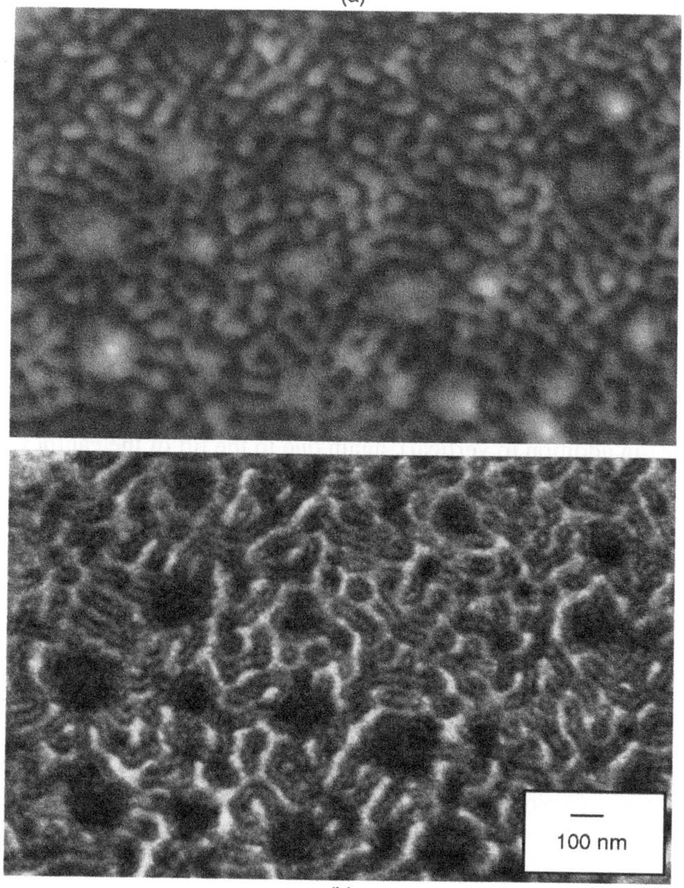

(b)

FIGURE 6.35 (a) Height and (b) adhesion images of a coating with a mixture of arborescent block copolymer and drug Rapamycin.

linear block copolymer (at similar weight fractions) [104]. There are moreover some very fine specks of the lowest adhesion, suggesting that a multiscale drug dispersion may have resulted from multiscale polymer phase segregation. Our purpose here is to simply illustrate that such *fine details are accessible to force-curve mapping*; the presumed "finer touch" of dynamic mode is not needed to reveal ultrafine structures within such multicomponent, complex systems.

6.5.5 Comments on "Force Modulation" Mode

It should be apparent to the reader that for some sufficiently small *Z*-cycling amplitude, an adequately large pulling force will not be reached and thus the break of contact will not take place. Instead, the tip will remain in contact with sample but with a strongly varying or "modulating" applied force. One could further reduce the Z-cycling amplitude such that the tip remains pinned at a point location (i.e., not sliding via cantilever tilt); then the response amplitude can gauge the elastic stiffness of the contact and thus the modulus of the material. Different from fast force-curve mapping, one can imagine tracking surface topography in contact mode—driving the Z feedback with the *time-averaged vertical deflection* of the cantilever—while analyzing the higher-frequency response to this oscillatory force (i.e., faster than the response of the feedback circuit), perhaps with a lock-in amplifier to give both amplitude and phase, or in-phase and out-of-phase response. This concept historically has been dubbed *force modulation mode*. To increase the signal/noise of these measurements, one can consider driving at cantilever resonance in contact, more commonly known as contact resonance force microscopy, the focus of Section 9.2.3. But one must contend with the possible effects of a sliding contact even at very small amplitudes, and moreover the role of *lateral* contact stiffness (i.e., in addition to vertical), again due to the tilt of the cantilever [22]. For these and other reasons, vertical force modulation mode has seen relatively little usage compared to the modes of principal focus in this book.

REFERENCES

[1] Greenwood, J.A., Adhesion of elastic spheres. *Proc. R. Soc. A*, 1997, **453**:1277–1297.

[2] Haiat, G., M.C. Phan Huy, and E. Barthel, The adhesive contact of viscoelastic spheres. *J. Mech. Phys. Solids*, 2003, **51**:69–99.

[3] Greenwood, J.A. and K.L. Johnson, Oscillatory loading of a viscoelastic adhesive contact. *J. Colloid Interface Sci.*, 2006, **296**:284–291.

[4] Giri, M., D.B. Bousfield, and W.N. Unertl, Dynamic contacts on viscoelastic films: Work of adhesion. *Langmuir*, 2001, **17**:2973–2981.

[5] Johnson, K.L., K. Kendall, and A.D. Roberts, Surface energy and the contact of elastic solids. *Proc. R. Soc. London A*, 1971, **324**:301–313.

[6] Derjaguin, B.V., V.M. Muller, and Y.P. Toporov, Effect of contact deformations on the adhesion of particles. *J. Coll. Interf. Sci.*, 1975, **53**(2):314–326.

[7] Bradley, R.S., The cohesive force between solid surfaces and the surface energy of solids. *Philos. Mag. Ser.*, 1932, **13**:853–862.

[8] Maugis, D., Contact, adhesion and rupture of elastic solids, in *Springer Series in Solid State Sciences*. ed. M. Cardona, et al. 2000, Berlin: Springer-Verlag.

[9] Maugis, D., Adhesion of spheres: The JKR-DMT transition using a Dugdale model. *J. Coll. Interf. Sci.*, 1992, **150**:243–269.

[10] Fischer-Cripps, A.C., *Introduction to contact mechanics*. 2nd ed., 2007, New York: Springer, 221.

[11] Carpick, R.W., D.F. Ogletree, and M. Salmeron, A general equation for fitting contact area and friction vs load measurements. *J. Colloid Interface Sci.*, 1999, **211**:395–400.

[12] Pietrement, O. and M. Troyon, General equations describing elastic indentation depth and normal contact stiffness versus load. *J. Colloid Interface Sci.*, 2000, **226**:166–171.

[13] Schwarz, U.D., A generalized analytical model for the elastic deformation of an adhesive contact between a sphere and a flat surface. *J. Colloid Interface Sci.*, 2003, **261**:99–106.

[14] Johnson, K.L. and J.A. Greenwood, An adhesion map for the contact of elastic spheres. *J. Colloid Interface Sci.*, 1997, **192**:326–333.

[15] Barthel, E., Adhesive elastic contacts: JKR and more. *J. Phys. D: Appl. Phys.*, 2008, **41**:163001.

[16] Mesarovic, S.D. and K.L. Johnson, Adhesive contact of elastic-plastic spheres. *J. Mech. Phys. Solids*, 2000, **48**:2009–2033.

[17] Sauer, R.A. and S. Li, An atomic interaction-based continuum model for adhesive contact mechanics. *Finite Elements Anal. Des.*, 2007, **43**:384–396.

[18] Radhakrishnan, H. and S.D. Mesarovic, Adhesive contact of elastic spheres revisited: Numerical models and scaling. *Proc. R. Soc. A*, 2009, **465**:2231–2249.

[19] Israelachvili, J.N., *Intermolecular and Surface Forces*. 2nd ed., 1991, San Diego: Academic Press, 450.

[20] Fowkes, F.M., Additivity of intermolecular forces at interfaces. I. Determination of the contribution to surface and interfacial tensions of dispersion forces in various liquids. *J. Phys. Chem.*, 1963, **67**(12):2538–2541.

[21] Noy, A., D.V. Vezenov, and C.M. Lieber, Chemical force microscopy. *Annu. Rev. Mater. Sci.*, 1997, **27**:381–421.

[22] Mazeran, P.-E. and J.L. Loubet, Force modulation with a scanning force microscope: An analysis. *Trib. Lett.*, 1997, **3**:125–132.

[23] Fisher, R.A., On the capillary forces in an ideal soil; correction of formulae given by W. B. Haines. *J. Agric. Sci.*, 1926, **16**:492–505.

[24] Orr, F.M., L.E. Scriven, and A.P. Rivas, Pendular rings between solids: Meniscus properties and capillary force. *J. Fluid Mech.*, 1975, **667**:723–742.

[25] Bisschop, F.R.E. and W.J.L. Rigole, A physical model for liquid capillary bridges between adsorptive solid spheres: The Nodoid of Plateau. *J. Colloid Interface Sci.*, 1982, **88**(1):117–128.

[26] de Gennes, P.-G., F. Brochard-Wyart, and D. Quere, *Capillarity and Wetting Phenomena*. 2004, New York: Springer.

[27] Rowlinson, J.S. and B. Widom, *Molecular Theory of Capillarity*. 1982, Oxford: Clarendon Press.

[28] Starov, V.M., M.G. Velarde, and C.J. Radke, Wetting and spreading dynamics, in *Surfactant Science Series*. Vol. **138**, 2007, Boca Raton, FL: CRC Press.

[29] Rudnick, L.R., ed. *Synthetics, Mineral Oils and Bio-Based Lubricants*. 2006, Boca Raton, FL: CRC Press.

[30] Butt, H.-J. and M. Kappl, Normal capillary forces. *Adv. Colloid Interface Sci.*, 2009, **146**:48–60.

[31] van Honschoten, J.W., N. Brunets, and N.R. Tas, Capillarity at the nanoscale. *Chem. Soc. Rev.*, 2010, **39**:1096–1114.

[32] Charlaix, E. and M. Ciccoti, Capillary condensation in confined media, in *Handbook of Nanophysics*, K. Sattler, ed., 2010, Boca Raton, FL: CRC Press, p. 1–17.

[33] Willett, C.D., et al., Pendular capillary bridges, in *Granulation*, Salman, A.D., M.J. Hounslow, and J.P.K. Seville, eds., 2007, New York: Elsevier, p. 1317–1351.

[34] Verdaguer, A., et al., Molecular structure of water at interfaces: Wetting at the nanometer scale. *Chem. Rev.*, 2006, **106**:1478–1510.

[35] Nosonovsky, M. and B. Bhushan, Phase behavior of capillary bridges: Towards nanoscale water phase diagram. *Phys. Chem. Chem. Phys.*, 2008, **10**:2137–2144.

[36] Feiler, A.A., P. Jenkins, and M.W. Rutland, Effect of relative humidity on adhesion and frictional properties of micro- and nano-scopic contacts. *J. Adhesion Sci. Technol.*, 2005, **19**(3–5):165–179.

[37] Eastman, T. and D.-M. Zhu, Adhesion forces between surface-modified AFM tips and a mica surface. *Langmuir*, 1996, **12**:2859–2862.

[38] Thundat, T., et al., Characterization of atomic force microscope tips by adhesion force measurements. *Appl. Phys. Lett.*, 1993, **63**(15):2150–2152.

[39] Thundat, T., et al., Role of relative humidity in atomic force microscopy imaging. *Surf. Sci. Lett.*, 1993, **294**:939–943.

[40] Berger, C.E.H., et al., Functional group imaging by adhesion AFM applied to lipid monolayers. *Langmuir*, 1995, **11**:4188–4192.

[41] Fisher, L.R. and J.N. Israelachvili, Direct measurement of the effect of meniscus forces on adhesion: A study of the applicability of macroscopic thermodynamics to microscopic liquid interfaces. *Colloids Surf.*, 1981, **3**:303–319.

[42] Xu, L., et al., Wetting and capillary phenomena of water on mica. *J. Phys. Chem. B*, 1998, **102**:540–548.

[43] Sedin, D.L. and K.L. Rowlen, Adhesion forces measured by atomic force microscopy in humid air. *Anal. Chem.*, 2000, **72**:2183–2189.

[44] Farshchi-Tabrizi, M., et al., On the adhesion between fine particles and nanocontacts: An atomic force microscope study. *Langmuir*, 2006, **22**:2171–2184.

[45] Butt, H.-J., M. Farshchi-Tabrizi, and M. Kappl, Using capillary forces to determine the geometry of nanocontacts. *J. Appl. Phys.*, 2006, **100**:024312.

[46] Fuji, M., et al., Effect of wettability on adhesion force between silica particles evaluated by atomic force microscopy masurement as a function of relative humidity. *Langmuir*, 1999, **15**:4584–4589.

[47] Xiao, X. and L. Qian, Investigation of humidity-dependent capillary force. *Langmuir*, 2000, **16**(21):8153–8158.

[48] He, M., et al., Critical phenomena of water bridges in nanoasperity contacts. *J. Chem. Phys.*, 2001, **114**(3):1355–1360.

[49] Asay, D.B. and S.H. Kim, Effects of adsorbed water layer structure on adhesion force of silicon oxide nanoasperity contact in humid ambient. *J. Chem. Phys.*, 2006, **124**:174712.

[50] Rabinovich, Y.I., et al., Kinetics of liquid annulus formation and capillary forces. *Langmuir*, 2011, **27**(22):13514–13523.

[51] McFarlane, J.S. and D. Tabor, Adhesion of solids and the effect of surface films. *Proc. R. Soc. A*, 1950, **202**:224–243.

[52] Chen, L., et al., Effects of humidity and sample surface fre energy on AFM probe-sample interactions and lateral force microscopy image contrast. *Langmuir*, 2009, **25**:3494–3503.

[53] Binggeli, M. and C.M. Mate, Influence of capillary condensation of water on nanotribology studied by force microscopy. *Appl. Phys. Lett.*, 1994, **65**(4):415–417.

[54] Major, R.C., et al., Viscous water meniscus under nanoconfinement. *Phys. Rev. Lett.*, 2006, **96**:177803.

[55] Ogata, S., et al., Molecular dynamics simulation of ultra-thin liquid bridging between surfaces having two-dimensional roughness. *Microsyst. Technol.*, 2005, **11**:1146–1153.

[56] Pakarinen, O.H., et al., Towards an accurate description of the capillary force in nanoparticle-surface interactions. *Modelling Simul. Mater. Sci. Eng.*, 2005, **13**:1175–1186.

[57] Choi, H.J., et al., Molecular simulation of the nanoscale water confined between an atomic force microscope tip and a surface. *Molecular Simulation*, 2009, **35**(6):466–472.

[58] Chau, A., et al., Three-dimensional model for capillary nanobridges and capillary forces. *Modelling Simul. Mater. Sci. Eng.*, 2007, **15**:305–317.

[59] Men, Y., X. Zhang, and W. Wang, Rupture kinetics of liquid bridges during a pulling process: A kinetic density functional theory study. *J. Chem. Phys.*, 2011, **134**:124704.

[60] Jang, J., G.C. Schatz, and M.A. Ratner, How narrow can a meniscus be? *Phys. Rev. Lett.*, 2004, **92**(8):085504.

[61] Jang, J., M.A. Ratner, and G.C. Schatz, Atomic-scale roughness effect on capillary force in atomic force microscopy. *J. Phys. Chem. B*, 2006, **110**:659–662.

[62] Sprakel, J., et al., Capillary adhesion in the limit of saturation: Thermodynamics, self-consistent field modeling and experiment. *Langmuir*, 2008, **24**:1308–1317.

[63] Braunschweig, A.B., F. Huo, and C.A. Mirkin, Molecular printing. *Nature Chem.*, 2009, **1**:353–358.

[64] Salaita, K., Y. Wang, and C.A. Mirkin, Applications of dip-pen nanolithography. *Nature Nanotech.*, 2007, **2**:145–155.

[65] Weeks, B.L., et al., Effect of dissolution kinetics on feature size in dip-pen nanolithography. *Phys. Rev. Lett.*, 2002, **88**(25):255505.

[66] Riedo, E., F. Levy, and H. Brune, Kinetics of capillary condensation in nanoscopic sliding friction. *Phys. Rev. Lett.*, 2002, **88**(18):185505.

[67] Riedo, E., et al., The 2/3 power law dependence of capillary force on normal load in nanoscopic friction. *J. Phys. Chem. B*, 2004, **108**:5324–5328.

[68] Szoszkiewicz, R. and E. Riedo, Nucleation time of nanoscale water bridges. *Phys. Rev. Lett.*, 2005, **95**:135502.

[69] Szoszkiewicz, R. and E. Riedo, Nanoscopic friction as a probe of local phase transitions. *Appl. Phys. Lett.*, 2005, **87**:033105.

[70] Zheng, W. and Y.-P. Zhao, Growth of liquid bridge in AFM. *J. Phys. D: Appl. Phys.*, 2007, **40**:4368–4375.

[71] Feiler, A.A., et al., Effect of capillary condensation on friction force and adhesion. *Langmuir*, 2007, **23**:517–522.

[72] Sirghi, L., R. Szoszkiewicz, and E. Riedo, Volume of a nanoscale water bridge. *Langmuir*, 2006, **22**:1093–1098.

[73] Kohonen, M.M., N. Maeda, and H.K. Christenson, Kinetics of capillary condensation in a nanoscale pore. *Phys. Rev. Lett.*, 1999, **82**(23):4667–4670.

[74] Wang, J., J. Qian, and H. Gao, Effects of capillary condensation in adhesion between rough surfaces. *Langmuir*, 2009, **25**(19):11727–11731.

[75] Halsey, T.C. and A.J. Levine, How sandcastles fall. *Phys. Rev. Lett.*, 1998, **80** (14):3141–3144.

[76] Mayer, R.P. and R.A. Stowe, Nodoids and toroids: comparison of two geometries for the meniscus profile of a wetting liquid between two touching isolated spheres and extensions to the model of a collection of packed spheres. *J. Coll. Interf. Sci.*, 2005, **285**:781–788.

[77] Rabinovich, Y.I., M.S. Esayanur, and B.M. Moudgil, Capillary forces between two spheres with a fixed volume liquid bridge: Theory and experiment. *Langmuir*, 2005, **21**:10992–10997.

[78] Kusumaatmaja, H. and J.M. Yeomans, Modeling contact angle hysteresis on chemically patterned and superhydrophobic surfaces. *Langmuir*, 2007, **23**:6019–6032.

[79] Gao, L. and T.J. McCarthy, Contact angle hysteresis explained. *Langmuir*, 2006, **22**:6234–6237.

[80] de Lazzer, A., M. Dreyer, and H.J. Rath, Particle-surface capillary forces. *Langmuir*, 1999, **15**:4551–4559.

[81] Miranda, P.B., et al., Icelike water monolayer adsorbed on mica at room temperature. *Phys. Rev. Lett.*, 1998, **81**(26):5876–5879.

[82] Choe, H., et al., Formation, manipulation, and elasticity measurement of a nanometric column of water molecules. *Phys. Rev. Lett.*, 2005, **95**:187801.

[83] Granick, S. and H.-W. Hu, Nanorheology of confined polymer melts. 1. Linear shear response at strongly adsorbing surfaces. *Langmuir*, 1994, **10**:3857–3866.

[84] Williams, K.J., et al., Strong electronic coupling in two-dimensional assemblies of colloidal PbSe quantum dots. *ACS Nano*, 2009, **3**:1532–1538.

[85] Garcia, R., R.V. Martinez, and J. Martinez, Nano-chemistry and scanning probe nanolithographies. *Chem. Soc. Rev.*, 2005, **35**(1):29–38.

[86] Day, H.C. and D.R. Allee, Selective area oxidation of silicon with a scanning force microscope. *Appl. Phys. Lett.*, 1993, **62**:2691.

[87] Haugstad, G., et al., Probing biopolymers with scanning force methods: Adsorption, structure, properties, and transformation of gelatin on mica. *Langmuir*, 1994, **10**:4295–4306.

[88] Yannas, I.V., Collagen and gelatin in the solid state. *J. Macromol. Sci. C*, 1972, **7** (1):49–104.

[89] Djabourov, M., Architecture of gelatin gels. *Contemp. Phys.*, 1988, **29**(3):273–297.

[90] Rotsch, C. and M. Radmacher, Mapping local electrostatic forces with the atomic force microscope. *Langmuir*, 1997, **13**:2825–2832.

[91] Dufrene, Y.F., et al., Nanometer-scale surface properties of mixed phospholipid monolayers and bilayers. *Langmuir*, 1997, **13**:4779–4784.

[92] A-Hassan, E., et al., Relative microelastic mapping of living cells by atomic force microscopy. *Biophys. J.*, 1998, **74**:1564–1578.

[93] Domke, J. and M. Radmacher, Measuring the elastic properties of thin polymer films with the atomic force microscope. *Langmuir*, 1998, **14**:3320–3325.

[94] Haugstad, G. and R.R. Jones, Mechanisms of dynamic force microscopy on polyvinyl alcohol: Region-specific non-contact and intermittent contact regimes. *Ultramicroscopy*, 1999, **76**:77–86.

[95] Buehler, A. and J. McGowen, *Polyacrylamide Hydrogels and Hydrogel Arrays Made from Polyacrylamide Reactive Prepolymers.* 2002, United States: Motorola, Inc., p. 18.

[96] Laughlin, R.G., *The Aqueous Phase Behavior of Surfactants.* 1996, New York: Academic Press.

[97] Haugstad, G. and A. Avery. Probing the morphology and tribo-response of nanostructured fluid films for personal care applications. in Technical Proc. 2005 NSTI Nanotechnol. Conf. Trade Show. 2005, Anaheim, CA: Nano Sci. Technol. Inst.

[98] Gigler, A.M. and O. Marti, Quantitative measurement of materials properties with the (digital) pulsed force mode, in *Applied Scanning Probe Methods.* Tomitori, M.B. Bhushan, and H. Fuchs, eds. 2008, New York: Springer, p. 23–54.

[99] Dong, J., et al., Multimodal dynamic imaging of therapeutic biomedical coatings in aqueous medium. *Langmuir*, 2009, **25**:5442–5445.

[100] Rico, F., C. Su, and S. Scheuring, Mechanical mapping of single membrane proteins at submolecular resolution. *Nano Lett.*, 2011, **11**:3983–3986.

[101] Schon, P., et al., Quantitative mapping of elastic moduli at the nanoscale in phase separated polyurethanes by AFM. *Eur. Polym. J.*, 2011, **47**:692–698.

[102] Brandrup, J., E.H. Immergut, and E.A. Grulke, eds. *Polymer Handbook.* 4th ed., 1999, New York: Wiley-Interscience.

[103] Puskas, J.E. and Y. Kwon, *Polym. Adv. Technol.*, 2006, **17**(9–10): p. 615.

[104] Puskas, J.E., et al., The effect of hard and soft segment composition and molecular architecture on the morphology and mechanical properties of polystyrene–polyisobutylene thermoplastic elastomeric block copolymers. *Eur. Polym. J.*, 2003, **39**:2041–2049.

7

PROBING MATERIAL PROPERTIES III: LATERAL FORCE METHODS

As previewed in Section 1.2, lateral force measurements in contact mode can be highly sensitive to material composition and structure, and thereby provide *materials contrast* [1–9]. (This naturally includes the ability to image *surface contamination.*) As demonstrated in Figures 1.8, 1.9 and 1.12, sliding friction can be sensitive to spatial variations in *crystallinity* [6–9] and *viscoelasticity* [5,6,9], and can be used to probe changes due to gaseous environment (e.g., polymer *plasticization*) [6] as well as sample temperature (e.g., *glass-to-rubber transition*) [6,9–11]; of course, lateral force measurements are useful for studying *tribology per se*: [9,12,13] controlled, quantitative studies of friction and wear, and down to the nanometer scale.

7.1 COMPONENTS OF LATERAL FORCE SIGNAL

Lateral forces unavoidably result during sliding or shear modulation measurements. The latter are small-amplitude lateral oscillations with tip pinned by static friction (Section 7.7). In general, these measured lateral forces derive not only from pure shear—that is, force tangential to the tip–sample interface—but also from the lateral component of force that is *normal* to the tip–sample interface, given a surface with *nonzero slope*. This is depicted in Figure 7.1. In this section, we discuss how to identify and separate these physically distinct contributions to the overall lateral

Atomic Force Microscopy: Understanding Basic Modes and Advanced Applications, First Edition.
Greg Haugstad.

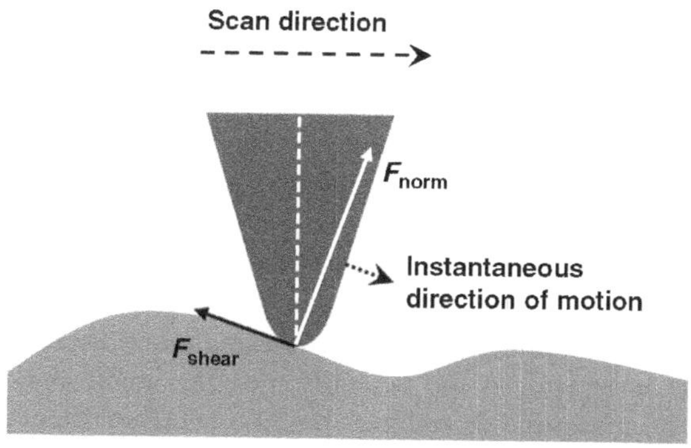

FIGURE 7.1 Illustration of a tip sliding tangentially to a surface under *Z* feedback. Solid force vectors normal and tangential (in shear) to the tip–sample interface are indicated. Both produce torque with respect to the tip axis are thus measured as lateral forces.

force. Importantly, the lateral force derived from the tip–sample normal force is not truly frictional in the scientific sense of the term (energy dissipative), because it is not directed to oppose the instantaneous (tangential) motion. The lateral force derived from the tangential force is frictional because it always resists motion; thus over a closed cycle, net work is done.

Let us first examine the means by which we measure lateral force. As previewed in Chapter 1, the quad photodiode in conjunction with the cantilever's *twisting* degree of freedom enables quantification of lateral forces via the torque on the tip, as depicted in Figure 7.2. This twist of the cantilever displaces the

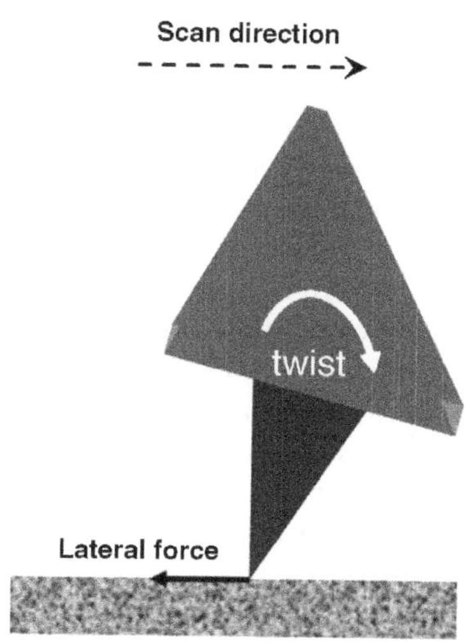

FIGURE 7.2 Illustration of an AFM cantilever (looking down the principal axis) twisting due to a lateral force exerted at the end of the tip.

reflected laser beam horizontally at the quad photodetector (see Figure 1.4). Similar to the vertical deflection measurements, the difference in the output signals produced in the left and right halves of the quad photodetector, normalized by the sum of all four outputs and multiplied by a constant (typically 10 V), is the first step toward quantitation: a voltage signal derived from lateral force. Usually, this measurement scheme is quite *linear*, which first means that the twist angles are very small. Thus, a given force produces some very small cantilever twist angle, which displaces the laser spot at the detector; doubling the force would double the (still very small) twist angle and thereby double the displacement of the laser spot at the detector (i.e., the small-angle approximation tan(angle) $\approx$ angle, measured in radians, holds true). But the overall linearity of the measurement further requires that the laser spot displacement be small compared to the spot size; otherwise, nonlinear contributions will arise due to the intensity profile of the laser spot, as was the case for vertical cantilever deflection measurement discussed in the context of Figures 3.11 and 3.12. To ensure linearity, lateral force signals should not near the limits of the measurement apparatus, typically ± 10 V. With commonly used tips one typically encounters lateral force signals of order 1 V or smaller in magnitude, consistent with this criterion. (If using a microsphere attached to the cantilever, i.e., colloid probe AFM, a lateral force signal may approach 10 V in magnitude and thus the photodetector measurement may become nonlinear.)

Quantification of lateral force also requires knowledge of the *zero* of the lateral force scale. The photodetector is never perfectly centered on the laser spot under zero force; thus for the most precise work, one would have to disengage tip and sample to determine the signal under zero lateral force. But we will see later that the zero of lateral force is moreover a function of position on the surface, due to an optical artifact (derived from the same interferometry concept as discussed in Section 3.4). One would have to actually *image* the zero lateral force signal, with tip and sample barely out of contact, and at all image sizes and locations of interest; thermal drift in the optical components would require this to be done repeatedly. Obviously, this is not a realistic approach for routine use. Fortunately, the need to measure the zero of the lateral force scale is obviated in most cases by taking measurements while scanning in each of two opposing directions (left-to-right/right-to-left or "trace"/"retrace"), the so-called *friction loop*, as was schematically shown in Figure 1.6 and here with experimental data (silicon tip/cantilever, $k \sim 3$ N/m, on diamond-like carbon film) in Figure 7.3. In the case of measuring or imaging friction, one can compute half the height of the friction loop as the quantitative gauge of sliding friction force (i.e., *kinetic* friction). Second, one usually leaves out the data near the turnaround points where static friction is dominant, that is, where the sign of the cantilever twist transitions from clockwise to counterclockwise or vice versa.

Exceptions do exist where the actual zero of lateral force signal must be measured. On certain crystalline molecular systems, anisotropy in the lattice (e.g., variations in the polar tilt of short-chain molecules) can result in different magnitudes of trace and retrace friction [8]. In this case, measuring the change in friction force

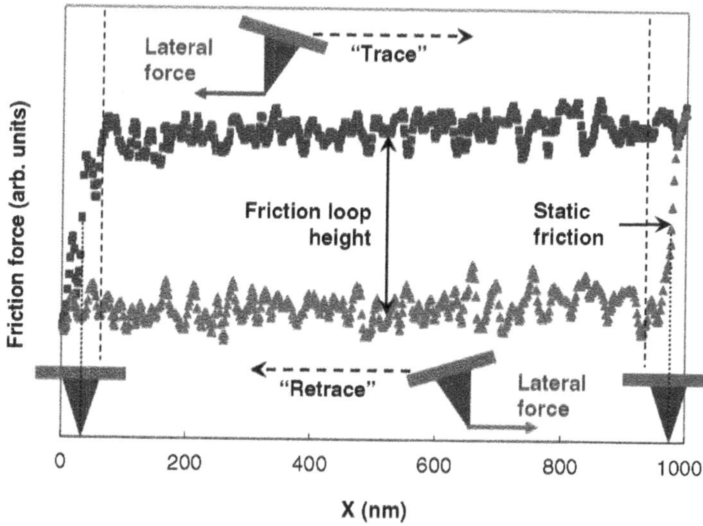

FIGURE 7.3 Example of "friction loop" data, with end view of cantilever depicted at certain locations. The loop height quantifies the magnitude of sliding friction, point by point along the scan line, as the trace-minus-retrace value at each X location (accounting for X hysteresis). Static friction outside of the vertical dashed lines is ignored.

signal upon changing scan direction cannot provide the individual magnitudes of friction for trace or retrace.

The essence of a sliding friction loop on an ideally smooth and horizontal two-material surface is illustrated in Figure 7.4. Here, we treat friction within the concepts of introductory physics. For coefficients of (kinetic) friction of $\mu_1 = 0.3$ and $\mu_2 = 0.6$—that is, the ratio of the magnitudes of horizontal shear force vector $\boldsymbol{F}_{\mathbf{shear}}$ and vertical normal force vector $\boldsymbol{F}_{\mathbf{norm}}$ during sliding on materials 1 and 2—the height of the depicted friction loop correspondingly varies by a factor of two. The direction of $\boldsymbol{F}_{\mathbf{shear}}$ is such that it opposes motion. For a constant $\boldsymbol{F}_{\mathbf{norm}}$, the relationship between the magnitudes of these forces is $F_{\mathrm{shear}} = \mu \cdot F_{\mathrm{norm}}$. Thus, the height of the friction loop on material 2, $2\mu_2 \cdot F_{\mathrm{norm}}$, is double that on material 1, $2\mu_1 \cdot F_{\mathrm{norm}}$.

Note that we do not make any assumptions, first, about the physical nature of F_{norm}; in general, it can include both the applied normal force due to the cantilever, which pushes or pulls on the tip depending on the sign of its deflection, and any intrinsically attractive or repulsive forces due to the tip–sample interaction (i.e., derived from van der Waals, DLVO, etc., forces), which act to push or pull the tip from or to the surface. In typical operation, a pulling (attractive) tip–sample interaction *adds* to a pushing cantilever force (applied load). In principle, this multi-component normal force F_{norm} can be described within the adhesive contact mechanics models of Sections 6.2 and 6.3. Second, we do not assume that the relationship between shear and normal force is described by *Amontons' Law*, meaning μ a *constant*. In general, μ may be a function of loading force, $\mu = \mu(F_{\mathrm{norm}})$ and indeed other parameters like temperature and scan rate.

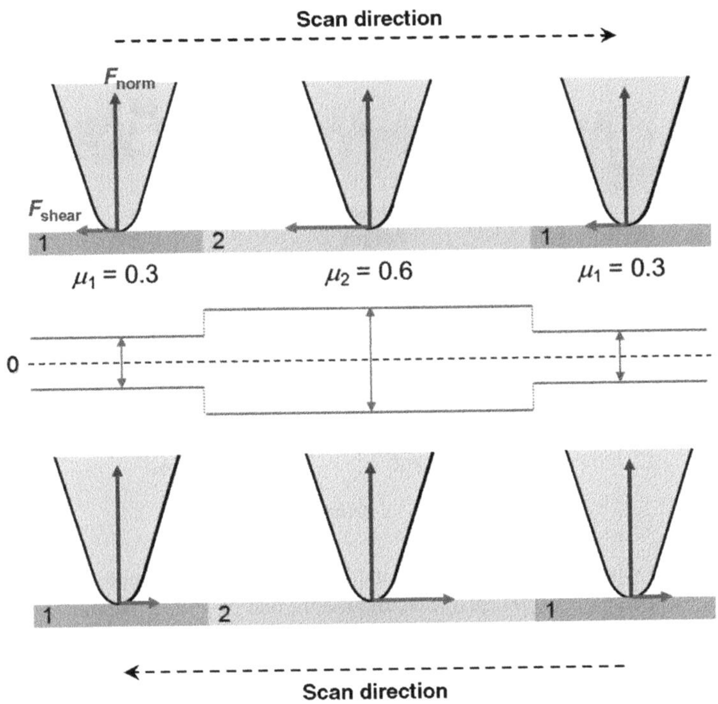

FIGURE 7.4 Illustration of shear (friction) and normal forces between tip and sample for two materials having coefficients of friction differing by a factor of two, for left-to-right and right-to-left scan directions. In middle is a depiction of a measured "friction loop" showing the change in shear force magnitude (per material) and sign (per scan direction).

In the presence of an inclined surface ($dZ/dX <> 0$, where X is transverse to the cantilever's principal axis), the picture is much more complicated, even on a single material and assuming tangential motion (perfect Z feedback). As seen in Figure 7.5, on a tilted surface there is not only a lateral component of the normal force but also a vertical component of the shear (friction) force. Significantly, the vertical component of the shear force *affects the Z feedback*. In order to keep the total vertical force constant, the feedback displaces Z so as to offset the effects of the vertical component of shear force. This Z displacement alters the vertical cantilever deflection and thus the normal force, which in turn changes the shear force via the physical relationship $\boldsymbol{F}_{\mathbf{shear}} = \mu \cdot \boldsymbol{F}_{\mathbf{norm}}$. The upshot is that the normal force—how strongly the tip is pressed to the surface—depends on whether the tip is sliding up or down an incline, as a result of Z feedback.

In further detail, Figure 7.5 illustrates tip scanning in an overall left-to-right direction on a single material but with steep inclines, slopes (dZ/dX) of magnitude $m = 0.55$ (m is again used to denote a dimensionless number but of totally different meaning compared to its usage in Section 6.2.4), and with coefficient of friction a

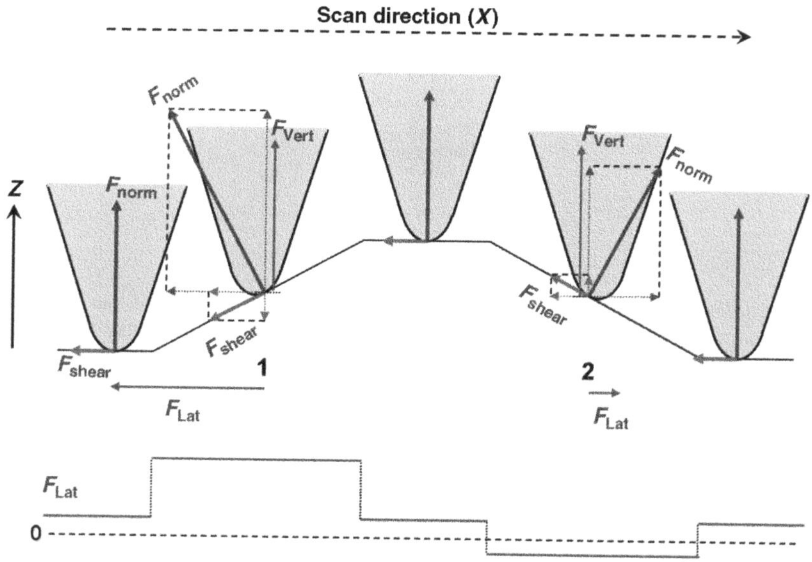

FIGURE 7.5 Illustration of vector forces on tip sliding left-to-right up, over and down a hill with steep, constant-slope inclines. The vertical component of shear force adds to or subtracts from the vertical component of normal force. Perfect Z feedback keeps the total vertical force F_{vert} constant, same as F_{norm} on horizontal terraces, thereby ensuring that the magnitude of the normal force is *not* constant. This in turn makes the magnitude of F_{shear} dependent on local slope. The sum of lateral components of F_{shear} and F_{norm} is the total lateral force F_{lat} depicted as a function of position, at bottom.

constant $\mu = 0.3$ at all locations. Dependent on the sign of slope of inclined regions 1 or 2, the vertical component of $\boldsymbol{F}_{\mathbf{shear}}$ either subtracts from (1) or adds to (2) the vertical component of $\boldsymbol{F}_{\mathbf{norm}}$. The net vertical force $\boldsymbol{F}_{\mathbf{Vert}}$ is held constant, the same as $\boldsymbol{F}_{\mathbf{norm}}$ on the horizontal locations, for the depicted case of perfect Z feedback (as driven by $\boldsymbol{F}_{\mathbf{Vert}}$). The lateral component of normal force either adds to or subtracts from the lateral component of the shear force to produce the aggregate lateral force $\boldsymbol{F}_{\mathbf{Lat}}$. We see that $\boldsymbol{F}_{\mathbf{Lat}}$ is much greater going up the incline (1) and much lesser going down (2). Indeed, the net lateral force going down is opposite in sign (values below the dashed line zero). Note that a slope dependence of lateral force would be present even in the absence of Z feedback; had $\boldsymbol{F}_{\mathbf{norm}}$ remained constant whether sliding up or down each incline, its lateral component still would have added to the lateral component of $\boldsymbol{F}_{\mathbf{shear}}$ going up and subtracted going down. So the impact of Z feedback is to *enhance* this effect.

One can start with the vector relationship $\boldsymbol{F}_{\mathbf{shear}} = \mu \cdot \boldsymbol{F}_{\mathbf{norm}}$ and write scalar equations relating F_{Lat} to that of F_{norm}, F_{shear}, F_{Vert}, and the sine or cosine of the angle of surface inclination, then solve to eliminate F_{norm} and F_{shear}. Also, replace the trigonometric functions with the slope m, for both left-to-right (trace) and right-to-left (retrace) scanning [14].[1] The following expressions relate the measurable quantities F_{Lat} (in trace and retrace), F_{Vert} and m to the generally unknown coefficient of friction μ:

$$F_{\text{Lat}}^{\rightarrow} = \frac{m+\mu}{1-m\mu} F_{\text{Vert}} \tag{7.1}$$

$$F_{\text{Lat}}^{\leftarrow} = \frac{m-\mu}{1-m\mu} F_{\text{Vert}}. \tag{7.2}$$

Here, F_{Lat} and m are treated as signed quantities, whereas μ and F_{Vert} are positive by definition. In the case of F_{Lat} positive corresponds with clockwise torque in the illustration of Figure 7.5. In the case of m, positive is defined as being upward for scanning from left to right. We see that a nonzero slope m effectively increases or decreases the coefficient of friction additively as $m \pm \mu$, while increasing or decreasing the AFM-controlled loading force F_{Vert} by the factor $(1 - m\mu)^{-1}$.

Typically, m and μ are fractional quantities, and m (and, in many cases, μ) often of order 0.1 or smaller. Thus $m\mu$ is often of order 0.01 (unlike Figure 7.5, where $m\mu = 0.165$) such that the factor $(1 - m\mu)^{-1}$ represents only a very fine correction. In this case, we may write the approximations

$$F_{\text{Lat}}^{\rightarrow} \approx (m+\mu) F_{\text{Vert}} \tag{7.3}$$

$$F_{\text{Lat}}^{\leftarrow} \approx (m-\mu) F_{\text{Vert}}. \tag{7.4}$$

Trace minus retrace lateral force eliminates m to give $2\mu F_{\text{Vert}}$, twice the friction force, as is the case of Figure 7.4 on horizontal surfaces. Trace plus retrace lateral force instead eliminates μ to give $2mF_{\text{Vert}}$, twice the slope-derived lateral force. Thus, these simple operations can be used to isolate material-specific friction—derived from intrinsic resistance to motion—from slope-derived lateral forces or what one might call extrinsic ("sandpaper") friction [3,15]. But we reiterate that these additive and subtractive operations can only be approximated as above if the *coefficients of friction and/or slopes across the surface are small.*

7.2 APPLICATION OF LATERAL FORCE DIFFERENCE

Figure 7.6 contains an example of the above mathematical operations applied to LFM data, on a discontinuous monolayer of the cationic surfactant behenyl trimethyl ammonium chloride (BTAC, with 22-carbon alkyl chain) adsorbed to cleaved mica, a fairly high surface energy (72 mJ/m^2) and atomically flat mineral

[1] There is a tiny correction due to the fact that the lateral force components when scanning up or down a slope are exerted slightly higher up on the tip, thereby reducing the torque by an amount that scales with the reduced effective length of the tip (moment arm); this reduction is normally of order 0.1% or less. A much more significant correction is needed if the angled side of a (pyramidal) tip touches the top of a tall vertical sidewall, micrometers above the end of the tip.

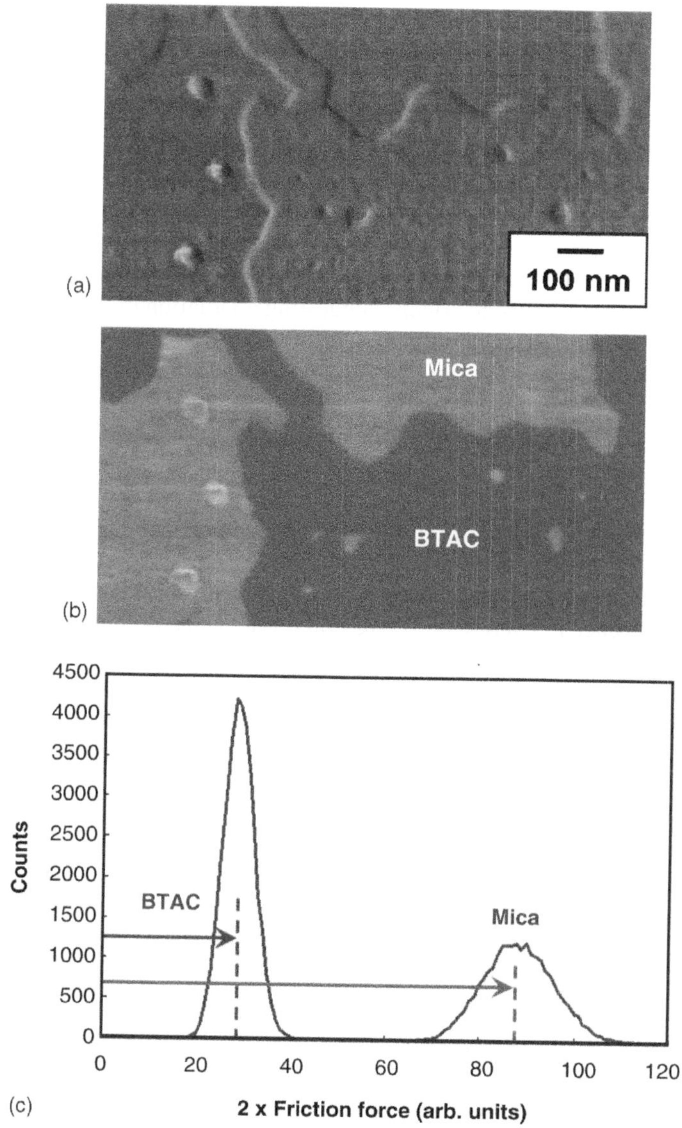

FIGURE 7.6 Sum (a) and difference (b) of trace and retrace lateral force images of a submonolayer of BTAC and bare mica substrate. (c) Histogram of (b).

surface that interacts relatively strongly with the polar (SiO_2) tip. The cation is expected to adsorb to the negatively charged mica surface during deposition (in this case from a *lamellar gel phase* mixture with cetyl alcohol and water). The short-chain molecules stand largely upright and close-pack together (due to interchain attractive van der Waals forces), producing a molecularly flat surface dominated by apolar methyl groups that interact relatively weakly with the AFM tip. The images in Figure 7.6 are (a) the sum signal $F^{\rightarrow}_{\mathrm{Lat}} + F^{\leftarrow}_{\mathrm{Lat}}$ and (b) the difference signal $F^{\rightarrow}_{\mathrm{Lat}} - F^{\leftarrow}_{\mathrm{Lat}}$, each operation performed after the retrace data were offset by a few pixels in X to account for scanner hysteresis. (Even if using a closed-loop scanner, one often finds a need for an X shift, to account for the tilt of the tip as discussed below.) The slope-derived character is apparent in (a), in comparison with the simultaneously acquired height image (revealing a 2-nm-thick BTAC film, not shown), whereas strong contrast between film and substrate is observed across large domains in (b). There are also three small particles at left whose difference signal seems to be dominated by edge effects, suggesting a violation of the approximations in Equations 7.3 and 7.4. Quantification of the film–substrate friction contrast is contained in Figure 7.6c, a histogram of the difference signal in (b). One finds that the friction force on mica is nearly three times greater than on BTAC (i.e., via peak position), and more heterogeneous (greater peak breadth). The peak areas quantify the relative surface areas of each component in the image of Figure 7.6b.

Examining several films and using multiple tips, it was found that friction forces on BTAC were roughly 1/4 to 1/3 of that on mica, whereas on films containing only a 16-carbon alkyl chain (cetyl trimethyl ammonium chloride, or CTAC, on mica), friction forces between 1/3 and 1/2 of that on mica were observed. Thus, there is a significant difference in friction on BTAC and CTAC when referenced to mica, even though the nominal chemistry at the surface of the film (methyl groups) is the same whether C-22 or C-16. It is generally known that longer alkyl chains tend to pack into more highly ordered crystalline monolayers than do shorter chains. More disorder in CTAC would translate into more internal degrees of freedom for submolecular motions such as *trans-gauche* conformational transitions (i.e., turnstile-type rotational motions) to dissipate strain energy [13,16,17]. But more disorder within the film probably results in more disorder at the surface, such that the nature of stick-slip bonding/debonding events on the atomic scale may be affected. The relative importance of internal-to-film versus purely tip–sample interfacial dissipation in such simple tribological systems is an area of ongoing research, including molecular dynamics simulations as well as experimental work. In this body of work, it seems clear, however, that the presence of atomic- to molecular-scale degrees of freedom in less tightly packed molecular systems has a strong impact on the observed frictional dissipation [17]. Nevertheless, one must reiterate that in Figure 7.6 it is the high surface-energy mica, an inorganic single crystal system with no internal conformational degrees of freedom, which exhibits the highest friction force. Presumably, differences in the strength of interfacial bonding between tip and sample are the principal sources of contrast in Figure 7.6b.

The preceding findings suggest that even after the removal of slope-derived lateral force, atomic-to-molecular scale surface roughness may still play a role in "material-specific" friction. During sliding, atomic-to-molecular scale stick-slip motions are likely taking place *within* the tip–sample contact zone, with the strength of each sticking point, and thus the energy dissipation of each slip event, affected by an at least partially topographic energetic barrier on the atomic or molecular scale, particularly on rigid materials [16,18]. Presumably, the Z feedback is not sensitive to such small structural effects within a contact zone, unless the contact zone itself is so small that individual stick-slip events can be observed within the *spatial and temporal* resolution of the technique. (Resolved stick-slip motion on atomic and molecular scales was among the earliest notable observations in frictional AFM [19,20].) In addition, or alternatively, material may creep under shear force and thereby dissipate mechanical energy [21,22], with the degree of creep being perhaps dependent on topography on these small scales. Understanding these very fundamental and atomic- or molecular-scale mechanisms of friction is the subject of ongoing investigation within the nanotribology community.

It turns out that the subtraction of trace and retrace lateral force is useful not only to isolate material-specific friction from topographic-slope derived contributions but also to remove a wavy background signal from large LFM images of reflective samples. This is exemplified in Figure 7.7 for a silicon calibration grating containing square pits, using a closed-loop scanner. The wavy background derives from

FIGURE 7.7 Raw lateral force images of a contaminated silicon grating, acquired in (a) trace and (b) retrace. (c) Lateral force difference image. (d) Corresponding height image.

interferometry of laser light reflected from the sample with incident light and that reflected off the cantilever, similar to the artifact discussed in Section 3.4; and is reduced if the AFM is equipped with a low-coherence laser. The waviness presumably results from a change in the relative optical path length of the interfering components as a function of lateral position of the *X*–*Y* scanner. (Images in Figure 7.7 were acquired with a tip-scanned AFM, which leads to an additional optical artifact: a "tilt" in the background due to slight movement of the laser spot position on the cantilever.) The optics of course does not know in which direction the scanner is moving, thus is independent of scan direction, apart from possible *X*-scanner hysteresis. So optical artifacts are largely removed by subtracting trace and retrace lateral force data, provided that one can account for *X*-scanner hysteresis by invoking a several pixel shift between trace and retrace (5 pixels in Figure 7.7). With an open-loop scanner, one often finds that the optimal shift in the center of the image is not the same as that near the edges of the image. (It is difficult to precalibrate the scanner to remove this difference.)

One also should note that the *X*-shift used to best account for scanner hysteresis may not be identical to the shift that best aligns slope-derived features (e.g., Figure 7.6b), because of differences in *tip tilt* between trace and retrace. That is, the optical artifact depends on the cantilever *X* position, which also determines the *X* position of the *topmost* part of the tip, at its connection to cantilever, whereas the imaged surface topography, and thus slope, depends on the *X* position of the *bottommost* part of the tip interacting with sample, which depends on the cantilever twist that tilts the tip. A closed-loop scanner with optimal *X* gain settings should produce an essentially unshifted LFM background in trace versus retrace, but a small hysteresis will remain in the slope-derived LFM features due to tip tilt. The *X* shift invoked to produce the lateral force difference image in Figure 7.7c was chosen to best align real surface features, meaning the location of the bottommost part of the tip. Nevertheless, it works to adequately remove the optical artifact.

Removing the optical artifact as described above can be essential to reveal contamination on highly reflective surfaces. Often the topographic signal of contamination is weak, and the frictional signal atop the optical background can be weak as well. This is demonstrated by comparing (a) and (b) with (c) in Figure 7.7. The latter reveals patches of both relatively high and low friction force signal. Some of the high-friction patches are seen to correlate with deposits visible in height, but others would require finer-scale images to find the corresponding height features, if they exist (i.e., a liquidy contaminant may not be revealed as a higher *Z* patch). If the point is to collect a survey image to look for contamination, then lateral force imaging is the more sensitive; but removal of the optical background is essential for high-quality results.

Another example of the importance of removing the optical artifact is in attempting to contrast *inorganic* materials. Often these materials exhibit only weak friction contrast: the surfaces may be oxidized and thus not greatly different in surface energy, unlike the corresponding unoxidized materials (e.g., in vacuum); internal degrees of freedom are not expected to play a role in the sense that they do in polymers and other organic materials (although dissipation via *phonons*, that is atomic lattice vibrations, is generally known to be appreciable). On top of these meager

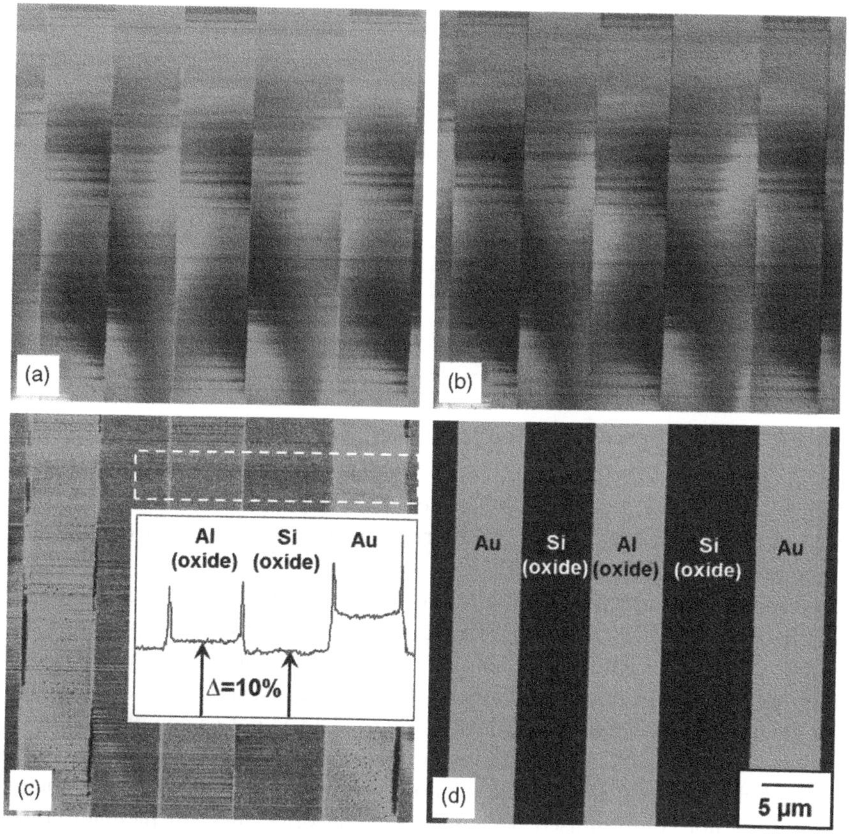

FIGURE 7.8 Raw lateral force images of aluminum and gold lines on silicon, acquired in (a) trace and (b) retrace. (c) Lateral force difference image with inset graph showing data averaged over 60 scan lines within the dashed rectangular region, easily distinguishing only 10% differences. (d) Corresponding height image.

differences, inorganic surfaces can be strongly reflective, further suppressing materials contrast. Thus, the variation of the optical background signal may be an order of magnitude larger than the variations in friction between materials. Figure 7.8 exemplifies such a case for lithographically grown aluminum (with native surface oxide) and gold strips, and bare silicon (native oxide) substrate. Here, the optical background (not using a low-coherence laser) overwhelms any materials contrast in the raw lateral force images (a) and (b); but measureable contrast is obtained between all three materials after subtracting trace and retrace lateral force signals (c) and averaging 60 friction loops to compute the difference data in (c) inset. In particular, aluminum and silicon oxides exhibit a 10% difference in the lateral force difference signal. Also in this example, an open-loop scanner was employed to demonstrate that adequate results can be obtained even if the X hysteresis between trace and retrace is not a constant across the image. A 3-pixel shift was chosen to

FIGURE 7.9 (a) Height and (b) trace lateral force images of silver rods on silver bromide.

optimize the alignment of the two data sets as best possible within the rectangular region to be analyzed, extending to the right edge of the image as denoted in (c), leaving residual background at left.

One should note, in passing, a useful trick. The optical artifact in the preceding two examples (Figures 7.7 and 7.8) can be significantly lessened if one positions the tip near the *edge* of the patterned part of the sample, and second positions the laser spot farther back on the cantilever away from the pattern, say, half way to the cantilever base. Then the interferometry depicted in Figure 3.15 is not as complicated by large variations in optical path length with respect to the reflective sample surface (i.e., laser spillover light reflecting at high and low points across the grating); yet the photodiode sensitivity is not compromised because the cantilever's angle of inclination at this location varies nearly as much as at the tip end, as discussed in Section 3.2.3.

As a final example, it is important to note that *some* inorganic materials can be imaged with ample friction contrast even prior to subtracting the optical background and slope-derived contributions to lateral force. Figure 7.9 contains height and left-to-right lateral force image of rod-like silver grains that grow out of silver bromide crystals hundreds of micrometers in size (as previewed in Figure 1.1a), induced by exposure to the red laser light of the AFM (i.e., with lab lights off) [23]. These silver domains exhibit a friction force only $\sim$1/3 as great as that on AgBr (111) [3], a dramatic contrast indeed. Notwithstanding a somewhat variable background as well as significant slopes across the silver grains, the overall strong contrast is clear. In this particular case, there is, of course, adequate information in the height image to identify the materials, but the point is that even in the absence of conclusive topographic identification, these particular materials are easily distinguished via raw lateral force data.

At least part of this huge contrast presumably derives from a larger tip–sample contact area on AgBr; it is a soft salt with a Young's modulus 55–60% that of Ag. In the simple Hertzian model of a spherical tip on a flat half-space, contact radius is approximately proportional to the $-1/3$ power of modulus, and thus contact area is proportional to the $-2/3$ power. This (quasistatic and nonadhesive) model therefore predicts friction force on Ag to be 70% of that on AgBr in the absence of differences in shear strength (i.e., intrinsically different shear force per unit area), certainly a measurable effect but a much lesser difference than observed. Small differences in pull-off force (~10%) also were observed, meaning adhesive contact mechanics models would predict a greater disparity in contact area than the Hertzian model considered above, but not enough to explain the experimental results. In addition, it is observed that AgBr wears under the tip (seen by zooming out) if the loading force is not kept minimal. It is possible that some degree of wear is always occurring, increasing the frictional dissipation. The fact that the AgBr is sensitive to red light indicates the presence of defects [24], which might be the source of additional atomic mobility and anharmonic vibrational effects that produce dissipation. It is not our purpose here to undertake a detailed analysis of mechanistic friction differences between these two materials, but the preceding statements provide a flavor of the considerations one would examine to better understand frictional data.

7.3 CALIBRATION OF LATERAL FORCE

Figures 7.6 and 7.8 exemplify that it is not necessary to convert lateral force data to force units to do semiquantitative work, that is, to determine *ratios* of friction force. By performing friction measurements on a standard reference material (e.g., the above example of mica), these relative numbers can become even more meaningful in terms of cross-experiment studies using multiple tips and samples. However, just as with normal force calibration (Section 3.7 and Appendix 1), methods have been identified for calibrating absolute lateral force in force units. This is desirable especially when the magnitude of friction force and/or coefficient of friction is being studied *per se*, usually within tribological research. Here, we describe a general strategy that neither requires calibration of the photodetection scheme (i.e., voltage output per unit angular twist of cantilever or lateral tip displacement) nor a determination of the torsional spring constant of the cantilever. The strategy follows directly from the concepts of Section 7.1: making use of variable sample slope m in order to measure the coefficient of friction μ, and thereby lateral force via a calibrated normal force.

The simplest usage of the method, call it Treatment #1, applies to cases where the approximate Equations 7.3 and 7.4 hold (i.e., $m\mu \sim 0.01$ or smaller). Within a friction loop, one may find locations where the data in trace or retrace passes through the zero of the measurement, thanks to a variable slope m. We see that with Equation 7.3 or 7.4, this condition implies $m = -\mu$ (trace) or $m = \mu$ (retrace), respectively. In both cases, this corresponds to a *downward* slope in the direction of scanning. Thus, the measured downward slope at which the net lateral force vanishes *is* the coefficient of friction [3]. Provided that the user has calibrated the vertical force measurement to force units (e.g., per the methods of Section 3.7 or Appendix 1), multiplication of the vertical force on a horizontal surface region by the determined value of μ calibrates the lateral force in force units.

One must exercise caution in using this method. First, it is imperative to assess how well the Z feedback tracking system is working, because a relatively steep surface location is being used to obtain the condition $m = -\mu$. One should operate with gains set high enough and scan velocity low enough (low scan-line frequency and scan size) to produce at most a small deviation of the vertical cantilever deflection from its setpoint value (corresponding to <5% change of vertical force), while avoiding feedback oscillations. Second, one should recognize that the portion of the tip that is touching the surface at this relatively steep location (i.e., not the endmost point of the tip) may not be the same as the portion of the tip touching during the measurements of interest. Thus, it is a good idea to probe lateral force over a *continuum* of slopes and check for consistency in the friction loop height (for a given material region), as a check of tip uniformity. For this reason, the use of a surface with vertical drops, or containing only specific values of slope (e.g., a triangular waveform profile grating or a stepped crystal surface) is probably not the best choice. Of course, in many such cases, the value of $m\mu$ may be much greater than 0.01 as well, such that the above approximation is not valid anyway. The most experimentally robust case is a sample with a range of gradually varying slopes across a given material, including locations where $m\mu \sim 0.01$ and the lateral force signal vanishes, $|m| \approx 0.1 \approx \mu$. Not only is the approximation criterion valid but also the uniformity of the tip can be confirmed as consistent friction loop heights at different surface slopes.

More generally, one may not be able to invoke Treatment #1, that is, the approximation of Equations 7.3 and 7.4, when $m\mu$ is considerably greater than 0.01 at locations of net vanishing lateral force (whether due to larger m, μ or both). Thus, it is useful to solve Equations 7.1 and 7.2 exactly for both the sum and difference of trace and retrace lateral force in terms of m, μ, and the vertical force F_{Vert}. After some algebra one obtains

$$F_{\text{Lat}}^{\rightarrow} + F_{\text{Lat}}^{\leftarrow} = \frac{2m(1+\mu^2)}{1-m^2\mu^2} F_{\text{Vert}} \tag{7.5}$$

$$F_{\text{Lat}}^{\rightarrow} - F_{\text{Lat}}^{\leftarrow} = \frac{2\mu(1+m^2)}{1-m^2\mu^2} F_{\text{Vert}}. \tag{7.6}$$

Note that one can determine the zero of the lateral force scale from half-height point of the friction loop at a location of zero slope m; thus, $F_{\text{Lat}}^{\rightarrow} + F_{\text{Lat}}^{\leftarrow}$ is measurable as a voltage signal without containing an arbitrary voltage offset. The denominator in Equations 7.5 and 7.6 can be replaced by unity for almost all cases, given that m and usually μ are fractional quantities. Even if m and μ are quite large as these quantities go, say around 0.3, this gives $m^2\mu^2 \approx 0.009$, an order of magnitude smaller than either m^2 or μ^2. Thus, we may to good approximation write

$$F_s \equiv F_{\text{Lat}}^{\rightarrow} + F_{\text{Lat}}^{\leftarrow} \approx 2m\left(1+\mu^2\right) F_{\text{Vert}} \tag{7.7}$$

$$F_d \equiv F_{\text{Lat}}^{\rightarrow} - F_{\text{Lat}}^{\leftarrow} \approx 2\mu\left(1+m^2\right) F_{\text{Vert}}. \tag{7.8}$$

Assuming a constant μ across a given material region, Equation 7.7 indicates that the sum signal $F_{\text{Lat}}^{\rightarrow} + F_{\text{Lat}}^{\leftarrow}$ will linearly increase with positive slope and decrease

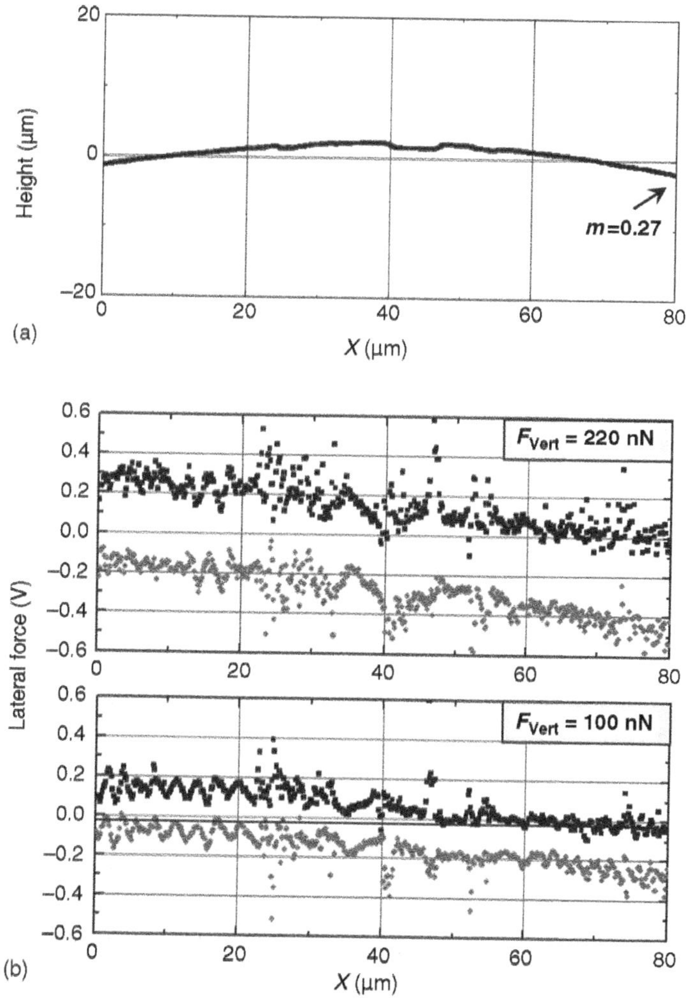

FIGURE 7.10 (a) Cross-sectional height data on a copper wire, with maximum magnitude of slope $m = 0.27$ at right. (b) Corresponding trace and retrace lateral force data obtained at two loading forces F_{vert}.

with negative slope. The difference signal will be constant until m become significant, whereupon it will increase quadratically with m, meaning the friction loop will further open on steep locations.

The veracity of this analysis is exemplified in Figure 7.10 on a curved surface, a copper wire 0.5 mm in diameter lying parallel to the principal axis of the cantilever.[2] Slow (~0.1 lines/s) 80-μm X scans were performed and the error signal

[2] This was snipped from an ordinary multiwire electronic cable, cleaned with acetone and methanol, and adhered to double-stick tape, being sure not to touch the operative surface to the tape during preparation.

monitored to verify good tracking. A representative height trace (a) and two representative friction loops (b) at two deflection setpoints corresponding to total vertical forces of 100 nN (lower plot) and 220 nN (higher plot) is shown. One observes that both friction loops contain a tilt, as the surface slope of the wire continuously changes versus X. The relative heights and tilts of the two friction loops are consistent with the relative vertical forces, as predicted in Equations 7.7 and 7.8 for a constant coefficient of friction. Last, although rather obscured by noise, the height of each friction loop near the right end is ~10% larger than near the middle portion (determined with data smoothing), consistent with the factor $(1+m^2)$ in Equation 7.8 for the measured slope of $m = 0.27$.

One can utilize Equations 7.7 and 7.8, to obtain a lateral force calibration, for the more general case where approximate Equations 7.3 and 7.4 do not hold. We refer to this as Treatment #2. The left side of Equations 7.7 and 7.8, the lateral force sum F_s, and difference F_d, respectively, are measured quantities but raw voltage signals. The slope m at various points on the surface is measureable from height data. Thus, one can divide the two Equations 7.7 and 7.8, making the left-hand side F_s/F_d a dimensionless number, and solve for the unknown μ. Then the lateral force can be converted to force units using either of Equation 7.7 or 7.8, inserting the known values of μ and m as well as F_{Vert} (the latter calibrated to force units via the method in Section 3.7 or those in Appendix 1, as done in Treatment #1).

7.4 LOAD-DEPENDENT FRICTION

7.4.1 Motivations

The most common variable, in analyzing frictional behavior, is loading force. Often a linear or very nearly linear relationship between friction force and load is observed with AFM, provided that wear does not result (e.g., Figure 1.10). In some very precise studies with smooth rounded tips, a *sublinear* dependence has been observed: often friction force increasing as the 2/3 power of load [25]. Within a model of uniform *shear strength* τ_0—the force per unit contact area required to make interfacial "bonding" yield and thus enable sliding—acting across the contact area, a 2/3 power relationship is consistent with adhesive contact mechanics models as discussed in Section 6.2. These models predict that contact radius a scales with the 1/3 power of load. Contact area is πa^2 for a smooth tip described as a surface of revolution, implying a contact area and thus friction force that increases as the 2/3 power of load.

Such consistency with adhesive contact mechanics models suggests that one could use friction–load measurements to extract a sample's material constants such as Young's modulus and surface energy, similar to the discussion in Section 6.2 in the context of force–Z measurements. But a problem in AFM is the ill-defined nature of tip–sample contact. The frictional phenomena within this contact zone may (1) involve multiple nanometer- or subnanometer-scale points of interaction, at which different stresses are exerted, (2) produce plastic deformation, that is, cohesive instead of adhesive yield to enable sliding, and (3) induce viscoelastic deformation including creep. Soft materials such as rubber polymers may reduce the likelihood of multiple individual contact points, instead producing a contiguous, *conformal* contact between tip and sample. But soft materials such as polymers are more likely to be viscoelastic, in which case the response may be explicitly rate- and scan-history dependent [26], complicating the extraction of material constants. The shear can be a highly nonlinear phenomenon; high stresses in the contact zone can greatly modify conformational states, such that the response is not representative of bulk response to the stresses typical of macroscale phenomena (e.g., bending). Indeed, many tribology studies have modeled a pressure-dependent shear strength [27],

$$\tau = \tau_0 + c_1 p, \tag{7.9}$$

where c_1 is a positive constant and p is normal force per unit contact area (i.e., contact pressure). This suggests that conformational states or interfacial "bonds" in the contact zone are affected by contact pressure, such that more shear force is needed to make a given interfacial "bond" yield. Thus, AFM nanotribology should not be viewed as a nonperturbative, external probe of material properties. Interestingly, the multiplication of the shear strength as expressed in Equation 7.9 by the contact area produces a load dependence that includes both a 2/3-power term (proportional to τ_0) and a linear term (proportional to c_1). Clearly, a rich amount of tribological detail can be contained, in principle, in the load dependence of friction.

7.4.2 Load Stepping and Ramping Methods

The simplest protocol for investigating the load dependence of friction force is to manually step the setpoint in one of three ways: (i) between individual friction loop measurements, (ii) between images of trace and retrace lateral force, or (iii) within such images. (Note that some AFM software by default does not allow the user to capture an image wherein the setpoint has been changed; an additional "capture" command must be issued after the last change of this parameter, before the raster scan is completed, to "force" the capture.) By doing so in image mode, as opposed to individual friction–loop cycles, one can improve signal/noise. At each setpoint, one can compile the friction loops within each 2D surface domain and examine the statistics, given a large number of data points.

Figure 7.11 shows an example of such analysis. The sample is a variable-crystallinity film of polyvinyl alcohol (PVA) [28] tens of nanometers thick, adsorbed on mica. The lower height regions in (a) are the predominant semicrystalline film surface, and the higher are ~4-nm-thick crystalline islands exhibiting greatly reduced friction force seen in (b). The large-scale height image in (c) inset exemplifies the overall shape of many of these crystalline islands. The reduced friction is due to fewer available molecular-scale degrees of freedom (e.g., *gauche-trans* transitions in amorphous domains). That is, the translational kinetic energy of the sliding tip is being transferred to myriad molecular motions, and the possibilities for this to occur are enhanced in amorphous domains where molecular subunits are less tightly locked

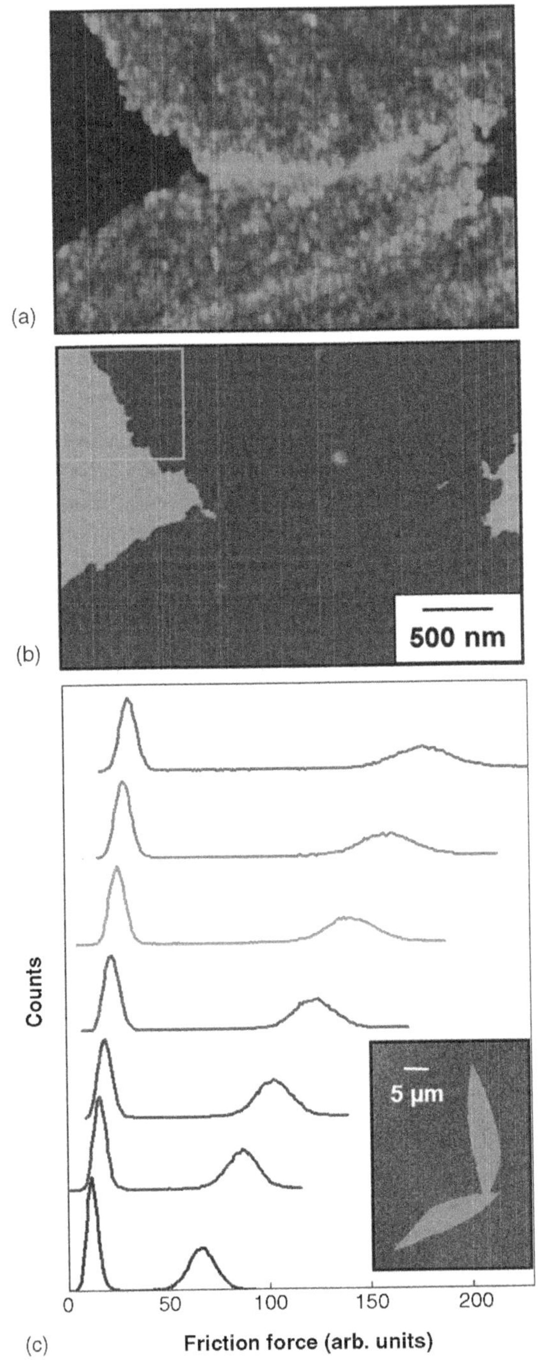

FIGURE 7.11 (a) Height and (b) friction force images of crystalline PVA islands on more amorphous PVA. (c) Friction force difference histograms at seven different loading forces.

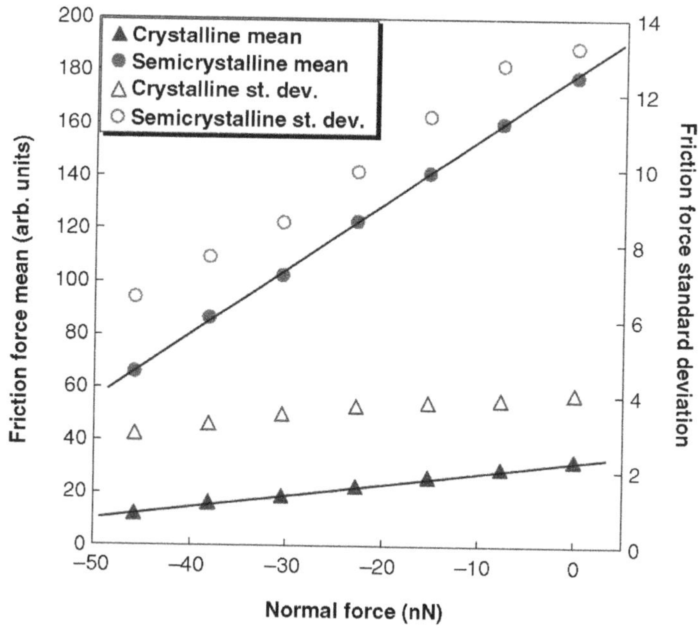

FIGURE 7.12 Load dependence of mean and standard deviation of friction force from histogram peak data in Figure 7.11c.

into particular conformational states. From trace minus retrace lateral force images, acquired at seven values of normal force within the tensile regime (pulling), a 200×230-pixel subregion (46,000 data points) located within the rectangular region designated in (b) is analyzed. The data are first compiled into histograms shown in (c); the vertical force was increasingly negative from top to bottom. These data are then statistically analyzed; mean and standard deviation values are plotted versus normal force in Figure 7.12. The smoothness of the linear data trend for mean value is excellent, the result of analyzing such a large quantity of data per histogram. The standard deviation is also seen to increase quite linearly with normal force. The slopes of the mean value (coefficient of friction) have a ratio of about 5 : 1, a huge difference derived from differences in crystallinity [6,29].

Friction–load measurements also can be performed with the slow-scan axis (Y) disabled (i.e., a 1D scan), such that consistently sized and located regional segments are being examined along each scan line at the various loads. One must carefully check for wear if repeatedly scanning the same scan line (i.e., conduct before/after imaging). However, if incrementing the setpoint, the scan line necessarily moves slightly in the lateral direction parallel to the cantilever, in the direction of the downward tilt of the cantilever (see Figure 3.7); if the setpoint increments are large, then with each change of setpoint there may be an immediate displacement in this direction followed by a smaller creep displacement, such that the scanning “track” is not really fixed along Y. If wear or other scan–history-dependent phenomena relating to repeated passes of the tip are being targeted for study [26], then this creep can be a problem; the method is poorly controlled.

In many cases, one benefits from a finer sampling of the loading force than in Figure 7.12, and down to the lowest friction forces measureable, that is, at the largest tensile (pulling) forces at which the contact persists. This may enable one to discern deviations from linear behavior, Amontons' Law, such as predicted by adhesive contact mechanics models. Thus, one is giving up high signal/noise statistics in the measurement of friction force at each load in order to obtain higher load resolution. One can implement this strategy manually, simply by invoking smaller load steps within an image or between images; but the reduction of such data sets can be tedious. A much more automated approach is to linearly ramp the setpoint, and moreover in both increasing and decreasing directions to examine physical hysteresis. One way to do this is during force-curve mode, by externally ramping the *X*-scanner voltage while *Z* is extending and retracting. This is normally done with a BNC breakout box, substituting a sinusoidal or triangle waveform for the AFM controller's applied *X* voltage signal. In this case, however, the system is out of deflection feedback, meaning the normal deflection can vary during each lateral oscillation cycle as the tip moves back and forth over a sloped surface.

A better-controlled and more robust scheme is to ramp the setpoint *during imaging*, such that *Z* feedback is maintained. Most AFM systems do not include set–point-ramping capability in software. One may, however, implement a setpoint ramp by breaking out the deflection signal with a BNC interface and adding a triangle waveform, then using this summed signal to drive the feedback circuit. Thus, one is "fooling" the AFM controller to think that the deflection is linearly changing, such that it displaces *Z* to oppose this change, and thereby ramps the deflection. Two key additional components are needed: a common function generator capable of millihertz-regime frequencies (as found, e.g., in many teaching labs) and an adder circuit. A homemade adder circuit can be constructed for almost negligible expense using simple op amp chips, resistors, capacitors, and a circuit board (example circuit diagrams are easily found on line) [30].

An example utilization of such an adder circuit is shown in Figure 7.13, for a silicon tip/cantilever ($k \sim 3$ N/m) scanning a diamond-like carbon film. The triangle-waveform set–point-ramping signal (solid line in (a)) was selected to have an amplitude of 200 mV and a frequency in the millihertz regime, such that three full load–unload cycles executed during the 512-line image. Conventional lateral force images were acquired on trace and retrace, and subtracted line-by-line and averaged across the cycle; each thus-integrated friction loop produced one square datum in (a). Lateral force signal after the loss of tip–sample contact at the largest pulling forces are excluded from the plot. In (b), these same data from the decreasing load portion of each of the three cycles are overlaid versus the applied vertical force signal, the latter zeroed according to the undeflected signal determined in force-curve mode with the adder circuit disconnected. Data were omitted from the portion of each cycle that tip–sample contact was lost following strong tensile loading (negative applied vertical force). One easily perceives a nonlinear data trend at negative applied vertical force.

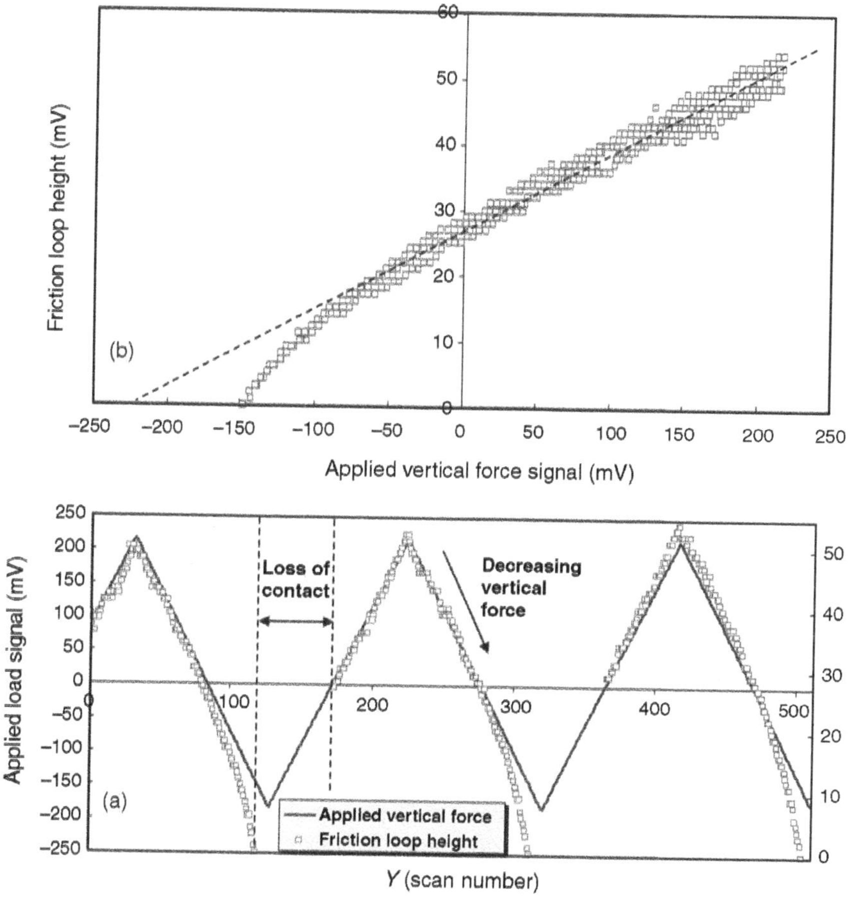

FIGURE 7.13 (a) Overlaid triangular load ramping signal (line) and resulting variation of friction loop height (squares, each datum representing a complete friction loop) during 2D imaging as a function of *Y* scan line number, omitting data during loss of contact under negative loading. (b) Load-dependent values of the friction loop-height data from (a), overlaid from the three segments of decreasing vertical force.

The friction–load data from Figure 7.13b were averaged and plotted in Figure 7.14. Superimposed is a curve with a functional form derived using the DMT contact mechanics model (Section 6.2.2), assuming friction force is proportional to contact area: the 2/3 power of the total loading force, the latter being the sum of the applied load of the cantilever (positive or negative), and the attractive tip–sample interaction force. The shape of this curve nearly matches that of the data trend. The slight inconsistency probably relates to the geometry of the real tip–sample system, that is, departures from an ideally spherical-ended tip on a flat half-space. Another possibility is a pressure-dependent shear strength that causes a steeper trend at the highest loads.

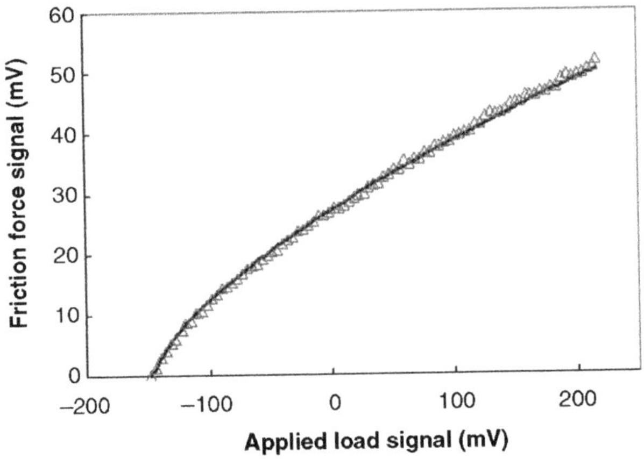

FIGURE 7.14 Average of data from Figure 7.13b and superimposed mathematical curve describing contact area versus applied load within the DMT adhesive contact mechanics model.

7.5 VARIABLE RATE AND ENVIRONMENTAL PARAMETERS IN AFM FRICTION AND WEAR

7.5.1 Motivations

Friction and wear are inherently *rate dependent* because they are nonequilibrium, irreversible processes. Any real stress–strain process in which force is imparted, and material or interface deforms, does not take place as a succession of equilibrium states placed infinitesimally apart. Thus, some potential energy imparted to the material and/or interface as strain—via the torsional spring force of the cantilever in the case of sliding friction—is not recovered upon the release of stress. The amount of dissipated energy can be specific to the type of material or interface. Indeed, this leads to the utility of phase, adhesion, and friction as imaging modes. Energy dissipation during shear processes is usually measureable, trivially so considering the ubiquity of friction. But measuring the inherent rate dependence can require exploring decades of rate, essentially for two reasons as follows.

To understand the first reason, one must recognize that friction and wear are *kinetic* phenomena, activated processes at finite temperature. Shear forces act to lower activation energy barriers for slip [31]. The longer an interfacial "bond" (usually van der Waals in character) is stressed, the *more likely* the bond will fail via thermal activation over the energetic barrier. Thus, there is an inherently statistical–mechanical nature to interfacial sliding friction. This is analogous to the concepts of Section 2.6.2 and Figure 2.27, but here pertaining to shear. As in Figure 2.27, the probability of transitioning over an energetic barrier ("slipping") in one direction is much greater than the reverse, as determined by the scanning direction. In considering such kinetic concepts for an idealized single-atom tip scanning over an atomically corrugated surface, the *Tomlinson model* plus thermal activation concepts can be used to predict a friction force that increases logarithmically with velocity [18,31,32]. Some studies on simple crystalline systems have measured this functional dependency, but requiring a wide range of velocities. Other studies on simple systems have obtained more complicated results, such as rate dependencies that depend strongly on relative humidity (RH) and surface chemistry (hydrophilic versus hydrophobic) [33]. Further work has explored thermal activation via variable temperature in conjunction with variable rate [33,34]. At this writing, understandings of the kinetics of friction on simple crystalline surfaces are emerging, but the field is still young compared to many others in statistical physics.

As discussed in Section 2.6.1 in the context of distance-dependent forces, a second source of rate-dependent friction is viscoelasticity [21,22,35]. A shear force at a surface or interface can deform a polymer (or other soft materials) such that internally nonequilibrium responses result. Moreover, the appearance and release of stress can leave the system in a state containing voids and disentangled polymer chains: wear. The tip perturbation time for a sliding contact—the most obvious being the contact diameter divided by the scan velocity—can be faster, slower, or about the same as the characteristic *relaxation times* of polymers [21,22,35–38]. These relaxation times describe how long it takes for a component of the imparted stress to relax away. In classic experiments, pulling on a bulk polymer can produce as many as three plateau values of strain over widely different timescales, meaning three classes of relaxations [22,37]. These are observed with conventional *dynamic mechanical analysis* (DMA) as well; oscillatory stress produces oscillatory strain that is variably in phase (δ by convention), dependent on the operational frequency relative to inverse relaxation times. The tangent of δ is also known as the *loss tangent* and is equal to the ratio of loss modulus E'' to storage modulus E', which comprise the *complex modulus* [21,22,35] (with i representing the square root of -1),

$$E_{\text{complex}} = E' + iE''.$$

The complex modulus relates the dynamic stress to the dynamic strain via a frequency-dependent Hooke's Law,

$$\sigma(\omega) = E_{\text{complex}}(\omega)\varepsilon(\omega).$$

Both the loss tangent and the loss modulus exhibit (broad) dissipative peaks in certain frequency windows, corresponding to different inverse relaxation times [22,37].

It has been known for more than a half century at this writing that all of the major classes of relaxations in polymers, often labeled alpha, beta, or gamma (slow to fast), are manifest in the rate and temperature dependence of sliding friction on polymers [39]. The alpha, being related to the *glass–rubber transition* and thus *segmental motion* along the main polymer chain, is usually, but not always, the strongest peak [22,37]. As with the preceding thermal activation concepts, one may need to explore many decades of rate to see these effects. In traditional polymer characterization, it is experimentally difficult to mechanically probe a wide enough frequency scale to probe all three classes. Often temperature has been varied in conjunction with frequency in order to explore a broad enough dynamic range [22]. Heating a sample by ~10°C can be equivalent to decreasing the frequency by an order of magnitude, reflecting an exponential (or *stretched* exponential [37]) relationship between rate and inverse temperature as is common to many kinetic phenomena. *Dielectric spectroscopy* probes similar polymer motions as does DMA (provided that the motional unit bears a net dipole moment) but usually can explore a wider range of frequency. Indeed, dielectric spectroscopy has in some cases revealed three broad peaks within single spectral sweeps spanning ~12 decades in frequency [22].

AFM-based friction force measurements at a given scan size are limited in the range of rate (i.e., scan-line frequency, resulting tip velocity) to a similar degree as DMA. By varying both scan size and scan-line frequency, however, many decades of rate can be explored in principle (barring difficulties with surface tracking at the highest rates, i.e., the *Z*-feedback circuit, and plastic deformation at the lowest rates, where stress may be repeatedly imparted to a small area). In some cases, as many as two classes of polymer relaxation have been probed within a single "spectrum" via variable scan size, scan-line frequency, and sample temperature. High loads also have caused the rate dependence of sliding friction to shift in some AFM studies [26,40], although presumably by changing the response function of the polymer rather than exploring its intrinsic rate-dependent spectrum. Scan velocity and sample temperature are considered better "external" parameters to explore, inert in the sense that they probe intrinsic polymer dynamics within different kinetic windows. Highly compressive loads might instead change the polymer-free volume or modify the distribution of conformational states in some other way, which can slow or hasten the polymer response function in principle.

7.5.2 Interplay of Rate, Temperature, Humidity, and Tip Chemistry in Friction

An example of AFM sliding friction measurements at variable rate and temperature, on a polymethyl methacrylate (PMMA) film, is shown in Figure 7.15a. The four scan velocities span two decades; the sample temperature was varied from 25°C to 110°C at each scan velocity. The observed date trend, combined with the known temperature dependence of the viscoelastic spectrum of PMMA [22], indicates the presence of both alpha and beta relaxations in Figure 7.15a [10]. Moreover, the temperature dependence shifts substantially among the four scan velocities at the lower temperatures, but little at the higher temperatures. This is consistent with the relative rate–temperature relationships for known alpha and beta relaxations in PMMA, derived from different activation energies for these dissipative motions. The data in Figure 7.15a are extensive enough to perform *time–temperature superposition* analysis of the beta relaxation [22,37]. One can plot the same data as friction force versus log rate, at each of six temperatures spanning 25–83°C, but multiplying the rate value by a variable *shift factor* a_T among each of five data sets so as to align with the sixth, thus creating a *master curve* [10]. By then plotting ln (a_T) versus $1/T$ and measuring the slope of this trend, one can extract a value of

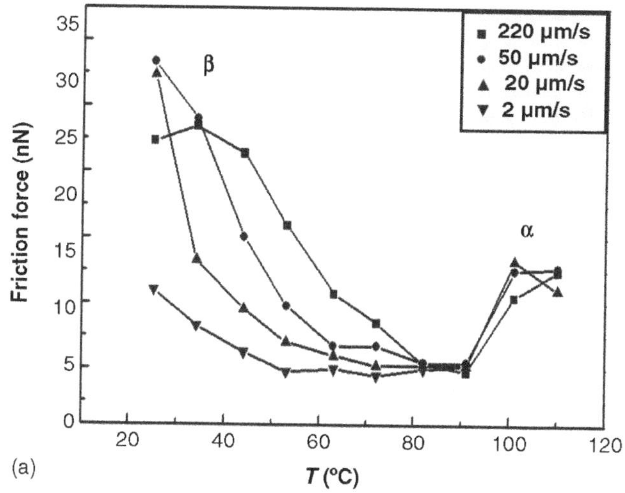

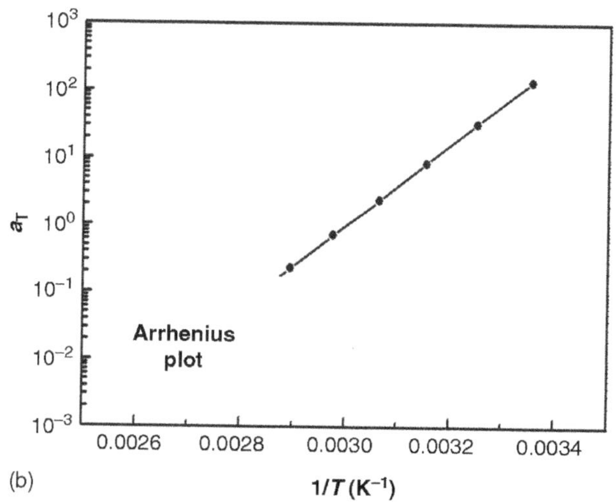

FIGURE 7.15 Time–temperature superposition analysis of frictional data collected on thin PMMA. (a) Temperature dependence of friction force at four scan velocities. (b) Log–linear plot of horizontal shift factors a_T versus inverse temperature (Arrhenius analysis) as explained in the text, with overlaid linear fit.

activation energy [22]. The shift factors determined from the data in Figure 7.15a are plotted on a semilog scale in Figure 7.15b. The linearity of this plot implies Arrhenius kinetics, meaning a single relaxation time, as is commonly observed for beta motions in DMA and dielectric spectroscopy of many bulk polymers [22,37]. This implies a small, well-defined motional unit; for PMMA, this beta motion has been assigned to turnstile-like rotations of the side chain [10]. Alpha relaxations, as stated earlier, involve larger segmental motions, usually along the main chain, and

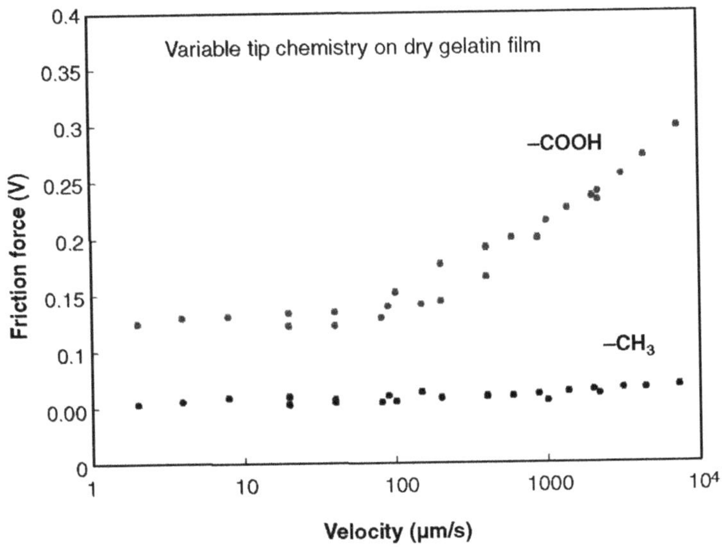

FIGURE 7.16 Comparative velocity-dependent friction force on gelatin for hydrophilic versus hydrophobic tip chemistries.

can include motions in concert between chains [22,36,37]. Such *cooperatively rearranging regions* may include subdomains with different relaxation times, and thus do not generally exhibit Arrhenius kinetics. In such cases, plots of log shift factor versus $1/T$ produce nonlinear curves [22,37], unlike that of Figure 7.15b.

In sliding friction of polymers, one unavoidable induces energy dissipation due to both phenomena described earlier: (1) the making and breaking of adhesive bonds at the tip–sample interface, likely stick-slip phenomena at some local level, and (2) viscoelastic response of the polymer (i.e., intrinsically dissipative shear and normal deformations). At this writing, the relative strengths of these processes are not well understood in AFM measurements. Interfacial chemistry can have an impact on dissipation both at the tip–sample interface (i.e., the strength of each "stick") and within the polymer. (Higher internal dissipation can result from larger stresses reached prior to slip, as a result of greater adhesion.) Thus, (1) and (2) are not independent phenomena. Such a coupling is suggested in Figure 7.16, a comparison of rate-dependent AFM friction force on nominally dry gelatin [41] employing tips with hydrophilic versus hydrophobic terminal chemistry. A rise in friction at velocities exceeding 100 μm/s on such films has been interpreted as due to beta relaxations [26]. In Figure 7.16, we see that this is true for a carboxylic–acid-terminated tip but not for a methyl-terminated tip, suggesting that more dissipative beta motions are only stimulated by a polar interaction. The higher friction force sensed by the polar tip at all velocities is more likely due to phenomenon (1) described above, the breaking of adhesive bonds. Gelatin, derived from the protein collagen, includes amine groups that form strong hydrogen bonds with carbonyl groups, such

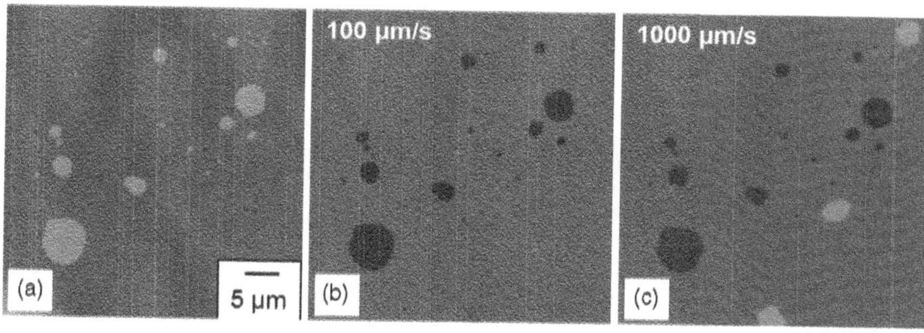

FIGURE 7.17 (a) Height and (b) friction force images of dry gelatin film containing crystalline islands on a more amorphous sublayer. (c) Additional high-friction domains are visible at high scan velocity using a gelatin-coated tip.

as present on the carboxylic acid tip. *Stereochemical* characteristics may be important as well (i.e., how well the molecular units "fit together"); but one is not surprised to find an overall stronger interaction than for methyl tips.

The examples of Figures 7.15 and 7.16 indicate that rate and/or temperature can be used as imaging parameters or "knobs" to affect materials contrast, as can tip chemistry. An example of the rate knob is given in Figure 7.17, again a gelatin film but containing islands of highly crystalline gelatin that exhibit low friction force independent of scan velocity within the accessible range of rates [42]. At scan velocities of 1000 μm/s or greater, however, additional high-friction domains become visible within the primary film when scanning with a gelatin-coated tip. In this case, as with the preceding example, there is a propensity for tip–sample coupling via hydrogen bonding and perhaps complementary stereochemistry, and for this coupling to enhance beta motion and thus increase frictional dissipation. One interpretation is that during film casting from aqueous solution and subsequent drying, some micrometer-scale domains retain higher water content or lower physical cross-linking density [41,43]; perhaps, these are the sites of final microdroplet evaporation during drying. In any case, this example demonstrates the keen sensitivity of sliding friction to subtle heterogeneities in polymeric materials and the ability to assess the physical nature of these heterogeneities by exploring scan rate.

The next example of rate-dependent friction begins by examining humidity-dependent friction on highly crystalline PVA moieties atop the surface of an otherwise semicrystalline film (~100-nm thick on mica) containing substantial amorphous content [6]. This sample was prepared by solvent–vapor (water) annealing immediately after the initially cast film was dried and imaged, in a humidity-controlled AFM system. The discontinuous, low-friction crystalline "skin" is apparent in Figure 7.18a and b. The graph in Figure 7.18c compares the subsequent, humidity-dependent friction on the highly crystalline (2) and more amorphous (1) domains. This dependence is quite flat on the highly crystalline domains up through 75% RH, whereas it exhibits a steep increase above 60% on the more amorphous domains. Elevated humidity causes water molecules to be absorbed into amorphous

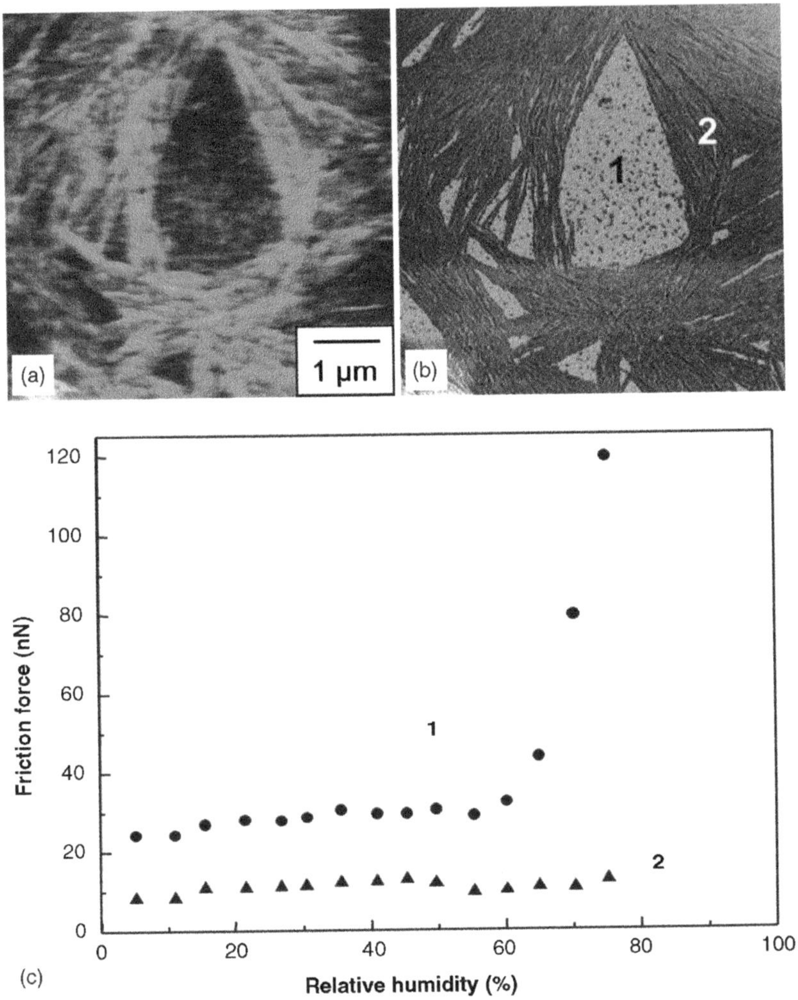

FIGURE 7.18 (a) Height and (b) friction force images of a polyvinyl alcohol film with numbered domain types differing in crystallinity. (c) Humidity dependence of friction for each domain type.

domains in (water-soluble) polymer matrices. This *plasticizes* the polymer to enhance molecular motion, and thus elevate friction [6]. Indeed, it is well known that a *solvent-induced* glass-to-rubber transition (enabling alpha motions) can result, whereby the solid transitions from a rigid plastic to an easily deformable rubber. The glass-to-rubber transition temperature (T_g) for anhydrous PVA is quite high, around 80°C, meaning that the absorbed water in domains #1 at 75% RH has effectively decreased T_g to around room temperature, a dramatic change in material properties. The rate dependence of the friction force measurements on domain 1,

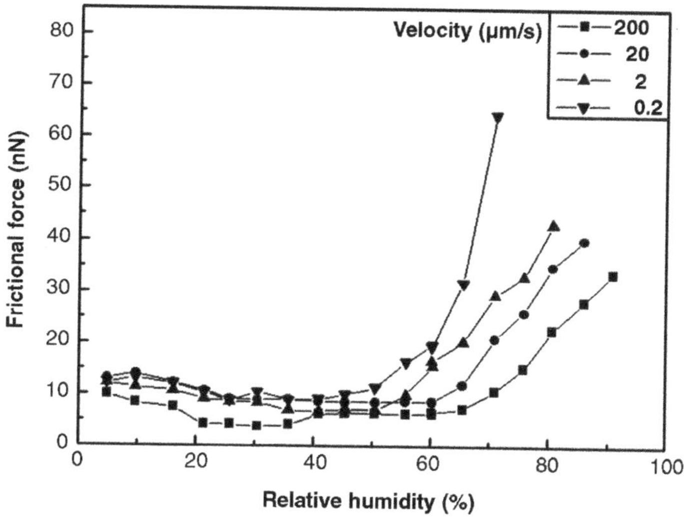

FIGURE 7.19 Humidity dependence of friction force on region #1 of Figure 7.18b, at four values of scan velocity.

the more amorphous component, is examined in Figure 7.19. At increasing scan velocities spanning three decades of rate, the onset of rubbery behavior shifts to higher humidity. In other words, scanning at high velocity preserves glassy response at high water content. Presumably, the small, dissipative amorphous regions do not have time to experience cooperative rearrangements during the brief time that the tip exerts shear stress.

Taken together, the methodological example of Figures 7.18 and 7.19 indicates that one can use friction force AFM as a gauge of small-molecule absorption into polymeric matrices, or use this absorption to generate contrast between domains that do not produce contrast otherwise, and moreover use rate to probe the visco-elastic nature of dissipative motions.

7.5.3 Wear Under Variable Rate and Temperature

Another example of the effects of variable rate and temperature is in the well-known phenomenon of polymer patterning, a type of wear, via shear forces in AFM raster scanning in contact mode. This intriguing phenomenon has received significant attention from several research groups [44–48]. It is beyond the scope of this section to summarize all of the studies to date or delve into proposed tribological mechanisms. Instead, we look at one empirical example, simply to underscore the broad relevance of rate and temperature in the response to an AFM tip including wear processes. Figure 7.20a contains five example height images obtained on a polystyrene film (20-nm thick on silicon wafer) at indicated temperatures in degrees Celsius, spanning from 39°C to 112.5°C. Ridges tend to form, perpendicular to the fast-scan direction,

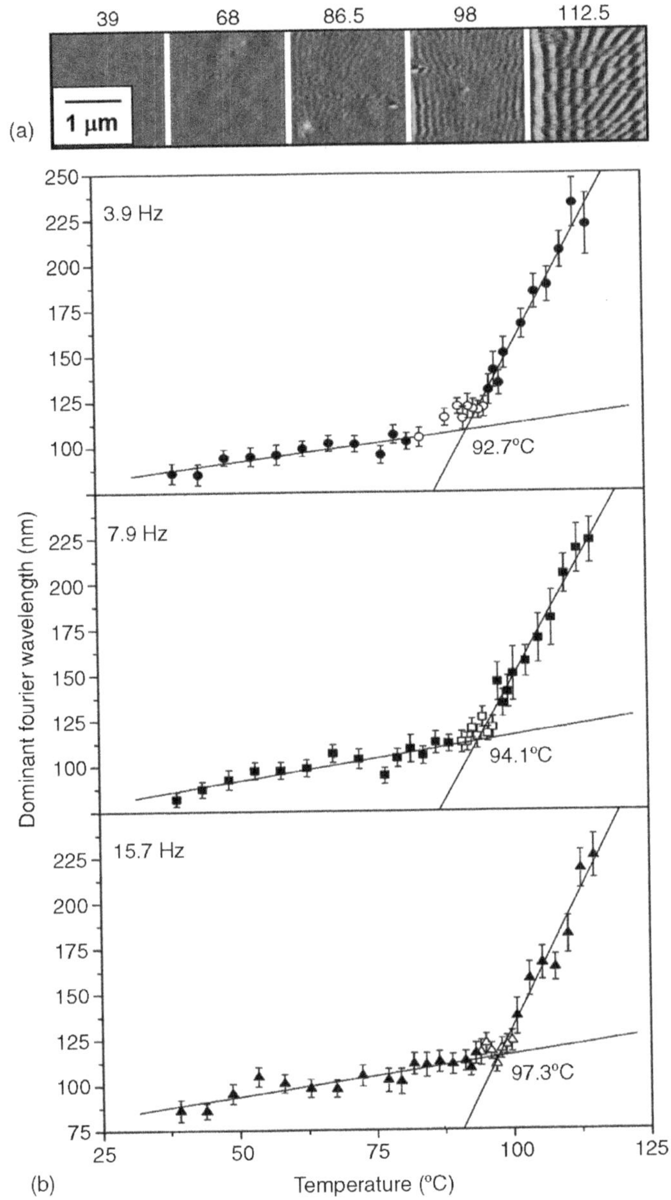

FIGURE 7.20 (a) Height images of polystyrene after one raster scan at the indicated temperature. (b) Spacing of ripples versus temperature and at three scan-line frequencies as shown, via Fourier analysis along the *X* axis.

during the first raster scan (shown) and strengthen with repeated raster scans (not shown) [44,45]. By Fourier analyzing these images using an *X* power spectral density routine (Section 8.3.2), a periodic component or dominant Fourier wavelength can be quantified from each image. This analysis was performed for a large set of images acquired at finer increments of temperature within the 39–112.5°C range, and at three scan-line frequencies as shown in Figure 7.20b (each image taken at a fresh location). Thus, each datum derives from a complete height image. The most notable observation is a marked change in the *slope* of approximately linear relationships between spatial wavelength and temperature. The crossing points of linear fits identify a temperature that is consistent with the glass-transition temperature of polystyrene. This temperature changes slightly between the three rates; the rate–temperature relationship is consistent with the known bulk behavior of polystyrene in the context of alpha activation energies [22,45].

It is worth considering the *utility*, from a microscopy standpoint, of wear processes such as the preceding example. One can imagine a case of blends or interfaces where two polymers are similar in mechanical properties, dielectric constant, and friction coefficient, and thus difficult to contrast via measurements of friction, tip–sample adhesion, stiffness, phase, or electromagnetic response. Such a case is shown in Figure 7.21, a hot-press laminate interface between polystyrene–acrylonitrile

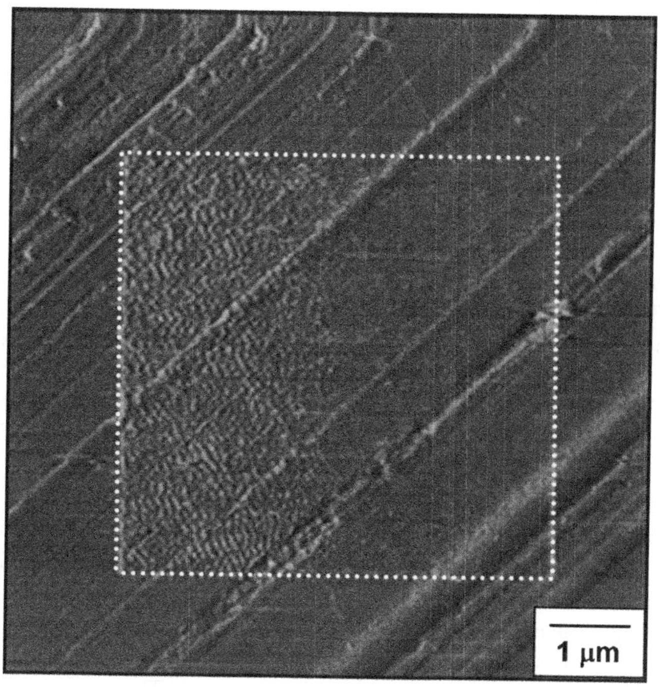

FIGURE 7.21 Deflection image (contact-mode error signal) of hot-press laminate interface between a PS-PAN copolymer and PMMA following high-force scanning inside of the square-designated region, to selectively roughen the PS-PAN copolymer (left) side of the interface.

copolymer (MW 92,000, 29% acrylonitrile) and PMMA (MW 108,000). Both polymers are glassy at room temperature. Phase and friction imaging in the absence of wear did not successfully contrast these polymers. A contact-mode deflection image (to highlight texture) is shown of the cryo-microtomed surface. The horizontal center of the image is the estimated location of the laminate interface. The image was taken following smaller-scale raster scanning at elevated load, within the marked square region. Some wear is detected throughout this square region, but it is dramatic only on the styrene–acrylonitrile side of the interface. The concept here is that a strong perturbation that produces *whole-chain movement* [22] can be sensitive to molecular differences (e.g., entanglement), even if these differences do not manifest at the more local scale (e.g., dissipation due to alpha or beta motions).

7.5.4 Musings on the Spectroscopic Nature of Friction and Other Measurements

We have now seen that several AFM imaging modes for distinguishing materials and investigating properties probe tip–sample *energy dissipation*. Examples from Section 7.5 as well as Chapters 5 and 6 include sliding friction force, phase lag in dynamic mode, and the vertical break of contact during force-curve mapping, wherein irreversible molecular rearrangements produce adhesion hysteresis. The principal experimental differences among these methods are the (i) direction of motion and (ii) interaction time per measurement. In all cases, the irreversible energetic excitations are inherently *rate dependent*. On viscoelastic materials, rate strongly impacts the degree to which any stress–strain cycle departs from reversible in the compression/decompression of vertical or shear motion [21,22,35]. Rate also affects the number of attempts to surpass activation energy barriers in stick-slip processes, as is commonly treated in sliding friction [11,16,31,32], but also as applies to activated transitions *internal* to a material (e.g., gauche/trans transitions along polymer chains as fibrils are under tension or matrices are sheared or indented) [10,11,22,31,37]. As seen in the case of sliding friction in Section 7.5.2, a conjugate of this rate dependence is the role of *temperature*: the higher the temperature, the greater the likelihood that activation barriers will be surpassed via thermal fluctuations [13,16]. Thus, the AFM measurement of friction, phase, or adhesion *as a function of rate and/or temperature* represents a fundamental probe of *dissipative molecular or atomic motions*.

There are traditional methods for measuring material responses as a function of rate and temperature that are often dubbed *spectroscopy* (i.e., dielectric, dynamic mechanical [22]). But the term spectroscopy is more conventionally applied to measurements as a function of *energy*, *frequency*, or *wavelength* [49,50]. Many of these latter techniques measure *quantum* excitations; a fundamental particle (usually an electron or photon) is incident on a material sample and the energy, frequency, or wavelength of single-quantum excitations is measured. The excitation information may be quantified by the same particle (detected after transmission or scattering) or by an ejected particle that was initially present in the sample (e.g., a photoelectron). For our purposes, it is instructive to consider the nature of excitations probed by these *quantum spectroscopies* in comparison with the excitations probed by traditional *mechanical* or *dielectric spectroscopies* as well as the energy-dissipative "force spectroscopies" of AFM. The quantum spectroscopies may quantify, for example, absolute or relative binding energies for stable electronic states within the atoms of a sample; or, as another example, the ground-to-first-excited chemical bonding (vibrational) states between atoms in a sample. A set of sharp and often isolated peaks are measured versus energy, frequency, or wavelength, a so-called "spectral fingerprint," which can identify different elements or bonding types (e.g., C–H) present in a sample. The discrete and energy-resolved nature of these spectroscopies derives from the fact that they involve *single-quantum* transitions; although a large number of quantum particles are incident on the material sample and thousands of particles are detected to construct a spectrum, each emission or absorption peak derives from repeated measurements of a *single type of quantum transition.*

Traditional mechanical and dielectric spectroscopies, and the nanoscale force spectroscopies of AFM, also produce a large number quantum excitations, as probed at a given measurement rate or sample temperature, but these excitations will generally span a variety of types (e.g., vibrations of different bond types, rotational isomer transitions with variable degrees of steric hindrance). The large number, and hence diversity, of these excitations holds true even if measured over only a nanoscale spatial domain and a seemingly short time interval. For example, even a small, 1-nN average friction force (e.g., 10-nN normal load times 0.1 coefficient of friction) integrated over a 500-nm sliding length divided into 512 sampling points (i.e., ~one measurement per 1 nm of sliding) translates into an average energy dissipation per sampling measurement of

$$\sim (1\,\text{nN}) \cdot (1\,\text{nm}) = 10^{-18}\,\text{J} \approx 6\,\text{eV}.$$

Thus, the number of individual molecular vibrations or rotational isomeric transitions contributing to this single datum number in the tens or even hundreds. So our "detector" is *not counting individual dissipative excitations* as a function of their energies, as do single-quantum transition spectroscopies. Instead, it is summing the aggregate energy dissipation within some short time interval, whether embodied in a force of resistance of motion, a force needed to break contact, or a phase shift in an oscillating cantilever. Nevertheless, the rate and temperature dependences of these measurements *may contain peaks*: an energetically narrower subset of all possible excitations may be preferentially excited *within certain rate and/or temperature windows*, just as seen in DMA or dielectric spectroscopy, or even differential scanning calorimetry. (Indeed, energy dissipation measurements are perhaps most analogous to heat conduction measurements with the understanding that the term "heat" may include more than a classic picture of bond vibrational motion, i.e., may include rotational isomeric transitions as understood in relaxational spectroscopies [36,37].) When combined with the extreme *spatial resolution* of AFM, these spectroscopies—although not as energetically "precise" as single-quantum methods like X-ray photoelectron spectroscopy or infrared absorption spectroscopy—nevertheless can be exquisite materials characterization tools.

7.6 TRANSVERSE SHEAR MICROSCOPY (TSM) AND ANISOTROPY OF SHEAR MODULUS

A much newer utilization of a special type of lateral force requires the measurement of cantilever twist during *longitudinal* scanning; with the fast direction of the raster scan oriented *parallel* to the principal axis of the cantilever. Thus, the detected force is *transverse* to the scan direction rather than opposed to it. This immediately suggests that the detected force is not truly frictional in nature. There are, of course, slope-derived forces that act to twist the cantilever regardless of the choice of fast-scan direction, indeed even in the absence of scanning; the force normal to the tip–sample interface at a point of nonzero slope (whether dZ/dX or dZ/dY) will, in general, have a lateral component that in turn can be resolved into components parallel and perpendicular to the principal cantilever axis. During lateral scanning and under an assumption of ideal surface tracking, this slope-derived force would be independent of the fast-scan direction. *Transverse shear microscopy* (TSM) instead concerns itself with *hysteretic* transverse forces perpendicular to the scan direction; within a closed scan cycle parallel to the principal cantilever axis, the cantilever's twist signal *reverses sign* between forward and backward scans, as depicted in Figure 7.22. But the force does zero net work because force and displacement are perpendicular to one another.

How can any material exert such a force? The key concept is an *anisotropic shear modulus*. For many crystalline molecular systems, the crystal axes are not equivalent, and thus the modulus depends on the direction of strain. It is not uncommon for a crystalline molecular system to have nonequivalent crystal axes within the scanning plane of an AFM image. An example is a self-assembled monolayer of short chain molecules that stand on end but with a finite polar tilt angle. The existence of a particular azimuthal tilt direction within a crystal grain necessarily means that the grain is anisotropic in-plane. Thus, different grains may exhibit

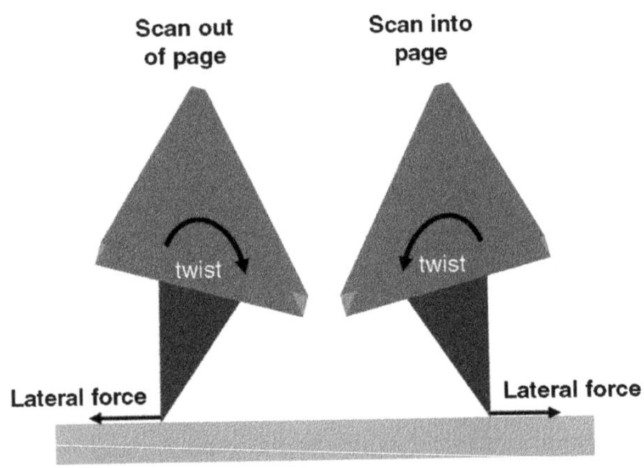

FIGURE 7.22 Illustration of an AFM cantilever (looking down the principal axis) twisting due to lateral forces exerted *transverse* to the fast-scan direction, which is in or out of the page.

different shear-force responses, assuming different tilt orientations from grain to grain. *Friction anisotropy* in AFM has been demonstrated on a number of systems dating back to the mid 1990s [8,51]. There is nothing methodologically different in that work from the topics considered in the preceding sections on lateral force measurements (except the need to measure the zero of lateral force independent of scanning the surface [8]). But the observation of transverse shear behavior as defined here, and its utilization as an analytical tool, is a newer development.

The system investigated in detail at this writing is pentacene deposited on silicon from the vapor phase (useful in semiconductor devices), previewed in Figure 1.8 of Section 1.2 in the context of friction imaging [7,52]. The pentacene molecule consists of five benzene-like rings bonded in a linear and planar geometry. The molecules pack into a van der Waals solid of triclinic lattice structure, in a "herring bone" motif parallel to substrate [53]. Individual grains within a polycrystalline first layer have been starkly revealed with TSM; an example is shown in Figure 7.23 [7].

FIGURE 7.23 Pentacene thin film containing an essentially complete first monolayer plus isolated dendritic islands of second layer. (a) Height, (b) trace transverse shear, and (c) retrace transverse shear. The fast-scan axis is vertically oriented.

The TSM images contain a mosaic-like contrast pattern, each monocolor domain being a single grain. The brightest grains in trace become darkest in retrace and vice versa; these are grains where a [110]-type crystallographic axis is nearly parallel or antiparallel to the fast-scan axis. The most intermediate-color grains, neither bright nor dark, remain consistently so whether imaged in trace or retrace; these are grains where a [100] or [010]-family crystallographic axis is nearly parallel or antiparallel with the fast-scan axis.

These findings have been systematized to allow precise mapping of grain orientations based on the magnitude and sign of the TSM hysteresis loop [54]. The relationship between the fast-scan direction and the cantilever twist response is depicted in Figure 7.24a. The molecules in the known herring-bone lattice are seen from above, the topmost benzene-ring being viewed on edge. Along each of the [010] and [100] crystallographic directions, the molecules encountered by the tip are oriented in two equal but opposite senses. Thus, the effects of molecular bending transverse to the scan direction intuitively cancel. Along the diagonal [110] family of directions, the molecules encountered by the tip are in two nonequivalent orientations, such that molecular bending would produce different elastic responses transverse to the scan direction. An example of grain mapping is shown in Figure 7.24b. The arrows indicate the orientations of the [110] directions specific to each single-crystal grain of order 1 μm in lateral size, determined by examining the magnitude and sign of the hysteresis loop signal per grain [54].

The preceding implies that TSM is sensitive to *elastic* rather than dissipative response. Consistent with this interpretation, TSM measurements on pentacene show no dependence on rate and temperature within the ranges examined, whereas friction force measurements show marked dependence within this same thermokinetic window [55]. This interpretation is also consistent with shear modulation results on this system, discussed in the following section.

7.7 SHEAR MODULATION METHODS

7.7.1 Motivations and Terminology

Under quasistatic contact, one can sinusoidally oscillate X at very small amplitude and measure the resulting oscillations in cantilever torsion (i.e., lateral force). This is equivalent to moving slightly up and down the slanted part of a friction loop (see Figure 7.3) near either end point, that is, within the static friction regime. Thus, the measured response amplitude is expected to be sensitive to the *shear stiffness* of the tip–sample contact, after accounting for torsional cantilever compliance. The possible advantages of shear stiffness characterization compared to normal stiffness characterization are twofold.

1. X displacement does not produce parasitic normal displacement (or a change of normal stress) in the sense that Z displacement produces lateral

displacement (or a change of lateral stress if contact is pinned) to lowest order in the driving displacement due to cantilever tilt (i.e., Figure 3.7).

2. In principle, shear stiffness measurements can be easier to analyze within adhesive contact mechanics models than normal stiffness because the

(a)
(b)

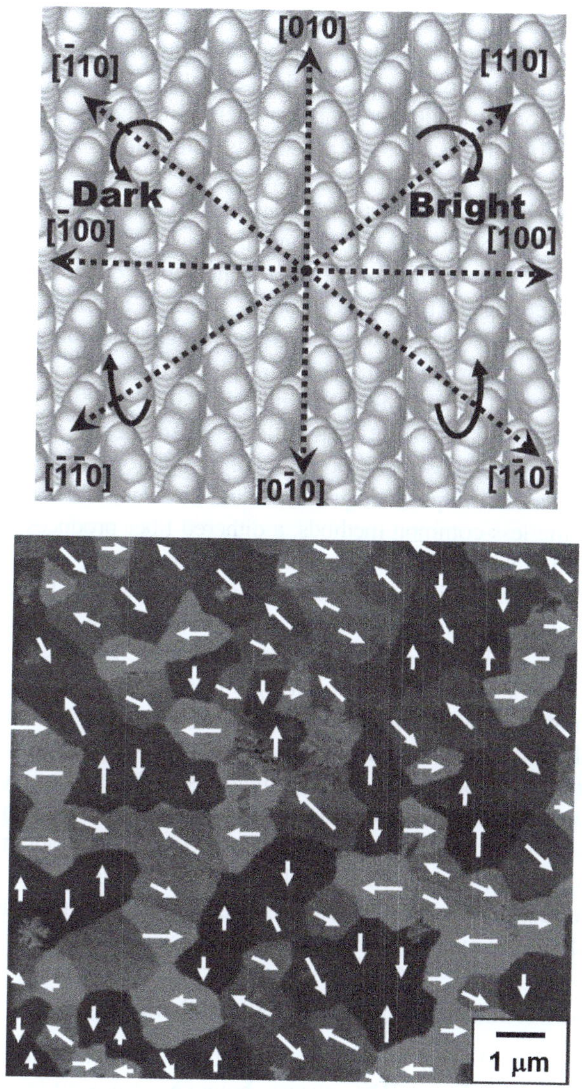

FIGURE 7.24 (a) The “herring bone” crystalline molecular lattice, the structure of pentacene, which exhibits finite transverse shear along the [110] family of crystalline axes but zero transverse shear along the [100] and [010] directions. (b) Example of grain-orientation mapping of the [110] directions for most of the grains in the image. The fast-scan axis, parallel to the cantilever principal axis, is horizontal.

relationship between lateral contact stiffness and modulus is independent of the choice of model (e.g., DMT, JKR). In particular, the relationship

$$S = Ka$$

generally holds, where K is given by Equation 2.12 and a is contact radius [56].

A clarification of terminology may be appropriate at this stage. The use of the term *modulation* within the topics of this section as well as related topics in the literature, such as vertical force modulation, is inconsistent with the usage of the term in other AFM methods. In this section, we are simply referring to the measured oscillatory force on the tip due to an oscillatory drive signal. This time-varying force signal is the sought measurement, as a route to some physical characteristic, such as shear stiffness or rheological properties. More commonly, however, one encounters the technical term 'modulation' in reference to *feedback* schemes in dynamic AFM methods, that is, to *track* something. In these latter usages, the term modulation is in (weak) analogy with radio systems wherein a high-frequency signal is more slowly varied (modulated), potentially in amplitude, frequency, or phase, to achieve some desired result. In the case of AM-AFM and FM-AFM—meaning amplitude or frequency-modulated dynamic AFM—a measurement of amplitude or frequency is the input signal to a feedback circuit that varies the Z position (of sample or cantilever base) in order to keep the amplitude or frequency signal constant, thereby *tracking surface topography*. (Indeed, in some less-common methods, a dithered fiber produces an oscillating *shear* force so as to track surface topography.) Another usage of 'modulation' within dynamic AFM methods is in long-range (e.g., electrostatic) force or force gradient (dF/dZ) characterization. One example is in Section 9.2.4, wherein alternative AM and FM methods of Kelvin probe force microscopy are discussed. These usages refer to feedback schemes for *tracking surface potential*. The point of our disclaimer here is that within the broad field of AFM-based methods, the term modulation may simply refer to oscillating signals that are imposed and correspondingly measured, including possible spatial variations in this measurement (as in this section), or modulation may refer to an electronic feedback system that *implements* an oscillating signal in order to measure spatial variations of some other signal. Rather than adopt a single usage of the term in this book—which might make better pedagogical sense—we instead wish to inform the reader of the variegated usages encountered in the scientific literature as well as AFM vendor manuals and application notes.

7.7.2 Shear Modulation During 1D Lateral Scanning

Most commercial AFM systems do not include the option to sinusoidally modulate X (or Y) in software. Given BNC access, however, it is simple to replace the AFM controller's X-scanner signal with an external sinusoidal signal from a commercial function generator. In some published implementations [9,56–59], the response signal (lateral force) is input to an external lock-in amplifier whose DC output is proportional to the amplitude of the response. This DC signal is then measured and potentially imaged by the AFM data acquisition system, via an auxiliary BNC input channel. The phase output of the lock-in amplifier also can be input to the AFM system via a second BNC channel for simultaneous measurement and imaging. This is particularly useful for polymeric or other viscoelastic systems that generally exhibit some phase lag between the drive and response signals. Provided a sufficient number of data acquisition channels, the driving signal and both amplitude and phase lag of the response could be measured.

The preceding measurements can be performed in a 1D imaging mode, wherein *Y* is controlled by the usual *Y*-scanner signal while *X* is externally modulated. For 2D imaging under *X* modulation, one can use an adder circuit such as that described in Section 7.4.2 to add the small-amplitude *X* modulation to the AFM controller's *X*-scanning signal. The limitation is that a common op-based adder circuit clips around ±13 V. Thus, the *X*-scan range will be substantially limited. For example, an AFM controller that drives a 100-μm range *XY* scanner using a ±200 V output will be limited to $(13/200) \cdot (100\ \mu\text{m}) = 6.5\ \mu\text{m}$ scan size along *X*. (Of course, there is also the question of contact mechanics; the tip is being scanned along *X* and thus is in a sliding mode (kinetic friction), whereas shear modulation measurements are usually interpreted under the assumption of a tip pinned by static friction.) Even if one chooses to disable the adder circuit to retain large *X*-scan capability, one cannot go to an arbitrary location within this imaged region to perform a point-measurement under shear modulation; the latter still requires the addition of a small signal to a potentially large signal, which the adder circuit cannot handle.

If *X* scanning is not needed, one can choose to modulate *X* at *low frequency* and thereby examine the real-time sinusoidal response at the AFM system's data sampling rate. This obviates the need for a lock-in amplifier and utilizes frequencies that are more akin to those in conventional DMA. Measurements can be conducted under 1D (*Y*) scanning or within force-curve acquisition (i.e., *Z* scanning). Under *Y* scanning with the *X*-scanner signal removed, the AFM software may still render a 2D "image" (because the software does not "know" that the *X* signal has been removed). The usual data sampling is being performed at the frequency of virtual pixel locations along the fast-scan axis, even though *X* is not being ramped up and down. Thus, in the acquired "image," *Y* is *Y* but *X* is *time*.

An example of this type of data acquisition is shown in Figure 7.25 as follows. The location and direction of the 1D domain of interest, a 1-μm *Y* scan, is marked with an arrow on 2D height and friction difference "images" of an *autophobically dewetted* (i.e., from itself) ultrathin PVA film (on mica substrate, not visible) in (a) and (b). During this 1D scan, a 25-nm, 500-Hz *X* modulation was being executed in time, expected to produce sliding. The 1-μm *Y* scan traverses two different components in an ultrathin PVA film (c): a *precursor* layer that is strongly adsorbed to mica and a second layer that contains roughly circular gaps micrometers in size, exposing the first layer (resulting from disentanglement between the precursor layer and additionally deposited polymer). The first layer is in a conformationally arrested state resulting in a friction force reduced by a factor of two relative to the second layer (quantified from (b)). The lateral force signal rendered as an "image" in (d) is graphically plotted in (e), at the two select *Y*-values indicated with horizontal lines in (d) above and below the edge of the second layer (dashed line). The

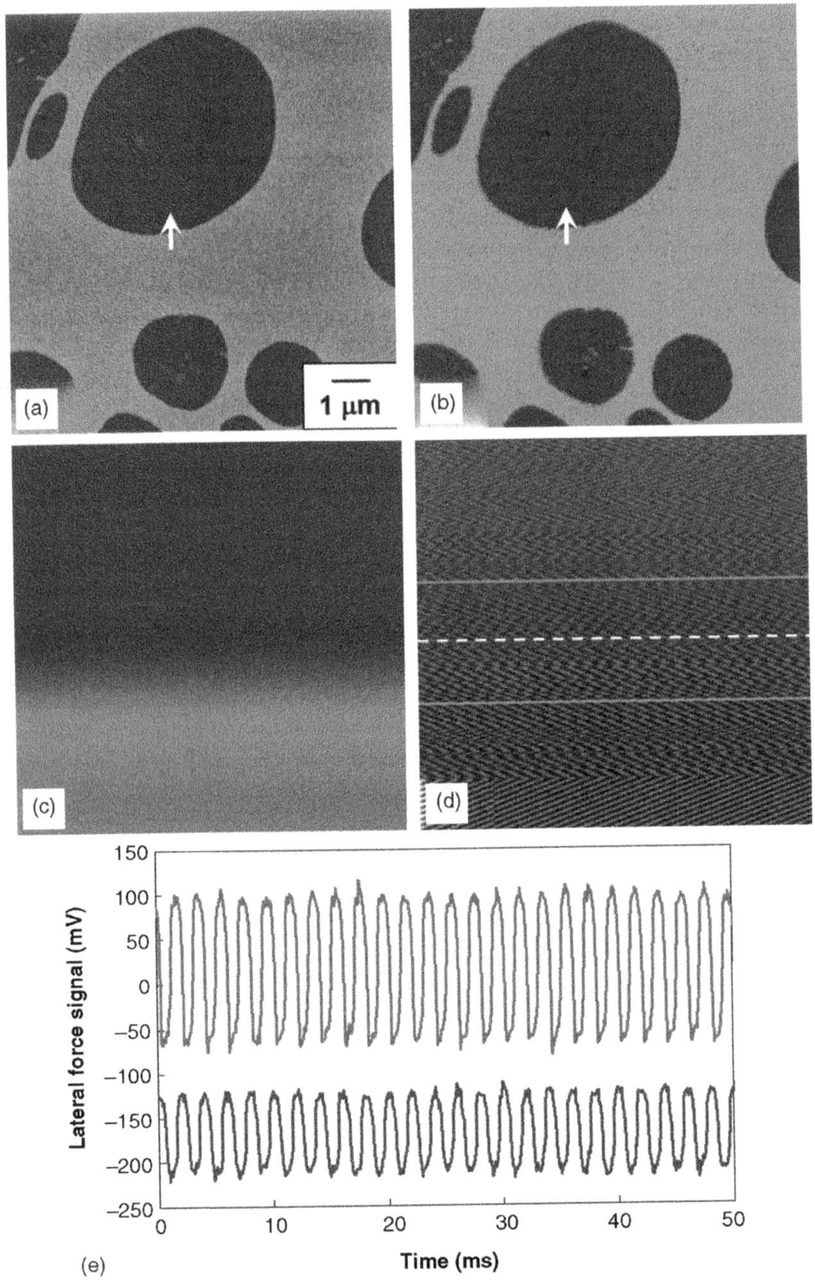

FIGURE 7.25 (a) Height and (b) friction difference images of an autophobically dewetted ultrathin PVA film. Arrows depict 1D (*Y*) domain of (c) height and (d) lateral force data acquisition under 500-Hz *X*-modulation. (e) Comparison of one timeline of data on first (top) and second (bottom) layers of PVA.

amplitudes of these shear modulation responses in (e) have a 2 : 1 ratio that corresponds well with the 2 : 1 ratio of friction forces in (b). This suggests that *sliding* is indeed taking place, and thus shear *stiffness* is not being characterized.

7.7.3 Diagnostics of Sliding Under Shear Modulation

The key to shear *stiffness* characterization is to avoid the onset of sliding by keeping the oscillation truly small. Given a contact diameter that may be only nanometers in size per the contact mechanics models of Section 6.2, one intuitively anticipates the need to set the X-modulation amplitude to a fraction of this contact diameter, say $\sim$1 nm, to avoid sliding. In this case, the torsional signal would change by only small amounts, meaning possibly poor signal/noise ratio. On softer materials, the lateral force threshold for sliding should be greater due to a larger contact diameter; on stickier materials, the lateral force threshold for sliding should be greater due to greater tip–sample adhesion. Either case should allow use of larger modulation amplitudes, thereby improving the signal/noise ratio. Of course, the torsional spring constant of the cantilever plays an essential role. First, a particular lateral displacement (drive amplitude) of sample or cantilever produces a net lateral displacement of the tip end reduced by the amount of lateral contact deformation, assuming a pinned (nonsliding) contact. This net lateral tip displacement determines the cantilever twist angle for a given tip length, and thus determines the applied shear force via the torsional cantilever spring constant. The shear force in turn determines the amount of lateral contact deformation. But a greater magnitude of applied shear force will result in the onset of sliding at a smaller lateral displacement. So a softer cantilever is expected to enable larger amplitudes without inducing sliding.

It is important to carefully examine the time-dependent lateral force signal, such as in Figure 7.25e, as a gauge of physical nonlinearities such as sliding or yield. These phenomena produce a departure from pure sinusoidal force response under sinusoidal X drive (i.e., a departure from linear response). The response waveform is expected to have a clipped or truncated appearance; force will grow with lateral displacement only up to a point, then the onset of sliding or yield causes the force to flatten out or even drop. This kind of response can be quantified by examining the Fourier transform of the signal: whereas fully sinusoidal, linear response would be described by only the fundamental sinusoid, a truncated sine wave generates strong *odd harmonics* [60]. (Indeed, it is well known that the limiting case of a pure square wave is produced by an infinite series of odd harmonics.) In Figure 7.25e, the drive amplitude of 25 nm was chosen to be high enough to cause sliding and thus odd harmonics in the response. These real-time waveforms on both first and second layers of PVA, examined over a narrower time interval in Figure 7.26a, have truncated appearances. The Fourier transforms of these responses, in Figure 7.26b, contain pronounced odd harmonics at 1500 Hz and 2500 Hz for a 500-Hz driving signal. Interestingly, the response on the second layer also contains weak but measurable even harmonics. This probably derives, in the real-time waveform, from a hint of sawtooth character, which in turn may result from different static and kinetic coefficients of friction. For these purposes, Fourier transforms are seen as revealing form of signal analysis to assess differences in contact behavior under shear modulation.

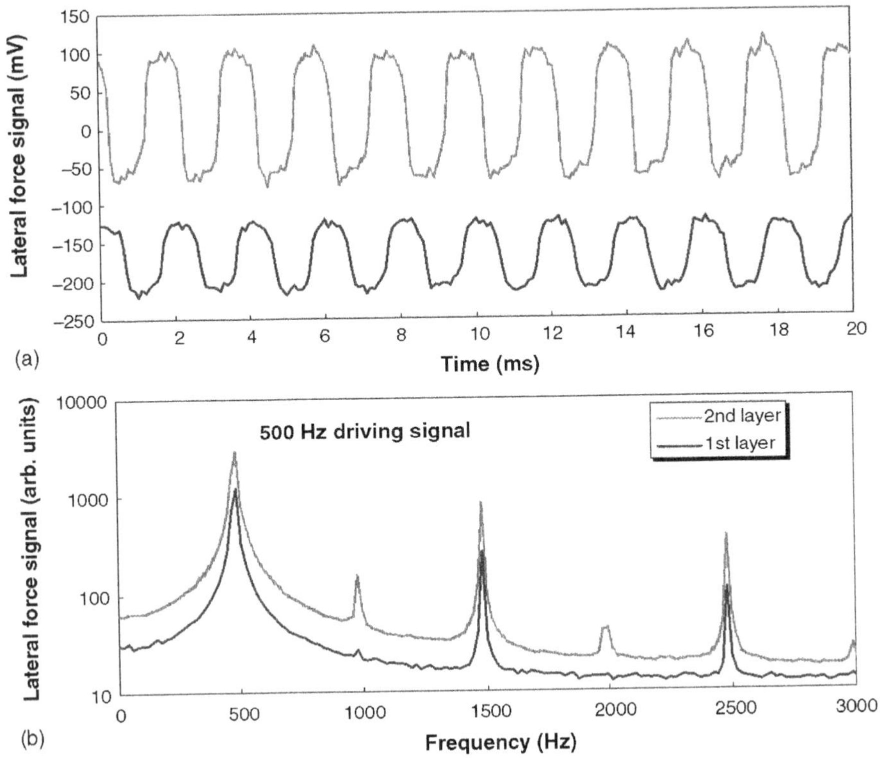

FIGURE 7.26 (a) Higher time resolution of data from Figure 7.25e, revealing square-wave response (to 500-Hz sinusoidal drive) characteristic of dominant sliding motion on each film layer. Response amplitudes differ by a factor of two, reflecting differences in coefficient of friction. (b) Fourier transforms of data sets in (a), showing fundamental response and strong odd harmonics in both cases, again indicating a predominantly sliding interaction, not a pinned contact.

7.7.4 Complementarity of Shear Modulation Methods to TSM

In Section 7.6, we saw that TSM is sensitive to anisotropic elastic shear response, dependent on the scan direction relative to the in-plane crystallographic axes of pentacene. A corroborative result is expected in shear modulation measurements performed with *driving modulation along the principal cantilever axes* (Y) while monitoring the *periodic twisting signal*, again, due to force transverse to the displacement direction. Armed with the Fourier diagnostic method of Section 7.7.3, we choose a small enough drive amplitude to avoid higher odd harmonics and thus pin the tip in the static friction regime within a given grain, so as to probe *elastic* (i.e., nonsliding) shear response perpendicular to the modulation axis. In Figure 7.27a and b are overlaid the drive and response signals acquired within two different crystal grains as illustrated in (c): one which exhibits strong clockwise twisting under shear ([110] direction), meaning a positive TSM hysteresis loop, and the other strong

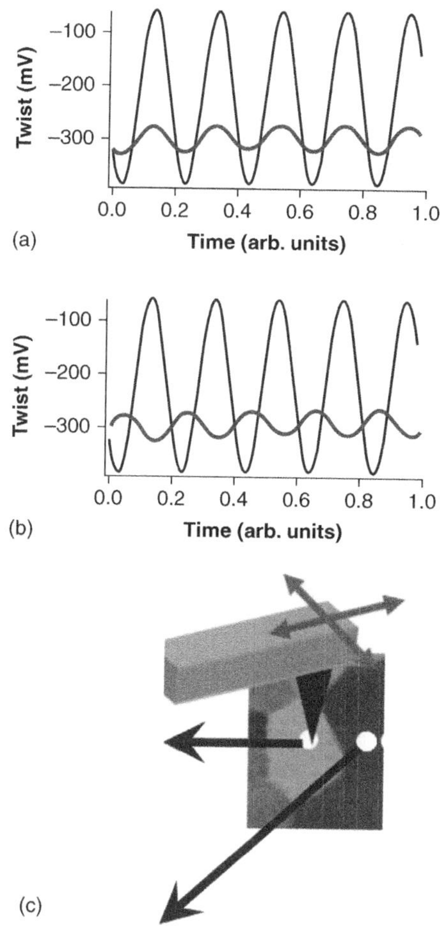

FIGURE 7.27 Twist responses (thick sine waves) due to shear forces transverse to the principal cantilever axis under longitudinal driving displacement (thin sine waves), meaning parallel to principal cantilever axis. Time-domain data in (a) and (b) derive from point-contact locations denoted in (c), within each of two adjacent pentacene grains that exhibit strong contrast in TSM imaging (longitudinal sliding).

counterclockwise twisting ([−110] direction), meaning a negative TSM hysteresis loop. Consistent with this, the shear-modulation response signal in (a) is in-phase with the drive signal along the [110] direction and in (b) is 180° out of phase with the drive signal along the [−110] direction.

7.7.5 Shear Modulation Within Force Curves: Material Creep

Shear modulation during force curves can be used to quickly explore the load dependence of shear stiffness (using small-amplitude drive, with fundamental harmonic response only) or sliding friction (using large-amplitude drive, with fundamental plus odd harmonic response). In the small-amplitude case, one can additionally examine in-phase and 90° out-of-phase components to assess viscoelasticity (in which case variable frequency is also of interest). The force-curve ramping time (usually via the *Z*-scan frequency) also can be explored to assess time-dependent creep.

An example of creep is shown in Figure 7.28. Raw vertical deflection as well as raw twisting signal under 1-nm *X*-drive amplitude is plotted as a function of time in

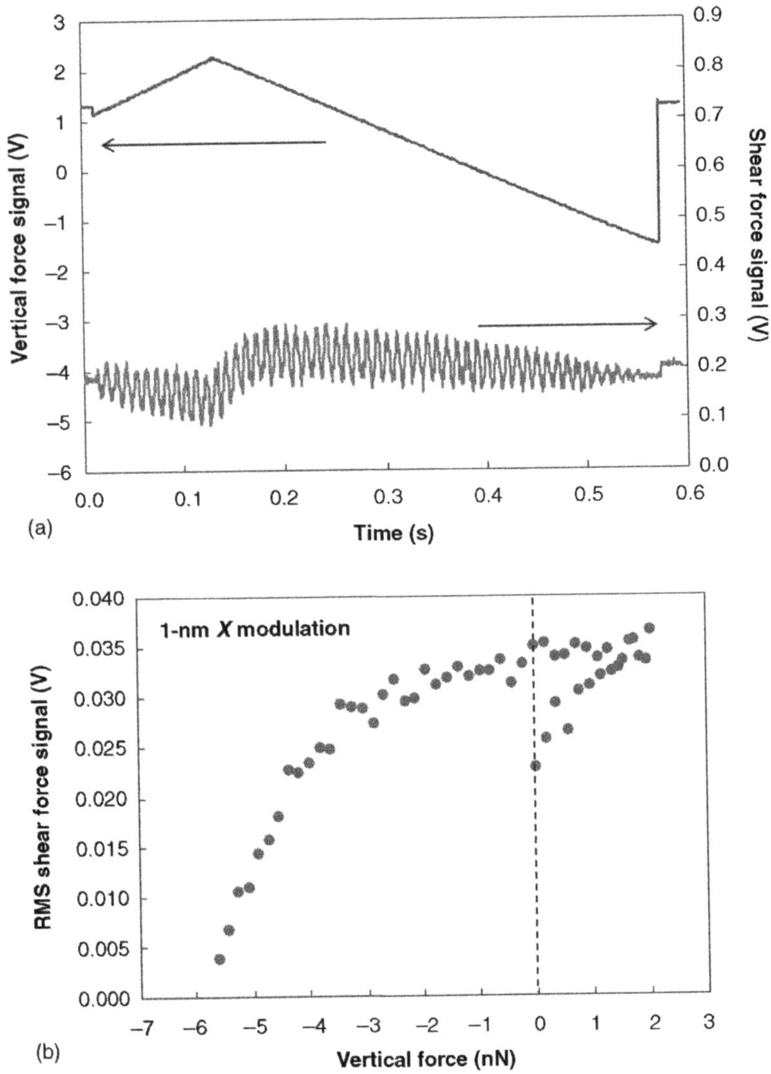

FIGURE 7.28 (a) Vertical and shear force signals as a function of time during the *Z* ramping of a force curve, on PVA. (b) Corresponding shear force signal, per period of modulation, plotted against average vertical force signal during each period.

Figure 7.28a during the *Z* ramp of a force curve on PVA (the dominant component in the inset of Figure 7.11c). Fourier analysis indicated a pinned contact: no discernable odd harmonics, which would indicate a sliding interaction [60] (or some other nonlinear response such as plastic deformation [61]). In Figure 7.28b, the root-mean-square sinusoidal twisting response was calculated per period of motion and plotted against the vertical force signal. The shear signal goes as approximately the 1/3 power of vertical force. This makes some sense; in adhesive contact mechanics treatments, shear stiffness is expected to be proportional to contact radius (Section 7.7.1), the latter scaling as the 1/3 power of vertical force (Section 6.2). What is not predicted by these models is the observed irreversibility of shear force signal upon reversal of the vertical force ramp, seen in Figure 7.28b. This implies a larger contact radius during the downward load ramp compared to the preceding upward load ramp, in turn suggesting that the tip is "sinking" into the polymer, even though glassy response is expected in these room-temperature measurements. One possibility is a more mobile behavior at the surface compared to the bulk.

A more dramatic onset of creep due to softening of polymer at a characteristic elevated *sample temperature* has been used by Overney and coworkers to identify the glass-to-rubber transition [59]. Thus, shear-modulation AFM can be a particularly useful probe of nanoscale polymer properties.

REFERENCES

[1] Meyer, E., et al., Friction and wear of Langmuir-Blodgett films observed by friction force microscopy. *Phys. Rev. Lett.*, 1992, **69**(12): 1777.

[2] Overney, R.M., et al., Friction measurements on phase-separated thin films with a modified atomic force microscope. *Nature*, 1992, **359**: 133–135.

[3] Haugstad, G., W.L. Gladfelter, and E.B. Weberg, Friction force microscopy of AgBr crystals: Ag_0 rods and adsorbed gelatin films. *Langmuir*, 1993, **9**(12): 3717–3721.

[4] Frisbie, C.D., et al., Functional group imaging by chemical force microscopy. *Science*, 1994, **265**: 2071–2074.

[5] Haugstad, G., W.L. Gladfelter, and R.R. Jones, Scanning force microscopy characterization of viscoelastic deformations induced by precontact attraction in a low cross-link density gelatin film. *Langmuir*, 1998, **14**: 3944–3953.

[6] Haugstad, G., J.A. Hammerschmidt, and W.L. Gladfelter, Viscoelasticity in nanoscale friction on thin polymer films, in *Interfacial Properties on the Submicrometer Scale*. Frommer J. and R.M. Overney, eds. 2001, Washington, D.C.: Oxford University Press.

[7] Puntambekar, K., et al., Structural and electrostatic complexity at a pentacene/insulator interface. *Adv. Funct. Mater.*, 2006, **16**: 879–884.

[8] Liley, M., et al., Friction anisotropy and asymmetry of a compliant monolayer induced by a small molecular tilt. *Science*, 1998, **280**: 273–275.

[9] Tsukruk, V.V. and K.J. Wahl, eds. *Microstructure and Microtribology of Polymer Surfaces*. ACS Symposium Series. Vol. **741**, 2000, ACS Books, 526.

[10] Hammerschmidt, J.A., G. Haugstad, and W.L. Gladfelter, Probing polymer viscoelastic relaxations with temperature controlled friction force microscopy. *Macromolecules*, 1999, **32**: 3360.

[11] Sils, S. and R.M. Overney, Creeping friction dynamics and molecular dissipation mechanisms in glassy polymers. *Phys. Rev. Lett.*, 2003, **91**(9): 1–4.

[12] Carpick, R.W. and M. Salmeron, Scratching the surface: Fundamental investigations of tribology with atomic force microscopy. *Chem. Rev.*, 1997, **97**: 1163–1194.

[13] Baumberger, T. and C. Caroli, Solid friction from stick–slip down to pinning and aging. *Adv. Phys.*, 2006, **55**(3–4): 279–348.

[14] Ogletree, D.F., R.W. Carpick, and M. Salmeron, Calibration of frictional forces in atomic force microscopy. *Rev. Sci. Intrum.*, 1996, **67**: 3298.

[15] Haugstad, G., et al., Probing biopolymers with scanning force methods: Adsorption, structure, properties, and transformation of gelatin on mica. *Langmuir*, 1994, **10**: 4295–4306.

[16] Persson, B.N.J., Sliding friction: Physical principles and applications, in *Nanoscience and Nanotechnology*. 2nd ed. von Klitzing K. and R. Wiesendanger, eds. 2000, Berlin: Springer-Verlag, 462.

[17] Harrison, J.A., et al., Elucidating atomic-scale friction using molecular dynamics and specialized analysis techniques. *J. Phys. Condens. Matter*, 2008, **20**: 354009.

[18] Gnecco, E. and E. Meyer, eds. Fundamentals of friction and wear on the nanoscale, in *Nanoscience and Technology*. 2007, Berlin: Springer-Verlag, 714.

[19] Mate, C.M., et al., Atomic-scale friction of a tungsten tip on a graphite surface. *Phys. Rev. Lett.*, 1987, **59**(17): 1942.

[20] Liu, Y., T. Wu, and D.F. Evans, Lateral force microscopy study on the shear properties of self-assembled monolayers of dialkylammonium surfactant on mica. *Langmuir*, 1994, **10**(7): 2241–2245.

[21] Ferry, J.D., *Viscoelastic Properties of Polymers*. 1980, New York: Wiley.

[22] McCrum, N.G., B.E. Read, and G. Williams, *Anelastic and Dielectric Effects in Polymeric Solids*. 1967, London: Wiley.

[23] Haugstad, G., et al., Atomic force microscopy of AgBr crystals and adsorbed gelatin films. *Langmuir*, 1993, **9**(6): 1594–1600.

[24] Hamilton, J.F., The silver halide photographic process. *Adv. Phys.*, 1988, **37**(4): 359.

[25] Schwarz, U.D., et al., Quantitative analysis of the frictional properties of solid materials at low loads. I. Carbon compounds. *Phys. Rev. B*, 1997, **56**(11): 6987–6996.

[26] Haugstad, G., et al., Probing molecular relaxation on polymer surfaces with friction force microscopy. *Langmuir*, 1995, **11**(9): 3473–3482.

[27] Briscoe, B.J. and D.C.B. Evans, The shear properties of Langmuir-Blodgett layers. *Proc. R. Soc. A*, 1982, **380**: 389–407.

[28] Finch, C.A., *Polyvinyl Alcohol: Properties and Applications*. 1973, London: Wiley.

[29] Strawhecker, K.E. and E. Manias, AFM of poly(vinyl alcohol) crystals next to an inorganic surface. *Macromolecules*, 2001, **34**: 8475–8482.

[30] Horowitz, P. and W. Hill, *The Art of Electronics*. 2nd ed. 1989, Cambridge: Cambridge University Press, 1125.

[31] Riedo, E., et al., Interaction potential and hopping dynamics governing sliding friction. *Phys. Rev. Lett.*, 2003, **91**(8): 084502.

[32] Gnecco, E., et al., Velocity dependence of atomic friction. *Phys. Rev. Lett.*, 2000, **84**: 1172–1175.

[33] Szoszkiewicz, R. and E. Riedo, Nucleation time of nanoscale water bridges. *Phys. Rev. Lett.*, 2005, **95**: 135502.

[34] Schirmeisen, A., et al., Temperature dependence of point contact friction on silicon. *Appl. Phys. Lett.*, 2006, **88**: 123108.

[35] Aklonis, J.J. and W.J. MacKnight, *Introduction to Polymer Viscoelasticity*. 2nd ed. 1983, New York: Wiley, 295.

[36] Johari, G.P., Secondary relaxations and the properties of glasses and liquids, in *Molecular Dynamics and Relaxation Phenomena in Glasses*. Dorfmüller, T. and G. Williams, eds. 1987, Berlin: Springer-Verlag, p. 90–112.

[37] Matsuoka, S., *Relaxation Phenomena in Polymers*. 1992, New York: Oxford University Press, 322.

[38] Meier, D., ed. *Molecular Basis of Transitions and Relaxations*. 1978, New York: Gordon and Breach, 429.

[39] Lee, L.-H., ed. Advances in polymer friction and wear, in *Polymer Science and Technology*. Vol. 5 1974, New York: Plenum Press, 405.

[40] Haugstad, G., W.L. Gladfelter, and R.R. Jones, Nanotribology on a polymer network film. *J. Vac. Sci. Technol. A*, 1996, **14**(3): 1864–1869.

[41] Yannas, I.V., Collagen and gelatin in the solid state. *J. Macromol. Sci. C*, 1972, **7**(1): 49–104.

[42] Haugstad, G., et al., Probing biopolymer films with scanning force methods. *Mat. Res. Soc. Symp. Proc.*, 1995, **355**: 253–258.

[43] Djabourov, M., Architecture of gelatin gels. *Contemp. Phys.*, 1988, **29**(3): 273–297.

[44] Schmidt, R.H., G. Haugstad, and W. Gladfelter, Scan-induced patterning in glassy polymer films: Using scanning force microscopy to study plastic deformation at the nanometer length scale. *Langmuir*, 2003, **19**: 898–909.

[45] Schmidt, R.H., G. Haugstad, and W.L. Gladfelter, Scan-induced patterning and the glass transition in polymer films: Temperature and rate dependence of plastic deformation at the nanometer scale. *Langmuir*, 2003, **19**: 10390.

[46] Schmidt, R.H., G. Haugstad, and W.L. Gladfelter, Correlation of nanowear patterns to viscoelastic response in a thin polystyrene melt. *Langmuir*, 1999, **15** (2): 317–321.

[47] Jing, J., et al., An atomic force microscopy study of polyester surfaces. *J. Mat. Sci.*, 1995, **30**: 5700–5704.

[48] Woodland, D.D. and W.N. Unertl, Initial wear in nanometer-scale contacts on polystyrene. *Wear*, 1997, 203–204, 685–691.

[49] Feldman, L.C. and W. Mayer, *Fundamentals of Surface and Thin Film Analysis*. 1986, New York: North-Holland, 354.

[50] Harris, D.C. and M.D. Bertolucci, *Symmetry and Spectroscopy: An Introduction to Vibrational and Electronic Spectroscopy*. 1978, Oxford: Oxford University Press.

[51] Overney, R.M., et al., Anisotropy in friction and molecular stick-slip motion. *Phys. Rev. Lett.*, 1994, **72**(22): 3546–3549.

[52] Kalihari, V., et al., Observation of unusual homoepitaxy in ultrathin pentacene films and correlation with surface electrostatic potential. *Adv. Mater.*, 2009, **21**: 1–7.

[53] Schwoerer, M. and H.C. Wolf, *Organic Molecular Solids*. 2007, Weinheim: Wiley-VCH Verlag, 438.

[54] Kalihari, V., et al., Grain orientation mapping of polycrystalline organic semiconductor films by transverse shear microscopy. *Adv. Mater.*, 2008, **20**: 1–7.

[55] Kalihari, V., G. Haugstad, and C.D. Frisbie, Distinguishing elastic shear deformation from friction on the surfaces of molecular crystals. *Phys. Rev. Lett.*, 2010, **104**: 086102.

[56] Carpick, R.W., D.F. Ogletree, and M. Salmeron, Lateral stiffness: A new nanomechanical measurement for the determination of shear strengths with friction force microscopy. *Appl. Phys. Lett.*, 1997, **70**: 1548–1550.

[57] Colchero, J., M. Luna, and A.M. Baro, Lock-in technique for measuring friction on a nanometer scale. *App. Phys. Lett.*, 1996, **68**: 2896–2898.

[58] Lantz, M.A., S.J. O'Shea, and M.E. Welland, Atomic-force-microscopy study of contact area and friction on $NbSe_2$. *Phys. Rev. B*, 1997, **55**(16): 10776–10785.

[59] Ge, S., et al., Shear modulation force microscopy study of near surface glass transition temperatures. *Phys. Rev. Lett.*, 2000, **85**: 2340–2343.

[60] Haugstad, G., Contrasting static-to-kinetic friction transitions on layers of an autophobically dewetted polymer film using Fourier-analyzed shear modulation force microscopy. *Trib. Lett.*, 2005, **19**: 49–58.

[61] Wilhelm, M., D. Maring, and H.-W. Spiess, Fourier-transform rheology. *Rheol. Acta*, 1998, **37**: 399–405.

8

DATA POST-PROCESSING AND STATISTICAL ANALYSIS

As with all techniques that generate images, post-processing may comprise a large portion of the analyst's effort. For AFM, the potential to extract quantitative information in post-processing is especially great, because the third dimension—whether height, phase, or other property-sensing modes—is not some scattering intensity or hue of transmitted light, as with electron- or photon-based microscopies; rather, it is a quantitative physical measurement. This chapter briefly describes several common operations used to manipulate 2D-domain AFM data sets, that is, conventional images. Commercial and custom methods for processing 3D-domain data cubes, such as generated by force-curve mapping routines, were specifically discussed in Section 6.4.

8.1 PRELIMINARY DATA PROCESSING

In order to calculate meaningful metrics from image data, one generally needs to ensure that extraneous contributions to the data—for example, due to noise or spurious changes in tip state—are removed or minimized. If significantly present in the image data, these deleterious contributions can measurably affect, even dominate, the computed metrics discussed in the following sections. Arbitrary tilts and perhaps artificial curvatures in height images, for example, also need to be removed in

Atomic Force Microscopy: Understanding Basic Modes and Advanced Applications, First Edition. Greg Haugstad.

order to extract meaningful metrics of surface roughness. Scanner nonlinearities as well as some important first-line data-processing procedures were discussed in Section 4.2.5: the removal of Z nonlinearities by subtracting curved mathematical surfaces $Z(X,Y)$ (higher-order "plane fitting" or "leveling") or by subtracting polynomials from individual fast-scan lines of data $Z_i(X)$, to make the average of the image quantity consistent from line to line ("flattening"). Such line-by-line adjustments are more commonly employed to remove the effects of spurious changes in the state of the tip (often dramatic) or other, subtler noise that occurs at frequencies ranging from on the order of the scanning frequency (lines per second), at the high end, down to three orders of magnitude lower, the inverse time for acquiring a complete image. Zeroth-order flattening means that the average value of the image quantity (height, lateral force, phase, etc.) within each scan line is subtracted out, whereas first and second order would individually remove the unique tilt and curvature, respectively, of the image quantity as a function of X along each scan line (i.e., linear and quadratic relationships).

An example of the zeroth-order flattening is seen by comparing raw and flattened phase images of a polybutyl/lauryl methacrylate blend, Figure 8.1a–c. This mathematical operation largely removes the extraneous streakiness seen in the initial image (a) but also induces false effects as seen in (b), bright bands along sets of scan lines containing a greater amount of dark phase regions (as also discussed with Figure 5.11). Thus, one could say that this operation both "cleans" and "dirties" the image. The presence of the darker ("lower") patches results in altered average values along these scan lines, and thus the subtraction of the average produces falsely higher (brighter) data away from the patches. To remove this effect, the data within numerous rectangular domains, as denoted by squares in Figure 8.1c, was removed from the calculation of average along each scan line (sometimes called *masking*); then each of these more "correct" averages was subtracted from the corresponding data within each scan line, yielding the image in Figure 8.1c.

Other artifacts that can be manifested in images are periodic signals of extraneous origin. Section 7.2 discussed a wavy periodic background in lateral force images, arising from optical interference (laser reflected off of cantilever and sample). Subtracting lateral force data during retrace scan from that during trace scan, with an appropriate X offset to account for scanner X hysteresis, was seen to remove most of the wavy background, given that it derived from intrinsic spatial variations (i.e., in X–Y). But other periodic signals may derive from variations *in time* that are manifest along the image X and Y-axes as determined by the data sampling rate and scan-line frequency. An example is contained in the LFM image (in TSM mode, of pentacene, Section 7.6) of Figure 8.2. This signal was determined to be 120 Hz and presumably of electrical origin; it was removed by covering the instrument with a metal enclosure. The diagonal character of the artifact in the image is typical of such a signal because scan frequency along the fast-scan axis is not an integral multiple of the 120-Hz signal.

On occasion, there is a need to remove *meaningful* signal—say, real variations of height, as opposed to extraneous signal—from an image before analyzing a finer signal of greater interest. A good example is the statistical analysis (Section 8.2) of

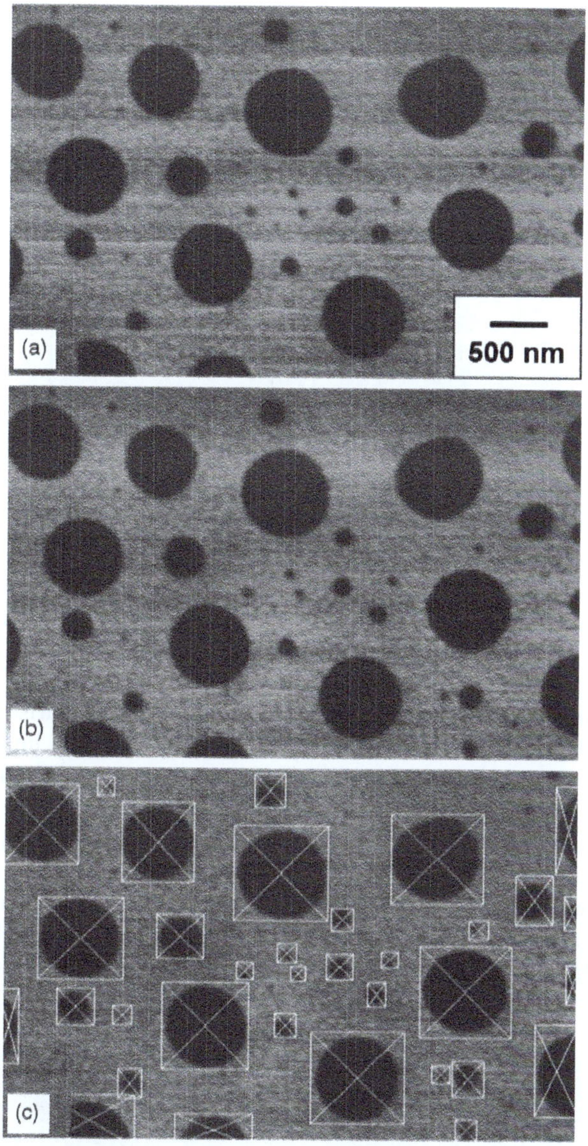

FIGURE 8.1 Example of flattening to remove streakiness. (a) Initial phase image of a blend of PBMA/PLMA. (b) Result of zero-order flattening applied over the entire image. (c) Result of flattening that excludes the data in square or rectangular regions (enclosing circular PLMA regions) from the calculation of average phase along each scan line.

nanoscale granular structures that are superimposed on larger-scale rolling or rippled topographies. In this case, higher polynomial subtraction routines can be very useful. As an extreme example, a 12th-order polynomial surface subtraction (using freeware program Gwyddion) was performed on a 50×50-μm height image

FIGURE 8.2 Example of 120-Hz extraneous background signal in a lateral force image of pentacene.

of a polycrystalline Cu surface as shown in Figure 8.3. Comparison of the initial image, the subtracted variable polynomial surface, and the final image exemplifies the profound leveling that can be produced with this numerical manipulation. In this particular case, the overall range of the subtracted Z variations is one order of magnitude greater than that due to scanner nonlinearity and cross coupling as characterized on a flat silicon wafer over the same 50-μm scale (i.e., mainly *real* variations of height are being removed from the height image in Figure 8.3). One way to avoid this gross data manipulation is by utilizing *Fourier analysis* (Section 8.3.2) to mathematically isolate the smaller features of interest from the larger-scale topographies (i.e., in *inverse space*).

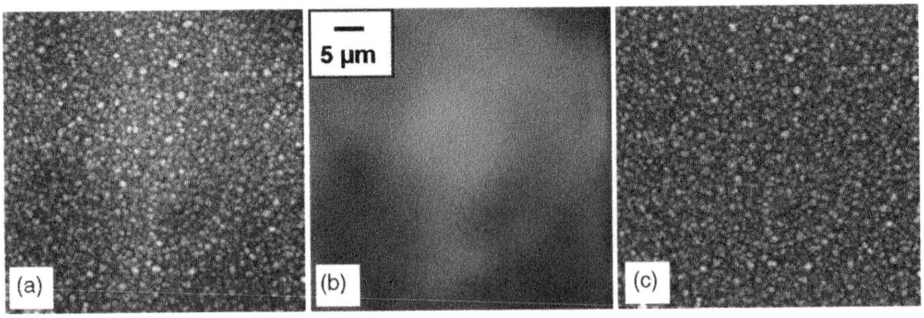

FIGURE 8.3 (a) Height image of polycrystalline copper. (b) Twelfth-order surface fit of image (a). (c) Image (a) following subtraction of image (b).

High-order line-by-line polynomial subtractions may be useful to remove noise occurring at frequencies up to an order of magnitude higher than the scan-line frequency. Or one may employ Fourier *filtering* (also Section 8.3.2), wherein an image is Fourier transformed then reverse transformed after excluding the portions of Fourier space that produce the extraneous periodic signal. (Of course, this latter method could also be employed to remove real rolling topography as done above via polynomial surface subtraction.)

8.2 1D ROUGHNESS METRICS

Probably, the most common computations from AFM images of various quantities Ω (Ω representing height Z, friction force F_f, phase ϕ, etc.) are *statistical metrics of the distribution* of Ω, which we denote $P(\Omega)$ [1]. The total number of measurements N contained in a conventional image extends well into the thousands (e.g., $N = 512^2 = 262{,}144$), providing for very well-determined distribution functions $P(\Omega)$. Common statistical metrics include the *mean* value

$$\bar{\Omega} = \frac{1}{N}\sum_{n=1}^{N} \Omega_n \tag{8.1}$$

as well as various *moments* about this mean, which quantify *deviations* from the mean. Such deviations are the most common ways to quantify surface *roughness*, for example, by analyzing Z data. The ith moment about the mean of the distribution is

$$\mu_i = \frac{1}{N}\sum_{n=1}^{N} \left(\Omega_n - \bar{\Omega}\right)^i. \tag{8.2}$$

The second moment μ_2 is also known as the *variance*.[1] The square root of the variance is the commonly employed *root-mean-square* (RMS) or *standard* deviation σ. The third and fourth moments are used to calculate the *skewness* and *kurtosis*:

$$\text{Skewness} = \frac{\mu_3}{\sigma^3} = \frac{1}{\sigma^3 N}\sum_{n=1}^{N} \left(\Omega_n - \bar{\Omega}\right)^3 \tag{8.3}$$

$$\text{Kurtosis} = \frac{\mu_4}{\sigma^4} = \frac{1}{\sigma^4 N}\sum_{n=1}^{N} \left(\Omega_n - \bar{\Omega}\right)^4. \tag{8.4}$$

[1] Equation 8.2 is sometimes called the biased variance, whereas replacing $1/N$ by $1/(N-1)$ provides the unbiased variance. This minor change is known as Bessel's correction, a statistical fine point named after Friedrich Bessel. For essentially all AFM data analysis, N is so large that the correction is unimportant.

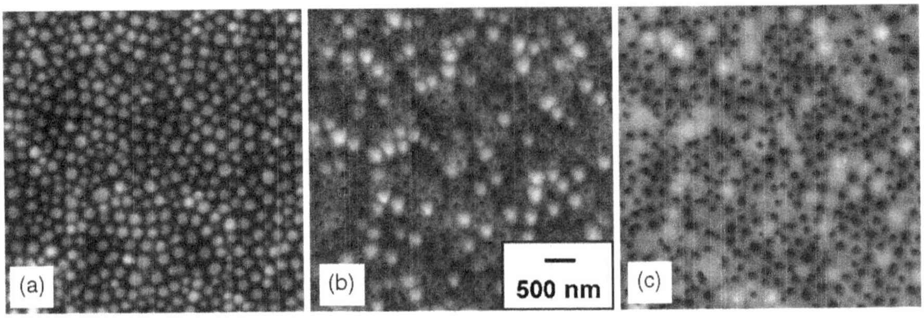

FIGURE 8.4 Height images used to compare roughness metrics. (a) Initial dry coating of PBMA and drug. Immersed coating after (b) 10 min and (c) 2.5 h in water.

Skewness provides a measure of the asymmetry of the distribution $P(\Omega)$, that is, weighted ("skewed") to the low or high side of the mean value. A large positive skewness results from a surface containing tall protrusions and a large negative skewness results from a surface laden with deep holes or canyons, within an otherwise relatively smooth surrounding surface. Kurtosis more strongly weights the tailings of the distribution, extremes of high and/or low (which might, e.g., be important features on tribological surfaces), but is not affected by the sign of these extremes whether tall or deep. A Gaussian distribution of height yields a kurtosis of 3, but some definitions of kurtosis (such as used by the program Gwyddion) subtract the number 3 to yield what is sometimes called the excess kurtosis; the sign of this latter quantity indicates the shape of the distribution *relative to a Gaussian distribution*, positive having greater tailings and negative having lesser tailings.

An example of comparative calculations of RMS roughness, skewness, and kurtosis is shown in Figure 8.4. Three height images were acquired on a $\sim$1-μm-thick coating (spun onto a silicon wafer) containing a mixture of the drug dexamethasone and glassy polymer polybutyl methacrylate (PBMA), initially in air (a), then after 10 min (b), and 2.5 h (c) in distilled or deionized water. Elution of drug from the coating and into the aqueous medium results in the appearance of holes or depressions at most locations that initially contained circular protrusions. These holes become deeper with elapsed time. A smaller number of the initial dry protrusions become taller protrusions upon water immersion, but with elapsed time collapse back down to flatter domains. The computed values of RMS roughness, skewness, and kurtosis, given in Table 8.1, quantify the above changes.

TABLE 8.1 Computed Values of RMS Roughness, Skewness, and Kurtosis

	RMS	Skewness	Kurtosis
(a)	12.8 nm	0.19	−0.62
(b)	12.3 nm	0.58	0.75
(c)	13.9 nm	−0.42	0.36

The RMS value does not change remarkably among the three images. The skewness changes from positive to negative for images (a) and (c), as expected given the dominance of deeper holes in (c); but in image (b), the skewness is actually more positive than in (a), because the effect of a few very tall protrusions outweighs those of a large number of shallow holes. The (excess) kurtosis changes from negative in (a), meaning a lack of longer tails in the height distribution, to strong positive in (b), meaning significant tailing in the distribution (due to the protrusions), to a smaller positive value in (c). The latter is due to the fact that the holes in (c) are not nearly as deep (~nm) as the protrusions are tall (~nm) in (b).

8.3 2D-DOMAIN ANALYSIS

8.3.1 Slope and Surface Area Analysis

The same set of statistical metrics discussed in Section 8.2 can be applied to the analysis of the *slope* of the image quantity Ω rather than the quantity Ω itself. This usually applies to height images, thus, dZ/dX or dZ/dY, or the steepest gradient at each location (i.e., along an arbitrary direction in the X–Y plane). In doing so, the X and Y values (i.e., the distance between pixel locations) play a role, whereas in the 1D analysis of Section 8.2, X and Y play no role whatsoever. Cases certainly arise where two surfaces exhibit similar RMS roughness—variations of Z—but very different variations of, say, dZ/dX. One should note that if the magnitudes of these slopes become quite large, then the shape of the tip may affect the measured slopes (as in Figure 4.31). In this case, one may perceive a preponderance of three- or four-sided pyramidal shapes in the image (depending of course on the type of tip used).

Again, by computing slope, one is examining not only the magnitude of changes in height but also, effectively, the lateral distance over which the changes are occurring. One can imagine two surfaces varying by the same amount in Z but with these changes occurring over wide versus narrow lateral scales, as depicted in cross section in Figure 8.5. One would tend to call the bottom cross-sectional surface "rougher," although the RMS variations of Z are identical for the two cases. What varies by greater amounts in this case is, of course, the local slope of the surface. Another, related calculation over the cross sections in Figure 8.5 that would differ for the two cases is a *path integral*, the actual distance traversed if tracking over the hills. Indeed, arguably, the most objective definition of "roughness" in an image is the actual surface area (i.e., how much paint is needed to cover it), computed by

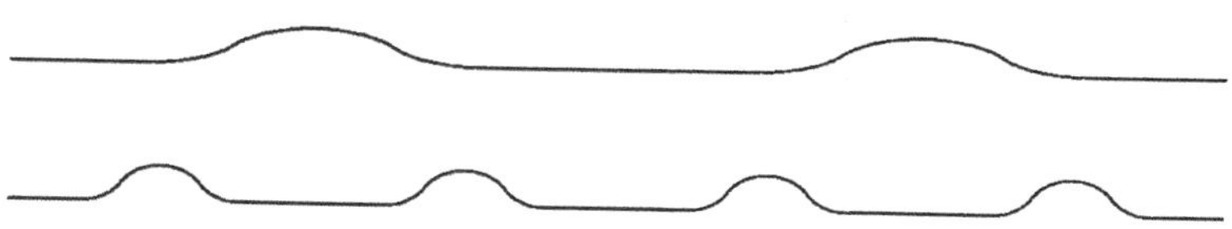

FIGURE 8.5 Cross-sectional depiction of bumps of a single height but more broadly spaced and broader in lateral extent (top) in comparison to more closely spaced and narrow in lateral extent (bottom).

essentially summing the path integrals along all scan lines in the image. The percentage by which this number exceeds the image area (i.e., the "plan view" area) is, then, the "excess area" that quantifies roughness. For the three 5×5-μm images of Figure 8.4a–c, for example, the excess areas are 4.6%, 3.0%, and 3.9%, respectively. Note that the relative values of excess area do not correlate well with the above-tabulated RMS roughness values (12.8, 12.3, and 13.9 nm, respectively). The lowest value of surface area corresponds to the lowest RMS value, but the other two numbers do not correlate in their relative values. So, indeed, the enumeration of "roughness" depends on the metric one uses.

8.3.2 2D-Domain Fourier Methods for Spatial Analysis

Fourier methods can be powerful in the hands of the AFM analyst [1]. An image can be mathematically reproduced by a superposition of sine waves of variable amplitude and wavelength by using a 2D fast Fourier transform (FFT) [2]. In some cases, this 2D pattern by itself is analytically useful, for example, to identify image periodicities, symmetries, or segregation length scales that are difficult to quantify in real-space images, and representative of the *ensemble* of structures in the image, not just a few selected locations. Three examples are shown in the insets of Figure 8.6: 2D FFTs of a (a) deflection image (gains zeroed) of the atomic lattice of a mica surface, bared via high-force abrasive scanning of a dry gelatin film as seen in upper left (within the shown square region roughly 0.5-μm across), (b) phase image of the block-copolymer poly(styrene-b-isoprene-b-styrene), and (c) phase image of the block copolymer poly(styrene-b-isobutylene-b-styrene). A hexagonal pattern of bright spots in Fourier space in (a) derives from the 2D periodic hexagonal lattice in real space. A bright ring in (b) derives from a single dominant polymer phase segregation length scale (styrene–isoprene) but lacking any long-range anisotropy. Two rings in (c) reflect two length scales: that between bright polystyrene-rich cylinders oriented parallel to the surface (inner Fourier ring), and that between cylinders oriented perpendicular to the surface (bright circles in image, producing outer Fourier ring). The inner Fourier ring in (c) also contains two spots, reflecting a predominant azimuthal orientation of the horizontal cylinders. The departure from a true circle in both (b) and (c) (i.e., slight elliptical shape) is due to scanner creep during image acquisition, a potentially useful diagnostic that is difficult to discern in the real-space image.

A fourth example usage of the 2D FFT is to quickly and precisely quantify symmetry angles in an image. This can free the analyst from more the laborious measurement of individual angles between real-space features, and moreover provide ensemble analysis. The inset of Figure 8.7, a 2D FFT of silver rods on a AgBr(111) surface, azimuthally oriented along the <112> family of crystallographic directions, contains three bright spokes at well-defined azimuthal angles, every 60° as is identifiable using the AFM software routine. Thus, the threefold symmetry of the AgBr(111) surface is readily apparent.

Another useful application of 2D FFT is image filtering. Commonly, streakiness along the horizontal fast-scan axis (e.g., due to changes in tip state) can be identified

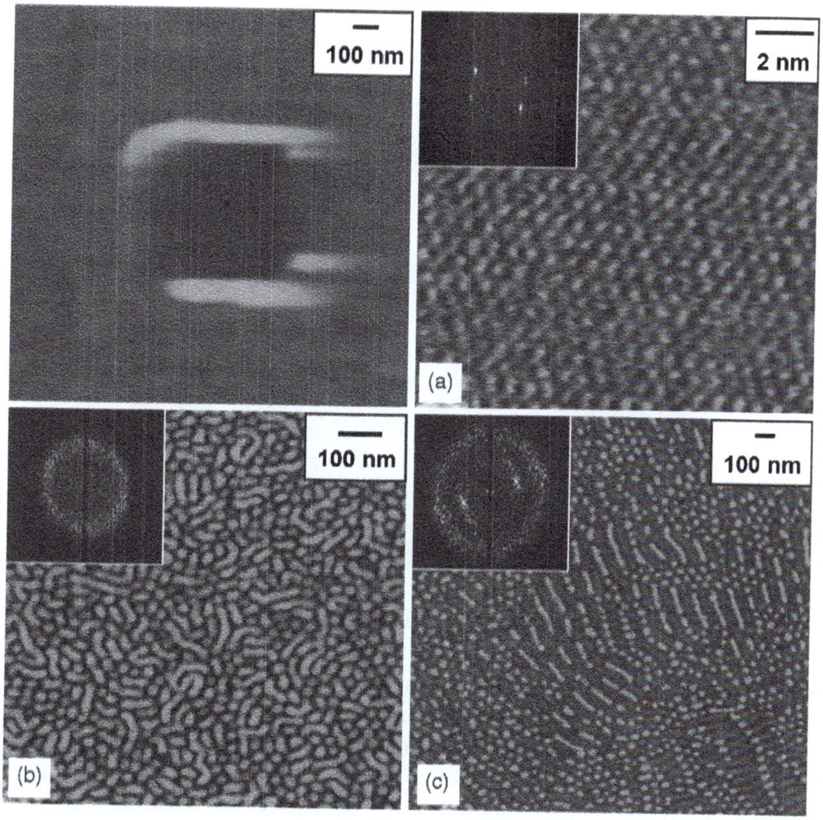

FIGURE 8.6 Three examples of 2D Fourier transform analysis (insets) of AFM images. (a) Deflection image of mica substrate bared by abrasive scanning of an ultrathin gelatin film (upper left). Phase images of block copolymers: (b) PS-PI-PS and (c) PS-PIB-PS.

in the 2D Fourier transform as a band along the vertical axis extending to large wave number (high spatial frequency, i.e., line-to-line). This portion of the transform can be removed, by editing in Fourier space, for the purpose of inverse transforming back to a filtered image. High wave number contributions along the horizontal axis in Fourier space can be removed as well, corresponding to the highest frequency noise quantified along the fast-scan direction. These are, in other words, customized implementations of low-pass filtering. An example is shown in Figure 8.8. Here, a height image of a dry gelatin film is low-pass Fourier filtered to remove streakiness as well as high-frequency noise, to better reveal the fine fibrous topology of the gelatin network [3,4]. The original and Fourier-filtered images are rendered with simulated overhead illumination to demonstrate the improvement.

Perhaps, the most common mathematical use of the FFT is in generating a power spectral density (PSD) function from a height image. In general, a PSD describes how the power of a signal in space or time is distributed with spatial or time frequency. In the case of electrical measurements, power is the voltage signal squared

FIGURE 8.7 2D Fourier transform analysis (inset) of AgBr(111) height image.

divided by the resistive load. But the term "power" in a PSD more generally means a squared signal of any kind, as if it were a voltage applied to a 1-Ω load. Thus, the power is the amplitude squared of each sine wave comprising the Fourier transform of the AFM image; the PSD is a plot of this power within increments of spatial frequency or wave number q ($=2\pi/\lambda$, where λ is wavelength, thus the PSD is sometimes plotted versus inverted wavelength). A 2D PSD can be performed, wherein the independent variable is radial wave number, or 1D PSD functions can be

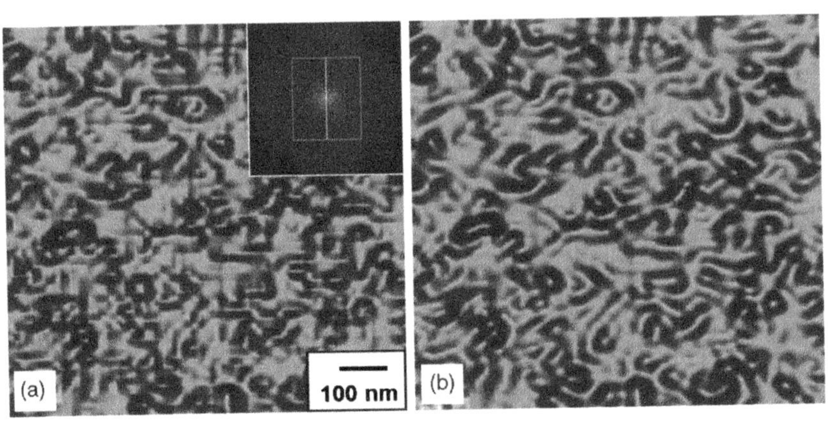

FIGURE 8.8 Height image of dry gelatin rendered in overhead illumination (to enhance the perception of fibrous morphology) (a) before and (b) after removal of horizontal streakiness via 2D Fourier transform filtering.

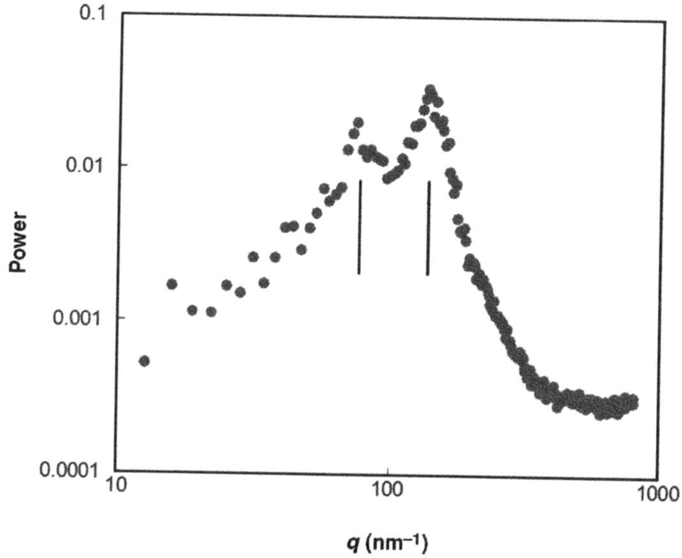

FIGURE 8.9 2D PSD plot produced by azimuthally integrating the 2D FFT in the inset of Figure 8.6c.

calculated for each pixel row or column and averaged across the image. Usually, the PSD is plotted on a log–log scale.

The 2D PSD azimuthally integrates the information contained in the 2D Fourier patterns (intensities squared), such as those in the insets of Figure 8.6 a–c. In particular, for case (c), the 2D PSD is double peaked, as shown in Figure 8.9. The low- and high-wave number components, denoted with vertical lines, correspond to the inner and outer rings in Figure 8.6c inset, respectively. Note that any azimuthally resolved information in the 2D FFT (i.e., spots) will contribute to the same signal as a ring located at the same radial value, upon integration over all azimuthal angles. In other words, azimuthal information is lost in the 2D PSD.

A second application of a 2D PSD examines what can be called *roughness scaling*, that is, how deviations in Z scale with X and Y. (Note that integrating the PSD over all values of wave number in the graph, and taking the square root, retrieves the RMS roughness.) Two PSDs are shown in Figure 8.10a, calculated from large and small height images (Figure 8.10b) of the same region of polycrystalline copper film. For such surfaces, one commonly finds at least a portion of the PSD, over a range of wave numbers, which is quite linear on a log–log scale (demarcated with dashed lines in Figure 8.10a). This is a domain of self-similar or fractal character: within some spatial size interval, zooming in or out produces the same kind of topography. There is a well-defined mathematical relationship between the range of heights and the image size. Thus, shorter hills and shallower valleys (small power) over a small lateral scale (high spatial frequency or wave number) become higher hills and deeper valleys (larger power) over a greater lateral scale (smaller spatial

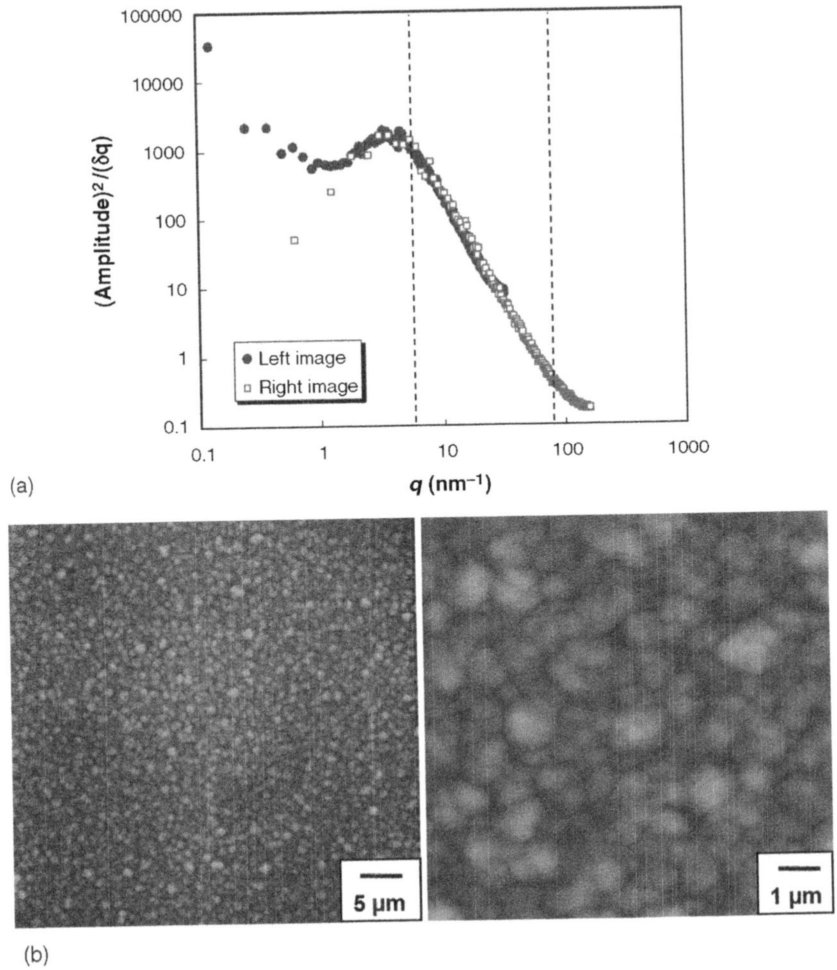

FIGURE 8.10 (a) 2D PSD plots derived from (b) left and right height images of polycrystalline copper. Fractal character (power law) is identified between the vertical dashed lines in (a).

frequency or wave number), by a simple multiplicative scaling relationship. An important metric is the fractal exponent, as determined from the linear slope of the PSD (on a log–log scale). Second, there is typically an approximate cutoff wave number below which the PSD become afractal, meaning the amplitude no longer changes with wave number via a simple scaling factor. Many applications relate the fractal exponent and cutoff wave number to kinetic parameters describing atom-by-atom or molecule-by-molecule growth of thin films, for example, in vacuum- or liquid-phase deposition [5]. Similarly, this methodology is applied to roughening processes during plasma treatments and other kinetic phenomena [6–8].

The overlay of the two PSDs in Figure 8.10a, deriving from images differing in size by a factor of five, requires two further comments. First, we note that two data points at the lowest wave numbers from the smaller image are inconsistent with those from the larger image. This reflects the fact that there are many more domains of this lateral size contained in the larger image, and this greater statistical sampling produces different PSD numbers. Indeed, the rising of power at the lowest wave numbers from the larger image reflects variations in film thickness sampled only over larger lateral scales, as seen in the larger image. The lack of any overall tilt in the smaller image is due to a plane fit operation performed prior to evaluating the PSD. Thus, monotonic variations in height, which would have produced lower-wave number Fourier components in the PSD, are absent and thus the PSD curve is lower. Second, we note that the power scale was normalized to the increment size for wave number δq, in order to get consistent PSDs from images of different sizes. Given the same number of pixels (512×512) contained in images of different size necessarily results in different physical distances between adjacent pixels, and thus different values of δq in the PSD. Such a difference in *bandwidth* results in a different power; normalizing to δq calculates power per unit bandwidth, which can be compared between images of different sizes.

As a last example of the 2D PSD, we compare surface regions that are nearly identical in RMS values of Z, but clearly "look different." This is the case of a polymer film (polyacrylamide) that has been scan-abraded to clear a patch of substrate (akin to that of Section 6.4.2.1). The first derivative (slope) image in Figure 8.11a clearly reveals qualitative differences in surface texture in comparing glass substrate and polyacrylamide film. But in the simultaneously acquired height image (not shown), Z RMS values within each marked rectangular domain are 0.48 nm and 0.55 nm for polymer and substrate, respectively, only a 15% difference. Correspondingly, the dZ/dX RMS slope values differ by 15%. The surface excess area values within these domains are 0.5% on polymer and 1.2% on substrate, small numbers but a more significant relative difference. Within the 2D PSDs of the height in Figure 8.11b, there are substantial differences *within certain intervals of wave number*. In particular, the glass substrate PSD curve lies well above (noting log scale) that for the polymer film within a wavelength range of ≈25–80 nm, whereas the polymer PSD curve lies well above that of the substrate at wavelengths greater than 100 nm. So which surface is "rougher?" Based on PSD analysis, one would say it depends on the lateral scale one considers. If demanding a simple answer, one might say the substrate is rougher because its excess area is higher.

8.3.3 Fourier Methods for Time-Domain Analysis

One should recognize that a 1D PSD can be useful for sampling *time-domain* periodic signals within the range of the sampling rate of the data acquisition system. Taken along X, the fast-scan direction (i.e., averaging PSD's over all scan lines) allows quantifying periodic signals up to frequencies equal to half the maximum data-sampling rate per the Nyquist theorem. For a maximum scan-line frequency

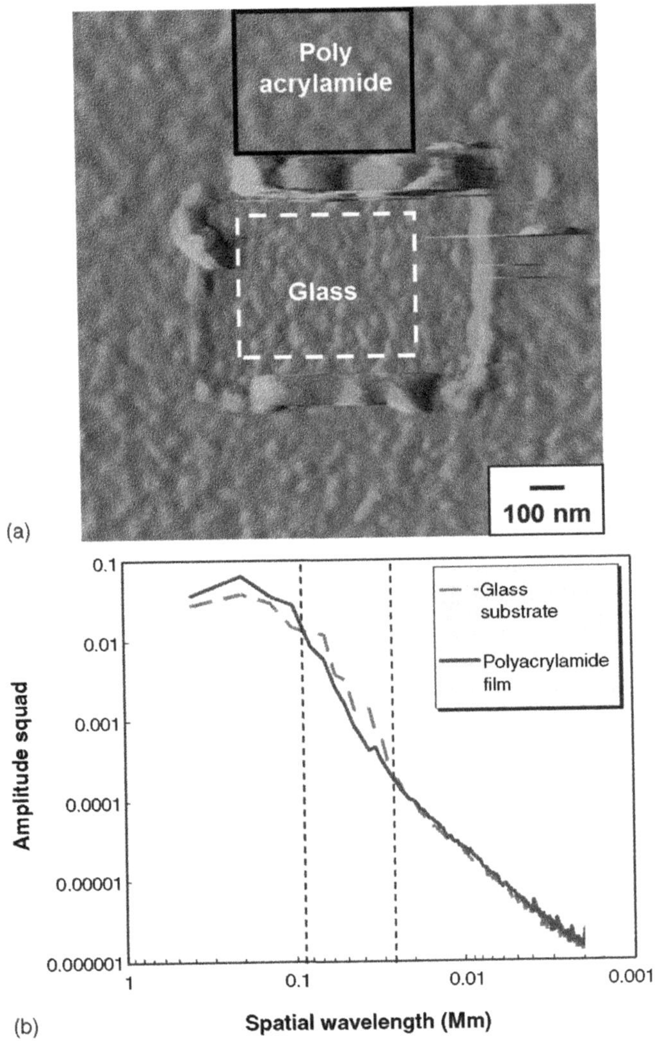

FIGURE 8.11 (a) Deflection image of a partially bared glass substrate otherwise covered with a polyacrylamide film. (b) PSD plots derived from corresponding height image data within the two denoted rectangular regions. Vertical dashed lines demarcate spatial wavelength domains in which the two plots are consistent (smallest spatial wavelengths) or differ in either sense: substrate "rougher" (middle wavelengths) or film "rougher" (lowest wavelengths).

of ~50 Hz, this amounts to

$$(50\,\text{scan cycles/s}) \times (1024\,\text{points/scan cycle})/2 \approx 25\,\text{kHz}.$$

(Some newer commercial AFM systems reach several megasamples per second, as do add-on systems with special high-rate data acquisition cards, such as used with

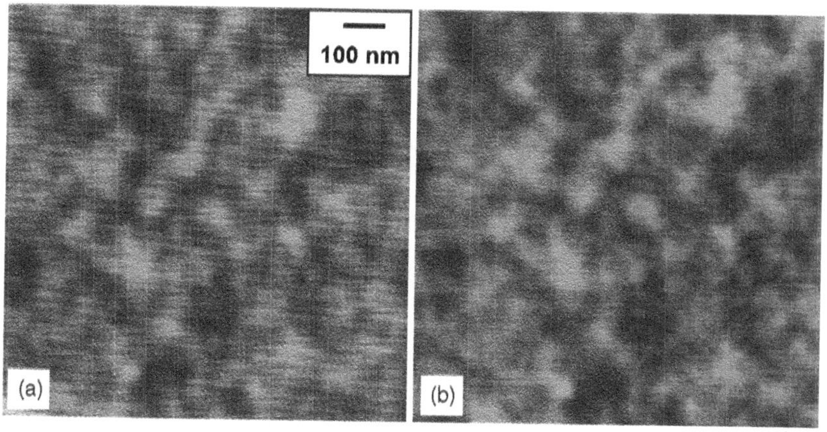

FIGURE 8.12 Height images of the native oxide of a silicon wafer with the AFM positioned (a) off and (b) on a commercial passive vibration isolation platform.

LabView or Witec Digital Pulsed Force Mode.) Taken along Y, the slow scan direction (i.e., averaging over all Y pixel columns) allows probing frequencies many decades lower, typically down to around

$$(0.1 \text{ scan lines/s}) \times (512 \text{ lines/image})/2 \approx 0.25 \text{ mHz}.$$

One use for time-domain analysis is the assessment of environmental vibrational signals that may degrade image quality. Typical building vibrations, such as generated by vehicle traffic, are primarily in the frequency range of tens of Hertz. Knowing the scan-line frequency and the number of measurements per scan line enables calculation of the data sampling rate, and thereby the frequency of vibrations, if analyzing PSD data in ASCII form. Equivalently, the AFM software's PSD routine can be used, converting from an identified spatial wavelength to a time period using the known scan size (accounting for possible overscan[2]) and scan-line frequency.

Figure 8.12 compares 1×1-μm height images of the native oxide of a silicon wafer with the AFM positioned (a) off and (b) on a commercial passive vibration isolation platform, acquired at a scan-line frequency of 1.5 Hz. Extra noise of a partially periodic character, on the order of several vibrational cycles per scan line (translating to a bit over 10 Hz), is clearly apparent in image (a), even with the AFM remaining atop a heavy granite slab on rubber pads in an underground laboratory. The determined RMS roughness and excess area in case (a) were 0.139 nm and 0.19%, whereas in case (b), these numbers increase to 0.146 nm and 0.23%, respectively. Although a small difference, environmental vibrations clearly can have both a qualitative and a quantitative impact.

[2] Most data acquisition systems allow the user to choose a scan size that is systematically larger than the acquired image. Often this is specified by an "overscan" parameter setting (e.g., 10%).

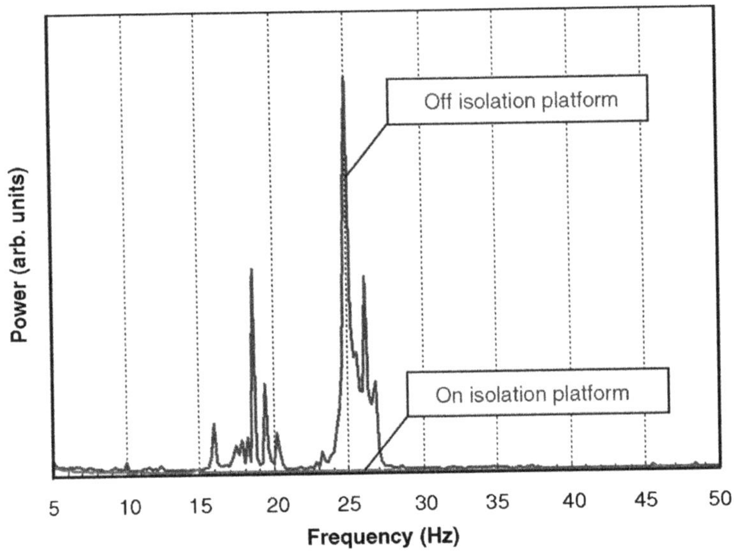

FIGURE 8.13 *X* power spectra density plots (converted from spatial frequency to time frequency domain) derived from the images in Figure 8.12 as indicated.

To best resolve the spectrum of environmental vibrations, the sampling time interval per PSD must be long enough to span many cycles of the lowest-frequency vibration. Second, one should image a topographically featureless sample (or choose zero scan size and hope to avoid location drift, a difficult task unless using a closed-loop scanner, which introduces additional noise as seen in Figure 4.24). In Figure 8.13, *X* PSDs computed from 100 × 100-nm images taken at 0.1 Hz scan-line frequency on a freshly cleaved, atomically flat mica surface, with and without the vibration isolation platform, are compared on a linear–linear scale. In the absence of the vibration isolation platform, a multi-peak vibrational spectrum is observed within two bands, 16–20 Hz and 25–27 Hz, but none of these vibrations are visible with the platform in place.

8.3.4 Grain or Particle Size Analysis

The analysis of the size distribution of grains or particles on a surface—not only lateral scale but also volume and surface area—is a unique capability of AFM. A conceptually simple, commonly available algorithm allows the analyst to choose a *height threshold*—a horizontal plane cutting through the image (following plane fitting)—so as to define grains or particles as islands rising above the plane, and the surrounding "sea" as regions below the plane. This can work well if the "sea" is truly flat, but often this is not the case. In such cases, a better, but less commonly available, scheme for demarcating grains and particles is the so-called *watershed algorithm*. It essentially finds the hill peaks and the polygonal valleys surrounding the peaks, and then draws meandering lines through the network of valleys to demarcate the grains. In either case, once the grains have been demarcated, the software can statistically analyze the distribution of domains in terms of their plan-view area, excess surface area, island volume, diameter, perimeter, etc., and generate histograms of these metrics. The example in Figure 8.14a shows a histogram of equivalent disk radius of island domains of the drug dexamethasone that phase segregate from the sea of polymer matrix (PBMA) as seen in the height image of Figure 8.14b (same as Figure 8.4a, a ∼1 μm-thick coating spun onto a silicon wafer). The demarcation of the islands as identified by the watershed algorithm (here using the freeware program Gwyddion) is shown in Figure 8.14c. Over 1000 drug islands were identified (colorized domains) and analyzed within a 10×10-μm area to produce the size histogram.

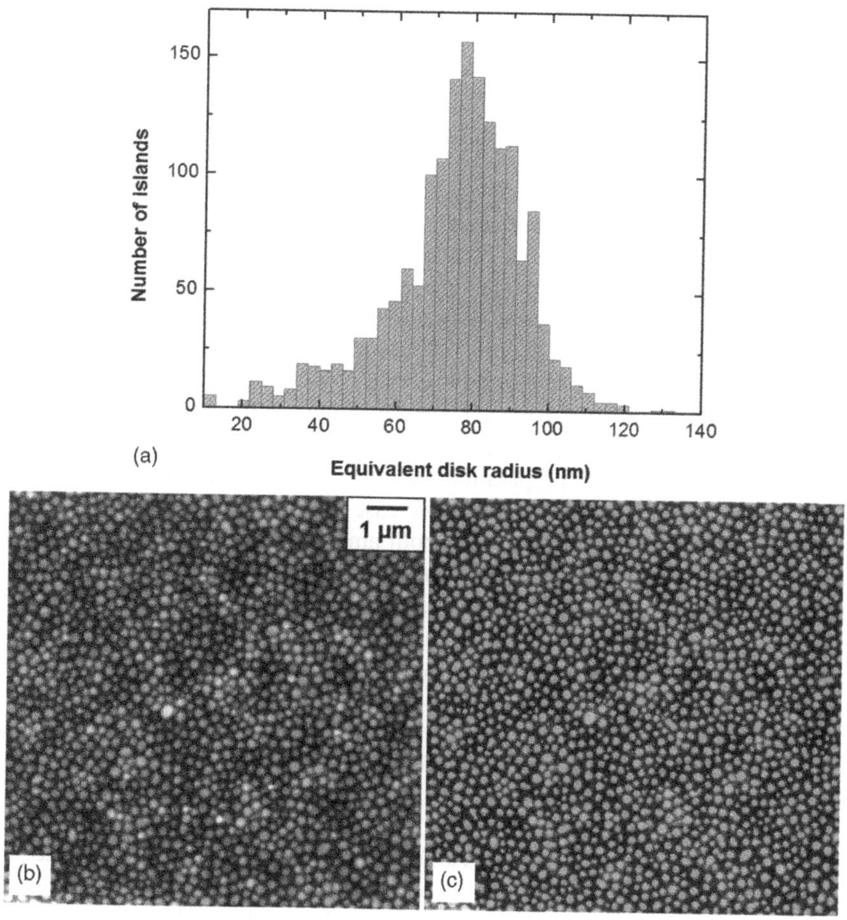

FIGURE 8.14 (a) Histrogram of island sizes determined using the watershed algorithm as applied to a (b) height image of PBMA plus drug mixture so as to (c) demarcate "island" locations (colorized or bright) from "sea."

8.4 "LINESHAPE" FITTING

In Figures 1.3, 4.10, 5.2, 5.39, 6.26–6.29, 6.33, 7.6, and 7.11, we saw ways of quantifying image or mapping information as *histograms*, akin to spectroscopic analysis. Although with an imaging technique one may choose to analyze subregions of an image in order to produce single-peak histograms (i.e., one peak per material type at the surface), there are many cases where this does not work. Material types may be too small and interdispersed, such as in the case of crystalline and amorphous regions that, for example, exhibit different dissipation. In such cases, one may instead obtain a histogram peak with a *shoulder*, or simply *broadened* or *asymmetric*, because more than one type of material or dominant surface elevation is contained in the region(s) that was converted to histogram. In such cases, there may be good reason to use mathematical curve fitting to reproduce a multicomponent, broad *lineshape* with a mathematical superposition of relatively narrow and symmetric peaks, say, each Gaussian forms

$$y_i(x) = A_i \exp\left[-\frac{(x - x_i)^2}{2\sigma_i{}^2}\right]. \tag{8.5}$$

Thanks to the need for lineshape analysis for myriad spectral techniques, there are numerous software packages available to perform multipeak curve fitting. Figure 8.15a shows one result for the case of a phase histogram (from Figure 5.2c) derived from the entire phase image in Figure 8.15b of a polyvinyl alcohol thin film on mica (as discussed in Sections 5.1, 5.3 and 6.4). Here, the breadth of each peak was allowed to be a free parameter (along with peak position and height) resulting in both narrow and broad lineshape subcomponent peaks. Broader peaks correspond to greater heterogeneity in tip–sample interaction, a potentially important finding, say as a function of sample preparation conditions such as deposition rate or temperature, annealing time, and others. In the case of Figure 8.15, Region 3 contains bulk-like semicrystalline polymer, with both amorphous and crystalline nanoscale domains, thereby producing a broad range of dissipative behavior ranging from highly dissipative (corresponding to phase ~90°) to mildly dissipative (~130°). An advantage of a peak-fitting analysis, as opposed to the region-specific histograms of Figure 5.2c, is the ability to determine the relative size of each material region by integrating each peak.

Other potential uses of curve fitting include PSD lineshape (Figure 8.10) and grain size distribution histograms (Figure 8.14). In the case of PSD, one is often attempting to locate the position of a peak (or peaks) on top of a strongly variable background (e.g., power law). This may, for example, reflect a common grain or particle size superimposed on otherwise fractal (self-similar) topography. The presence of the background, together with a possibly weak peak (i.e., some variation in particle or grain size) may make visual determination of the peak position (thus, the mean size of particles or grains) difficult, even if following image processing via flattening, filtering, etc. Moreover, the quantification of peak breadth (to gauge heterogeneity or objectively determine error bars) may be difficult atop a sloped

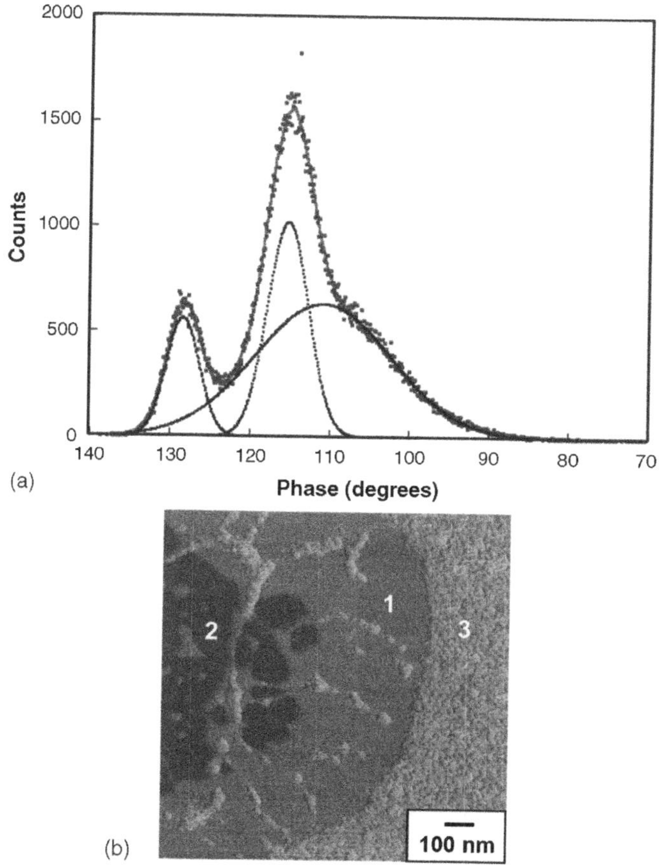

FIGURE 8.15 (a) Phase histogram of the entire image in (b), of a polyvinyl alcohol ultrathin film exhibiting three characteristically dissipative regions. Curve fitting achieves a good fit to three Gaussians.

background. An example is shown in Figure 8.16. The height image is of a block copolymer film of poly(styrene-b-isoprene-b-styrene). It was first-order plane fit and zeroth-order flattened (reflecting tip instability); the remaining, displayed image in Figure 8.16b, contains a variable underlying topography, atop of which the phase-segregated block copolymer domains can be faintly seen. The 2D PSD plot (linear–linear) in Figure 8.16b contains a short, broad peak atop a background that steeply increases toward lower wave number. A polynomial background was fit to the data outside of the vertical lines and subtracted from the entire data set; this enabled fitting of the remaining peak with a Gaussian function to objectively measure peak position and breadth. Thereby, the corresponding characteristic size of the phase-segregated polymer domains, in this ensemble analysis (i.e., from the entire image), was determined to be 30.0 ± 3.6 nm. In this particular case, the

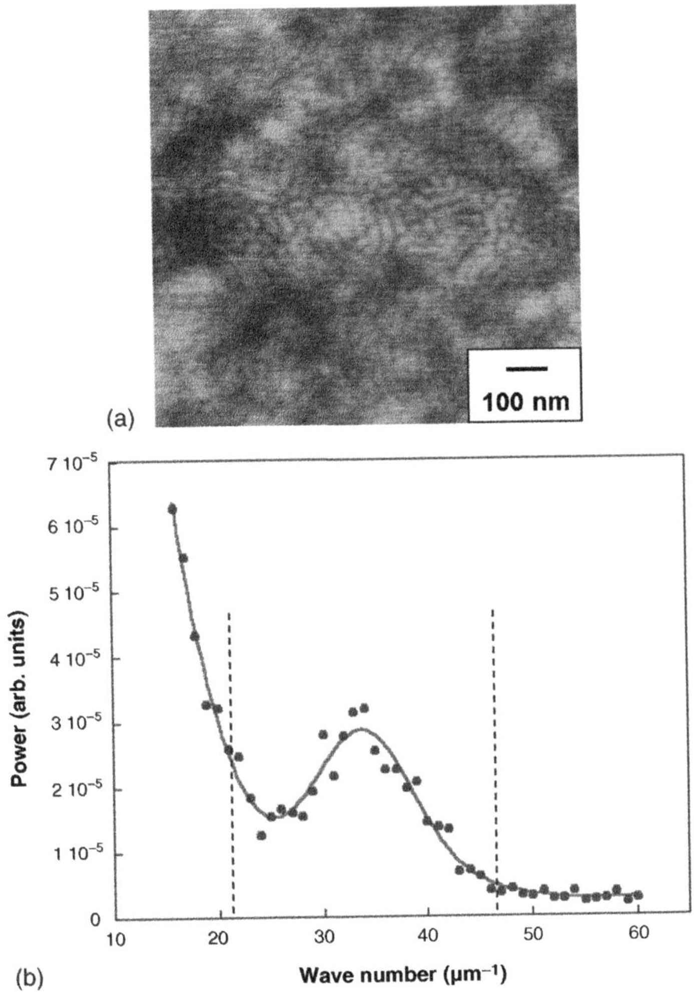

FIGURE 8.16 (a) Height image of block copolymer, PS-b-PI-b-PS. (b) PSD plot (dots) of the image in (a). Solid curve is a fit of the PSD to a polynomial plus Gaussian peak.

corresponding phase image (Figure 8.6c) may be the better choice for PSD analysis, but Figure 8.16 exemplifies the degree to which weak nanoscale morphologies can be quantitatively extracted from a coarser topography.

REFERENCES

[1] Gadelmawla, E.S., et al., Roughness parameters. *J. Mater. Proc. Technol.*, 2002, **123**: 133–145.

[2] Sikora, A., et al., Application of FFT transformation for correlation analysis of near field microscopy measurements. *J. Phys.: Conf. Ser.*, 2009, **146**: 012037.

[3] Haugstad, G., et al., Probing biopolymers with scanning force methods: Adsorption, structure, properties, and transformation of gelatin on mica. *Langmuir*, 1994, **10**: 4295–4306.

[4] Djabourov, M., Architecture of gelatin gels. *Contemp. Phys.*, 1988, **29**(3): 273–297.

[5] Barabasi, A.-L. and H.E. Stanley, *Fractal Concepts in Surface Growth*. 1995, Cambridge, MA: Cambridge University Press.

[6] Petri, R., et al., Silicon roughness induced by plasma etching. *J. Appl. Phys.*, 1994, **75**(11): 7498–7506.

[7] Constantoudis, V., et al., Characterization and simulation of surface and line-edge roughness in photoresists. *J. Vac. Sci. Technol. B*, 2001, **19**(6): 2694–2698.

[8] N. S. John, et al., Cellular network formation of hydrophobic alkanethiol capped gold nanoparticles on mica surface mediated by water islands. *J. Chem. Phys.*, 2010, **133**: 094704.

9

ADVANCED DYNAMIC FORCE METHODS

This chapter introduces further analytical methods in dynamic AFM, utilizing the resonant oscillation of the cantilever in additional ways.[1] We first consider methods long available in research-grade commercial AFM systems, which vibrate the cantilever at its fundamental flexural resonance as usual but under special acquisition schemes to sense longer-range forces (Section 9.1). We second introduce methods that more recently have become commercially available, utilizing cantilever vibrational frequencies *far from the fundamental flexural resonance* (Section 9.2). In general, sensitivity to long-range force gradient (see Section 2.4) enables one to probe variations in electrostatic charging, polarization, or magnetization [1–4], as well as differences in surface potential [3–7]. Section 9.2 touches on methods to probe these same properties, as well as those treated in earlier chapters (e.g., mechanical response), utilizing *higher vibrational modes* of AFM cantilevers [8,9]. A more home-built implementation of these latter methods is also possible provided that the user has a sufficient number of data acquisition channels and appropriate signal access (usually, a BNC interface), and sufficient funds for electronic hard-

[1] Sometimes these methods are categorized within the broader scanning probe microscopy family, rather than AFM. Force is still the quantity being sensed, and the sample and tip still consist of atoms imparting forces on each other, but unbound electrons and magnetic spins may play the dominant role in what is being probed. By some conventions, this goes beyond the definition of "atomic force" microscopy, but here we use the term more broadly.

Atomic Force Microscopy: Understanding Basic Modes and Advanced Applications, First Edition. Greg Haugstad.

ware attachments, such as high-frequency lock-in amplifiers, function generators, and signal-adding circuitry [10].

The methods of Chapter 9 are arguably a step up in complexity from those of the preceding chapters, and the undergirding science also can be very rich. It is not our purpose to fully review this literature or describe a comprehensive tool kit with detailed examples. This is especially true of electromagnetic property characterization, a large subfield within scanning probe methods, and the focus of edited volumes on research applications [3,4,7,11]. Instead, our purpose is to provide a basic understanding of several dynamic methods that apply to broad classes of materials, via a few phenomenological examples including discussion of parameter settings. We leave some of the scientific and methodological fine points to the aforementioned edited volumes, public-domain descriptions (e.g., web pages and downloadable presentations), and scientific journals including a number of review articles [6,12,13].

9.1 PRINCIPLES OF ELECTRONIC METHODS UTILIZING DYNAMIC AFM

Before considering methodological details of electronic methods, it is important to understand the principles of electron energy levels at the heart of electrostatic force microscopy (EFM) and Kelvin-probe force microscopy (KFM or KPFM; also SKPM is used, for scanning Kelvin-probe microscopy). Figure 9.1 depicts an idealized tip–sample system (ignoring the possible effects of a cantilever). If we bring together two dissimilar materials (tip and sample) that are not electrically grounded, their electronic vacuum (energy) levels become the same (Figure 9.1a). In general, the two *Fermi energy* levels—the highest energy of occupied electronic states—are different. (If tip and sample are metallic, the difference of vacuum to Fermi level is the *work function*.) Upon electrically connecting the materials and assuming an adequate number of mobile electrons, a substantial displacement of electron charge will act to *align Fermi levels* and thus offset vacuum levels (Figure 9.1b). This produces an electric field between tip and sample and thus a net force of attraction between opposite charges, as well as charge–dipole attractive forces. Applying an appropriate external bias between tip and sample can generate a larger total tip–sample voltage difference, enhancing the strength of attraction and its possible variation from point to point along the surface (e.g., if heterogeneous in dielectric constant). This additional external bias may be needed to generate adequate image contrast in EFM.[2] Alternatively, by connecting just the right dc voltage offset between the two materials, the vacuum levels can be brought into alignment and thus the force eliminated or *nulled*. To achieve this condition, the applied

[2] One should bear in mind that large biases, of order several volts, can change the electron energetics in the sample such that the method cannot be considered a "weakly perturbative" probe of sample properties. In physicists' language, perturbation or linear response theory does not apply, that is, the Hamiltonian of the system is being substantially modified by the strong electric field.

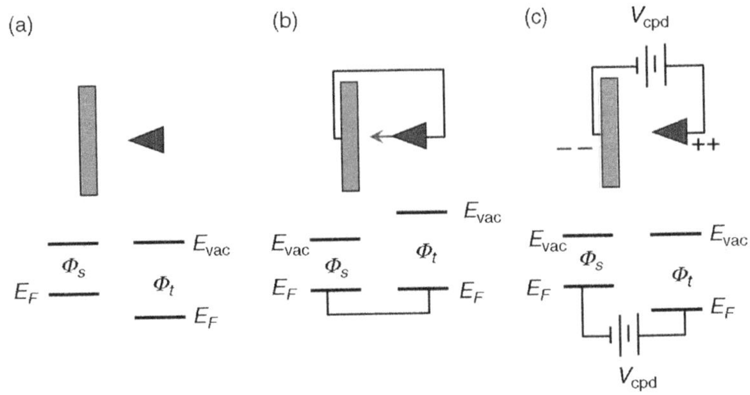

FIGURE 9.1 Idealized tip and sample electronic energies at the vacuum level and Fermi level, for the case of dissimilar work function. Materials that are (a) not electrically grounded, (b) electrically grounded (common), and (c) electrically connected with a battery in between, of the exact potential difference needed to match the work-function difference and thus align vacuum levels.

voltage must be equal and opposite to the surface potential difference of the (commonly grounded) materials, also known as the *contact potential* (or contact potential difference, V_{cpd}). If the work function of the tip ϕ_t is greater than that of the sample ϕ_s, then a positive tip voltage would be needed in order to drop the energy levels (of negatively charged electrons) to achieve the nulling state (Figure 9.1c). In KFM, one wishes to map this applied voltage across the surface to obtain a contact potential image. On a heterogeneous surface containing work functions variably *less* than that of a high work-function tip, the nulling condition on the *lowest* work-function surface locations, for example, would require the *most positive* tip bias. (We leave it as an exercise to show that for the case of sample work functions variably *greater* than that of the tip, the nulling condition on the *lowest* work-function surface locations would require the least negative tip bias. Thus, the relative contrast seen in KFM images would be the same, whether using a high or low work-function tip.)

9.1.1 Shifted Dynamic Response Due to Force Gradient

In any case, the quality of data from a signal-to-noise standpoint is enhanced by utilizing dynamic AFM methods. This also applies to the case of magnetic force microscopy (MFM), which is operationally similar to EFM except that no voltage bias is being applied, and of course, there are no static magnetic "charges," only dipole–dipole interactions. Most commonly, the phase signal is used to enable the measurement (and imaging) of electrostatic or magnetostatic force gradient from the surface, dF/dZ. In EFM or MFM, one usually measures shifts in either: (1) the phase signal itself or (2) the cantilever resonance frequency as determined by a dedicated feedback circuit that *alters the drive frequency to maintain a constant phase* value. This is depicted in Figure 9.2. In case (1), the phase measurement serves as a

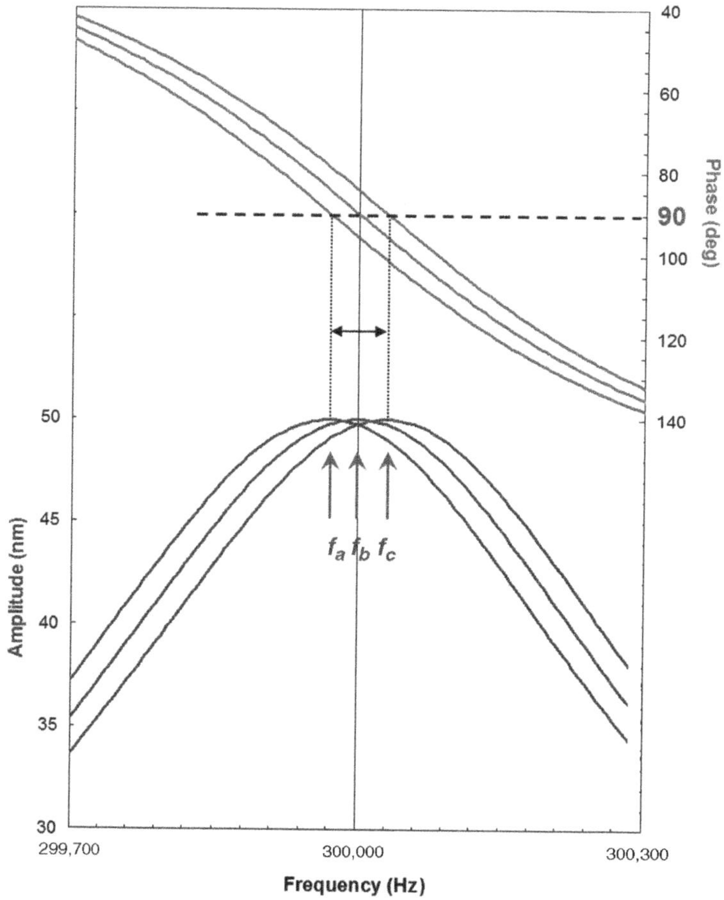

FIGURE 9.2 Theoretical frequency dependence of amplitude (bottom) and phase (top) at three sample locations a–b–c during *X–Y* scanning, at which the electrostatic force gradient achieves different values corresponding to decreasing strength of attraction. Feedback-adjusted drive frequency values, f_a, f_b, and f_c, keep the phase measurement constant at 90° as the cantilever resonance frequency shifts.

proxy for the oscillator resonance frequency; the conversion of phase to frequency shift requires measurement of the phase–frequency relationship (e.g., as in Figure 9.2, top). In case (2), the one-to-one (i.e., monotonic) relationship between phase and frequency enables a dedicated feedback circuit (i.e., *not* that used for topographic imaging) to adjust the drive frequency to keep the phase value constant. Moreover, the essentially linear relationship between phase and frequency very near resonance means that the feedback circuit sensitivity is a constant (i.e., variable gain is not needed). In either case, the shift in the resonance frequency is proportional to the electrostatic or magnetic force gradient sensed at a distance, with the understanding that this gradient may be integrated over a significant range of

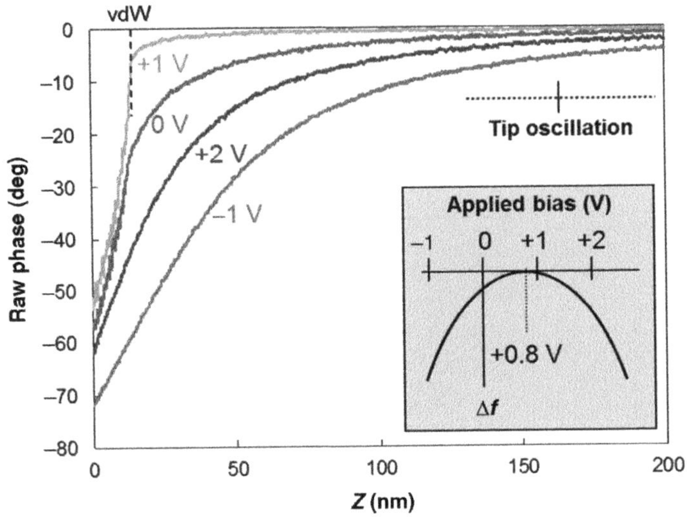

FIGURE 9.3 Measured raw phase-Z relationships for a metallic sample interacting with a variably biased metallic tip. Tip oscillation amplitude is depicted at right. *Inset*: Trend of frequency shift Δf versus externally applied tip bias, showing a minimum frequency shift at +0.8 V, at which the intrinsic tip–sample surface potential difference is nulled by the external tip bias.

distances depending on the oscillation amplitude of the cantilever. Case (2) is often called frequency modulation (FM) because it is the variation of the frequency of the oscillator under feedback that is providing the information about the tip–sample interaction.

Figure 9.3 exemplifies the Z dependence of phase in the presence of electrostatic forces. Raw phase (i.e., zeroed at free resonance, rather than set to 90°) versus Z curves at a metal surface are compared for four biases applied to a metallic tip: −1, 0, +1, and +2 V. Figure 9.3 displays: (a) the manifestation of long-range forces seen as variable phase shift versus distance beyond van der Waals (vdW) force engagement (compare to Figure 5.5b), (b) the influence of applied tip bias on the phase-Z relationship, and (c) the interplay of applied bias and intrinsic surface potential differences (of tip and sample) to produce the observed electrostatic interaction. Related to (c), the bias trend among these curves does not make sense at first sight; the weakest long-range interaction (i.e., least negative phase shift) is seen at a bias of +1 V rather than 0 V and the strongest interaction is at a bias of −1 V rather than +2 V. How can this be? As further discussed in Section 9.1, there is a *parabolic* relationship between tip–sample potential difference and attractive force (or force gradient and thus frequency shift Δf). The question is the *location of the vertex* of this parabola. Because the tip–sample potential difference is determined by two things, the applied bias *and* the difference in intrinsic surface potentials of tip and sample materials (i.e., work functions for metals), the vertex of this parabola is necessarily located at *nonzero* bias. A more detailed study of the system of Figure 9.3 found this value to be +0.8 V as depicted in the inset; examining this parabolic relationship, the ratio of frequency shifts at +2 V versus 0 V biases is then $(2-0.8)^2/(0-0.8)^2 = 2.25$, consistent with the approximate doubling of phase shift from 0 V to +2 V bias at large values of Z (beyond vdW forces).[3]

A remaining question is how to implement this kind of measurement in AFM, where the surface of interest is not ideally flat. Both interleave-based (two-pass) and single-pass methods are commercially available. We introduce the interleave case in the following subsection, reserving the single-pass case for Section 9.2.4 on multifrequency methods.

9.1.2 Interleave Methods for Long-Range Force Probing

Section 1.7 previewed a two-pass imaging scheme wherein the conventional height and perhaps phase are imaged on the first pass, whereas long-range (e.g., electrostatic) forces are sensed and imaged on the second interleave pass, as depicted in Figure 1.21. During the second, interleave pass, a scan trajectory $Z_{\text{interleave}} = Z(X) + Z_{\text{offset}}$ is followed, where $Z(X)$ is the topographic trajectory saved in memory from the first pass and Z_{offset} is a constant, user-selected vertical offset. Also, during the interleave pass, the cantilever may be driven exactly the same as, slightly different from, or very different from the first pass, depending on the specific application. In EFM and MFM, the vibration source during interleave is usually the same as in the first pass (usually mechanical), and at or near the same resonance frequency. In EFM, a dc bias may be applied to tip or sample to generate a tip–sample electric field and thus electrostatic forces relating to the capacitance of the system, or no bias may be applied such that *intrinsic* charging or surface potential differences are the source of electrostatic forces. Exploration of the bias parameter can vary the net strength of the electrostatic interaction and thus materials contrast, as mentioned in the previous section.

In interleave-based Kelvin-probe force microscopy (i.e., during a second line scan following the topographic contour collected in the first line scan), the mechanical vibration is *replaced by an ac electrical bias plus a dc bias* that is varied while scanning point to point across the surface.[4] This is illustrated in Figure 9.4. The ac bias generates a time-varying electrostatic force between tip and sample, producing a vertical cantilever oscillation. In the absence of a dc bias, the amplitude of this oscillation would scale with the magnitude of the difference in surface potentials of the tip and sample. The dc bias at any given X–Y location is chosen via feedback to zero or "null" the cantilever oscillation, by *compensating for the difference in*

[3] A more precise correspondence would require measuring frequency shift rather than phase, and require vanishingly small tip oscillation amplitude.

[4] The application of an ac bias is not to be confused with one of the names sometimes used for ordinary dynamic AFM, "AC mode," whether under mechanical or magnetic (MAC) drive. These latter methods certainly use an ac electrical signal to generate a mechanical or magnetic drive signal and thus an "AC" force, but the cantilever (or sample) is not itself being ac biased. To distinguish, lowercase ac is used herein to refer to an actual ac bias of tip or sample.

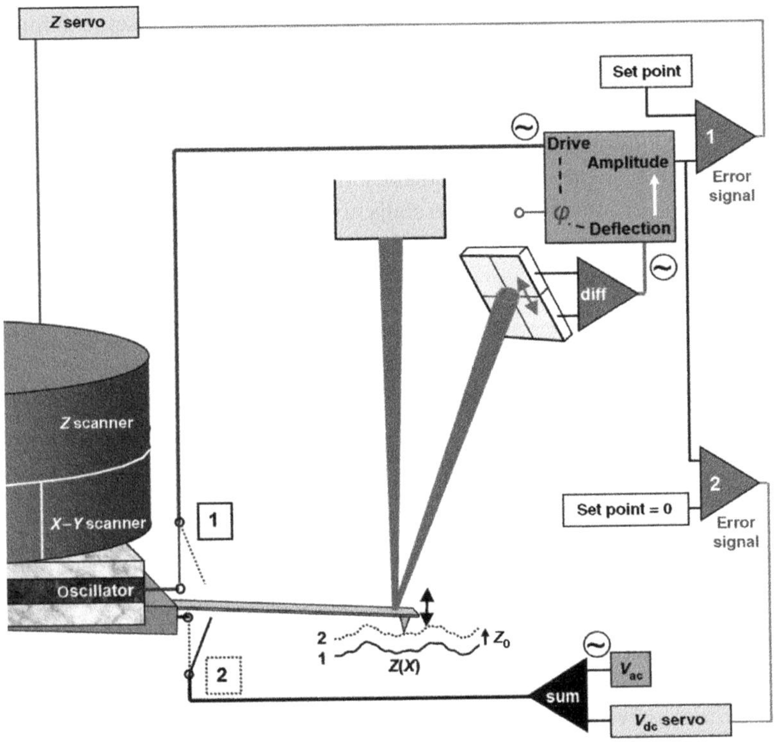

FIGURE 9.4 Inclusion of KFM feedback loop in dynamic AFM schematic. First-pass imaging is the same as in Figure 4.5, to obtain topography (and optionally phase) via Z servo control to maintain constant setpoint amplitude of the tip oscillation under a constant-amplitude drive of a mechanical oscillator, with no electrical bias to the tip. Second-pass or "interleave" imaging is performed over the $Z(X)$ scanner trajectory saved from the first pass but offset by a constant Z_0 to displace tip from sample. During interleave, the mechanical oscillator drive signal is removed while the cantilever/tip is biased by an ac voltage summed with a dc voltage that is varied by the servo so as to achieve zero amplitude of tip oscillation.

surface potentials. Thus, the determined dc bias equals the contact potential difference of sample and tip materials in the absence of this bias (including the possible roles of surface contamination, surface oxides, and/or trapped charge). The precision with which this is done is dependent on the quickness of the electronics that must: (1) quantify the amplitude of the cantilever oscillation and (2) vary the dc bias so as to minimize this amplitude. Often this feedback scheme is called *amplitude modulation* (AM) because it is the measured amplitude of cantilever response (usually chosen to be at resonance) that forces the applied bias to be varied so as to null the cantilever amplitude. In Section 9.2.4, we describe an alternative scheme, dubbed frequency modulation, where the applied bias is instead varied under feedback via a measurement that is sensitive to the frequency response, in parallel with topographic imaging (i.e., in a single pass, no interleave required). Another important distinction of these two cases is that AM is strictly sensitive to force, whereas frequency modulation is sensitive to force gradient. The latter falls off more steeply with distance and thus can provide contrast that is more localized (i.e., higher resolution images of surface potential).

Figure 9.5 bottom is a qualitative comparison of materials contrast obtained via imaging of: (a) EFM frequency shift and (b) KFM dc bias on a heterogeneous surface, a nominal ~50-nm-thick polycrystalline aluminum film on a n-doped silicon wafer (capped by native oxide), using a metallic-coated (Pt/Ir) tip. The Al surface is from the same sample as examined in Chapter 4 (Figures 4.7, 4.13, 4.17, 4.18, and 6.7), containing islands of a relatively soft contaminant that are larger than the metal grains visible between these islands (height image in Figure 9.5c). The sample was maintained at an overall ground at the underside of the Si wafer, and thus the aluminum was not directly grounded. For the EFM image, the tip was biased by 5 V relative to the sample ground and the interleave-scan lift height set to a small

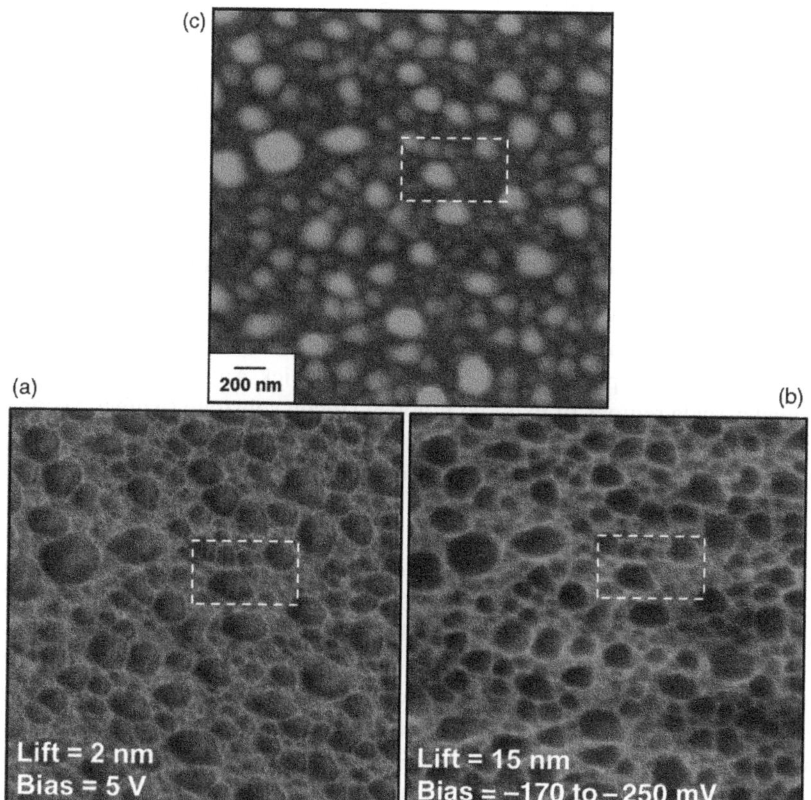

FIGURE 9.5 (a) EFM frequency shift and (b) KFM images of a polycrystalline aluminum surface with islands of likely organic contaminant. (c) Height image of the same, obtained in first pass with KFM data.

value, 2 nm, just enough to avoid regular engagement of tip to sample. These parameters were chosen to generate contrast in the frequency-shift image, which was inverted in contrast before displaying in Figure 9.5, such that a stronger attractive force gradient corresponds to brighter signal in (a). In the KFM image, Figure 9.5b, the applied tip bias (to produce the nulling condition) was 80 mV higher on the brighter regions. Although this value is small, the materials contrast obtained in the KFM image is somewhat stronger and less affected by surface slope than in the EFM case. The spatial resolution in EFM is marginally better, however, related to the sensitivity to force *gradient* in the case of EFM (Section 9.1.1). Both the EFM and the KFM images seem to give a more definitive demarcation and identification of islands of contaminant than does the height image. For example, this can be seen by comparing "grains" within the marked rectangular region.

If considered in the context of Figure 9.1, the darker contaminant regions seen in KFM would correspond to a higher "work function." But we are not dealing with a metal surface, rather, an organic contaminant. That a presumably nonconducting medium would have deeper electronic energies at the weakest bound states makes physical sense ("less reactive"). In EFM, these regions exhibit weaker attractive force gradient under a large bias. This also makes sense from a materials standpoint; the contamination is expected to act as a dielectric layer and thereby reduce electric field strength at the surface. More detailed and quantitative interpretations of these images, however, require more precise experimental control as well as careful variation of parameters (e.g., tip bias in EFM).

9.1.3 Interleave-Based EFM/KFM on Different Metals and Silicon

In this section, we more quantitatively examine the dependence of EFM electrostatic force gradient on applied tip voltage across a surface containing micrometer-scale strips of lithographically fabricated ~50-nm-thick Au and Al films, as well as micrometer-scale regions of exposed n-doped silicon substrate (wafer). Second, we examine variations in KFM contact potential difference across a similar sample, relating not only to different materials but also to variations in electronic-charge displacement within the silicon.

Figure 9.6 examines the effects of tip bias on EFM cantilever resonant frequency shift (i.e., attractive force gradient integrated over the cantilever oscillation amplitude). The bias was varied in discrete 0.25-V steps ranging from −1.5 V to +2.5 V on a sample at two alternative grounding locations: at the underside of the Si wafer and on the top of the wafer near the metal strips (using silver colloid paint). For this sample, a thick (~300 nm), insulating oxide was grown on the silicon before the metal strips were lithographically deposited on top of this oxide. For the topside-grounded case, the AFM frequency-shift EFM image is shown as Figure 9.6a. Interestingly, the sense of contrast on the metal strips changes from negative to positive tip bias; indeed, a +2 V bias produces contrast that is essentially inverted compared to a −1 V bias. How does one make sense of this? The key is to look more systematically at the voltage dependence of frequency shift on each material region. The mean frequency shift within each material domain at a given tip bias is plotted in

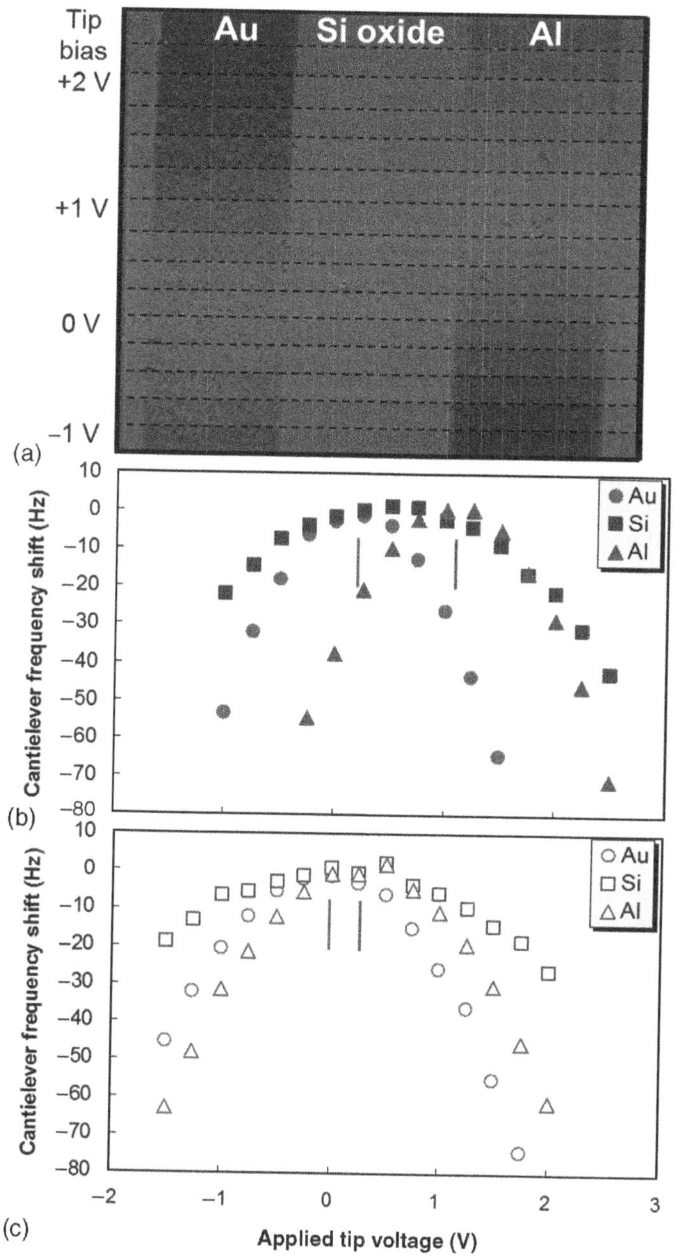

FIGURE 9.6 (a) EFM frequency-shift image obtained across Au and Al metal strips and intervening thick oxide of Si, while stepping tip bias, for topside sample ground. (b) Plots of data from (a) per region. (c) Similar plots for bottom-side sample ground. Vertical lines show potential difference between Au and Al.

Figure 9.6b; in each case, a parabolic trend is seen (as discussed with the inset of Figure 9.3). The most notable observations are: (1) the vertices of the parabolas are located at different voltages, shifted by 0.9 V between the two metals with Al exhibiting a more positive shift (lower work function; see Figure 9.1), and (2) the curvature of the parabolas is similar for the two metals but much lesser on the Si. It is immediately interesting to compare with the data in Figure 9.6c, acquired as in Figure 9.6b but with the sample grounded on the underside rather than topside. Here, the parabolic vertices are located closer together, with only a 0.3 V shift between the two metal cases and weaker parabolic curvatures for the metals.

Published tabulations consistently show a smaller work function for Al compared to Au, qualitatively consistent with Figures 9.6b and c (albeit on air-oxidized Al) and by a difference on the order of 1 V. Thus, the simplest interpretation of Figure 9.6 is that better grounding of the sample can produce more accurate values of relative work function. Stronger parabolic curvature in the voltage dependence of frequency shift indicates larger capacitive energy storage on the metals than the silicon, and even more so if the metal is better grounded. Our emphasis here is not on research findings but rather on *two methodologically important observations* for EFM: (1) complete inversions in qualitative contrast resulting from change of applied bias and (2) details of grounding affecting quantitative results (unlike some other material-contrasting modes such as friction or adhesion). As is common with property characterization and especially electronic methods, to obtain quantitatively meaningful AFM results, one may need to execute *careful sample preparation and data acquisition and analysis* procedures as well as control of tip state.

Figure 9.7 contains lateral 1D profiles of height and relative contact potential difference (i.e., variable KFM dc tip bias, arbitrarily zeroed) across alternating strips of Au and Al with intervening bare substrate. Here, the metal lines were

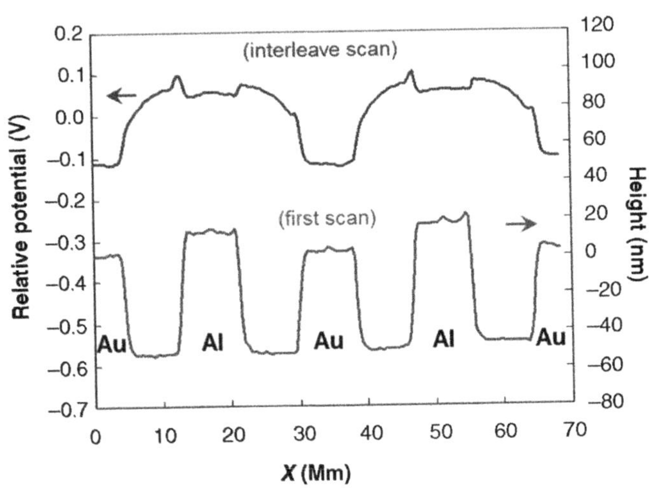

FIGURE 9.7 Contact potential difference and height profiles across strips of metal and silicon with native oxide. Sample is underside grounded.

lithographically grown on Si capped only with thin (~1 nm) native oxide and the sample was grounded on the bottom side. This grounding is likely not good enough to produce a reliable Al–Au work-function difference measurement; but what is of principle interest in this example is the strong variation of contact potential difference *across the semiconductor region*. This is most likely the result of energy *band bending* in the near-surface region of the silicon, as discussed below. Strong bias-induced charging in the metals produces carrier charge displacement within the semiconductor.

These observations can be semiquantitatively understood in the context of *Schottky barrier* theory. Our motivation here is not a study of Schottky barrier *per se*, rather to introduce important physical phenomenology underlying contrast mechanisms. A Schottky barrier strictly describes a metal–semiconductor interfacial system with an effectively infinite spatial extent parallel to the interface. The word "barrier" refers to the electrical *rectification* that can result across such an interface in an electronic device (e.g., at an electrode of a transistor) [10,14]. In our case, this would best apply to the metal–silicon interface centered under the metal strips. (But the presence of a native oxide capping the substrate prior to metal deposition invalidates a strict Schottky barrier treatment.) The variation of electrostatic potential within the semiconductor, perpendicular to the interface below the metal, would be described by the differential equation derived from the Poisson equation [14],

$$\frac{\partial^2 \Phi}{\partial z^2} = -\frac{n_c e}{\varepsilon \varepsilon_0} \tag{9.1}$$

where Φ is the electrostatic potential, n_c is the number density of charge carriers in the semiconductor, e is the single-electron charge, and ε and ε_0 are the permittivities of semiconductor and vacuum. This equation predicts a nonlinear variation of electrostatic potential with distance below the interface, into the semiconductor [14,15]. A result is nonlinear energy band bending perpendicular to the interface, as depicted in Figure 9.8 for two cases, relatively large and small work-function metals interfaced

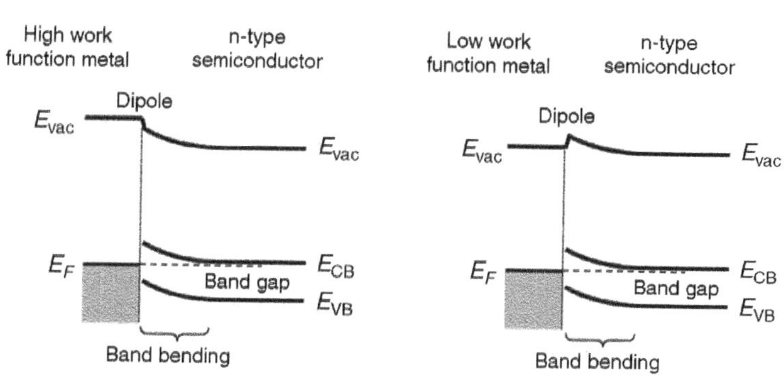

FIGURE 9.8 Illustration of electronic band bending for cases of high and low work-function metals contacting an n-type semiconductor.

with an n-type semiconductor (and including a possible localized *interface dipole* [14]). The rise of the conduction and valence band edges E_{CB} and E_{VB}, along with the vacuum level E_{vac}, upon nearing the interface may be driven by *midgap* electronic states at the interface, which *pin* the effective Fermi level of the semiconductor to the metal Fermi level [14,15]. The region near the interface in which the semiconductor band edges are *bending* is the locus of carrier charge displacement [14,15]. The displacement of charge toward the interface is another example of *Debye screening* (as in Figure 2.17, in which ions are displaced in solution to screen a charged surface). A parameter describing the distance over which charge is displaced by Debye screening is the Debye length L_D,

$$L_D = \sqrt{\frac{\varepsilon\varepsilon_0 k_B T}{e^2 n_c}} \tag{9.2}$$

where k_B is the Boltzmann's constant and T is the Kelvin temperature [14].

In the data of Figure 9.7, we are instead examining an electrostatic potential profile along the (oxidized) semiconductor surface, parallel to the metal–semiconductor interface and spanning micrometers beyond it. Nevertheless, the preceding concepts remain valid in relating variable electrostatic potential to charging at the metal–semiconductor interface and screening by carriers within the semiconductor. In other words, it is this physics that is the source of the observed variation of contact potential difference across the semiconductor. The length scale of this variation, several micrometers, is much greater than typical for Schottky barriers (Eq. 9.2), usually measured in nanometers for technological dopant concentrations (albeit normal to the interface). This suggests a very low carrier concentration in the case of Figure 9.7, possibly resulting in part from the oxidized (insulating) state of the silicon near the surface. Here again, we are only concerned with basic physical insight into contrast mechanisms; a precise treatment of this physical system would require numerical modeling with finite element methods.

9.1.4 KFM of Organic Semiconductor, Including Cross-Technique Comparisons

In this section, we consider interleave-based KFM imaging of ultrathin films of a crystalline organic semiconductor, pentacene (on silicon oxide), to exemplify variations in contact potential difference due to: (1) grain boundaries and (2) the presence or lack of *epitaxy* (molecular lattice "registry") between first- and second-layer domains [16]. KFM in #2 is compared with another property-sensitive AFM measurement, friction force (under sliding contact), which is also sensitive to the epitaxial relationship between first and second layer as previewed in Figure 1.8. We will see that the information from the two techniques is highly complementary, each imaging mode helping to elucidate what is being seen with the other mode.

Figure 9.9a contains height (top) and contact potential difference (bottom) images of approximately one monolayer of pentacene on silicon oxide. The height

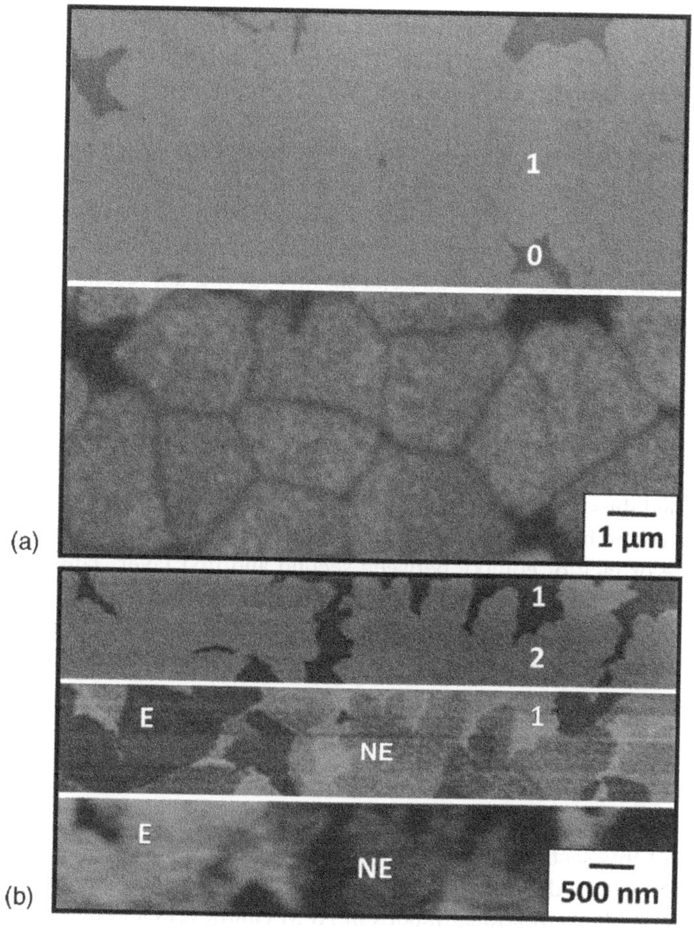

FIGURE 9.9 (a) Height (top) and contact potential difference (bottom) images of nearly complete pentacene monolayer on silicon oxide. (b) Height (top), friction force (middle), and contact potential difference (bottom) of nearly complete pentacene bilayer on silicon oxide with epitaxial (E) and nonepitaxial (NE) domains marked. Substrate, first, and second layers marked 0, 1, and 2.

image reveals several gaps in the film, indicating that there is slightly less than full monolayer coverage. The KFM image at bottom reveals much more, however: the presence of a mosaic of grains demarcated by grain boundaries approximately 40 mV lower in contact potential difference. The higher effective work function implies deeper energy states, including unoccupied conduction states, which could serve as traps impeding electron mobility and, therefore, electronic device (e.g., transistor) performance.

Figure 9.9b contains height (top), friction force (middle), and contact potential difference (bottom) images of a bit less than two monolayers of pentacene on silicon oxide. The height image reveals the presence of these two surface elevations, first and second layer. The friction force image is similar to that in Figure 1.8, the most interesting result being *two distinctly different levels of friction within the second monolayer*. Again, the reduced-friction domains are locations of initial second-layer grain growth atop underlying grains with a definitive epitaxial relationship, whereas the higher friction domains are locations where a given second-layer grain had crossed a first-layer grain boundary to grow onto a different underlying grain of arbitrary lattice orientation, adjacent to the initial epitaxially related grain. The lack of epitaxy upon crossing over to an adjacent first-layer grain results in much greater frictional (dissipative) response on the second layer, as well as other unique behaviors not shown in Figure 9.9: preferential chemical etching [17] and preferential wear under tip sliding at elevated temperature. The KFM image at bottom reveals variations in contact potential difference in inverse correlation with variations in friction force. The two types of second-layer domains differ by approximately 50 mV in the KFM image, the epitaxial domains being higher in contact potential difference and thus corresponding to effectively lower work function. Thus, there is a clear connection between electronic energy-level positions in the second layer and the presence or lack of epitaxy between second and first layer. This is a profound, revealing connection between physical attributes (frictional response, electrostatic potential) that at first glance would seem unrelated, and thus underscores the value of using multiple imaging modes to explore the same sample region with the same tip.

9.2 METHODS USING HIGHER VIBRATIONAL MODES

The topics of this section could be considered "newer methods." At this writing, it has been well over a decade, however, since the first journal papers were published investigating higher vibrational modes in AFM cantilevers and demonstrating their sensitivity to details of tip–sample interaction [8,9]. For many years, groups that utilized these modes (in conjunction with the fundamental mode, i.e., *multifrequency* methods) custom-assembled the necessary equipment, such as high-frequency lock-in amplifiers and signal adding circuitry. Only recently have AFM vendor companies begun to offer add-on electronics for this purpose. This section presents basic formulism relevant to these methods and provides examples obtained with commercial add-on electronics. More detailed analyses are available in numerous articles in the primary literature [8,9,18–32], as well as in review articles and edited volumes [33–35].

9.2.1 Mathematics of Beam Mechanics: The Music of AFM

To help understand the concepts of this section, we begin with the results of a simple experiment: using a digital oscilloscope to record the cantilever oscillation just after pull-off in a force-curve measurement, as previously considered in Section 6.5. Figure 9.10 contains 2.5 cycles of the damped sinusoidal motion during this

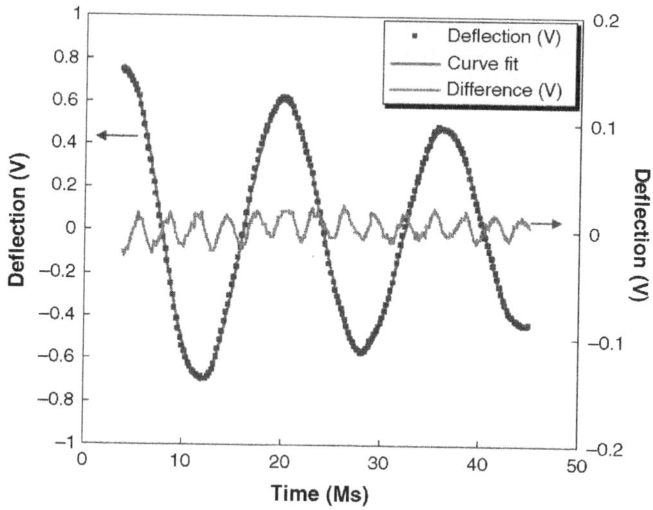

FIGURE 9.10 Digital oscilloscope trace of the decaying free oscillation (data points) of a V-shaped cantilever immediately following force–curve pull-off, plus an overlaid mathematical fit (exponentially damped sine function) and the difference of the two: a small-amplitude, high-frequency signal (expanded scale).

ring-down (of a gold-coated V-shaped silicon nitride cantilever). Upon mathematically subtracting a damped sine fitting function and expanding the vertical scale, one unambiguously sees an additional periodic component of much smaller amplitude and roughly 5.5 times higher frequency. What is this component? To this point, we have treated the tip–cantilever oscillator as an equivalent point mass m^* attached to a spring of force constant k, with damping due to the immersion medium (Section 2.4). An oscillating point mass–spring system with damping has only one natural frequency of motion (Eq. 2.20). What is the source of the higher frequency component so clearly present in Figure 9.10?

The key to this behavior is quite obvious in hindsight. The AFM tip–cantilever is not, of course, a point mass, rather, an extended 3D body. There is not just one mode of resonant wave motion in such a body. The physics is very much akin to that of musical instruments; in addition to the fundamental, lowest frequency mode, there are higher frequency modes—you could say "overtones." Moreover, the act of pulling an AFM tip out of adhesive contact confers some initial bent shape to the cantilever that is only close to, not exactly, the shape of the fundamental vibrational mode. In Section 3.2, we discussed the bent shape of an end-loaded, ideal rectangular cantilever (Eq. 3.5). To determine the possible bent shapes of such a cantilever under *free oscillation*, one needs to consider the *wave equation* describing this motion. We will see that the shape of the deformed cantilever as it oscillates can be described as a linear combination of many shapes, known as *eigenfunctions* of the differential wave equation. At the break of adhesive contact between tip and sample, the initial bent shape of the cantilever can be described as a *linear combination* of these various eigenfunctions. With passing time following release, each of these eigenfunctions undergoes oscillatory motion at its own characteristic eigenfrequency. The subtraction of the dominant sine wave in Figure 9.10, the motion of the fundamental eigenmode, reveals *at least* one of these other motions, a sine wave at higher frequency, which is contained in the overall cantilever motion, a sum of sine waves, both fundamental and overtones.

Let us examine the mathematics of these various eigenfunctions and eigenfrequencies in more detail. The time-dependent displacement in the "soft" direction $z(x,t)$ of each cross-sectional element of differential volume $(wt)dx$ located at x along the long axis of a rectangular cantilever, in the absence of damping or driving force, is most generally described by the partial differential equation

$$EI\frac{\partial^4 z}{\partial x^4} + \rho_m A\frac{\partial^2 z}{\partial t^2} = 0 \tag{9.3}$$

with the constants deriving from the cantilever's geometry and properties: Young's modulus E, cross-sectional area moment of inertia I $(=wt^3/12$, Eq. 3.4), mass density ρ_m, and cross-sectional area A $(=wt)$. That this wave equation is *fourth order* in spatial derivative results from the role of *curvature* in beam mechanics as follows. First, as mentioned in Section 3.2, the local curvature of the cantilever (see Figure 3.9, inset) relates to the cantilever's local bending moment M, Young's modulus E, and cross-sectional area moment of inertia I according to

$$\frac{\partial^2 z}{\partial x^2} = \frac{M}{EI}. \tag{9.4}$$

The local bending moment has energy units and is defined as

$$M = \int_A z\sigma dA \tag{9.5}$$

where σ is the stress acting on a differential area element dA, a distance z from the neutral surface (Section 3.2). M will be eliminated, however, in the following. Second, the local force F on a cross-sectional element arises from the local gradient of moment,

$$F = \frac{\partial M}{\partial x} = EI\frac{\partial^3 z}{\partial x^3}, \tag{9.6}$$

and Newton's second law for a differential shear force ∂F accelerating a cross-sectional element of differential mass $\rho_m A\partial x$ gives

$$\partial F = -\rho_m A\partial x\frac{\partial^2 z}{\partial t^2}. \tag{9.7}$$

Combining Equations 9.6 and 9.7 yield the wave equation, Equation 9.3, relating fourth-order spatial derivative to second-order time derivative.

For a simple rectangular bar of uniform cross section A and length L, the solution of Equation 9.3 under appropriate boundary conditions for rigid clamping at one end of the cantilever ($z=0$, $\partial z/\partial x=0$ at $x=0$) and free oscillation at the other (no moment or shear force, meaning $\partial^2 z/\partial x^2=0$, $\partial^3 z/\partial x^3=0$ at $x=L$) is a set of eigenfunctions

$$z_n = \cos(k_n x) - \cosh(k_n x) + \left(\frac{\sin(k_n L) - \sinh(k_n L)}{\cos(k_n L) + \cosh(k_n L)}\right)[\sin(k_n x) - \sinh(k_n x)], \tag{9.8}$$

with each eigenfunction z_n having a characteristic wave number κ_n, oscillating in time at eigenfrequency f_n as determined by the dispersion relation

$$\kappa_n = \left(\frac{12(2\pi f_n)^2 \rho_m}{Et^2}\right)^{1/4} \tag{9.9}$$

with allowed values of $\kappa_n L$ determined by the characteristic equation

$$\cos(\kappa_n L)\cosh(\kappa_n L) + 1 = 0. \tag{9.10}$$

Numerical solutions of the characteristic equation for $n=1$–5 are $\kappa_n L \approx 1.88$, 4.69, 7.85, 11.00, and 14.14. The corresponding five eigenmode shapes are depicted in Figure 9.11a. The eigenfrequencies f_n for $n=1$–10, normalized to the $n=1$ case, are shown in Figure 9.11b. (For $n>1$, these approximately obey the relation $f_n = f_1[(2n-1)/1.194]^2$.) Of particular note is the fact that the eigenfrequencies are not harmonics—integral multiples—of the $n=1$ case, unlike the case of a vibrating string pinned at both ends.

The theoretical eigenfrequency for $n=2$ is 6.27 times higher than the fundamental mode ($n=1$). This frequency ratio is a bit larger than observed in Figure 9.10 for a V-shaped cantilever, upon subtraction of the fundamental sine wave. The dynamics of cantilevers with shapes more complicated than rectangular, whether V, tapered ("picket fence"), or otherwise, and including the significant mass of the tip have been studied by finite-element methods. A similar spacing of eigenfrequencies has been found, but with normalized values somewhat lower than theoretical predictions for purely rectangular, tipless cantilevers as considered above. Of particular significance is the additional mass of the tip, which generally decreases these eigenfrequencies relative to those of tipless cantilevers, and decreases the higher eigenfrequencies relative to the fundamental. This is consistent with Figure 9.12, which compares the $n=2$–6 experimental eigenfrequencies of a real rectangular cantilever (i.e., including slight end taper and tip), determined by power spectral density analysis of the cantilever ringdown following force–curve pull-off, to theory for a purely rectangular tipless cantilever. Each eigenfrequency is normalized to the $n=1$ mode

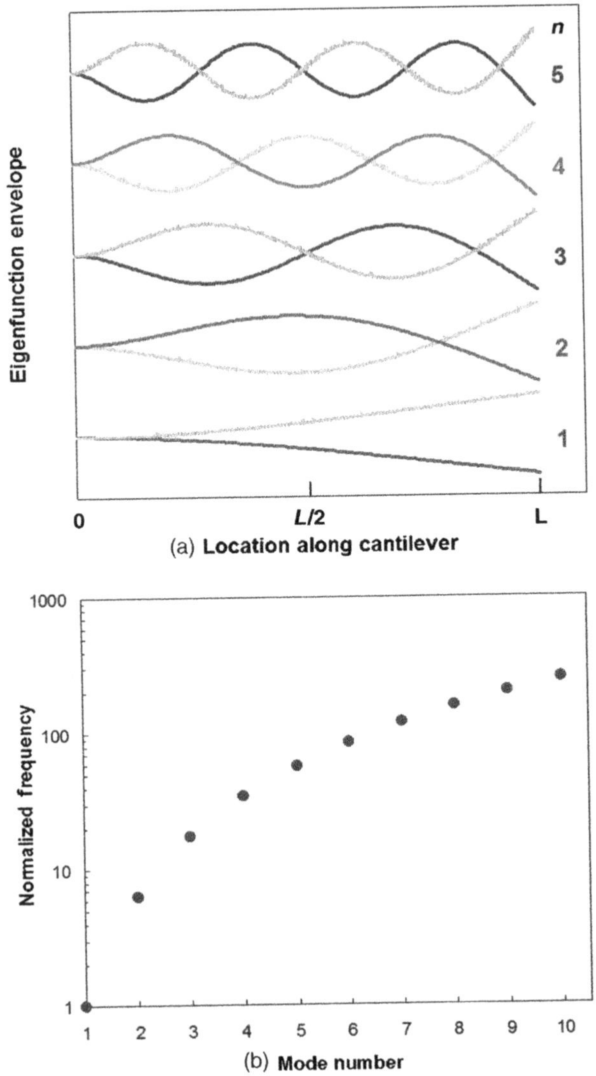

FIGURE 9.11 (a) Theoretical side-view shapes of five eigenmodes for a rectangular bar vibrating vertically (arbitrarily scaled). With passing time, these shapes represent the envelope of vibrational motion (i.e., its amplitude). (b) Resonant frequencies of the first 10 eigenmodes, normalized to that of the fundamental.

value of 68.4 kHz. (A time stream of raw deflection data was acquired at 5 megasamples per second with a LabView system, and then Fourier analyzed. The sixth eigenfrequency of approximately 2 MHz was the highest measurable, given the limit imposed by the Nyquist theorem.)

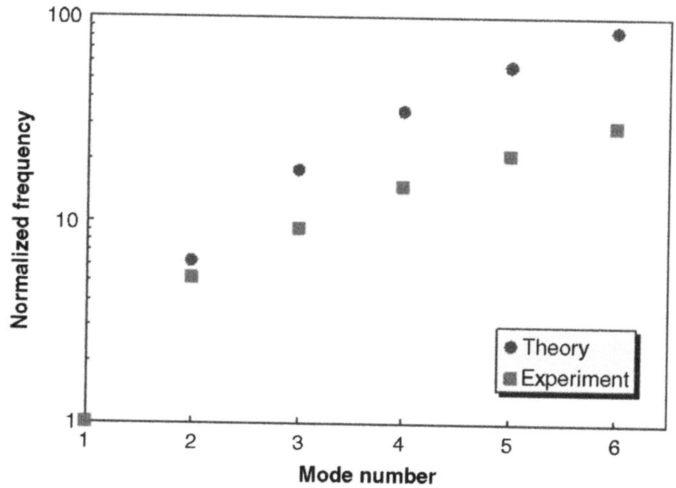

FIGURE 9.12 Comparison of experimental eigenfrequencies (squares) of first six vibrational modes of a rectangular AFM cantilever (i.e., with tapered shape and attached tip), normalized to the frequency of the fundamental mode, to those of the first six theoretical modes (circles) from Figure 9.11b.

9.2.2 Probing Tip–Sample Interactions Via Multifrequency Dynamic AFM

In this section, we introduce multifrequency cantilever excitation to generate higher-eigenmode imaging, for the purpose of enhancing materials contrast. It is known that with lock-in amplification, one can detect signals at higher eigenfrequencies in the absence of their external excitation. This phenomenon primarily results from a tip–sample interaction that effectively transfers some of the cantilever's kinetic energy from the (driven) fundamental mode to higher eigenmodes. But in recent years and with commercial packages, it has been found that explicitly driving the cantilever at additional, higher eigenfrequencies can produce stronger, and more useful, signal at the driven frequencies.

Experimentally, a first need is signal-adding circuitry in order to generate a multifrequency cantilever drive signal, and the second is multiple lock-in amplifiers to analyze the individual periodic signals contained within the cantilever response. In the following examples, a commercial add-on electronics[5] has been used for the purpose of analyzing two periodic signals in the case of bimodal excitation (but other commercial systems exist as do third-party lock-in amplifiers and signal adding circuitry). An additional operational consideration is the precise positioning of the laser spot on the cantilever. In Figure 9.11a, we see that for eigenmodes $n > 1$, there is at least one node away from the attached location, where the eigenmode amplitude goes to zero. Centering a focused laser spot position at these locations is

[5] Agilent MACIII, used with a model 5500 SPM.

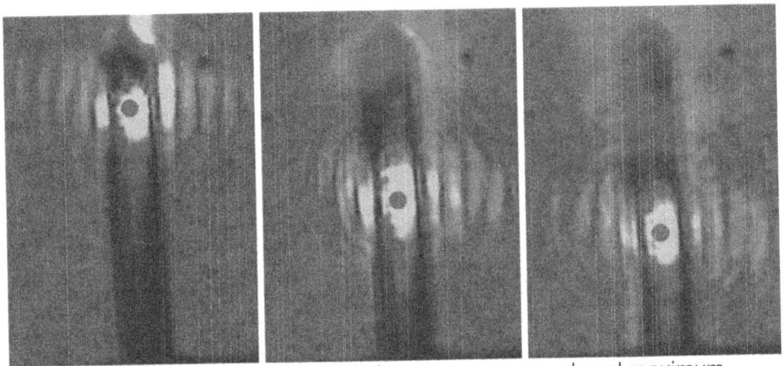

FIGURE 9.13 Experimental identification of the three laser spot positions on a nominally rectangular silicon cantilever at which the second eigenmode *signal* (i.e., generated by changes of *inclination*) reaches local extremes.

expected to provide *high* sensitivity to the eigenmode's motion because the *angle of inclination strongly varies* at these locations during oscillation. Near the free end of the cantilever, the slope appears to reach slightly greater values. Apart from the free end, positioning the laser at a local maximum of eigenmode amplitude is expected to *minimize* the sensitivity to this motion because the angle of inclination varies the *least*. This consideration is thus important to the operator's ability to detect higher eigenmode activation.

Figure 9.13 shows three laser spot positions on approximately rectangular silicon cantilever at which the second eigenmode amplitude signal reached maximum, minimum, and local maximum values. The minimum corresponds to approximately half way between the tip- and base-end of the cantilever, quite consistent with theoretical prediction for the $n=2$ case seen in Figure 9.11a (maximum amplitude and thus minimum change of angle of inclination). The endmost maximum was 10–20% greater than the local maximum, some 70% of the way back, toward the pinned end of the cantilever, also roughly consistent with the picture in Figure 9.11a. Laser positioning was performed very precisely on tapered-end-with-tip rectangular cantilevers by Rabe et al. [36], using a 5-μm-focused laser spot and interferometric methods to examine eigenmode shapes for $n=2$–6. Very good agreement with theory was found, except near the tip end of the cantilever for $n>3$. All things considered, an overall strategy for maximal sensitivity to higher eigenmode motion is to position the laser near the free end. But if laser spill-off becomes an issue, one should recognize that other locations along the cantilever are available while still maintaining good sensitivity to the $n=1$ mode (usually for driving the surface-tracking Z-feedback circuit).

In the following examples, a rectangular cantilever was excited at the two lowest eigenfrequencies $n=1, 2$. The amplitude of the second eigenmode signal was one order of magnitude smaller than the fundamental, producing an overall cantilever excitation signal similar to that in the top of Figure 9.14, data from a freely

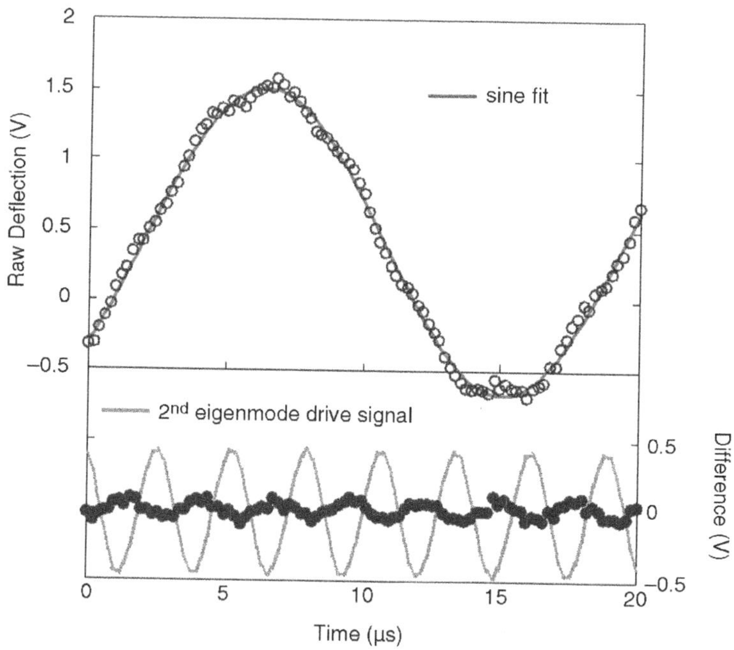

FIGURE 9.14 Top: Measured response of a freely oscillating rectangular cantilever (open circles) driven at first and second eigenfrequencies, and overlaid sinusoidal curve fit. Bottom: Difference (solid circles) of experimental signal and sine fit from top; overlaid second-mode driving signal (curve).

oscillating cantilever. Note that one cannot directly infer the relative spatial amplitudes of different eigenmodes, in nanometer units, from the relative raw amplitudes of eigenmode signal in volts, because of differences in sensitivity. Examining the modal profiles in Figure 9.11a, one sees that at the end of the cantilever, for similar modal amplitudes on a distance scale, higher eigenmodes exhibit greater slopes. Thus, the measured amplitude of the $n = 2$ signal should be *greater* than that for the $n = 1$ in the event that the actual amplitudes in distance units are the *same*. A second observation is the presence of a phase lag in the $n = 2$ signal relative to its drive signal, as shown in the bottom of Figure 9.14. Here, the $n = 2$ signal is extracted from the raw, composite deflection signal at top by subtracting a single sine-wave curve fit. Normally, the phase signal from each lock-in amplifier can be treated with a unique phase offset, specific to the mode being analyzed, such that a phase lag of 90° is achieved precisely at the eigenfrequency of interest.

Both the amplitude and the phase of each eigenmode are generally expected to change upon tip–sample interaction, but without necessarily the same relationship to the physics of tip–sample interaction as the $n = 1$ case (i.e., Chapters 4 and 5). Given the recent availability, at this writing, of multifrequency methods, the understandings of such relationships for higher eigenmodes are likely in their infancy

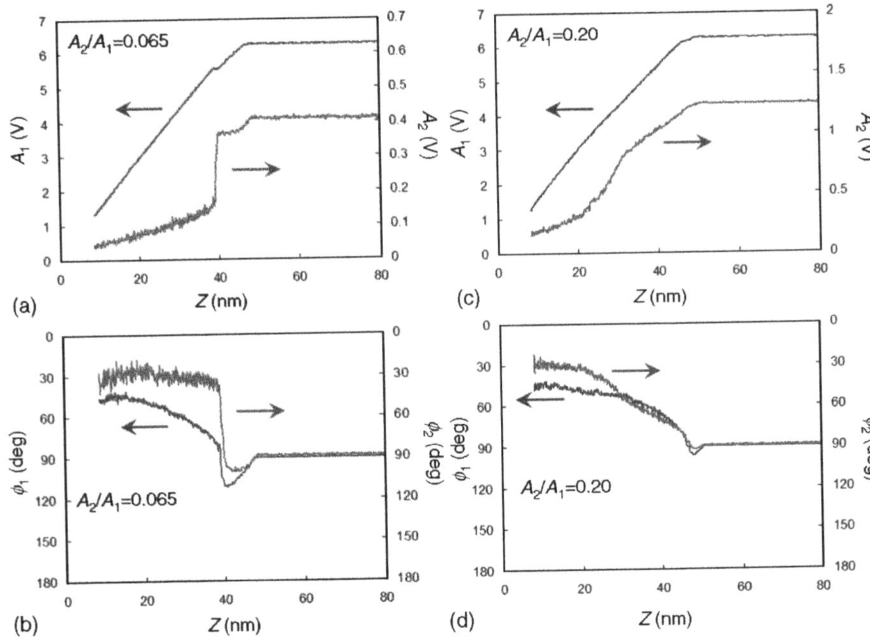

FIGURE 9.15 Overlaid amplitude (a) and (c) and phase (b) and (d) responses of first and second eigenmodes for two indicated values of excitation amplitude ratio, A_2/A_1.

compared to understandings of the $n=1$ case. In the following, we examine the Z dependence of the second eigenmode amplitude and phase simultaneous along with amplitude and phase of the first eigenmode, at two values of the ratio A_2/A_1, the relative magnitude of the two excitations, at a fixed value of A_1. Figure 9.15 shows the cases of $A_2/A_1=0.065$ and 0.20, respectively, for a tip interacting with a glassy polymer blend of polyethyl methacrylate and polymethyl methacrylate. In each case, the Z-axis was zeroed at approximately the location at which A_1 extrapolates to zero. For Figure 9.15, $A_1=48$ nm using a nominal $k=3$ N/m silicon rectangular cantilever.

In Figure 9.15a and b, for $A_2/A_1=0.065$, one observes in plots of A_1-Z and φ_1-Z the well-known transition from net attractive to repulsive regime, as discussed in Section 5.1. Most notably, the nearly linear reduction in A_1 with decreasing Z within both regimes, with similar slope (reduction of A_1 signal in volts per unit distance in Z displacement) is not reproduced by A_2 versus Z. Instead, the decrease of A_2 with decreasing Z plateaus just above the transition from net attractive to repulsive regimes. Below this transition, the decrease of A_2 with further decrease of Z is finite but less steep than the decrease of A_1, considering the respective values of A_2 and A_1 at free oscillation. These observations exemplify an important feature of the A_2 signal when kept small relative to A_1: a relatively weak variation over intervals of Z in which A_1 varies strongly. One important result is that *the A_2 signal is less affected by the slope of topography* along the fast-scan direction, dZ/dX, than is A_1 or φ_1.

One also notes that φ_1 is also only weakly variable with Z, and thereby A_1. Thus, in both cases, one has a signal that may provide materials contrast but contains little effect from surface slope, unlike conventional phase, φ_1, which can exhibit strong slope effects (e.g., Figure 5.42).

In Figure 9.15c and d, for a free-oscillation value of A_2 that is three times greater than for Figure 9.15a and b, one sees that A_2 and φ_2 vary more steeply with Z. Thus, these signals would be more strongly affected by surface slope than at the three times smaller value of A_2 in Figure 9.15a and b. One also observes that the distinct, jump transition between net attractive and repulsive regimes of Figure 9.15a and b has been replaced by a continuous change in A_1 and φ_1 of Figure 9.15c and d, and at a higher value of Z, such that the attractive regime almost completely disappears. This is consistent with both experiments and numerical simulations indicating that *the repulsive regime generally can be stabilized by the excitation of the second eigenmode at larger amplitudes* [31]. The nonlinearity of tip–sample interaction can cause problems such as chaotic motion in dynamic AFM [35]; driving at more than one frequency may aid a transition to chaos within some regions of parameter space. The exploration of parameter space should proceed cognizant of such possibilities.

Figure 9.16 demonstrates the sensitivity of A_2 as a materials-contrasting signal, in comparison with weak-at-best contrast simultaneously obtained in conventional phase imaging, φ_1. The ratio A_2/A_1 was set to 0.065 and the system was in the net

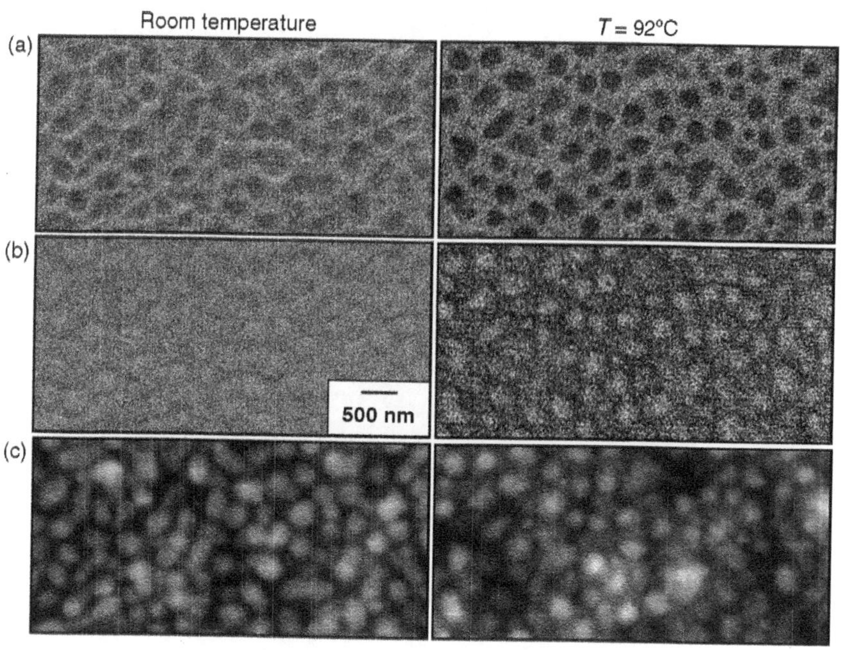

FIGURE 9.16 Comparison of contrast in images of (a) second eigenmode amplitude and (b) conventional phase on a PEMA/PMMA blend at room temperature (left) and 92°C (right). (c) Corresponding height images.

repulsive regime. At room temperature, there is slight contrast of PEMA- and PMMA-rich phase-segregated domains in A_2 while completely absent in φ_1. At an elevated temperature, 92°C, upon which PEMA has entered glass–rubber transitional behavior, the A_2 contrast has become stark while φ_1 contrast is meager at best. Although a fundamental understanding of the contrast mechanism is difficult to postulate in this example, the practical utility of A_2 imaging is unmistakable.

Figure 9.17 demonstrates the sensitivity of the second eigenmode phase, φ_2, as a materials-contrasting signal, similar to conventional phase φ_1 but exhibiting better

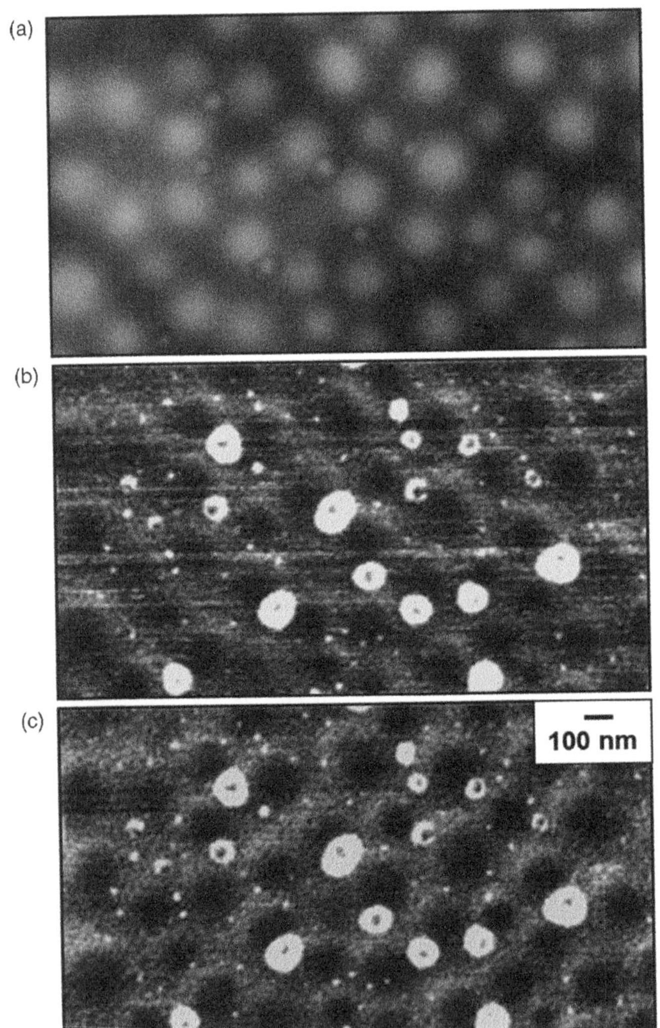

FIGURE 9.17 (a) Height, (b) conventional phase, and (c) second eigenmode phase images of a blend of PBMA, PLMA, and drug. Both phase images are raw.

signal/noise. The system was in the net repulsive regime but the φ_1 image inverted in contrast for better comparison to the φ_2 image. The sample is a blend of polybutyl methacrylate (PBMA), polylauryl methacrylate (PLMA), and the drug dexamethasone. The brightest regions in both phase images are PLMA domains, each apparently containing a drug particle (which appeared only after exposure to high humidity, presumably due to drug mobilization). The darkest domains in both phase images contain concentrated dexamethasone in the as-spin-cast coatings ($\sim$1 μm thick on a silicon wafer), confirmed in elution experiments during which most of these protruding domains become depressions in the height image, as drug diffuses into the aqueous medium [37]. The intervening regions of intermediate phase are predominantly PBMA. Both phase images represent *raw* data (e.g., unflattened); yet only the φ_1 image is laden with streaky manifestations of noise and/or instability. There is also generally a more distinct nanoscale texture visible in the φ_2 image within the PBMA/dexamethasone regions.

9.2.3 Contact Resonance Methods

Multifrequency cantilever excitation also can be utilized with the tip in continuous contact with the surface. When exciting eigenmodes, these methods go by the names of scanning acoustic force microscopy or, more commonly, contact resonance AFM [36,38–42]. When utilizing frequencies away from any resonances, the term ultrasonic force microscopy has been applied [43,44]. Here, we treat only the case of (resonant) eigenmodes. These modes are significantly modified under continuous contact with respect to the corresponding eigenmodes in dynamic AFM (Section 9.2.2). One can think of this contact as being intermediate between the free-oscillation case of Section 9.2.1, depicted in Figure 9.18a, and a completely pinned, infinitely rigid contact, depicted in Figure 9.18d. In this latter case, the boundary conditions at both ends of the cantilever would be $z=0$ and $\partial z/\partial x=0$ (see Section 9.2.1), and one can imagine a very different set of eigenmodes akin to those of a vibrating string (i.e., musical instruments). Thus, the more meaningful case of a contact with finite stiffness k^*, Figure 9.18c—meaning a tip indenting into a material with finite Young's modulus—is expected to produce eigenmodes having resonant frequencies *intermediate* between the cases of a free-end cantilever and a completely pinned cantilever (i.e., Figure 9.18a and d). Even more generally, one can consider a tip that is in contact with a material that additionally exhibits finite *damping*, whether *viscoelastic* in origin or otherwise (e.g., due to *adhesion hysteresis* within the contact periphery, with contact area continually expanding and receding under an oscillatory contact force). A simple model of such a contact is a spring/dashpot combination (Figure 9.18b). One expects a corresponding modification of both the characteristic resonance frequencies (stiffness) *and* line widths (damping) in the response spectrum of the composite system.

An example of a contact-modified resonance spectrum is shown in Figure 9.19, for a tip in contact with a glassy polymer, PBMA. Arrows denote the modified peak positions for the first and second eigenmodes of a rectangular silicon cantilever, compared to those for the same cantilever under free oscillation at the same drive

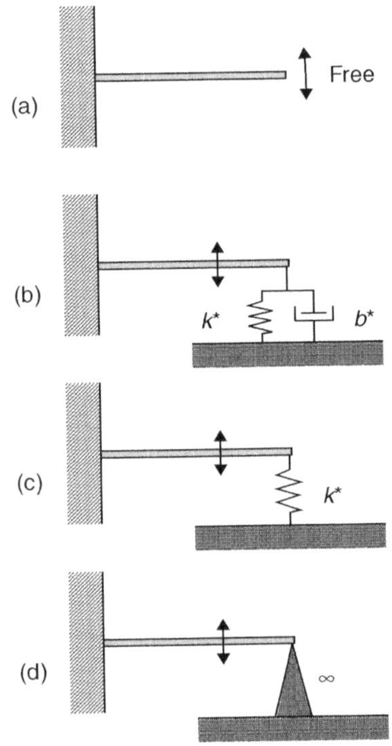

FIGURE 9.18 Cross-sectional depictions of AFM cantilever for four cases of tip-end freedom or constraint: (a) freely oscillating, (b) elastically coupled to sample with damping, (c) elastically coupled to sample without damping, and (d) rigidly fixed.

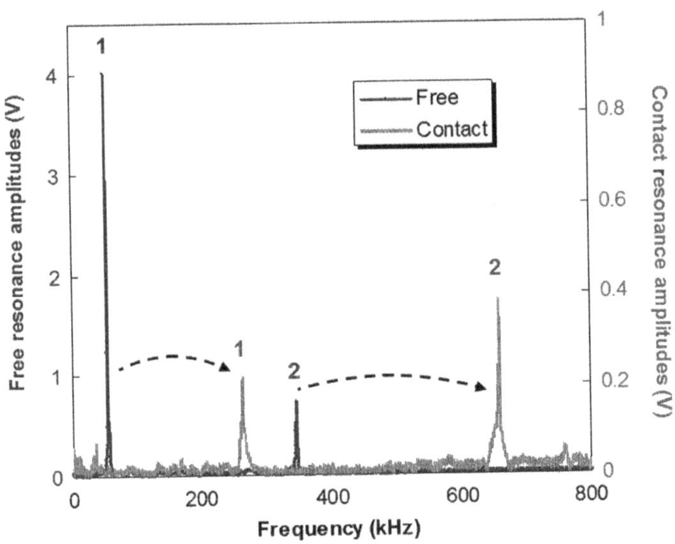

FIGURE 9.19 Frequency-dependent resonances of first and second eigenmodes for free cantilever (leftmost 1 and 2, left amplitude scale) and tip in contact with PBMA (right-shifted 1 and 2, right amplitude scale, expanded).

amplitude. In contact, the peaks are substantially reduced in amplitude relative to the corresponding free cases.

More meaningful analyses of peak position, as well as breadth, are made by comparing the right-shifted contact resonance spectra on different materials (Figure 9.20). Here, the first eigenmode response is compared between an (oxidized) aluminum polycrystalline film and organic contamination atop this film. Figure 9.20a and b shows the conventional topography and friction difference images. The latter contains only weak contrast between the metal surface and the mounds of contaminant. Figure 9.20c contains an image of the cantilever response amplitude at the driven frequency of 287.6 kHz, which corresponds to the resonant peak position f_1 of the first eigenmode *on the aluminum surface*. The contaminant appears darker in this image because the position of the resonance at this location has shifted down to 283.1 kHz, leaving a much smaller excitation at the drive frequency 287.6 kHz. The entire shape of the resonance on each material region is compared in Figure 9.20d.

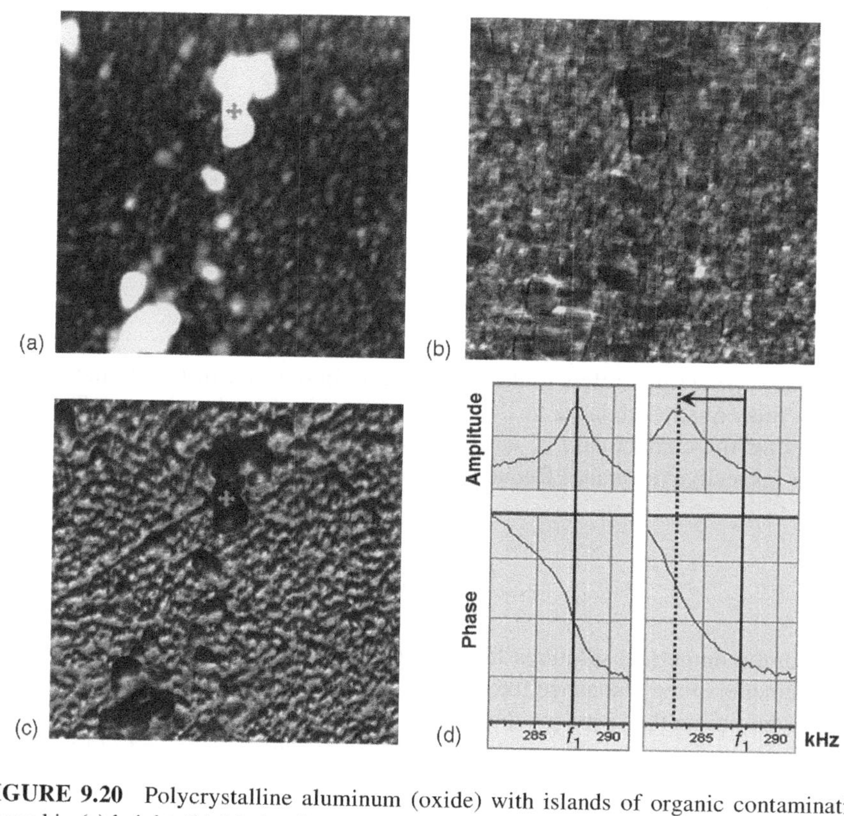

FIGURE 9.20 Polycrystalline aluminum (oxide) with islands of organic contamination imaged in (a) height, (b) friction force, and (c) cantilever response amplitude at a driven frequency $f_1 = 287.6$ kHz as denoted with vertical solid line in the (d) frequency response spectra. The left and right spectra were acquired at the left and right locations, respectively, marked (+) in (a)–(c). Vertical dotted line in (d) denotes the lower contact resonant frequency atop the contaminant compared to the aluminum.

In addition to the 4.5-kHz downward peak shift, the peak width is appreciably larger on the contaminant. A lower-frequency and broader resonance on the contaminant is consistent with a softer and more dissipative, presumably viscoelastic, material. That the friction force (Figure 9.20b) shows little contrast and no sign of wear suggests that the contaminant is not extremely compliant. This is a good example of material differences that are clear, both qualitatively and quantitatively, in contact resonance data, whereas uncertain in the friction data.

To extract more quantitative information from contact resonance data like that in Figure 9.20d, one must first utilize formalism akin to Section 9.2.1 but including the effects of the finite k^*, and potentially b^*, to include the effects of tip–sample contact (Figure 9.18b and c). Here, we treat purely elastic effects due to the spring k^*, positioned between the end of the cantilever and a rigid surface. Similar to the case of a freely oscillating cantilever (Section 9.2.1), the spring-coupled end can transfer no moment, $\partial^2 z/\partial x^2 = 0$ at $x = L$. But unlike the free case, the shear force at $x = L$, $EI\partial^3 z/\partial x^3$, is not zero; rather, it is given by $-k^* z$. Thus, the other boundary condition at $x = L$ can be written as

$$\frac{\partial^3 z}{\partial x^3} = \frac{k^* z}{EI} = \frac{3k^* z}{kL^3}. \tag{9.11}$$

This leads to the characteristic equation for allowed values of $\kappa_n^c L$,

$$\sinh(\kappa_n^c L)\cos(\kappa_n^c L) - \sin(\kappa_n^c L)\cosh(\kappa_n^c L) = \frac{(\kappa_n^c L)^3 k}{3k^*}\left[\cos(\kappa_n^c L)\cosh(\kappa_n^c L) + 1\right] \tag{9.12}$$

where the superscript c refers to the contact case. In the case of $k^* = 0$, meaning no contact, this equation reduces to Equation 9.10, with κ_n^c replaced by κ_n. The wave numbers for the contact case κ_n^c can be determined from those of the free case κ_n via their respective resonance frequencies,

$$\kappa_n^c = \kappa_n \sqrt{\frac{f_n^c}{f_n}}. \tag{9.13}$$

Given the numerical solutions for $\kappa_n L$ listed in Section 9.2.1, and experimental measurements of the resonance frequencies for both free (f_n) and contact (f_n^c) cases, one can determine the corresponding values of $\kappa_n^c L$. Then using Equation 9.12, one can determine the ratio of contact to cantilever stiffness, k^*/k, and ultimately k^* given a calibrated value of k (Section 3.7 or Appendix 1).

The relationship of contact stiffness k^* to Young's modulus for a given sample (the sought quantity) depends on the contact mechanics model invoked. Often the applied load in contact resonance measurements is much greater than the tip–sample adhesion, such that the Hertzian model is invoked. In other cases, the DMT model (Section 6.2) is used, that is, the adhesive force is added to the applied load to give the total contact force, and the latter treated the same as the applied load in the Hertzian model. In either case, the key relationship between contact stiffness k^* and modulus can be written as

$$k^* = 2aE^* \tag{9.14}$$

where a is the contact radius and E^* is the reduced elastic modulus of the tip–sample contact,

$$\frac{1}{E^*} = \frac{(1-\nu_1^2)}{E_1} + \frac{(1-\nu_2^2)}{E_2} \tag{9.15}$$

with $E_{1,2}$ and $\upsilon_{1,2}$ representing the Young's moduli and Poission's ratios for tip and sample. (Note that E^* is simply $3K/4$, with K as defined in Section 2.3. Recall that the Poisson's ratio does not vary much among sample types, and is typically assumed to be some appropriate value.) The need to know contact radius in Equation 9.14 is commonly circumvented by performing measurements on at least one reference material with known elastic modulus and reasonably flat surface. Having a second reference material can further eliminate the assumption of elastic modulus of the tip.

Finer treatments consider one or more further geometric or physical details, such as (i) a contact that is offset slightly from the end of the cantilever, reflecting the true location of the tip [36,38], (ii) a more complicated adhesive contacts mechanics model such as that of Maugis (Section 6.2.4), (iii) a viscoelastic contact (nonzero b^*) [39], and (iv) a nonspherical tip model, such as conical on a soft material or flat punch on a rigid material (i.e., accounting for the dulling of the tip at the high forces and pressures applied). Ultimately, the tip geometry affects the exponent in the power law relationship between measured contact stiffness and calculated elastic modulus, as discussed in Section 2.3. More complicated models result in the need for multiple measurements on *reference materials with moduli bracketing those of the samples of interest*, as well as the examination of multiple resonances both in contact and free of surface. This simply reflects the fact that a greater number of unknown quantities require a greater number of relationships to measurable quantities.

9.2.4 Single-Pass Electric Methods

Section 9.1 discussed the interleave procedure to image and characterize longer-range electrostatic interactions between metallic tip and sample in a two-pass imaging scheme. Historically, however, some of the first methods developed to image these long-range forces did *not* use interleave. Rather, *multifrequency* methods were utilized to *simultaneously* yet independently characterize different kinds of forces within each pass of a given scan line [1,2,5]. But as the community of users grew through the 1990s using single-frequency, commercial systems for long-range force characterization (i.e., methods of Section 9.1), multifrequency methods did not similarly flourish, presumably due to a lack of available commercial systems. More recently at this writing, however, there has been a flurry of activity utilizing multifrequency schemes for long-range force applications. (An exhaustive treatment is beyond the scope of this book.) Importantly, some of these methods have become commercially available; thus, we expect the methods to become increasingly commonplace. In this section, we describe two commercial variants.

The electric methods most commonly utilize ac biasing at either low frequency, well below the fundamental flexural resonance, or at higher eigenmode frequencies. In some other cases, the mathematics of ac interactions results in cantilever excitation at higher *harmonics* (integral multiples) of the excitation frequency. However, because nonlinear tip–sample interactions of any sort (electrically induced or otherwise, e.g., those due to vdW or mechanical interactions) can yield higher harmonic excitations [8], these latter methods must be used with much caution to avoid coupling electrical and topographic information. Here, we focus on methods using below- and on-eigenfrequency ac excitations.

Before getting into measurement details, it is important to understand some theoretical underpinnings of many variants of AFM that utilize both dc and ac biasing. We consider the case of a metallic tip at voltage V relative to sample, consisting of both dc and ac *applied* voltages, V_{dc} and $V_{ac}\sin(\omega_{elec}t)$, as well as an *intrinsic* difference in surface potential or *contact potential* V_{cp}, defined as sample relative to tip.[6]

For tip–sample capacitance $C(z)$, the force of electrostatic interaction is then

$$\begin{aligned} F_{elec} &= \frac{1}{2}\frac{\partial C}{\partial z}V^2 = \frac{1}{2}\frac{\partial C}{\partial z}\left(V_{dc} + V_{ac}\sin\omega_{elec}t - V_{cp}\right)^2 \\ &= \frac{1}{2}\frac{\partial C}{\partial z}\left[\left(V_{dc} - V_{cp}\right)^2 + 2\left(V_{dc} - V_{cp}\right)\left(V_{ac}\sin\omega_{elec}t\right) + V_{ac}^2\sin^2\omega_{elec}t\right] \\ &= \frac{1}{2}\frac{\partial C}{\partial z}\left\{\left[\left(V_{dc} - V_{cp}\right)^2 + {}^1\!/\!_2 V_{ac}^2\right] + \left[2\left(V_{dc} - V_{cp}\right)\left(V_{ac}\sin\omega_{elec}t\right)\right]\right. \\ &\quad \left. - \left[{}^1\!/\!_2 V_{ac}^2\cos 2\omega_{elec}t\right]\right\} \qquad (9.16) \end{aligned}$$

using the trigonometric identity $\sin^2(u) = [1 - \cos(2u)]/2$. The brackets designate three force terms: one constant and two periodic at frequencies ω_{elec} and $2\omega_{elec}$. Using a dedicated feedback circuit to vary V_{dc} from point to point during imaging, so as to null the cantilever amplitude at frequency ω_{elec} as measured with a lock-in amplifier, one then can measure V_{cp}. Provided a second lock-in amplifier (i.e., multifrequency detection), one can also measure the cantilever amplitude at $2\omega_{elec}$ and thereby image variations in the capacitance gradient $\partial C/\partial Z$.

But the reader will recall that these ac biasing signals are intended to be used simultaneous to topographic imaging in dynamic mode, that is, while the cantilever

[6] Note that if the sample is an operating electrical device, such as a transistor, there may be additional voltages superimposed on device elements, for example, a gate bias.

is being mechanically driven to oscillate near its resonance frequency. Clearly, one cannot have an ac bias frequency equal to the mechanical drive frequency, as this would inherently convolve electrical and topographic information: when the tip passes over a region of stronger electrical interaction, its time-dependent variation would affect the cantilever amplitude and thus the Z feedback system (i.e., the height image). We also do not want the electrical force variation to *slowly modulate* the cantilever's resonant amplitude (i.e., produce a time-varying envelope of the fundamental excitation) over such a timescale that the Z feedback can respond to it. Thus, one immediately conceives two strategies for choosing the ac bias frequency: (1) well below the mechanical drive frequency yet well above the *bandwidth* of the Z feedback system (say a few kilohertz, the inverse of its quickest response time), and (2) well above the mechanical drive frequency. For case 2, the availability of higher eigenmodes is appealing, as the signal or noise should be better than that at any arbitrary frequency. Case 1 implies the usage of large voltage biases and soft cantilevers such that adequate signal or noise is achieved. This can be problematic on semiconductors and potentially insulators, as large biases can produce substantial energy band bending, changing the intrinsic electronic structure of the sample.

The preceding schemes, as well as the interleave-based scheme of Section 9.1, are sometimes called AM-KFM because the amplitude of force variation (cantilever deflection) at ω_{elec} is causing the feedback to vary V_{dc} so as to null (or minimize) the amplitude of force variation, and thereby track V_{cp}. There is also a type of KFM known as frequency modulation (FM)-KFM, which is sensitive to electrically induced frequency shifts of the cantilever excitation, in order to vary V_{dc} and thereby track V_{cp}. Here, we discuss implementations of FM-KFM that still *track topography via the amplitude* of mechanical excitation of the fundamental flexural eigenmode, as usual. The overall operational mode is sometimes called AM/FM, denoting the feedback schemes for topography/surface potential [3,45]. (There are also FM/FM schemes used primarily in vacuum-based AFM systems, where frequency shifts are utilized for tracking both topography and surface potential [46–51]. FM/AM methods exist as well [51–53].) As discussed in Section 2.4, the presence of a long-range attractive force, with uniform gradient $\partial F/\partial Z$ over a tip oscillation cycle, results in a modified cantilever resonance frequency $\omega_0{}'$ according to

$$\omega_0{}' = \sqrt{\frac{k - \partial F/\partial Z}{m^*}} \approx \omega_0 \left(1 - \frac{1}{2k}\frac{\partial F}{\partial Z}\right), \tag{9.17}$$

where a power series expansion was retained only to first order by assuming the force gradient $\partial F/\partial Z \ll k$.

In one implementation of FM-KFM [54], the ac electrical bias is applied at frequency of the second eigenmode, but a weak version of this signal is also added to the mechanical excitation signal that is driving the cantilever at the fundamental resonance (i.e., with a much larger amplitude). The KFM feedback varies V_{dc} to maintain a fixed phase between the excitation and response signals at the second eigenmode as monitored by a second lock-in amplifier (the first being used as usual to track topography via the fundamental eigenmode amplitude). But there are important fine points in order to successfully implement this scheme. First, out of tip–sample engagement (withdrawn by approximately 100 nm), V_{dc} is approximately matched to the surface potential so as to minimally shift the fundamental eigenmode from its predetermined natural resonance frequency (say by monitoring phase). Upon approach to engagement (at the fundamental amplitude setpoint), the ordinary tip–sample interaction in the absence of electrical biasing produces a phase shift of the second eigenmode; the phase offset of the second lock-in amplifier is adjusted to match this phase shift. Then, upon biasing the tip at the same second eigenfrequency, variations in the phase of this signal derive from variations in tip–sample electrostatic potential, which then drive the KFM feedback to vary V_{dc} to null these phase variations. One measures and images these variations of V_{dc}.

In the other implementation of FM-KFM considered in this section [48], a scheme that is commercially available and also utilizing two lock-in amplifiers,[7] the applied ac voltage is varied at frequency ω_{elec} well below the fundamental resonance used to mechanically drive the cantilever for tracking topography. This ac voltage varies F, and thereby $\partial F/\partial Z$, per Equation 9.16. As a result, the mechanical resonance spectrum of the cantilever contains not only the fundamental resonance peak at ω_0 but also peaks at ω_{elec} and $2\omega_{elec}$, and moreover *sidebands* adjacent to the fundamental excitation due to indirect modulation at $\omega_0 \pm \omega_{elec}$ and $\omega_0 \pm 2\omega_{elec}$ [48]. In setting $\omega_{elec} \ll \omega_0$, the frequency-modulated cantilever deflection is equivalent to the sum of sine and cosine waves at ω_0 (i.e., 90° out of phase) with the latter's amplitude *varying in time at* ω_{elec}. With an appropriate setting of its phase offset, a first lock-in amplifier outputs an amplitude signal (i.e., used to track topography) extracted from its *in-phase* response (sometimes called the "X" signal in lock-in parlance), and a second output signal whose amplitude is time-varying at ω_{elec} from its *out-of-phase* response (the "Y" lock-in signal). The latter becomes the input signal for a second lock-in amplifier, which then outputs an amplitude signal proportional to

$$\frac{\partial^2 C}{\partial z^2}\left(V_{dc} - V_{cp}\right). \tag{9.18}$$

The KFM feedback varies V_{dc} to null variations in this latter signal; one measures and images these variations of V_{dc}. Thus, the FM-KFM designation can be confusing; although it is ultimately an amplitude signal driving the KFM feedback circuit, this amplitude derives from the *magnitude of a frequency shift that scales with electrostatic force gradient*. This last italicized point is the essential distinction.

Given the preceding descriptions of FM-KFM, it is exceedingly important to understand the significance of *force gradient detection as opposed to force detection*, for long-range interactions. As discussed in Section 2.2.2, long-range forces

[7] Agilent MACIII, used with a model 5500 SPM

depend on probe geometry well beyond the rounded end of the tip. The models employed in KFM studies usually include a three-part depiction of the entire probe: a hemispherical end cap attached to a truncated conical tip, in turn attached to a planar but tilted cantilever (as discussed in Section 3.6.4). Each of these elements contributes to the measured electrostatic force, but over very different distance intervals. The contributions of tip shank and cantilever are nearly constant over the vertical cycling range of the tip, contributing negligibly to a measurement that is sensitive to force gradient. Thus, the force gradient is primarily sensitive to interactions with the hemispherical end cap of the tip. This part of the tip is of course much more laterally localized, leading to much better spatial resolution in FM-KFM compared to AM-KFM [45]. (There are other advantages of AM-KFM in ultrahigh vacuum KFM, where a different interaction mechanism apparently yields atomic resolution [52]). Moreover, the determination of absolute surface potential is improved in FM-KFM; the portion of the electric field between the cantilever/tip and sample that contributes to the measured surface potential is only that localized at the sample region of interest, whereas in AM-KFM, forces exerted all along the cantilever, interacting with a broad area of sample underneath, can significantly affect the absolute surface potential measurement [3]. This latter issue is especially critical to the characterization of electronic device constructs having micrometer to submicrometer-scale devices or spacings between devices.

REFERENCES

[1] Stern, J.E., et al., *Phys. Rev. Lett.*, 1988, **53**: 2717.

[2] Terris, B.D., et al., *Contact electrification using force microscopy. Phys. Rev. Lett.*, 1989, **63**(24): 2669–2672.

[3] Girard, P. and A.N. Titkov, Electrostatic force and force gradient microscopy: Principles, points of interest and application to characterisation of semiconductor materials and devices, in *Applied Scanning Probe Methods II: Scanning Probe Microscopy Techniques*, Bhushan B. and H. Fuchs, eds. 2006, New York: Springer, p. 283–320.

[4] Nakamura, M. and H. Yamada, Electrostatic force microscopy, in *Roadmap of Scanning Probe Microscopy*, S. Morito, ed. 2007, New York: Springer, p. 43–51.

[5] Nonnenmacher, M., M.P. O'Boyle, and H.K. Wickramasinghe, *Kelvin probe force microscopy. Appl. Phys. Lett.*, 1991, **58**(25): 2921–2923.

[6] Fujihira, M., *Kelvin probe force microscopy of molecular surfaces. Annu. Rev. Mater. Sci.*, 1999, **29**: 353–380.

[7] Glatzel, T., et al., Principles of Kelvin probe force microscopy, in *Scanning Probe Microscopy: Electrical and Electromechanical Phenomena at the Nanoscale*, S. Kalinin and A. Gruverman, eds. 2007, New York: Springer, p. 113–131.

[8] Stark, R.W., T. Drobek, and W.M. Heckl, *Tapping-mode atomic force microscopy and phase-imaging in higher eigenmodes. Appl. Phys. Lett.*, 1999, **74**(22): 3296–3298.

[9] Hillebrand, R., M. Stark, and R. Guckenberger, *Higher-harmonics generation in tapping-mode atomic-force microscopy: Insights into the tip-sample interaction. Appl. Phys. Lett.*, 2000, **76**(23): 3478–3480.

[10] Horowitz, P. and W. Hill, *The Art of Electronics*. 2nd ed. 1989, Cambridge, MA: Cambridge University Press, 1125.

[11] Kalinin, S. and A. Gruverman, eds. *Scanning Probe Microscopy: Electrical and Electromechanical Phenomena at the Nanoscale*. Vol. **1**, 2007, New York: Springer, 565.

[12] Lhernould, M., et al., *Electrostatic forces in micromanipulations: Review of analytical models and simulations including roughness. Appl. Surf. Sci.*, 2007, **253**: 6203–6210.

[13] Oliver, R.A., *Advances in AFM for the electrical characterization of semiconductors. Rep. Prog. Phys.*, 2008, **71**: 076501.

[14] Sze, S.M., *Physics of Semiconductor Devices*. 2nd ed. 1981, New York: Wiley-Interscience, 868.

[15] Kittel, C., *Introduction to Solid State Physics*. 6th ed. 1986, New York: Wiley.

[16] Kalihari, V., et al., *Observation of unusual homoepitaxy in ultrathin pentacene films and correlation with surface electrostatic potential. Adv. Mater.*, 2009, **21**: 1–7.

[17] Puntambekar, K., et al., *Structural and electrostatic complexity at a pentacene/insulator interface. Adv. Funct. Mater.*, 2006, **16**: 879–884.

[18] Sahin, O. and A. Atalar, *Simulation of higher harmonics generation in tapping-mode atomic force microscopy. Appl. Phys. Lett.*, 2001, **79**(26): 4455–4457.

[19] Stark, R.W. and W.M. Heckl, *Fourier transformed atomic force microscopy: Tapping mode atomic force microscopy beyond the Hookian. Approximation. Surf. Sci.*, 2000, **457**: 219–228.

[20] Stark, M., et al., *Spectroscopy of the anharmonic cantilever oscillations in tapping-mode atomic-force microscopy. Appl. Phys. Lett.*, 2000, **77**(20): 3293–3295.

[21] Stark, M., *Spectroscopy of higher harmonics in dynamic atomic force microscopy. Nanotechnology*, 2004, **15**: 347–351.

[22] Stark, R.W., *Dynamics of repulsive dual-frequency atomic force microscopy. Appl. Phys. Lett.*, 2009, **94**: 063109.

[23] Lee, S.I., et al., *Nonlinear dynamics of microcantilevers in tapping mode atomic force microscopy: A comparison between theory and experiment. Phys. Rev. B*, 2002, **66**: 115409.

[24] Xin, X., et al., *Compositional contrast of biological materials in liquids using the momentary excitation of higher eigenmodes in dynamic atomic force microscopy. Phys. Rev. Lett.*, 2009, **102**: 060801.

[25] Crittenden, S., A. Raman, and R. Reifenberger, *Probing attractive forces at the nano-scale using higher-harmonic dynamic force microscopy. Phys. Rev. B*, 2005, **72**: 235422.

[26] Lee, S.I., et al., *Nonlinear dynamics of microcantilevers in tapping mode atomic force microscopy: A comparison between theory and experiment. Phys. Rev. B*, 2002, **66**: 115409.

[27] Proksch, R., *Multifrequency, repulsive-mode amplitude-modulated atomic force microscopy. Appl. Phys. Lett.*, 2006, **89**: 113121.

[28] Lozano, J.R. and R. Garcia, *Theory of multifrequency atomic force microscopy. Phys. Rev. Lett.*, 2008, **100**: 076102.

[29] Lozano, J.R. and R. Garcia, *Theory of phase spectroscopy in bimodal atomic force microscopy. Phys. Rev. B*, 2009, **79**: 014110.

[30] Rodriguez, T.R. and R. Garcia, *Compositional mapping of surfaces in atomic force microscopy by excitation of the second normal mode of the microcantilever. App. Phys. Lett.*, 2004, **84**: 449–451.

[31] Thota, P., S. MacLaren, and H. Dankowicz, *Controlling bistability in tapping-mode atomic force microscopy using dual-frequency excitation. Appl. Phys. Lett.*, 2007, **91**: 093108.

[32] Ghatkesar, M.K., et al., *Resonating modes of vibrating microcantilevers in liquid. Appl. Phys. Lett.*, 2008, **92**: 043106.

[33] Morita, S., et al., Noncontact atomic force microscopy. in *Nanoscience and Technology*, P. Avouris, et al. ed. Vol. **2**, 2009, New York: Springer-Verlag, 449.

[34] Raman, A., J. Melcher, and R. Tung, Cantilever dynamics in atomic force microscopy, in *Nano Today*. 2008, New York: Elsevier, p. 20–27.

[35] Stark, R.W., Bistability, higher harmonics, and chaos in AFM, in *Materials Today*. 2010, New York: Elsevier, p. 24–32.

[36] Rabe, U., K. Janser, and W. Arnold, *Vibrations of free and surface-coupled atomic force microscope cantilevers: Theory and experiment. Rev. Sci. Instrum.*, 1996, **67**(9): 3281–3293.

[37] Haugstad, G. and K. Wormuth, Digital pulsed force mode AFM and confocal Raman microscopy in drug-eluting coatings research, in MRS Proceedings. 2011, Materials Research Society.

[38] Huey, B.D., *AFM and acoustics: Fast, quantitative nanomechanical mapping. Annu. Rev. Mater. Res.*, 2007, **37**: 351–85.

[39] Yuya, P.A., D.C. Hurley, and J.A. Turner, *Contact-resonance atomic force microscopy for viscoelasticity. J. Appl. Phys.*, 2008, **104**: 074916.

[40] Hurley, D.C., Contact resonance force microscopy techniques for nanomechanical measurements, in *Applied Scanning Probe Methods XI*. 2009, New York: Springer.

[41] Killgore, J.P., et al., *Quantitative subsurface contact resonance force microscopy of model polymer nanocomposites. Nanotechnology*, 2011, **22**: 175706.

[42] Stan, G., et al., *Contact-resonance atomic force microscopy for nanoscale elastic property measurements: Spectroscopy and imaging. Ultramicroscopy*, 2009, **109**: 929–936.

[43] Kolosov, O. and K. Yamanaka, *Nonlinear detection of ultrasonic vibrations in an atomic force microscope. Jpn. J. Appl. Phys.*, 1993, **32**: L1095–L1098.

[44] Yamanaka, K., H. Ogiso, and O. Kolosov, *Ultrasonic force microscopy for nanometer resolution subsurface imaging. Appl. Phys. Lett.*, 1994, **64**(2): 178–180.

[45] Gil, A., et al., *Electrostatic force gradient signal: Resolution enhancement in electrostatic force microscopy and improved Kelvin probe microscopy. Nanotechnology*, 2003, **14**: 332–340.

[46] Glatzel, T., S. Sadewasser, and M.C. Lux-Steiner, *Amplitude or frequency modulation-detection in Kelvin probe force microscopy. Appl. Surf. Sci.*, 2003, **210**: 84–89.

[47] Nony, L., et al., *Understanding the atomic-scale contrast in Kelvin probe force microscopy. Phys. Rev. Lett.*, 2009, **103**: 036802.

[48] Zerweck, U., et al., *Accuracy and resolution limits of Kelvin probe force microscopy. Phys. Rev. B*, 2005, **71**: 125424.

[49] Kitamura, S. and M. Iwatsuki, *High-resolution imaging of contact potential difference with ultrahigh vacuum noncontact atomic force microscope. App. Phys. Lett.*, 1998, **72**(24): 3154–3156.

[50] Kikukawa, A., S. Hosaka, and R. Imura, *Silicon pn junction imaging and characterizations using sensitivity enhanced Kelvin probe force microscopy. App. Phys. Lett.*, 1995, **66**(25): 3510–3512.

[51] Burke, S.A., et al., *Determination of the local contact potential difference of PTCDA on NaCl: A comparison of techniques. Nanotechnology*, 2009, **20**: 264012.

[52] Kawai, S., et al., *Atomic contact potential variations of Si(111)-7 $\times$ 7 analyzed by Kelvin probe force microscopy. Nanotechnology*, 2010, **21**: 245704.

[53] Sadewasser, S. and M.C. Lux-Steiner, *Correct height measurement in noncontact atomic force microscopy. Phys. Rev. Lett.*, 2003, **91**(26): 266101.

[54] Ziegler, D., et al., *Compensating electrostatic forces by single-scan Kelvin probe force microscopy. Nanotechnology*, 2007, **18**: 225505.

APPENDICES

APPENDIX 1: SPECTRAL METHODS FOR MEASURING THE NORMAL CANTILEVER SPRING CONSTANT k

User measurement of cantilever spring constants, both normal k and torsional κ, is a rich subtopic in AFM. As is extensively reported in the literature, there are various *spectral* calibration methods wherein one examines either the externally or the thermally driven cantilever resonance spectrum. These methods vary in precision, ease of laboratory implementation, and simplicity of data reduction. It is not the purpose of this appendix to comparatively review different strategies or recap their historical progression. Several review-type articles have effectively served this purpose [1–7]. Instead, we describe three methods that have passed the scrutiny of different reviewers and seem by consensus to be quite accurate (within 10%, as with the reference cantilever method of Section 3.7) and relatively free of difficulties or pitfalls.

Some of the following, follows from the method of Section 3.7. Recall that in this method the user collects a force–Z curve on a compliant surface of "known" stiffness—a tipless, single-material, rectangular reference cantilever—and compares the contact slope ($\delta s_Z/\delta Z$) to that on a surface of (nearly) zero compliance, the chip to which the cantilever is attached. The determination of spring constant k of this purely rectangular reference cantilever proceeds via $k = Ewt^3/4L^3$ (Eq. 3.6), given Young's modulus E and cantilever dimensions w, t, and L for width, thickness, and length, respectively. Width and length can be microscopically measured to acceptable accuracy, whereas thickness generally cannot. But one can also

Atomic Force Microscopy: Understanding Basic Modes and Advanced Applications, First Edition.
Greg Haugstad.

develop an expression for k wherein thickness has been eliminated [4],

$$k = \frac{(59.3) \cdot w\left(f_0 L \sqrt{\rho}\right)^3}{\sqrt{E}}, \tag{A1.1}$$

thus requiring only the plan-view dimensions of length L and width w and the material constants of density ρ and Young's modulus E, plus a measurement of cantilever resonance frequency f_0 in air or vacuum. This is how manufacturers compute spring constants (typically claiming three significant digits) for the above reference cantilevers, and thus how the user can check these values.

But the reference cantilever method of Section 3.7 suffers one obvious drawback: the need to push the tip against surfaces other than the sample(s) of interest—reference cantilever and rigid surface such as the cantilever chip—with enough force to establish the force–Z trend (slope) in contact. In some cases, the user may prefer to never touch a solid surface in this way, say if only studying DLVO or steric forces (and if measurements of the deflection sensitivity S are found to be adequately consistent from one cantilever mounting to the next, for the sought precision of force and distance calibration). Of course, one can wait until all measurements of interest have been completed before performing force curves on the reference cantilever surface, depending on successful calibration without damaging the tip or cantilever. But in some cases, one would prefer to obtain k with a *no-touch* method, whether in air or in a liquid medium of interest. This is one of the attractive features of the following spectral methods.

A1.1 Plan-View/Resonance Frequency Method

In this method, one applies the preceding formalism to the *cantilever used for AFM operation*, that is, assuming that the inclusion of the mass of the tip as well as small deviations from purely rectangular geometry (e.g., angled corners) have only a minor effect on the determined value of k [2]. Plan-view cantilever dimensions (i.e., X–Y, viewed from above the sample) are usually specified to acceptable accuracy and precision by the cantilever manufacturer (i.e., within the goal of 5–10% overall calibration accuracy). Otherwise, the cantilever length and width can be determined using a good light microscope, certainly within 5% error and approaching 1% in the case of length. In air or vacuum, the determination of resonance frequency is most easily achieved within a frequency-sweep measurement ("cantilever tune"), as performed to select the driving frequency in dynamic mode. If, however, one is calibrating in a heavy damping medium of operational interest (e.g., water) with a relatively low-mass cantilever, the damped-driven oscillator (DDO) resonance frequency may be significantly different from what is truly sought, the intrinsic resonance frequency, as per Equation 2.18 (i.e., if b/m^* is not $\ll \omega_0$),

$$\omega_0^{\mathrm{DDO}} = \sqrt{\omega_0^2 - \frac{1}{2}\left(\frac{b}{m^*}\right)^2}. \tag{A1.2}$$

In this latter case, it is better to determine the *thermally* driven—that is, by k_BT energy, meaning Brownian motion—resonance frequency. To do this, one examines *the power spectral density* (PSD) of the cantilever's thermal fluctuations in deflection (a rapid succession of measurements versus time) and finds the dominant resonance peak. This is similar to the use of PSD to measure environmental vibrations, as discussed in Section 8.3.3, but examining much higher frequencies and with the tip away from the sample. In a heavy damping medium, these thermally driven motions are typically slow enough (for soft cantilevers with low resonance frequencies, as used for high force sensitivity in liquids) to characterize using standard AFM data acquisition systems (i.e., at sampling rates in the tens of kilosamples/s); indeed, for very soft cantilevers, the thermally driven motion may be slow enough even in air to characterize in this fashion. One can simply capture a deflection signal "image" out of tip–sample engagement (zeroing Z-feedback gains) at the highest achievable scan rate, then perform a 1D PSD along the fast-scan (X) direction using the instrument software or a third-party AFM analysis program, as exemplified in Figure A1.1 for a nominal $k = 0.2$ N/m rectangular silicon cantilever. This averages all the 1D PSDs corresponding to each raw scan line of deflection data. Within such programs, it is typically the dominant spatial wavelength or wave number that the software reads out for a given image size. Multiplying by the scan velocity, one determines resonance frequency in cycles/s:

$$[\text{Spatial wave number (cycles}/\mu\text{m)}] \cdot 2 \cdot [\text{Scan size } (\mu\text{m})] \cdot [\text{Scan frequency } (\text{s}^{-1})].$$

One must also note whether a nonzero overscan parameter (e.g., called "Rounding" in Bruker scanner calibration software) is used. If, for example, the overscan is 10% (Rounding 0.1), then the actual scan size to insert into the above equation is 10% greater than the displayed scan size. Alternatively, one may simply zero the overscan parameter prior to acquiring the image. The latter was done for the image in Figure A1.1a, resulting in the following calculation of resonance frequency:

$$f_0 = [188\,\text{cycles}/\mu\text{m}] \cdot 2 \cdot [0.5\,\mu\text{m}] \cdot [61.0\,\text{s}^{-1}] = 11.4\,\text{kHz}.$$

The plan-view dimensions provided by the manufacturer of this silicon cantilever were $w = 50\,\mu$m and $L = 450\,\mu$m; using a density $\rho = 2.33$ g/cm^3 and Young's modulus $E = 179$ GPa, k is determined by Equation A2.1,

$$k = \frac{(59.3) \cdot (50 \times 10^{-6}\text{m}) \left[(11{,}500\,\text{s}^{-1})(450 \times 10^{-6}\,\text{m}) \sqrt{2330\,\text{kg}/\text{m}^3}\right]^3}{\sqrt{179 \times 10^9\,\text{N}/\text{m}^2}}$$
$$= 0.11\,\text{N}/\text{m}.$$

This example of a user-determined k differing from the manufacturer's nominal value by nearly a factor of two is not uncommon.

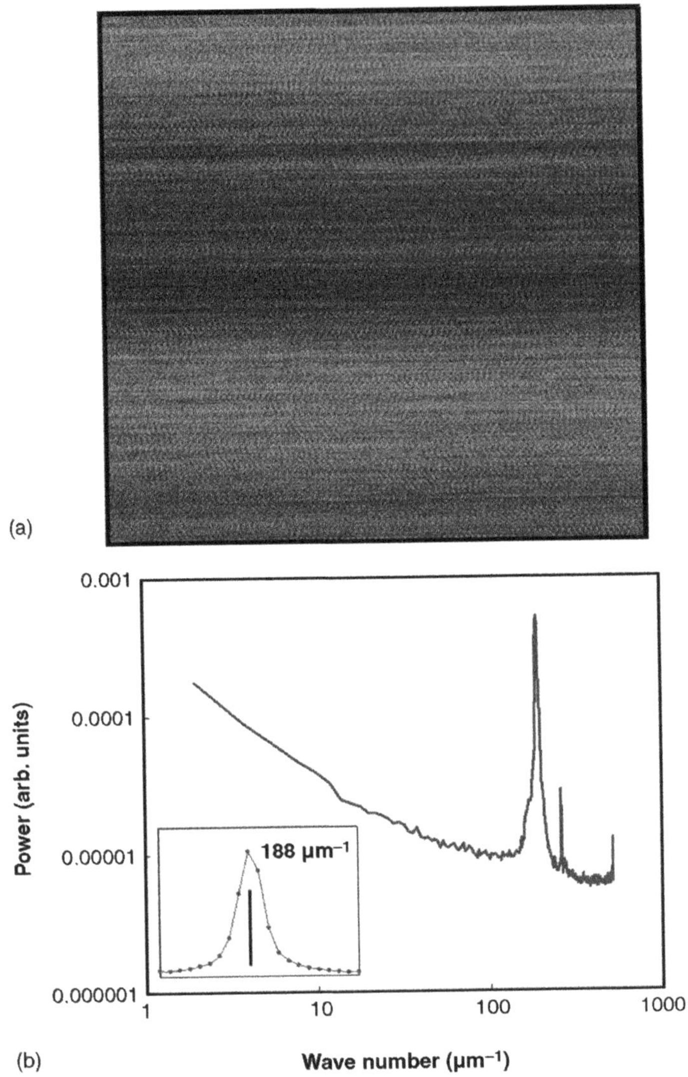

FIGURE A1.1 (a) Cantilever deflection "image" taken out of contact at 61 lines/s, at a scan size of 0.5 μm with no overscan, following first-order plane fitting (for viewing clarity). (b) Resulting *X* PSD (along fast-scan direction) and identification of spatial wave number (cycles/μm).

It is very important to recognize that for many applications in air or vacuum, the cantilever's thermal motion is too fast (resonance frequency too high) to track versus time with a standard AFM data acquisition system. If one nevertheless attempts to measure the power spectral density of cantilever deflection as described above, one often obtains a peak and with a Lorentzian lineshape akin to that in

Figure A1.1b. But in this case, the cantilever motion is being *aliased*, meaning sampled at a rate below the rates of fluctuational motion in such a way as to produce a virtual periodic signal. To understand aliasing, think of each successive measurement moving slightly further in phase along a sine wave but *skipping several periods* between each measurement, thereby producing a *virtual* sine wave of much longer time period (lower frequency) than the actual motion being sampled. In such cases, a faster measurement device is needed, such as a spectrum analyzer or simply a fast data acquisition card (e.g., as part of a LabView system), and of course BNC access to the cantilever deflection signal. Some newest-generation AFM systems, at this writing, are standard-equipped with data acquisition cards extending into the megahertz regime.

The plan-view/resonance frequency method of this section introduces sources of error that are absent, in one way or another, from the methods of the next two sections. These include the values of cantilever density and Young's modulus and/or the measurements of plan-view dimensions.

A1.2 Sader Method

This method is named after the author of several papers proposing and refining procedures whereby cantilever resonance frequency and damping are analyzed to determine spring constant [8,9]. Plan–view-dimensional measurements are required as in the preceding section, but not the density and Young's modulus of the cantilever. What is additionally required is the density and viscosity of the immersion fluid, typically air, and strongly different cantilever dimensions $L \gg w \gg t$ (but in practice $L/w > 3$ has been found to be acceptable [9]). The spring constant expression is

$$k = 7.52 \cdot \rho_f w^2 L Q f_0^2 \Gamma_i(Re) \tag{A1.3}$$

with

$$\text{Re} = \frac{2\pi \rho_f f_0 w^2}{4\eta_f}, \tag{A1.4}$$

where ρ_f and η_f are the mass density and viscosity of the fluid medium, Q is the oscillator quality factor, and Γ_i is the *imaginary component of the hydrodynamic function*, which is in turn a function of the *Reynolds number* Re via Equation A1.4.

The hydrodynamic function is shown graphically in reference [8] and at this writing can be evaluated via Sader's online calculator or downloadable Mathematica files [10].

One might attempt to measure values of f_0 and Q from frequency sweep measurements (sometimes called "cantilever tune"). As discussed in Section A, this is acceptable for f_0 provided that the damping is not too high (the case in air) and the cantilever mass not to low (usually true). But the Q value is dependent on the *shape* of the resonance peak. In Section 5.2.6, extraneous contributions to this shape for an externally driven cantilever were discussed, in terms of the frequency-dependent coupling of the driving signal to the cantilever base. A more reliable way to measure Q with high accuracy is to mathematically fit the peak in the PSD of the cantilever's thermally driven motion with a Lorentzian function plus background:

$$A_{\text{white}} + A_f/f + \frac{A_1 Q}{\sqrt{(f/f_0)^2 + Q^2\left(1 - (f/f_0)^2\right)^2}}, \tag{A1.5}$$

where the constant A_{white} is the white noise (frequency independent), A_f is the $1/f$ noise, and $A_1 Q$ is the peak height. Using this expression to fit the spectrum in Figure A1.1b produces $Q = 35$ and $f_0 = 11.5\,\text{kHz}$. Applicable values of fluid density and viscosity for air were $1.17\,\text{kg/m}^3$ and $1.86 \times 10^{-5}\,\text{kg/m/s}$. Together with the values of w and L from Section A, one obtains $k = 0.094\,\text{N/m}$ with the Sader method, within roughly 10% of the value computed using the method of Section A1.1, and consistently about half of the manufacturer's nominal value of 0.2 N/m.

A1.3 Thermal Method

The name for this method often includes an additional noun, for example, "thermal tune," "thermal k," or "thermal noise." Alternatively, it is sometimes designated by the authors of the original paper, Hutter and Bechhoefer [11], and/or the authors of later papers refining the method (e.g., Butt and Jaschke [12]). Obviously, the measurement of thermal noise is not unique to this method. What is unique is the use of the *amplitude* of thermally driven cantilever motion as the gauge of spring constant. This requires, of course, that the magnitude of thermal fluctuations is significantly greater than the noise floor of the deflection (cantilever inclination) measurement electronics; thus, the method is often not applicable to the stiffest cantilevers such as those often used in dynamic AFM in air. The fluctuations are characterized across a range of frequencies via the power spectral density, but the measurement is dominated by thermal noise in the vicinity of the fundamental resonance frequency. The precise value of this *frequency*, however, does *not* enter the computation.

The key relationship in the thermal method derives from a fundamental principle of physics, the *equipartition theorem* for a single oscillatory degree of freedom, which relates the cantilever's thermal motion to $k_B T$,

$$\frac{1}{2}k\langle s_z^2\rangle = \frac{1}{2}k_B T,$$

or

$$k = \frac{k_B T}{\langle s_z^2\rangle}. \tag{A1.6}$$

The quantity in brackets, the mean squared vertical displacement of the tip measured over a sufficiently long time, is approximately evaluated in the frequency domain by integrating under the power spectral density peak for the fundamental mode (i.e., subtracting the background) such as that in Figure A1.1b. For good accuracy of integration, this requires a high point density in the frequency domain, which is not the case in Figure A1.1b; it would result if sampling the data for longer time intervals. But there are further details that introduce corrective factors in the ultimate expression for k. One factor derives from the fact that not all the cantilever's kinetic energy is contained in the fundamental mode as if an ideal spring. Rather, approximately 3% of the energy is contained in higher eigenmodes as can be determined by solving a fourth-order differential equation within the context of beam theory (see Section 9.2.1). A second consideration is the fact that the actual linear tip displacement s_z in length units is not being measured with the position-sensitive photodetector, rather, the cantilever's *angle of inclination*, not far from the tip end of the cantilever. The relation between this angle of inclination and the vertical tip displacement is not the same for a thermally driven, free cantilever as it is for a statically loaded cantilever (Section 3.2), whether the latter is in sample contact or experiencing some other equilibrium net force such as DLVO. To account for both of these corrective factors (which act in the same direction, to decrease the determined value of k), in the case of a rectangular cantilever, Equation A1.6 is replaced by

$$k = 0.817\frac{k_B T}{\langle s_z^{*2} \rangle}, \tag{A1.7}$$

where the asterisk indicates the *measured* displacement of the freely fluctuating cantilever, as calibrated via the sensitivity S discussed in Section 3.2, which is determined via the force–Z slope in static loading measurements (force curve on rigid surface).

For the case of a V-shaped cantilever, in comparison with a rectangular cantilever of the same length and leg width, it has been numerically shown that the prefactor in Equation A1.7 is replaced by 0.764, a 7% reduction [13]. In general, one anticipates a similar, slightly reduced value for other V-shaped compared to rectangular cantilevers of similar length and leg width.

The above corrective factors can require further modification if the laser spot diameter is considerable relative to the length of the cantilever. In this case, a range of inclination angles may be contributing to the position of the reflected laser beam spot on the photodetector. This issue is only expected to matter for very small cantilevers, such as recently developed at this writing.

APPENDIX 2: DERIVATION OF VAN DER WAALS FORCE–DISTANCE EXPRESSIONS

The problem of deriving the distance dependence of van der Waals force between a tip of axisymmetric geometry and a flat sample (Section 2.1.1) is central to AFM modeling and simulations. First, one integrates over the dipole–dipole interaction between a single-point dipole (inside the tip), at a distance z from the sample surface, and the continuum of point dipoles of number density n_1 throughout the semi-infinite half-space of the sample, as shown in the lower portion of Figure A2.1. Assuming a nonretarded dipole–dipole potential

$$V(r) = -Cr^{-6}, \tag{A2.1}$$

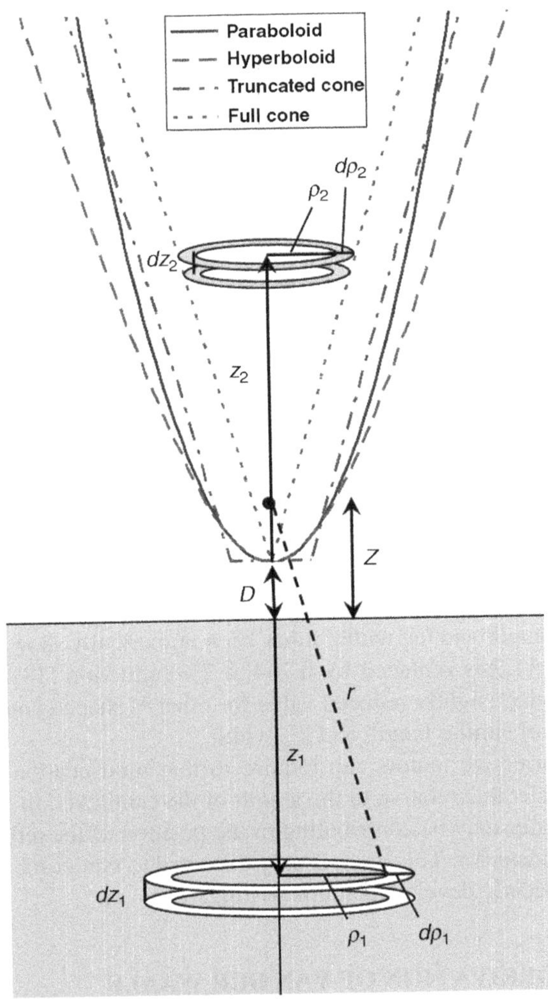

FIGURE A2.1 Illustration of two spatial integration domains in cylindrical coordinates: first, an infinite half-space interacting with a point dipole on the principal axis of the tip, and second, over an axisymmetric tip with bottommost point, a distance D from the surface of the half-space.

where C is a constant and r is distance, one may integrate over the semi-infinite half-space cylindrical coordinates,

$$V(z) = \int_0^\infty dz_1 \int_0^\infty \frac{-C(2\pi\rho_1 n_1)d\rho_1}{\left(\sqrt{\rho_1^2 + (z+z_1)^2}\right)^6} \tag{A2.2}$$

and thus [14]

$$V(z) = -2C\pi n_1 \int_0^\infty \frac{1}{4(z+z_1)^4} dz_1 = -\frac{C\pi n_1}{6z^3}. \tag{A2.3}$$

Second, one may integrate in cylindrical coordinates over all point dipoles of number density n_2 in the tip, extending the z limit to infinity by assuming negligible van der Waals interactions at large distances (and neglecting retardation effects). Given a distance D between the end of the tip (at $z_2 = 0$) and the sample surface, we have

$$V(D) = -\frac{C\pi n_1}{6} \int_0^\infty dz_2 \int_0^{\rho(z_2)} \frac{(2\pi\rho_2)n_2 d\rho_2}{(D+z_2)^3}$$

or

$$V(D) = -\frac{A_H}{6} \int_0^\infty \frac{[\rho(z_2)]^2 dz_2}{(D+z_2)^3}, \tag{A2.4}$$

where the Hamaker constant is $A_H = \pi^2 C n_1 n_2$, and $\rho(z_2)$ describes the functional shape of the tip. Here, we consider the model tip geometries

$$\rho(z_2) = \begin{cases} \sqrt{2Rz_2} \\ a\sqrt{\left(\left(\frac{z_2}{b}+1\right)^2 - 1\right)} \\ \tan(\beta)\cdot(z_2+z_0) \\ \tan(\beta)\cdot z_2 \end{cases} \tag{A2.5 – A2.8}$$

for the cases of paraboloid of revolution, hyperboloid of revolution (offset such that the end of tip is located at $z_2 = 0$), and truncated and full cones, respectively, as depicted in top portion of Figure A.2. (z_0 is the distance of the end of the truncated cone from the mathematical end point of a virtual full cone.) For the purpose of consistent tip aspect ratios in the latter three models, we choose $\tan(\beta) = a/b$, β being the tip half-angle; for consistent radius of curvature in the first two models, we choose $R = a^2/b$.

For any tip geometry, one may evaluate force by taking the derivative with respect to D prior to evaluating the integral, given a smooth, continuous, and finite integrand over the integration interval in Equation A2.4. Thus, we have

$$F = -\frac{dV}{dD} = -\frac{A_H}{2} \int_0^\infty \frac{[\rho(z_2)]^2 dz_2}{(D + z_2)^4}. \quad \text{(A2.9)}$$

The four tip models $\rho(z_2)$ produce integrals of the form

$$I_n(D) = \int_0^\infty \frac{z_2^n dz_2}{(D + z_2)^4} \quad \text{(A2.10)}$$

with $n = 0, 1, 2$. The solutions are [14]

$$I_0(D) = -\frac{1}{3(D + z_2)^3}\Bigg|_{z_2=0}^{\infty} = \frac{1}{3D^3} \quad \text{(A2.11)}$$

$$I_1(D) = \left[-\frac{1}{2(D + z_2)^2} + \frac{D}{3(D + z_2)^3}\right]\Bigg|_{z_2=0}^{\infty} = \frac{1}{6D^2} \quad \text{(A2.12)}$$

$$I_2(D) = \left[-\frac{1}{D + z_2} + \frac{D}{(D + z_2)^2} - \frac{D^2}{3(D + z_2)^3}\right]\Bigg|_{z_2=0}^{\infty} = \frac{1}{3D}. \quad \text{(A2.13)}$$

Thus, for the four tip geometries and using Equations A2.5–A2.13, we obtain the following expressions:

$$\text{Paraboloid of revolution} \quad F(D) = -\frac{A_H R}{6D^2} \quad \text{(A2.14)}$$

$$\text{Hyperboloid of revolution} \quad F(D) = -\frac{A_H R}{6D^2}\left(\frac{D}{b} + 1\right) \quad \text{(A2.15)}$$

$$\text{Truncated cone} \quad F(D) = -\frac{A_H \tan^2\beta}{6D}\left(1 + \frac{z_0}{D} + \frac{z_0^2}{D^2}\right) \quad \text{(A2.16)}$$

$$\text{Full cone} \quad F(D) = -\frac{A_H \tan^2\beta}{6D}. \quad \text{(A2.17)}$$

The magnitude of these four $F(D)$ expressions are plotted on a log–log scale in Figure 2.2, using $A_H = 6.5 \times 10^{-20}$ J, $R = 10$ nm, $b = 140$ nm, $z_0 = 50$ nm, and $\beta = 15°$. We see that the paraboloid or hyperboloid models produce nearly identical distance dependences, whereas the cone and truncated cone produce force–distance trends very different from the other two models.

APPENDIX 3: DERIVATION OF ENERGY DISSIPATION EXPRESSION, RELATIONSHIP TO PHASE

One can derive a mathematical expression relating the energy dissipated per tip–sample interaction in dynamic AFM to the measured value of phase (Section 2.4.4). In dynamic equilibrium, the power imparted by the driving force at the base of cantilever, at any driving frequency, must be equal to the power loss due to the cantilever motion (viscous medium and any intrinsic cantilever anelasticity) plus the power loss due to tip–sample interaction [15,16]:

$$P_{\text{driver}} = P_{\text{cantilever}} + P_{\text{tip-sample}}. \tag{A3.1}$$

In the general case, one must integrate the energy imparted over a full driving cycle, multiplied by the number of driving cycles per second, and then subtract the energy loss due to cantilever damping over its oscillation cycle, multiplied by the number of cantilever oscillations per second, to solve for the tip–sample power dissipation. For the driving expression, one first notes that the instantaneous power delivered is the force on the driver times its velocity [17], given by

$$k[s_z(t) - z_{dr}(t)]\frac{dz_{dr}}{dt}$$

where

$$s_z(t) = A\cos(\omega t + \varphi)$$

and

$$z_{dr}(t) = A_{dr}\cos(\omega t)$$

are the time-dependent vertical positions of tip and cantilever base, respectively, and k is the cantilever spring constant. This assumes harmonic cantilever response, with negligible energy in higher or subharmonics; the tip–sample interaction simply affects tip amplitude A as well as phase shift φ between drive and response. Integrating this power over a time of one driving period, and multiplying by the driving frequency $f = \omega/2\pi$, determines P_{driver},

$$P_{\text{driver}} = \frac{\omega}{2\pi}\int_0^{2\pi/\omega} k[s_z(t) - z_{dr}(t)]\frac{dz_{dr}}{dt}dt = \tfrac{1}{2}kA_{dr}A\omega\sin\varphi. \tag{A3.2}$$

For the cantilever damping expression, one integrates the velocity-dependent resistive force

$$-b\frac{ds_z}{dt} = -\left(\frac{k}{Q\omega_0}\right)\frac{ds_z}{dt},$$

where it is assumed that the damping velocity is approximately equal to the tip velocity. Here, the point is to express the viscous damping coefficient b in terms of experimentally measurable quantities: the cantilever spring constant k, the free resonance frequency ω_0, and quality factor Q of the *freely oscillating* cantilever at a distance just barely beyond tip–sample interaction (see Section 5.2.3). This does *not* imply that the effective quality factor of the cantilever under dynamic tip–sample interaction remains that of the free oscillator (certainly not the case) or that the operational drive frequency must be ω_0. Integrating this damping over one cycle period determines $P_{\text{cantilever}}$,

$$P_{\text{cantilever}} = -\frac{\omega}{2\pi}\int_0^{2\pi/\omega}\left(\frac{k}{Q\omega_0}\right)\left(\frac{ds_z}{dt}\right)^2 dt = \frac{1}{2}\frac{k}{Q\omega_0}A^2\omega^2. \tag{A3.3}$$

Taking the difference of the expressions for P_{driver} (Eq. A3.2) and $P_{\text{cantilever}}$ (Eq. A3.3) and using the relationship $A_0 = QA_{dr}$ yields

$$P_{\text{tip-sample}} = \frac{1}{2}\frac{kA^2\omega}{Q}\left(\frac{A_0\sin\varphi}{A} - \frac{\omega}{\omega_0}\right).$$

The energy dissipation per cycle is the power times a time period of $2\pi/\omega$ or

$$E_{\text{tip-sample}} = \frac{\pi kA^2}{Q}\left(\frac{A_0\sin\varphi}{A} - \frac{\omega}{\omega_0}\right).$$

Typically, $\omega \approx \omega_0$ to a very good approximation, resulting in the following useful expressions, solving in terms of the energy dissipation per cycle normalized to the total oscillator energy during imaging,

$$\frac{E_{\text{tip-sample}}}{1/2kA^2} = \frac{2\pi}{Q}\left(\frac{\sin\varphi}{A/A_0} - 1\right), \tag{A3.4}$$

the right side importantly being *independent of cantilever spring constant* and dependent on the *amplitude reduction* A/A_0 rather than the magnitude of A or A_0. A disadvantage of this expression is that it diverges as A/A_0 goes to zero. A similar expression, which avoids such a divergence, normalizes the energy dissipation per cycle to the total oscillator energy when free of the surface,

$$\frac{E_{\text{tip-sample}}}{1/2kA_0^2} = \frac{2\pi}{Q}\left(\sin\varphi - \frac{A}{A_0}\right). \tag{A3.5}$$

One can also solve for the phase angle,

$$\varphi = \sin^{-1}\left(\frac{QE_{\text{tip-sample}}}{\pi k A A_0} + \frac{A}{A_0}\right), \tag{A3.6}$$

with the inverse sine function providing *two possible values of phase* for a given value of energy dissipation (i.e., within each of the attractive and repulsive regimes).

APPENDIX 4: RELATIONSHIPS IN MENISCUS GEOMETRY, CIRCULAR APPROXIMATION

Figure A.4.1 describes the relationships between geometric parameters needed to describe a capillary meniscus in solid sphere–flat geometry (Section 6.3.3), within

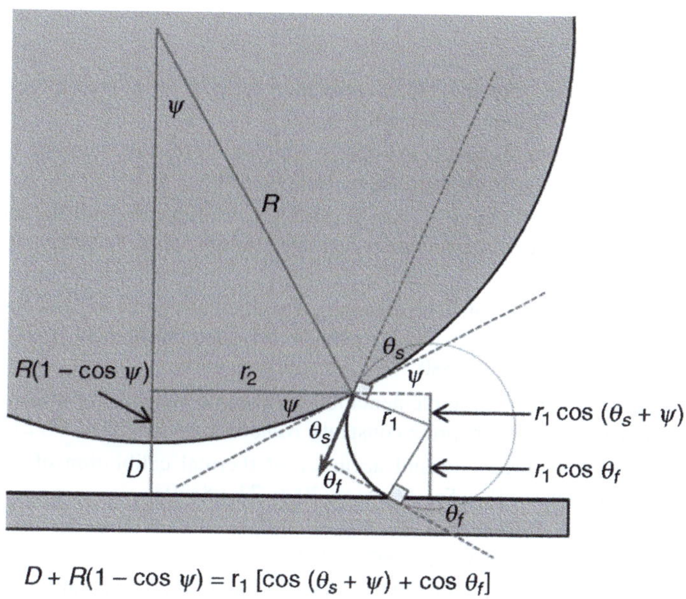

FIGURE A4.1 Cross-sectional geometry of capillary meniscus between a sphere and a flat surface in the model known as the *circular approximation.* The curvature of the liquid–vapor interface is described by two radii: r_1 is that of the shown circular arc, in a plane perpendicular to the flat surface, and r_2 is that of a circular cross section parallel to the flat surface (extending in and out of the diagram). Thus, the 3D meniscus geometry is a toroid. Geometric relationships allow one to express r_1 in terms of the *filling angle* ψ. A surface tension vector is shown at the contact circle (of radius r_2) between the meniscus and the sphere. This and the Laplace pressure over the horizontal meniscus cross section act to exert an attractive force between the sphere and the flat surface (Section 6.3).

the *circular* (toroidal) *approximation* of meniscus shape (referring to the circular-arc liquid–vapor interface shown in the plane of the diagram). These include three angles: the *filling angle* ψ and the *contact angles* between the meniscus and the solid surfaces of sphere, θ_s, and flat, θ_f; three *radii of curvature*, the sphere radius R and the radii describing the circular meniscus cross sections perpendicular to the flat solid surface, r_1, and parallel to the flat solid surface, r_2 (cutting through the diagram), and the distance D between the lowest point of the solid sphere and the solid flat surface. The principal sought relationship is obtained by equating a vertical length, the sum of distance D and spherical cap height $R(1 - \cos\psi)$, to the height of the meniscus $r_1\cos(\theta_s + \psi) + r_1 \cos(\theta_f)$. (The latter requires identification of complementary angles.) A second needed expression is the vertical component of the surface tension (per unit length) around the contact circle between meniscus and solid sphere, which is the fraction $\sin(\theta_s + \psi)$ of the total force (per unit length), the latter directed as shown along the tangent of the liquid–vapor interface at the three-phase line (i.e., the contact circle), the point of contact with solid sphere.

REFERENCES

[1] Burnham, N.A., et al., Comparison of calibration methods for atomic-force microscopy cantilevers. *Nanotechnology*, 2003, **14**: 1–6.

[2] Gibson, C.T., D.A. Smith, and C.J. Roberts, Calibration of silicon atomic force microscope cantilevers. *Nanotechnology*, 2005, **16**: 234–238.

[3] Butt, H.-J., B. Cappella, and M. Kappl, Force measurements with the atomic force microscope: Technique, interpretation and applications. *Surf. Sci. Rep.*, 2005, **59**: 1–152.

[4] Clifford, C.A. and M.P. Seah, The determination of atomic force microscope cantilever spring constants via dimensional methods for nanomechanical analysis. *Nanotechnology*, 2005, **16**: 1666–1680.

[5] Cook, S.M., et al., Practical implementation of dynamic methods for measuring atomic force microscope cantilever spring constants. *Nanotechnology*, 2006, **17**: 2135–2145.

[6] Matei, G.A., et al., Precision and accuracy of thermal calibration of atomic force microscopy cantilevers. *Rev. Sci. Instrum.*, 2006, **77**: 083703.

[7] Ohler, B., *Practical Advice on the Determination of Cantilever Spring Constants. Bruker Application Note #AN94*. 2007, p. 1–12.

[8] Sader, J.E., J.W.M. Chon, and P. Mulvaney, Calibration of rectangular atomic force microscope cantilevers. *Rev. Sci. Instrum.*, 1999, **70** (10): 3967–3969.

[9] Chon, J.W.M. and J.E. Sader, Experimental validation of theoretical models for the frequency response of atomic force microscope cantilever beams immersed in fluids. *J. Appl. Phys.*, 2000, **87** (8): 3978–3988.

[10] Sader Research Group, *Atomic Force Microscope Cantilevers*, www.ampc.ms.unimelb.edu.au/afm/index.html.

[11] Hutter, J.L. and J. Bechhoefer, Calibration of atomic-force microscope tips. *Rev. Sci. Instrum.*, 1993, **64**: 1868–1873.

[12] Butt, H.-J. and M. Jaschke, Calculation of thermal noise in atomic force microscopy. *Nanotechnology*, 1995, **6**: 1–7.

[13] Stark, R.W., T. Drobek, and W.M. Heckl, Thermomechanical noise of a free v-shaped cantilever for atomic-force microscopy. *Ultramicroscopy*, 2001, **86**: 207–215.

[14] Beyer, W.H., *Standard Mathematical Tables*. 26th ed., 1981, Boca Raton, FL: CRC Press.

[15] Cleveland, J.P., et al., Energy dissipation in tapping-mode atomic force microscopy. *Appl. Phys. Lett.*, 1998, **72**: 2613–2615.

[16] Garcia, R., et al., Phase contrast in tapping-mode scanning force microscopy. *Appl. Phys. A*, 1998, **66**: S309–S312.

[17] French, A.P., *Vibrations and Waves*. 1971, New York: W. W. Norton, 316.

INDEX

Abrasive scanning, 13–16, 21–23, 158–160, 299–302, 304, 307, 386, 387
AC electrical bias, 405, 406, 430, 431
Acoustic force microscopy, *see* Contact resonance
Activation, 7, 25, 85–87, 116, 259, 261, 352–362, 420. *See also* Kinetics
Adhesion
  force, *see* Pull-off force
  hysteresis, *see* Hysteresis, adhesion
AgBr (silver bromide), 2, 152, 153, 342, 343, 386, 388
Aging (polymer), 7, 16, 114
Alexander - de Gennes theory (steric forces), 76
Algorithms
  data acquisition, 297, 298
  data processing, 22, 156, 254, 298, 299, 311, 394, 395
  modeling, 37
  scanning, 21, 144, 147, 163, 171, 179, 180, 220
Aliased signal, *see* Signal, aliasing
Aluminum (sample), 2, 29, 129, 130, 149, 155, 164, 165, 174, 288, 289, 341, 407, 408, 427, 428
Amontons' law, 240, 333–336, 346, 349, 350
Amorphous (materials), 6, 9, 13, 40, 41, 194, 216, 242, 252, 347, 348, 357–359, 396
Amplitude
  free oscillation, 145, 160, 167, 182, 190, 191, 194–201, 208–214, 219–225, 227, 228, 234–236, 240, 243–247, 307–309, 415, 422, 423, 425
  modulation (AM), *see* Modulation, amplitude
  ratio, *see* Setpoint, amplitude ratio
  reduction, *see* Setpoint, amplitude ratio
  second mode, *see* Eigenmode(s)
  setpoint, *see* Setpoint, amplitude. *See also* Deviation from setpoint
Anelasticity, 57, 447
Angle, contact, *see* Contact angle
Angle of inclination, *see* Inclination

Atomic Force Microscopy: Understanding Basic Modes and Advanced Applications, First Edition.
Greg Haugstad.

Anisotropy, 35, 332. *See also* Friction anisotropy and Modulus, shear
Annealing, *see* Polymer, solvent-vapor annealing
Arctangent function (phase versus frequency), 57, 230–232
Arrhenius, *see* Kinetics; Activation
Artifacts
  optical interferometry, *see* Interferometry, artifact
  piezoscanner, *see* Nonlinearity, scanner; Hysteresis, scanner; Creep; Cross coupling
Atomic motion, *see* Molecular or atomic motions
Autophobic dewetting (of polymer film), 369, 370
Axisymmetric (geometry), 76, 77, 278, 282, 443, 444

Baseline of force, *see* Zero, of vertical force scale
Beam mechanics (theory), 414–419
Behenyltrimethyl ammonium chloride (BTAC), 336
Bell–Evans model (rate-dependent adhesion), 87
Bins (binary increments), 3, 153, 192. *See also* Histogram
Biofilm(s), 14, 15, 304
Biomolecule(s), 1, 6, 21, 24, 33, 72
Bistability, 63, 64, 141, 142, 196–200, 242, 423
Blend, *see* Polymer blend
Block copolymer, 19, 20, 27, 159, 172, 173, 216, 235, 311, 322, 323, 386, 387, 397, 398
Blunting, tip, 106, 132, 181, 194, 260, 291, 292, 295
Bocek (degrees in Extender electronics), 222
Boundary conditions (mathematical, of cantilever), 417, 425
Bridging
  capillary meniscus, 54, 103, 108, 260, 280–289
  macromolecular, *see* Extension, macromolecular
Brittle (crystalline) polymer, 13
Calibration
  deflection, *see* Sensitivity, in calibration
  lateral force, 343–346
  normal spring constant, 133–135, 437–443
  phase (Extender), 221–230
Cantilever
  coating, 110, 203, 204
  rectangular, 101–104, 129–131, 133–135, 415–422, 425, 426, 437–439, 443
  V-shaped, 110, 111, 415
Cantilever ring-down, 315–319, 415
Cantilever tune, 438, 441. *See also* Driven oscillator
Capacitance, 43, 146, 195, 202, 261, 405, 408–410, 430
Capillary,
  condensation, 275–278, 283
  necking, 281, 313. *See also* Bridging, capillary
  pick-up (of nanoparticles), 293, 294
  polymer melt, *see* Polymer melt
  transport, 15. *See also* Lithography
Carpick–Ogletree–Salmeron approximation, 273. *See also* Piétrement–Troyon approximation
Cationic surfactant, 19, 313, 336
Cetyltrimethyl ammonium chloride (CTAC), 313, 314, 318, 319
Chain molecules
  long (macromolecular), 14, 16, 19–23, 72–76, 78–83, 85, 125, 126, 287, 303, 353–362
  short, 19, 72, 82, 275, 332, 364
Chain movement, whole, 74, 75, 362
Characteristic equation, 417, 428. *See also* Eigenmodes
Chip (cantilever-attached), 5, 17, 18, 62, 94, 133, 134, 144–146, 184, 185, 202, 209, 210, 231–233, 437, 438
Circular approximation (capillary), 282, 283, 449
Closed-loop (scanning)111, 114, 167–173, 180, 338–340, 394
Coefficient (calibration)
  deflection, 131. *See also* Sensitivity
  scanner, 114, 144, 169, 180

Coefficient (physical)
of damping, 56, 67, 129, 146, 167, 202, 210–214, 219, 231, 234, 235, 317, 415, 416, 425, 426, 438–441, 447, 448
of dissipation in dynamic mode, 242
offriction, 160, 240, 242, 343, 361, 363, 371, 372. *See also* Amontons' law
Cohesion, 14, 15, 304–311
Co-ion, 68. *See also* Counter-ion
Collagen, *see* Gelatin
Colloid probe, 76–78, 87, 332
Complex dielectric function, *see* Dielectric function, complex
Complex modulus, *see* Modulus, complex
Compliance, 9, 44–51, 95, 106, 108, 137, 143, 160, 182, 198, 240, 263, 269, 276, 319, 366, 437
Computer simulation, *see* Simulation
Condensation, 275–278, 283, 292
Conical tip (model), 48, 283, 289, 429. *See also* Truncated conical tip
Confinement, 26, 27, 76, 87, 277, 292. *See also* Squeeze film damping
Contact angle (liquid–solid), 216, 217, 253, 274, 281, 283–285, 450
Contact (tip-sample)
conformal, 263, 275, 276, 347
potential (difference, CPD), 402, 406–414, 430
size, 47, 264–267, 271–273, 275, 342, 346, 368, 375, 429
Contamination
sample, 26, 29, 54, 103, 149, 157, 159, 164, 165, 330, 339, 340, 406–408, 427, 428
tip, 45, 54, 103, 109, 181, 219–221, 260, 276, 313
Continuum (material modeling), 45, 70, 281–284, 444
Contrast mechanisms, 12, 188, 411, 412, 424
Controller, 137–139, 143–147, 162–166, 350, 368, 369. *See also* Algorithm, scanning Convolution, phase imaging, 18, 21, 296
Copper, 345, 382, 389–391
Cooperative motion (chain molecules), 25, 354–362
Counter-ion, 68, 131. *See also* Co-ion
Creep, 114, 143, 167–172, 178, 180, 184, 349, 350, 386
Cross coupling, 114, 143, 177, 178, 382
Crosslinking (polymers), 76, 123. *See also* Network, of macromolecules
Crystalline organic thin film, 13, 14, 269, 332, 338, 353, 364–367, 412–414
Crystalline polymer, 14, 15, 189–194, 216, 240, 242, 252, 307–311, 347–349, 357, 358, 396, 397,
Crystallographic orientations (in images), 152, 172, 273, 364–366, 386. *See also* Anisotropy

Damping, *see* Coefficient, damping
Dashpot (viscous model element), 56, 84, 85, 425
Data acquisition (system, rate), 60, 133, 146, 148, 156, 203, 219, 229, 258, 261, 296–299, 315, 368–370, 391–393, 439–441
Data cube, 21, 220, 221, 258, 296–298, 302, 309, 379
Debye forces, 38
Debye screening length, 68, 131, 412
Deflection (cantilever)
basic concepts, 93–96
setpoint, *see* Setpoint, deflection. *See also* Deviation from setpoint
time-averaged (in dynamic mode), 220, 304, 309, 310
Degrees of freedom
molecular, 7, 242, 259, 302, 338, 340, 347
stepper motor, 113
Depth location (of objects), *see* Sensitivity, to depth location
Derjaguin approximation (relating force to interfacial energy), 76, 77
Deviation from setpoint, 140–143, 150, 156, 157, 162, 163, 165, 167, 214–216, 248–254, 315, 344
Deviation, standard, *see* Standard deviation
Dewetting, *see* Autophobic dewetting
Dexamethasone (drug), 320, 321, 384, 395, 425
Diamond-like carbon, 332, 350

Dielectric
constant, 41, 78, 195, 361, 401
function (complex), 40
spectroscopy, *see* Spectroscopy, dielectric
Differential equation
of electrostatic potential (in semiconductor), 411
of end-loaded cantilever (deflection), 102
of wavemotion (cantilever oscillation), 57, 166, 416, 443
Digitization (force or signal), 132, 133, 148, 149
Dilation (image), 181
Dipole interactions, 7, 23, 24, 28, 29, 36–44, 67, 68, 77–79, 108, 109, 196, 274, 354, 401, 402, 412, 443–445
Disengagement, *see* Engage(ment) and disengagement
Disjoining (pressure), 279, 280, 287, 289
Dislocations (defects in crystalline organic thin films), 12
Dispersion forces, 38, 40, 62, 274
Dispersion relation (cantilever eigenfrequencies), 417
Dissipation (energy)
in dynamic mode (phase), 234–255, 447–449
in contact mode, *see* Friction, loop
in force curves and force-curve mapping, *see* Hysteresis
Distribution function, *see* Gaussian; Lorentzian
DLVO forces, 50, 69, 70, 72, 83, 125, 131–133, 438, 443
DMT contact mechanics model, 50, 263–269, 271–273, 275, 276, 351, 352, 368, 428
DNA (biophysical AFM measurements), 80, 87
Double layer (electric), 68, 69
Drift, 113, 114, 137, 143, 149, 166, 168, 182–184, 202, 207–210, 213, 214, 231, 254, 307, 309, 332, 394
Drive
amplitude, 188, 207–209, 219, 231–233, 238, 247, 371–374
frequency, 60, 141, 160, 182, 188, 191, 200–203, 207, 208, 213, 214, 231–233, 236, 402, 403, 427, 431, 448
Driven oscillator, 56–60, 64, 146, 182, 209, 438
Driving power, 234, 447
Drug–polymer mixtures (morphology, segregation, elution), 2, 3, 320–323, 384, 395, 424, 425
Dynamic stiffening, 87, 159

Eigenfrequencies, 416–421, 430, 432
Eigenfunction(s) (and their linear combination), 415–417
Eigenmode(s), 231, 416–433, 443
Elastic modulus, *see* Modulus
Elastic recovery (following indentation), 118, 119, 121
Electron microscopy, 1, 150, 276
Electron volts (eV, in dissipation), 66, 67, 85, 242, 245, 363
Electronic absorption, *see* Lifshitz theory. *See also* Spectroscopy, quantum
Electronic states, 29, 363, 401, 412
Electronic transport (rate and temperature dependence), 24
Energy dissipation, *see* Dissipation
Energy landscape (stress modification), 86
Engage(ment) and disengagement, 6, 18, 146–148, 167, 185, 200, 202, 203, 211, 213, 219, 222, 227, 234, 235, 332, 404, 408, 432, 439
Ensemble (of entities or measurements), 7, 75, 86, 386, 397
Entanglement (molecular), 78, 82, 362, 369. *See also* Network, of macromolecules
Enthalpy, 72, 76, 81
Entropy, 72–76, 81, 281
Envelope
of cantilever oscillation, 418, 431
of zero dissipation in phase versus amplitude, 222, 227, 238
Epitaxial (relationship of film layers), 412–414
Equipartition theorem, 442
Error signal, 144–146, 150–152, 163–167, 171, 194, 195, 214, 215, 219, 251–253, 345, 361
Exponential
damping in time, 166, 315–317, 415
distance dependence, 34, 37, 68–72, 76, 124, 125, 131, 132

relationship of rate and inverse temperature, 354
stretched, 354
Extender electronics (Bruker), 191, 221–232, 234, 235, 240, 243
Extension (between tip and sample)
of capillary meniscus, *see* Bridging, capillary meniscus
macromolecular, 73, 78–83, 87, 116, 126–129, 294–297, 303
Extensive quantity, 62, 76, 248, 259, 260

Fermi level, 29, 401, 402, 412
Fibrous image, 295, 387, 388
Filling angle (of capillary about tip), 284, 449, 450
Finite element methods, 261, 273, 412, 417
First generation AFMs, 110, 113, 114
Flattening (in post-processing), 156, 207, 238–240, 302, 306, 380, 381, 396, 397, 425
Flexural (cantilever resonance), 17, 400, 430, 431
Fluctuations
addition in quadrature, 127
cantilever thermal motion, 79, 85, 362, 439, 442
electromagnetic (in materials), 40
inlaser intensity, 95, 143
inscanner position signal, 116
Fluid cell, 184, 232. *See also* Immersion
Fly fishing (via molecular tethering), 24
Fly height, 54, 138, 140, 189, 193, 195, 196, 219
Force balance, 102, 175
Force instability, *see* Jumps
Force curve, fast, 159, 161, 183, 279, 290, 322, 324
Force gradient, *see* Gradient, force
Force modulation, *see* Modulation, vertical force
Force spectroscopy, 21, 78, 81
Force volume (Bruker force curve mapping routine), 21, 139, 220, 297–300
Fourier (analysis, filtering), 2, 360, 361, 371, 372, 382, 383, 386–391, 418
Free energy (Gibbs), 259, 275, 281
Free oscillation (of tip/cantilever), *see* Amplitude, free oscillation
Freely jointed chain (model), 73–75, 79–81, 127
Freeware, *see* Gwyddion
Frequency modulation (FM), *see* Modulation, frequency
Friction
anisotropy, 365
kinetic, 135, 332, 333, 369
loop, 9, 10, 332–334, 341, 343–347, 350, 351, 366
static, 45, 100, 159, 260, 330, 332, 333, 366, 369, 372

Gain (of feedback system)
integral, 163–167
look-ahead, 163–165
proportional, 163–165
Gaussian (function, distribution), 104, 105, 384, 396–398
Gel, 83, 123–125, 268, 294
Gelatin, 14–16, 25, 81–83, 108, 109, 123–125, 127, 128, 294–296, 356, 357, 386–388
Geometric factor (of tip), 42, 43, 131
Geometry (experimental), 143, 182, 184–186
Glass (silica), 21–23, 72, 95, 98, 99, 131, 132, 253, 268, 298, 299–303, 391, 392
Glassy polymer, 9, 17, 20, 26, 28, 46, 48, 85, 100, 118, 123, 183, 235, 266–270, 304, 319–322, 360–362, 375, 384, 422, 425
Glass–rubber transition, 26, 27, 330, 353, 358–361, 375, 424
Gold, 2, 4, 29, 110, 155, 171–174, 194, 195, 228, 229, 245, 246, 341
Gradient (with tip-sample distance)
of capacitance, 43, 430
of energy, 37, 39
of force, 28, 35, 50–54, 67, 68, 71, 72, 107–109, 139–142, 196, 221, 313, 368, 400, 402–404, 407, 408, 431–433
Grain size analysis, 389–391, 394–396
Granular (polycrystalline) images, 12, 149, 152, 153, 158, 195, 295, 296, 320, 321, 342, 364–367, 372, 373, 382, 384, 390, 392, 395, 407, 412–414
Gravitational force, 35, 36, 280

Gwyddion (post-processing freeware), 177, 381, 384, 395

Hair conditioner film (images), 313, 317
Half-space (model sample), 38, 39, 42, 47–49, 68, 76–78, 98, 107, 125, 265, 342, 351, 444, 445
Hamaker constant, 24, 39–44, 53, 54, 61, 78, 107, 242, 263, 275, 279, 286, 287, 293, 445
Harmonic oscillator, 51–53. *See also* Driven oscillator
Harmonics, 371, 372, 375, 417, 430
Helical (triple), 15. *See also* Gelatin
Helmholtz layer, 68
Herring-bone molecular packing motif (anisotropic forces), 365–367
Hertzian contact (model), *see* Contact mechanics
Heterogeneous surface, 34, 92, 137, 160, 187–194, 198, 253, 283, 338, 357, 396, 401, 402, 407
Histogram
  of AFM measurements, 3, 4, 86, 143, 152–154, 191, 192, 239–242, 249–251, 308–314, 319, 320, 337, 338, 348, 349, 395–397
  of chain-molecule end-to-end distances, 74
Hookean force (Hooke's law), 35, 45, 53, 81, 84, 103, 353
Humidity (experimental), 7, 14, 15, 24, 25, 27, 28, 33, 103, 104, 110, 187, 200, 210, 216, 253, 260, 273, 279, 280, 283–295, 353, 354, 357–359, 425
Hummer–Szabo model (rate dependent adhesion), 87
Hydration, 72, 83
Hydrogen bonding, 15, 24, 71, 72, 87, 108, 260, 274, 295, 356, 357
Hydrophilic, 24, 28, 72, 216, 277, 282, 287, 299, 353, 356
Hydrophobic, 23, 24, 28, 72, 216, 253, 353, 356
Hyperboloid tip (model), 43
Hysteresis
  adhesion, 23, 27, 62, 106–109, 141, 142, 188, 237, 242, 247, 259, 275, 301–304, 310, 319, 350, 362, 425
  amplitude versus Z, 198, 199, 310, 311
  contact (plastic deformation), 119–121, 304–306
  lateral force, 9, 366, 373
  scanner, 114–117, 119–121, 126, 127, 143, 168, 173, 174, 199, 216, 298, 333, 338, 340, 341, 380
  viscoelastic, 297

Immersion (in fluid medium), 21, 40, 41, 94–96, 123, 127, 166, 187, 222, 233, 274, 299, 322, 384, 415, 441
Inclination (angle of cantilever), 93, 94, 101–104, 110, 131, 135, 144, 231, 342, 420, 442, 443
Indentation, 7, 8, 45–49, 62, 85, 87, 98–101, 106, 113, 118–126, 160–162, 182, 183, 191, 235, 240, 246, 262–273, 296, 297, 303–306, 316, 362, 425
Index of refraction, 35, 40, 41, 94–96, 110
Inertia, *see* Moment, of inertia
In-phase (response), 324, 373, 374, 432
Integral gain, *see* Gain, integral
Integral multiples (of frequency), 70, 73, 380, 417, 430
Intensive quantity, 76, 248, 259, 260
Interaction stiffness, 64, 202
Interfacial
  bond, 85, 86, 338
  energy, 76, 77, 273, 274, 281, 282
  force, 67, 72, 131
Interferometry
  artifact, 110, 111, 126, 306, 332, 340, 342
  asa sensor, 34, 171
Interleave, 28–30, 189, 261, 405–408, 412, 429, 431
Intermittent (contact or interaction), 17, 18, 28, 51, 63, 139, 141, 159–162
Inverse optical lever sensitivity, *see* Sensitivity
Inverse tangent, *see* Arctangent function
Ionic (forces, screening), 33, 38, 68, 69, 87, 132,
Irreversible processes, 159, 188, 242, 259, 319, 352, 362. *See also* Hysteresis

JKR contact mechanics model, 121, 263–273, 315, 319, 368
Jumps (as a function of *Z*)

between dynamic interaction regimes, *see* Bistability
into contact, 8, 72, 103, 104, 106–109, 141, 142, 157, 183, 259, 287, 313–318
as decreased adhesion, 8, 83, 105–109, 141, 142, 259, 270, 284, 287, 289, 304, 313–318

Keesom forces, 38
Kelvin equation or radius, 282–284
Kelvin probe (force microscopy, KFM), 29, 68, 368, 401, 405
Kinetic energy, 7, 66, 212, 230, 234, 247, 347, 419, 443
Kinetics, 4, 116, 277, 280, 352–362. *See also* Activation
Kuhn length, 79, 80. *See also* Persistence length
Kurtosis, 383–385

Laminate interface (revealing via wear), 361, 362
Langevin function (in statistical physics), 79
Laplace pressure, 278, 281–284, 449
Laser, 5, 6, 35, 51, 93–96, 101, 103–106, 110–113, 123, 126, 127, 129, 133, 144, 152, 153, 163, 166, 185, 214, 216, 231, 332, 340–342, 380, 419, 420, 443
Laser mode competition, 111
Lennard-Jones interaction, 37, 107, 262, 271
Lifshitz theory, 40. *See also* Hamaker constant
Lift height, 407
Ligand-receptor, *see* Receptor
Linear combination (of eigenfunctions), *see* Eigenfunctions
Linear response (of material), 45, 84, 240, 242, 262, 276, 344–350, 355, 361, 401, 423, 430
Lineshape (mathematical), 202, 227, 233, 234, 396, 397, 440
Liquidy
contaminants, 54, 109, 340
sample domains, 19, 38, 219–221, 313, 317
Liquid, imaging in, *see* Immersion
Lithographic (structures imaged), 2, 29, 129, 155, 164, 171, 341, 408, 411
Lithography (via tip), 277, 293
Load dependence
of adhesion (memory), 319
of friction, 346–352
Lock-in amplifier, 146, 221, 225, 228–230, 324, 368, 369, 401, 414, 419, 421, 430, 432
London forces, 38
Long-chain (macro-) molecule, 19, 20, 23, 72, 73, 81, 82
Lorentzian (function), 57, 58, 104, 105, 183, 440–442
Loss modulus and tangent, 248, 353
Lossymaterial, 198, 237
Lubrication, 277, 292, 293

MAC mode (Agilent), *see* Magnetic excitation
Magnetic excitation (of cantilever), 71, 72, 129, 145, 233, 405
Magnetic force (microscopy), 29, 30, 108, 400–403
Magnetization, 187, 400
Many-body interaction (in dielectric media), *see* Lifshitz theory
Mass, equivalent point-, 35, 51–57, 60, 415
Master curve (of viscoelastic response), *see* Time–temperature superposition
Maugis–Dugdale model, 261, 268, 270–273, 429
Maxwell element (in viscoelasticity), 84, 85
Mean (statistical), 383, 384
Melt, *see* Polymer melt
Meniscus, *see* Capillary condensation
Meridional curvature (of meniscus), 281, 282
Mica, 19, 41, 71, 72, 189, 100, 124, 125, 277, 287–291, 294, 307, 313, 317, 336–338, 343, 347, 357, 369, 386, 387, 394, 396
Mode (statistical), 249
Modulation
amplitude, 140, 368, 406
frequency, 139, 368, 404, 406, 407, 431
shear, 85, 330, 366–375
vertical force, 323, 324, 368

Modulus
complex, 353
loss, 248, 353
shear (anisotropic), 364–366, 372, 373
storage, 62, 248, 353
Young's, 45, 84, 98, 100, 102, 103, 133, 262, 266, 270, 342, 346, 416, 425, 428, 437–439, 441
Molecular bridging, *see* Bridging, chain molecules
Molecular dynamics (simulations), 277, 280, 338
Molecular or atomic motions, 7, 12, 24, 25, 85–87, 259, 338, 347, 357, 362
Moment
bending (cantilever), 102, 416, 417, 428
dipole, *See* Dipole interactions
of inertia, 102, 336, 416
statistical, 383
Multiasperity, 263, 276
Multifrequency methods, 189, 405, 414, 419–433
Multimode (Bruker), 196, 220

Nanofibril, 109, 294
Nanoindentor, 45, 125, 262. *See also* Indentation
Nanolithography, *see* Lithography via tip
Nanoparticle(s), 19, 181, 200, 201, 293, 294
Nanoscope (Bruker), 196, 204, 207, 221, 234
Network
hydrogen bonded, 71, 72
of macromolecules, 15, 123, 387. *See also* Crosslinking
Newton's second law, 57, 166, 416
Newtonian rheological response, 240. *See also* Rheology
Nodoid (capillary geometry), 280, 282
Noise limit or floor, 97, 133, 149, 442
Nonlinearity
of force measurement apparatus, 104–106, 108, 110, 119–121, 332
of material response, *See* Linear response of material
of scanner, 114–119, 126, 127, 143, 156, 167, 168, 171, 173–180, 184, 380, 382
Nonlinear relationship
infriction-load, 346, 347, 350–352, 356, 371–375. *See also* Amontons' law
in phase-dissipation, 249. *See also* Dissipation, in dynamic mode
Nonlinear mathematical treatment
in calibration of phase measurement, 227–229, 235, 238, 240
inpost processing, 306
Nonmonotonic relationship, 24, 56, 140, 208, 287, 291, 292
Non-Newtonian, *see* Newtonian rheological response
Nonspecific interactions, 24
Nulling
of Kelvin-probe feedback signal, 29, 401, 402, 404–406, 408, 430–432
of topographic error signal, 142
Nyquist theorem (in power spectral density analysis), 391, 418

Oliver-Pharr method (nanoindentation), 118, 119, 123, 266
Open loop (scanning), 167, 171–173, 180, 341
Operative spring constant, 93, 134, 135
O-ring (couplingof fluid cell to sample), 184
Oscilloscope readings, 225–228, 232, 414, 415
Out-of-phase (response), 324, 374, 432
Overtones (cantilever), 415, 416. *See also* Eigenmodes

Paraboloid tip (model), 38–39, 41, 48, 180, 445–447
Patterning, scan-induced, *see* Scanning effects, patterning
Pauli repulsion, 37, 48, 49
Peak force tapping (Bruker), 9, 21, 139
Pentacene, 12, 365–367, 372, 373, 380, 382, 412, 413
Periodic ripples, 2, 17, 360. *See also* Scanning effects, patterning
Persistence length (macromolecules), 80, 303. *See also* Kuhn length
Phase
as measure of dissipation, *See* Dissipation, dynamic mode
lag (general topic), 18, 56, 57, 60–65, 146, 182, 202, 203, 296, 362, 369, 421
of second eigenmode, 422, 424, 432

segregation, 19, 186, 235, 306, 311, 313, 322, 323, 386
shift, apparatus, 146, 203, 233, 254, 255
-Z curves, 188, 196–201, 211–215, 228, 229, 231, 234, 243, 245, 307–309, 311, 312, 404
Photodetector/photodiode, 5, 6, 9, 51, 93–96, 98, 99, 103–106, 110, 113, 119, 121, 123, 127, 133, 143–145, 150, 163, 166, 182, 185, 315, 331, 332, 342, 443
Piétrement–Troyon approximation, 273. *See also* Carpick–Ogletree–Salmeron approximation
Piezoelectric
bimorph, 145, 202, 231, 235
scanner, 30, 113, 114, 129, 143, 145, 177, 178, 202, 231, 235
Planefit, 156, 173–175, 380, 391, 394, 397, 440
Plan-view cantilever dimensions, 438, 439, 441. *See also* Calibration, normal spring constant
Plastic deformation, 7, 47, 100, 159, 262, 304, 306, 346, 354, 375. *See also* Wear
Plasticization (of polymer by absorbed molecules), 7, 330, 358, 359
Plateau force (for polymers), 81, 127, 128
Poisson's ratio, 46, 98, 262, 429
Polarizability, 24, 92. *See also* Dipole interactions
Polarization, 35, 36, 39, 187, 400. *See also* Dipole interactions
Poly(butyl methacrylate) (PBMA), 9, 11, 160, 161, 183, 203, 205–207, 214–216, 219–221, 227–230, 238–240, 245, 247, 253, 254 ,304–306, 319–322, 380, 381, 384, 395, 424–426
Poly(butylene terephthalate-ethylene oxide) triblock copolymer (PBT-PEO), 27, 28, 172, 173, 216–219
Poly(dimethyl siloxane) (PDMS), 19, 41, 292, 293, 311–314, 317–319
Poly(ethyl methacrylate) (PEMA), 26, 27, 422–424
Poly(lauryl methacrylate) (PLMA), 9, 11, 126, 160, 161, 203, 205–207, 214–216, 219–221, 228–230, 238–240, 245, 247, 253, 254 , 304–306, 319, 320, 380, 381, 424, 425
Poly(methyl methacrylate) (PMMA), 26, 27, 41, 54, 354, 355, 361–362, 422–424
Poly(styrene-acrylonitrile) (SAN) copolymer, 361, 362
Poly(styrene-isobutylene) triblock copolymer (SIBS), 20, 21, 235–237, 322, 323, 386, 387
Poly(styrene-isoprene) triblock copolymer (SIS), 386, 397
Poly(vinyl alcohol) (PVA), 13, 14, 100, 101, 189–194, 200, 201, 240–245, 307–311, 347–349, 357–359, 369–372, 374, 375, 396, 397
Polyacrylamide, 21–23, 78, 252, 299–302, 391, 392
Polycrystalline metals, *see* Aluminum; Copper; Gold
Polymer
block, *see* Block copolymer
crystalline, *see* Crystalline, polymer
drug mixture, *see* Drug–polymer mixture
melt, 292, 293, 313, 317
nanofibril on tip, *see* Nanofibril
solvent-vapor annealed, 20, 235, 357
Polypeptide, *see* Gelatin
Polysaccharide, 6, 14
Polystyrene (PS), 41, 54, 98, 99, 118, 119, 121, 122, 359, 360
Post processing, 152, 156, 173, 175, 189, 207, 261, 298–301, 311, 379–398
Potential energy of attraction, 37–39, 45
Potential energy of cantilever, 275, 319, 352
Potential, surface/electrostatic, *see* Surface potential
Power law (relationships), 75, 76, 124, 125, 390, 396, 429
Power of driver, *see* Driving power
Power spectral density (analysis), 439–443
Pressure
contact, 47, 48, 157, 194, 266, 347, 429. *See also* Shear strength
disjoining, *see* Disjoining pressure
Laplace, *see* Laplace pressure
Protein, 3, 6, 14, 15, 80, 81, 123, 127, 295, 356
Pull-off force, 26, 85, 182, 259, 265, 269, 279, 288–293, 315–319, 343, 414, 415, 417

Pulsed force mode (Witec), 9, 21, 139, 192, 315, 393
Pyramidal geometry (of tip), 48, 118, 125, 182, 336, 385

Q control, 167. *See also* Quality factor
Quad photodiode, 5, 9, 143–145, 150, 166, 182, 331, 332, 406
Quadratic (scanner nonlinearity), *see* Nonlinearity of scanner
Quality factor (of cantilever), 35, 57, 65, 166, 207, 210–213, 233, 237, 317, 441, 448. *See also* Q control
Quantum physics (of relevance in force measurements), 37, 362, 363
Quartz (as standard for mechanical measurements), 118–122
Quasistatic (interaction, imaging process), 30, 64, 72, 107, 137–144, 156, 161, 180, 187, 194, 196, 220, 258, 274, 275, 279, 296, 304, 305, 309, 313, 314, 316, 342, 366

Radius of curvature (tip, sample), 39, 43, 45, 46, 71, 72, 91, 121–123, 132, 181, 182, 194, 237, 238, 262–264, 271, 275, 276, 445
Radius of gyration (polymer random coil), 75, 293
Rapamycin (drug), 322, 323
Rate dependence (of material response), 25, 27, 84, 87, 114, 259, 262, 266, 352–358, 362
Rate, measurement or sampling, *see* Data acquisition
Receptor (biological), 3, 24, 87
Reduced elastic modulus, 46, 48, 118, 266, 429. *See also* Modulus, Young's
Relaxation (time), 159, 184, 276, 353–356
Relaxations (alpha, beta, gamma), 353–358, 361, 362. *See also* Kinetics
Resolution, 2, 3, 5, 19–21, 34, 44, 46, 47, 92, 114, 143, 157, 196, 221, 265, 275–277, 293–295, 322, 339, 350, 363, 372, 407, 408, 433
Retardation (of interaction in time), 36, 39, 77, 78, 444, 445
Rheology, 25, 240, 368
Ring-down, cantilever, *see* Cantilever ring-down; Damping
Root-mean-square (RMS), *see* Roughness
Rotational isomeric motions (chain molecules), 25, 363. *See also* Turnstile motions
Roughness (surface), 2, 78, 149, 152, 188, 261, 263, 276, 277, 279, 280, 286, 287, 291, 292, 338, 380, 383–386, 389, 393
Rubber(y) (sample), 7, 9, 20, 26, 27, 47, 48, 87, 118, 235, 269, 270, 306, 319, 322, 330, 347, 353, 358, 359, 375, 424

Sader method, 319, 322, *see* Calibration, normal spring constant
Sawtooth, force profile, 83, 303, 371
Scanner, piezotube, *see* Piezoelectric, scanner
Scanning algorithm, *see* Algorithm, scanning
Scanning effects
  history (dependence), 115, 116, 168, 319, 347, 349
  patterning (polymer), 17, 260, 359–362
Schottky barrier (metal-semiconductor interfaces), 411, 412
Second generation AFMs, 93, 110, 114
Segmental motion (polymer), 353–356. *See also* Relaxations
Self-assembled monolayer, 24, 269, 364
Sensitivity
  in calibration(cantilever deflection), 95, 98, 102, 103, 111, 129, 183, 342, 357, 400, 420, 421, 438, 439, 443
  of AFM (as an analytical tool), 9, 10, 16, 20, 21, 29, 71, 202, 233, 259, 291, 311, 414, 423, 424
  to depth, 20, 100, 101, 186, 235, 304, 307, 311, 312
  to frequency shifts, 403, 408
Servo, 141, 143–146, 406. *See also* Gain
Setpoint
  amplitude, 160, 161, 167, 188–191, 197, 198, 200, 208, 210, 214, 219, 227, 230, 234–240, 249–251, 309, 406, 432
  amplitude ratio (*A*/*A*0), 66, 67, 160–162, 196–198, 203, 208, 210, 219, 224, 228, 230, 234–251, 448, 449

deflection, 142, 150, 164, 182, 183, 346
ramping, 350–352
Shank of tip, 41–43, 131, 279, 280, 283, 289, 433
Shear
modulation, *see* Modulation, shear
force, 13–17, 45, 139, 159, 293, 304, 333–335, 343, 347, 352, 353, 359, 365–375, 416, 417, 428. *See also* Tangential force
motion, stress derived from, 139, 259, 362
stiffness, 366–368, 371, 373, 375
strength (in friction), 159, 343, 346, 347, 351
Shift factor (viscoelastic response analysis), *see* Time–temperature superposition
Signal (manipulation and issues)
adder circuit, 350, 369
aliasing, 441
-to-noise ratio, 28, 34, 60, 81, 133, 275, 289, 324, 347, 350, 371, 402
Silanized (glass), 21, 23, 253, 275, 299, 301, 302
Silica, *see* Glass
Silicon
cantilever/tip, 103, 108, 118, 120, 121, 131, 133, 268, 269, 287, 289, 290, 294, 317, 332, 350, 420, 422, 425
sample or substrate (with oxide), 2, 11, 12, 29, 103, 106, 110, 129, 148, 155, 160, 164, 168, 169, 171, 173–175, 191, 196–199, 203, 204, 228, 235, 238, 243, 268, 293, 304, 321, 339, 341, 359, 365, 382, 384, 393, 395, 407–415, 425
Silicon nitride
cantilever/tip, 110, 133, 287
sample, 124
Silicone, *see* Poly (dimethyl siloxane)
Simulation (computer), 75, 243, 262, 273, 278
Single molecule extension, 78, 81–83, 303
Sinusoid(-al motion), 18, 56, 57, 71, 126, 202, 225, 230, 315–317, 350, 366, 368–372, 375, 414, 421
Skewness, 383–385
Skin, polymer, 13, 14, 20, 237, 306, 357
Snell's law (laser beam path), 94, 95
Solvation forces, 69–72
Solvent
exclusion, 279
induced glass–rubber transition, 358
Specific biological adhesion, 24
Spectral fingerprint, 363
Spectroscopy
dielectric, 354, 355, 363
force, *see* Force spectroscopy
quantum, 363
Spherical tip (model), 78, 107, 123–125, 267, 272, 342
Spring constant, *see* Calibration, normal spring constant; Operative spring constant
SPS Imaging (Agilent force curve mapping routine), 21, 139, 298, 299
Squeeze film damping (effects on cantilever oscillation during coarse approach), 211
Standard deviation (statistical), 2, 75, 128, 149, 349, 350, 383, 389
Standard linear model (viscoelastic), 84
Steady state (dynamics), 57, 189, 194, 208, 260
Stereochemical (molecular freedom), 24, 76, 357
Steric (force, repulsion), 50, 72–76, 78, 82, 83, 87, 125, 279, 287, 297, 301, 302, 363, 438
Stern layer, 68
Stick-slip motion, 289, 338, 339, 356, 362
Strain, 44–46, 84, 85, 317, 338, 352, 353, 362–364
Stress, 7, 12, 16, 45, 84–87, 100, 159, 184, 188, 260–262, 266, 275, 306, 317, 352–354, 359, 362, 366, 367, 416
Structural forces, 72, 277, 284, 287, 289
Surface (sample)
charge, 24, 92, 131
energy, 21, 39, 92, 123, 262 187, 235, 237, 262, 263, 268, 269, 273–275, 281–283, 311, 313, 319, 336, 338, 340, 346
forces, 24, 67, 71, 76, 77, 110, 263, 277–279, 283, 287
potential, 29, 42–44, 61, 68, 129, 195, 196, 198, 260, 368, 400–407, 430–433
Surface tension (of capillary), 210, 273, 278–281, 283, 284, 449, 450
Swelling (of polymer matrix), 21, 279, 280

Tan delta measurement, *see* Loss tangent
Tangential force, 45, 139, 157, 330, 331, 333, 334. *See also* Shear force
Temperature (sample), 24–27, 33, 187, 320–322, 330, 352–363, 366, 375
Tensile force, 85, 159, 276, 284, 349, 350
Thermalactivation, 7, 85–87, 259–261, 352, 353
Thermal calibration, *see* Calibration, normal force
Thermal energy, 40, 85
Thermal fluctuation, *see* Fluctuations, thermal
Thermal k, *see* Calibration, normal force
Thermal noise, 103, 123, 163, 442
Thermally driven cantilever, 437, 439, 442, 443
Thermodynamics, 4, 20, 69, 72, 123, 187, 235, 259, 273, 274, 279, 282–286, 311, 319
Theta solvent (for polymers), 75, 76, 79
Three-phase contact line (capillary), 278, 450
Time domain (signal analysis in), 129, 225, 316–318, 373, 391, 393, 394
Time-temperature superposition, 354, 355. *See also* Viscoelasticity
Tomlinson model (in friction), 353
Toroid (meniscus geometry), 280–282, 449
Torque (of tip), *See* Twisting of cantilever
Tractions (intersurface), 45
Trans (conformation), 79–81
Trans-gauche (conformational transitions), 85, 338
Transverse shear microscopy, 364–367
Tribology (nano), 2, 12, 17, 33, 100, 181, 330, 338, 339, 343, 347, 359, 384
Trigger(ing), 182, 220, 221, 297, 304, 305, 307–313
Truncated conical tip, 41, 287, 371, 433, 445–447
Tune (cantilever), *see* Cantilever tune
Turnstile motions (in chain molecules), 25, 75, 338, 355. *See also* Rotational isomeric
Twisting (of cantilever), 5, 166, 331, 332, 336, 340, 343, 364–366, 371–375

Ultrahigh vacuum AFM, 139, 433
Ultrasonic force microscopy, 425
Underdamped motion, 57

Vacuum level, 29, 68, 401, 402, 412
Van der Waals, *See* Dipole interactions
Variance, statistical, 383
Variational calculus (meniscus shape), 280
Vibrational modes (of cantilever), *See* Eigenmodes
Vibrations (environmental isolation), 393, 394, 439
Video access, 148
Virial (theorem), 247, 248
Viscoelasticity, 25, 45, 60, 62, 84–87, 98, 116, 121, 129, 159, 187, 229, 233, 237, 242, 247, 248, 250, 259, 261, 262, 266, 269, 276, 297, 298, 317–319, 330, 346, 347, 353–359, 362, 369, 374, 425, 428, 429
Viscousimmersion medium, 87, 166, 233, 447, 448
Viscous tip–sample interaction, 21, 66, 84, 194, 240, 250, 259, 261, 319, 445

Water contact angle, *see* Contact angle
Watershed algorithm, *see* Grain size analysis, watershed
Wave equation (of cantilever), 415–417. *See also* Differential equation
Wave number, 387–391, 397, 417, 428, 439, 440
Wear, 2, 14, 15, 47, 160, 260, 262, 277, 330, 343, 346, 349, 352, 353, 359–362, 414, 428. *See also* Plastic deformation
Wetting (of tip), *see* Capillary
Work function, 29, 42, 195, 260, 401, 402, 404, 408–411, 413, 414
Worm like chain model, 80–81

Xenon (as spherical atom), 35, 36, 40

Yield point (for shear motion), 346, 347, 371

Zero of measurement scale
  amplitude, 235, 309
  distance, 39, 92, 124, 125, 212, 245
  lateral force, 9, 332, 335, 343, 344, 365
  phase, 207, 227, 229
  vertical force, 13, 53, 68, 96, 106, 110–113, 138, 157, 182, 183